HYDROMECHANICS

PART II

HYDRODYNAMICS

A TREATISE ON HYDROMECHANICS

PART II
HYDRODYNAMICS

BY

A. S. RAMSEY, M.A.

LATE FELLOW OF MAGDALENE COLLEGE, CAMBRIDGE

CBS

CBS Publishers & Distributors Pvt. Ltd.

New Delhi • Bengaluru • Chennai • Kochi • Kolkata • Mumbai
Hyderabad • Nagpur • Patna • Pune • Vijayawada

ISBN: 81-239-1191-2

First Indian Edition: 1988
Reprint: 1998, 2000, 2006

This edition has been reprinted in India by arrangement with
G. Bell & Sons Limited, UK

Published by **Satish Kumar Jain** and produced by **Varun Jain** for
CBS Publishers & Distributors Pvt. Ltd.,
4819/XI Prahlad Street, 24 Ansari Road, Daryaganj, New Delhi - 110002
delhi@cbspd.com, cbspubs@airtelmail.in • www.cbspd.com
Ph.: 23289259, 23266861, 23266867 • Fax: 011-23243014

Corporate Office: 204 FIE, Industrial Area, Patparganj, Delhi - 110 092
Ph: 49344934 • Fax: 011-49344935
E-mail: publishing@cbspd.com • publicity@cbspd.com

Branches:

- ***Bengaluru:*** 2975, 17th Cross, K.R. Road, Bansankari 2nd Stage, Bengaluru - 70 • Ph: +91-80-26771678/79 • Fax: +91-80-26771680 E-mail: cbsbng@gmail.com, bangalore@cbspd.com
- ***Chennai:*** No. 7, Subbaraya Street, Shenoy Nagar, Chennai - 600030 Ph: +91-44-26681266, 26680620 • Fax: +91-44-42032115 E-mail: chennai@cbspd.com
- ***Kochi:*** Ashana House, 39/1904, A.M. Thomas Road, Valanjambalam, Ernakulum, Kochi • Ph: +91-484-4059061-65 Fax: +91-484-4059065 • E-mail: cochin@cbspd.com
- ***Kolkata:*** 6-B, Ground Floor, Rameshwar Shaw Road, Kolkata - 700014 Ph: +91-33-22891126/7/8 • E-mail: kolkata@cbspd.com
- ***Mumbai:*** 83-C, Dr. E. Moses Road, Worli, Mumbai - 400018 Ph: +91-9833017933, 022-24902340/41 • E-mail: mumbai@cbspd.com

Representatives:

- Hyderabad: 0-9885175004
- Patna: 0-9334159340
- Vijayawada: 0-9000660880
- Nagpur: 0-9021734563
- Pune: 0-9623451994

Printed at:
Neekunj Print Process, Delhi (India)

PREFACE TO THE FIRST EDITION

Dr Besant's *Treatise on Hydromechanics* was first published in one volume in 1859. When a fourth edition was called for in 1882 the subject matter had grown sufficiently to warrant the sub-division of the book into two volumes. Part I on Hydrostatics appeared alone in 1883 and in the preface a hope was expressed that Part II on Hydrodynamics would follow shortly. Several chapters were written and materials for other chapters were collected but laid aside owing to pressure of other work. In 1904 Dr Besant kindly invited me to co-operate with him in bringing out a new edition—the sixth—of the Hydrostatics, and suggested that I should undertake to complete the Hydrodynamics. This latter task I was unable to perform until the present year. Dr Besant kindly placed all his materials at my disposal, but as modes of expression and analysis have altered somewhat in the last thirty years, it seemed desirable to write a new book *ab initio*....

In the matter of the sign of the velocity potential I have followed the precedent of Professor Lamb and Sir George Greenhill. It does not seem a matter of intrinsic importance which sign is used, but uniformity is desirable and as all serious students of the subject will ultimately read it in the classic work of Professor Lamb, there is good reason why his precedent should be followed....

I am indebted to Mr W. Welsh for advice and assistance, and most of all my thanks are due to Mr J. G. Leathem for reading the whole book in proof and making many valuable criticisms and suggestions.

A. S. R.

MAGDALENE COLLEGE
CAMBRIDGE
December 1912

PREFACE TO THE FOURTH EDITION

As stated in the preface to the first edition, the book was written in the first place for beginners, it does not profess to be an exhaustive treatise and it does not aim at taking the reader to the limits of knowledge in the subject. But since of late years the study of hydrodynamics has become increasingly important in view of the rapid developments of the kindred subject of Aeronautics, therefore the additions have been made with a view to the needs of students who desire to take their studies further in this direction. A chapter on viscosity has been added as well as some applications of contour integration to problems of two-dimensional motion and some discussion of the part played by 'circulation' in producing 'lift' and of the application of conformal transformation to aerofoil theory. The chapter on viscosity is placed at the end of the book but it contains direct references to so few of the other chapters that there is no reason why a student who prefers to do so should not read it at an earlier stage, e.g. before the chapters on waves.

In conclusion I wish to acknowledge my indebtedness and express my gratitude to Dr S. Goldstein of St John's College for reading the proofs of all the additions and making many valuable suggestions. The book owes much to his careful criticism.

A. S. R.

March 1935

CONTENTS

CHAPTER I: KINEMATICS

CHAPTER II: EQUATIONS OF MOTION

CHAPTER III: PARTICULAR METHODS AND APPLICATIONS

CHAPTER VI: THE USE OF CONFORMAL REPRESENTATION. DISCONTINUOUS MOTION. FREE STREAM LINES. AEROFOILS

CHAPTER VII: IRROTATIONAL MOTION IN THREE DIMENSIONS

CHAPTER VIII: MOTION OF A SOLID THROUGH A LIQUID

Chapter XI: VIBRATIONS OF STRINGS

Chapter XII: SOUND WAVES

HYDRODYNAMICS

CHAPTER I

KINEMATICS

1·1. In the introduction to Part I of this work it was explained that all propositions in Hydrostatics are true for all fluids whatever their degree of viscosity. A very little consideration will suffice to shew that the motion of fluids cannot be independent of such properties as viscosity, and the results obtained from a discussion of the motion of fluids which ignores their internal friction can only be regarded as an approximation to what actually takes place in nature, and in some cases are far from representing reality. In the 'classical' treatment of the subject of Hydrodynamics however it is usual, for the sake of simplicity, to regard the fluid medium as a 'perfect fluid', incapable of exerting shearing stress, and, whether at rest or in motion, such that the pressure it exerts on any surface in contact with it is always normal to the surface and consequently, as was shewn in Hydrostatics, the pressure at any point in such a fluid is the same in every direction.

In the present chapter we shall limit ourselves to the consideration of some properties of the motion of fluids which are independent of causation, that is with the kinematics of fluids, leaving the equations of motion, or equations connecting the acting forces with the motions arising therefrom, for a subsequent chapter.

1·11. There are two methods of treating the general problem of Hydrodynamics or motion of a continuous medium; in the one, any particle of the fluid is selected and observation is made of its particular motion—it is pursued throughout its course; in the other, any point in the space occupied by the fluid is selected and observation is made of whatever changes of velocity, density and pressure take place at that point. The two methods are commonly called the Lagrangian and the Eulerian methods respectively though both were used by Euler, but the former was used by Lagrange in the *Mécanique Analytique*. The latter is sometimes also called the flux method. Clerk Maxwell suggested the words Historical and Statistical as descriptive of the two methods. We shall obtain the equations requisite for the determination of fluid motions from both these points of view.

1·12. In the **Lagrangian Method** if x, y, z denote the coordinates of a particle at time t, then the components of its velocity are $\dot{x}$, $\dot{y}$, $\dot{z}$ and the components of its acceleration are $\ddot{x}$, $\ddot{y}$, $\ddot{z}$. Also x, y, z and the velocities and accelerations are functions of t and of three independent parameters a, b, c which define the position of the chosen particle at a particular instant, thus a, b, c may be the coordinates of the chosen particle at the instant of time from which t is measured. In using this method it is well to remember that it resembles that of Dynamics of a Particle only in so far as the coordinates x, y, z of the chosen particle are dependent on the time t; but in the case of fluid motion t is not the only independent variable, for the particle is *any* particle in the fluid, and three other variables a, b, c are needed to specify which particle has been chosen, so that there are altogether four independent variables a, b, c, t.

1·13. In the **Eulerian Method** velocity at a point is measured thus: if a small plane surface be placed at the point at right angles to the direction of flow, the velocity at the point is measured, when uniform, by the volume of fluid per unit area that flows across the surface in unit time; and when variable by the time-rate of flow of volume of fluid per unit area across the surface.

Thus if q be the velocity and ρ the density of the fluid at any point, the mass that in time δt flows across a small area A, the normal to which makes an angle θ with the velocity, is $\rho q A \cos\theta\, \delta t$, and the rate at which mass crosses the surface is $\rho q A \cos\theta$.

As stated in **1·11**, in the Eulerian Method a particular point in the space occupied by the fluid is selected; we shall denote this point by (x, y, z) so that in this case x, y, z and t are *independent* variables. And it is important to remember that in the use of this method, unless some further meanings are assigned to the symbols, such expressions as dx/dt, d^2x/dt^2 do not occur, for the simple reason that x and t are independent.

When the axes are rectangular we shall use u, v, w to denote the components of the velocity q at the point (x, y, z). In general u, v, w are functions of the four independent variables x, y, z and t.

If we regard (x, y, z) as a fixed point, then the values of u, v, w will tell us what happens at that point as t changes; and if we regard t as fixed, then since (x, y, z) may be any point of the fluid, u, v, w will tell us what is happening at every point of the fluid at the particular instant under consideration.

If we wish to connect the Eulerian and Lagrangian methods or combine both notations in any particular problem, we regard u, v, w as the components of velocity of the element of fluid at (x, y, z) and the relation between the two sets of symbols is then $u, v, w = \dot{x}, \dot{y}, \dot{z}$.

1·2. Acceleration. In considering the meaning of acceleration and how to obtain its value by the flux method, we have to take two facts into account. Firstly, if P denote the point (x, y, z), then inasmuch as u, v, w are functions of t a change of velocity of the fluid can take place at the fixed point P as time progresses without any variations in x, y, z. Secondly, in order to estimate correctly the acceleration of an elementary portion of the fluid it is not sufficient merely to note what change of velocity is taking place at the point P, but we must also pursue the element for a short space after it passes P, in order to observe whether as it moves onwards it does so with the velocity it had on reaching P or acquires any additional velocity.

Let $$u = f(x, y, z, t).$$

The particle which is at (x, y, z) at time t will after a short interval δt have moved to $(x + u\,\delta t,\ y + v\,\delta t,\ z + w\,\delta t)$ so that its velocity will become

$$u + \delta u = f(x + u\,\delta t,\ y + v\,\delta t,\ z + w\,\delta t,\ t + \delta t)$$
$$= f(x, y, z, t) + \left(u\frac{\partial f}{\partial x} + v\frac{\partial f}{\partial y} + w\frac{\partial f}{\partial z} + \frac{\partial f}{\partial t}\right)\delta t$$
$$+ \text{terms containing higher powers of } \delta t.$$

Hence the x component of acceleration, being Lt $\delta u/\delta t$, is equal to

$$\frac{\partial f}{\partial t} + u\frac{\partial f}{\partial x} + v\frac{\partial f}{\partial y} + w\frac{\partial f}{\partial z} \quad \ldots\ldots\ldots\ldots\ldots\ldots(1),$$

or

$$\frac{\partial u}{\partial t} + u\frac{\partial u}{\partial x} + v\frac{\partial u}{\partial y} + w\frac{\partial u}{\partial z} \quad \ldots\ldots\ldots\ldots\ldots\ldots(2);$$

and in this expression the first term is the rate at which the velocity increases at the point (x, y, z) regarded as a fixed point in space, and the other terms arise from the changing velocity of the element of fluid in its onward course.

We shall denote the operator

$$\frac{\partial}{\partial t}+u\frac{\partial}{\partial x}+v\frac{\partial}{\partial y}+w\frac{\partial}{\partial z}$$

by the symbol $\frac{D}{Dt}$, and speak of it as 'differentiation following the motion of the fluid'.

With this notation the components of acceleration are Du/Dt, Dv/Dt, Dw/Dt; and if $f(x, y, z, t)$ be *any* function of the position of a particle of the fluid and the time, the rate of change of this function following the motion of the fluid is Df/Dt.

1·21. As an illustration let us consider the flow of water through a pipe, which is filled by the water. Firstly, let the pipe be of uniform section, then the velocity u is the same at every point, but inasmuch as the water may be forced through the pipe at varying speeds there may be an acceleration $\partial u/\partial t$, which, in this case, will at any instant have the same value at all points in the pipe. Secondly, let the motion be *steady*, i.e. the velocity at any particular point of the pipe keeps the same value u for all time; also let the pipe be of variable section, then the velocity varies from one point to another inasmuch as the section is variable, for the total flow across each section must be the same. Hence if s denote distance measured along the pipe to a point where the velocity is u, the element of fluid which occupies this position at time t will at time $t+\delta t$ have moved to a point indicated by $s+u\,\delta t$, and if $u=f(s)$, its velocity in the second position is

$$u+\delta u=f(s+u\,\delta t)=f(s)+\frac{\partial f}{\partial s}u\,\delta t$$

to the first power of δt. Therefore

$$\delta u=\frac{\partial f}{\partial s}u\,\delta t,$$

and the element of fluid has therefore an acceleration

$$=\mathrm{Lt}\,\delta u/\delta t=u\,\partial u/\partial s.$$

Thus we see that even in steady motion there may be acceleration; and in the general motion of water through a pipe of variable section the acceleration is given by

$$\frac{\partial u}{\partial t}+u\frac{\partial u}{\partial s}.$$

1·3. The Equation of Continuity.

The motions that we shall have to consider will be, in general, continuous motions; that is, we shall assume that u, v, w are finite and continuous functions and that their space derivatives $\partial u/\partial x$, $\partial u/\partial y$, $\partial u/\partial z$ are also finite.

In continuous motion, if we consider any closed surface drawn in the fluid, it is clear that the increase in the mass of fluid within the surface in any time δt must be equal to the excess of the mass that flows in over the mass that flows out.

Let ρ denote the density of the fluid at (x, y, z), and with this point as centre construct a small parallelepiped with edges of lengths h, k, l parallel to the axes.

The mass of fluid which in time δt crosses a unit of area parallel to the yz plane at (x, y, z) is $\rho u \delta t$, or say, $f(x, y, z)\,\delta t$.

To find the flow across the face kl of the parallelepiped nearest to the origin, take a point $(x-\frac{1}{2}h,\ y+\eta,\ z+\zeta)$ on this surface. Then the mass of fluid which in time δt flows across a small area $d\eta\, d\zeta$ at this point is $f(x-\frac{1}{2}h,\ y+\eta,\ z+\zeta)\, d\eta\, d\zeta\, \delta t$, or, to the first power of the small quantities, h, η, ζ,

$$\left(\rho u - \tfrac{1}{2}h\frac{\partial \rho u}{\partial x} + \eta\frac{\partial \rho u}{\partial y} + \zeta\frac{\partial \rho u}{\partial z}\right) d\eta\, d\zeta\, \delta t.$$

The total flow in time δt into the parallelepiped across this face kl is obtained by integrating the last expression over the area kl, noting that ρu and its derivatives are values at the centre of the parallelepiped and η, ζ are the only terms which vary over the face kl. Hence the flow is

$$\int_{-\frac{1}{2}k}^{\frac{1}{2}k}\int_{-\frac{1}{2}l}^{\frac{1}{2}l}\left(\rho u - \tfrac{1}{2}h\frac{\partial \rho u}{\partial x} + \eta\frac{\partial \rho u}{\partial y} + \zeta\frac{\partial \rho u}{\partial z}\right) d\eta\, d\zeta\, \delta t,$$

or

$$\left(\rho u - \tfrac{1}{2}h\frac{\partial \rho u}{\partial x}\right) kl\,\delta t.$$

Similarly the flow out across the opposite face is

$$\left(\rho u + \tfrac{1}{2}h\frac{\partial \rho u}{\partial x}\right) kl\,\delta t.$$

Therefore the increase in mass inside the parallelepiped in time δt due to the flow across these two faces is $-\dfrac{\partial \rho u}{\partial x} hkl\,\delta t$. The other pairs of faces give like contributions so that the total increase in mass in time δt is

$$-\left\{\frac{\partial \rho u}{\partial x} + \frac{\partial \rho v}{\partial y} + \frac{\partial \rho w}{\partial z}\right\} hkl\,\delta t.$$

But the original mass inside the parallelepiped is ρhkl, and its increase in time δt is $\dfrac{\partial \rho}{\partial t} hkl\,\delta t$. Whence we get the equation

$$\frac{\partial \rho}{\partial t} + \frac{\partial \rho u}{\partial x} + \frac{\partial \rho v}{\partial y} + \frac{\partial \rho w}{\partial z} = 0 \quad \text{.................(1).}$$

This equation is clearly equivalent to

$$\frac{D\rho}{Dt}+\rho\left(\frac{\partial u}{\partial x}+\frac{\partial v}{\partial y}+\frac{\partial w}{\partial z}\right)=0 \quad \text{.................(2)},$$

and either of these may be called the equation of continuity.

It follows from above that the expression

$$\frac{\partial u}{\partial x}+\frac{\partial v}{\partial y}+\frac{\partial w}{\partial z} \quad \text{......................(3)}$$

measures the rate at which the volume of an element of fluid at (x, y, z) is expanding. It may be called the **dilatation** or the **expansion**.

The expression (3) when u, v, w denote components of any vector is called the **divergence** of the vector, and is often written div (u, v, w).

If the fluid is homogeneous and incompressible, ρ is constant and the equation reduces to

$$\frac{\partial u}{\partial x}+\frac{\partial v}{\partial y}+\frac{\partial w}{\partial z}=0 \quad \text{...................(4)}.$$

If the fluid is heterogeneous and incompressible, ρ is a function of (x, y, z, t) such that $\frac{D\rho}{Dt}=0$, i.e. the density of an element does not alter as that element moves about; hence in this case also (4) follows from (2).

1·31. We can also obtain the equation of continuity by following the motion of a small element $\rho\, dx\, dy\, dz$ of fluid, and expressing the fact that its mass remains unchanged during the short interval δt.

If x, y, z are the coordinates of a particle at time t, its coordinates at time $t+\delta t$ are $x+u\,\delta t$, $y+v\,\delta t$, $z+w\,\delta t$. Similarly the particle whose coordinates are $x+dx$, y, z, will move in time δt to

$$x+u\,\delta t+dx+\frac{\partial u}{\partial x}dx\,\delta t, \quad y+v\,\delta t+\frac{\partial v}{\partial x}dx\,\delta t, \quad z+w\,\delta t+\frac{\partial w}{\partial x}dx\,\delta t,$$

so that dx is changed to ds_1, whose projections on the axes are

$$dx\left(1+\frac{\partial u}{\partial x}\delta t\right), \quad dx\frac{\partial v}{\partial x}\delta t, \quad dx\frac{\partial w}{\partial x}\delta t,$$

with similar expressions for dy and dz. Therefore the new volume of the parallelepiped is

$$dx\,dy\,dz\begin{vmatrix} 1+\frac{\partial u}{\partial x}\delta t, & \frac{\partial v}{\partial x}\delta t, & \frac{\partial w}{\partial x}\delta t \\ \frac{\partial u}{\partial y}\delta t, & 1+\frac{\partial v}{\partial y}\delta t, & \frac{\partial w}{\partial y}\delta t \\ \frac{\partial u}{\partial z}\delta t, & \frac{\partial v}{\partial z}\delta t, & 1+\frac{\partial w}{\partial z}\delta t \end{vmatrix},$$

$$=dx\,dy\,dz+\left(\frac{\partial u}{\partial x}+\frac{\partial v}{\partial y}+\frac{\partial w}{\partial z}\right)dx\,dy\,dz\,\delta t \quad \text{..............(1)}$$

to the first power of δt, and the density ρ is changed to $\rho+\frac{D\rho}{Dt}\delta t$. Equating the product of these to the original mass $\rho\, dx\, dy\, dz$, we get

$$\frac{D\rho}{Dt}+\rho\left(\frac{\partial u}{\partial x}+\frac{\partial v}{\partial y}+\frac{\partial w}{\partial z}\right)=0 \quad\ldots\ldots\ldots\ldots\ldots\ldots\ldots\ldots(2)$$

as before.

1·32. We may also obtain the equation of continuity by making use of Green's Theorem—

$$\iint (lf+mg+nh)\,dS=\iiint\left(\frac{\partial f}{\partial x}+\frac{\partial g}{\partial y}+\frac{\partial h}{\partial z}\right)dx\,dy\,dz,$$

where f, g, h are functions of (x, y, z), which with their first derivatives are finite and continuous throughout a region bounded by a closed surface S, and (l, m, n) are direction cosines of the outward drawn normal at a point on the surface, the $\iint$ being taken over the surface and the $\iiint$ throughout the space enclosed.

By considering any region in the fluid bounded by a closed surface S we have

$$\begin{aligned}\frac{\partial}{\partial t}\left\{\iiint \rho\, dx\, dy\, dz\right\}\delta t &= \text{increase in mass inside the surface in time } \delta t\\ &= \text{excess of flow in over flow out across the surface in time } \delta t\\ &= -\iint (l\rho u+m\rho v+n\rho w)\,dS\,\delta t\\ &= -\iiint\left(\frac{\partial \rho u}{\partial x}+\frac{\partial \rho v}{\partial y}+\frac{\partial \rho w}{\partial z}\right)dx\,dy\,dz\,\delta t,\end{aligned}$$

by Green's Theorem.

Therefore $$\iiint\left\{\frac{\partial\rho}{\partial t}+\frac{\partial\rho u}{\partial x}+\frac{\partial\rho v}{\partial y}+\frac{\partial\rho w}{\partial z}\right\}dx\,dy\,dz=0$$

for all ranges of integration within the fluid.

Therefore $$\frac{\partial\rho}{\partial t}+\frac{\partial\rho u}{\partial x}+\frac{\partial\rho v}{\partial y}+\frac{\partial\rho w}{\partial z}=0$$

at every point of the fluid.

1·4. The Equation of Continuity in the Lagrangian Method. Let a, b, c be the coordinates of a particle P at a given epoch, and x, y, z the coordinates of the same particle after the lapse of time t. Take a small tetrahedron $PABC$ in the fluid with its edges PA, PB, PC parallel to the coordinate axes of lengths δa, δb, δc.

After time t, the element of fluid that occupied the space $PABC$ at the given epoch will form a differently situated tetrahedron $P'A'B'C'$, and x, y, z being the coordinates of P', the coordinates of A' relative to P' will be

$$\frac{\partial x}{\partial a}\delta a,\quad \frac{\partial y}{\partial a}\delta a,\quad \frac{\partial z}{\partial a}\delta a,$$

of B'

$$\frac{\partial x}{\partial b}\delta b,\quad \frac{\partial y}{\partial b}\delta b,\quad \frac{\partial z}{\partial b}\delta b,$$

and of C'

$$\frac{\partial x}{\partial c}\delta c,\quad \frac{\partial y}{\partial c}\delta c,\quad \frac{\partial z}{\partial c}\delta c.$$

Hence the volume of the tetrahedron $P'A'B'C'$

$$=\tfrac{1}{6}\begin{vmatrix} \frac{\partial x}{\partial a}, & \frac{\partial y}{\partial a}, & \frac{\partial z}{\partial a} \\ \frac{\partial x}{\partial b}, & \frac{\partial y}{\partial b}, & \frac{\partial z}{\partial b} \\ \frac{\partial x}{\partial c}, & \frac{\partial y}{\partial c}, & \frac{\partial z}{\partial c} \end{vmatrix} \delta a\,\delta b\,\delta c,$$

and its mass

$$=\tfrac{1}{6}\rho\frac{\partial(x,y,z)}{\partial(a,b,c)}\delta a\,\delta b\,\delta c.$$

But if ρ_0 be the initial density the mass is $\tfrac{1}{6}\rho_0\delta a\,\delta b\,\delta c$, and therefore

$$\rho\frac{\partial(x,y,z)}{\partial(a,b,c)}=\rho_0,$$

which is the equation of continuity.

1·41. We can prove, by a direct transformation, the equivalence of the two forms of the equation of continuity. Beginning with the Lagrangian form, let

$$J=\frac{\partial(x,y,z)}{\partial(a,b,c)};$$

then ρJ is constant, and

$$d(\rho J)/dt=0,$$

or

$$J\frac{d\rho}{dt}+\rho\frac{dJ}{dt}=0.$$

But these time-rates are variations due to the motion of a particle, or the variability of x, y, z; and we can change now from the Lagrangian to the Eulerian system of variables by either writing D/Dt instead of d/dt or by writing u, v, w for $\dot{x}$, $\dot{y}$, $\dot{z}$. And on this hypothesis we shall write

$$\frac{\partial u}{\partial a}\text{ for }\frac{d}{dt}\frac{\partial x}{\partial a},\text{ etc.}$$

Hence
$$\frac{dJ}{dt}=\frac{\partial u}{\partial a}\frac{\partial(y,z)}{\partial(b,c)}+\frac{\partial u}{\partial b}\frac{\partial(y,z)}{\partial(c,a)}+\frac{\partial u}{\partial c}\frac{\partial(y,z)}{\partial(a,b)}$$
$$+\frac{\partial v}{\partial a}\frac{\partial(z,x)}{\partial(b,c)}+\frac{\partial v}{\partial b}\frac{\partial(z,x)}{\partial(c,a)}+\frac{\partial v}{\partial c}\frac{\partial(z,x)}{\partial(a,b)}$$
$$+\frac{\partial w}{\partial a}\frac{\partial(x,y)}{\partial(b,c)}+\frac{\partial w}{\partial b}\frac{\partial(x,y)}{\partial(c,a)}+\frac{\partial w}{\partial c}\frac{\partial(x,y)}{\partial(a,b)}$$
$$=\frac{\partial(u,y,z)}{\partial(a,b,c)}+\frac{\partial(x,v,z)}{\partial(a,b,c)}+\frac{\partial(x,y,w)}{\partial(a,b,c)}.$$

But
$$\frac{\partial u}{\partial a}=\frac{\partial u}{\partial x}\frac{\partial x}{\partial a}+\frac{\partial u}{\partial y}\frac{\partial y}{\partial a}+\frac{\partial u}{\partial z}\frac{\partial z}{\partial a},$$
$$\frac{\partial u}{\partial b}=\frac{\partial u}{\partial x}\frac{\partial x}{\partial b}+\frac{\partial u}{\partial y}\frac{\partial y}{\partial b}+\frac{\partial u}{\partial z}\frac{\partial z}{\partial b},$$
$$\frac{\partial u}{\partial c}=\frac{\partial u}{\partial x}\frac{\partial x}{\partial c}+\frac{\partial u}{\partial y}\frac{\partial y}{\partial c}+\frac{\partial u}{\partial z}\frac{\partial z}{\partial c},$$

and by eliminating $\dfrac{\partial u}{\partial y}, \dfrac{\partial u}{\partial z}$ we get
$$\frac{\partial u}{\partial x}\frac{\partial(x,y,z)}{\partial(a,b,c)}=\frac{\partial(u,y,z)}{\partial(a,b,c)},$$

or
$$\frac{\partial(u,y,z)}{\partial(a,b,c)}=J\frac{\partial u}{\partial x};$$

and from this and similar expressions we get
$$\frac{dJ}{dt}=J\left(\frac{\partial u}{\partial x}+\frac{\partial v}{\partial y}+\frac{\partial w}{\partial z}\right),$$

and therefore
$$\frac{D\rho}{Dt}+\rho\left(\frac{\partial u}{\partial x}+\frac{\partial v}{\partial y}+\frac{\partial w}{\partial z}\right)=0.$$

1·5. Particular cases of the Equation of Continuity. The equation of continuity may be transformed to cylindrical and to polar coordinates by the ordinary processes of change of the independent variable, but it is simpler to obtain it directly in each case from the principle that the increase in the mass contained in an element of volume in any short time δt is equal to the excess of the mass that flows in over the mass that flows out.

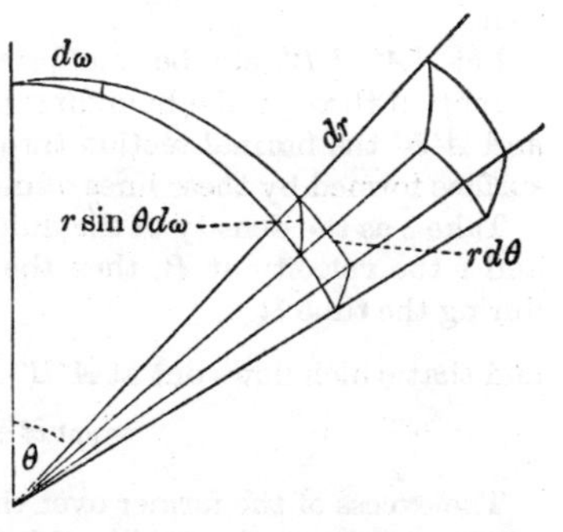

Thus in **polar coordinates** if q_r, q_θ, q_ω denote the components of velocity in the directions of the elements $dr, rd\theta, r\sin\theta d\omega$, the excess of flow in over the flow out arising from the face $r^2 \sin\theta\, d\theta\, d\omega$ and the opposite face is

$$-\frac{\partial}{\partial r}(\rho q_r r^2 \sin\theta\, d\theta\, d\omega)\, dr\, \delta t,$$

from the face $r \sin\theta\, d\omega\, dr$ and the opposite face

$$-\frac{\partial}{r\partial\theta}(\rho q_\theta r \sin\theta\, d\omega\, dr)\, r\, d\theta\, \delta t,$$

and from the face $r d\theta\, dr$ and the opposite face

$$-\frac{\partial}{r\sin\theta\,\partial\omega}(\rho q_\omega r\, d\theta\, dr)\, r\sin\theta\, d\omega\, \delta t,$$

and the increase in mass is

$$\frac{\partial}{\partial t}(\rho r^2 \sin\theta\, dr\, d\theta\, d\omega)\, \delta t.$$

Therefore

$$\frac{\partial\rho}{\partial t} + \frac{1}{r^2}\frac{\partial}{\partial r}(\rho q_r r^2) + \frac{1}{r\sin\theta}\frac{\partial}{\partial\theta}(\rho q_\theta \sin\theta) + \frac{1}{r\sin\theta}\frac{\partial \rho q_\omega}{\partial\omega} = 0 \quad \text{...(1)}.$$

Similarly if in **cylindrical coordinates** v_r, v_θ, v_z denote the components of velocity in the directions of the elements dr, $r d\theta$, dz, we can shew that

$$\frac{\partial\rho}{\partial t} + \frac{1}{r}\frac{\partial}{\partial r}(\rho v_r r) + \frac{1}{r}\frac{\partial}{\partial\theta}(\rho v_\theta) + \frac{\partial}{\partial z}(\rho v_z) = 0 \quad \text{.........(2)}.$$

1·51. Another form of the equation of continuity may also be given.

Let $PQ = \delta s$ be an arc of the line of motion passing through a point P; and let AB be a small area normal to the arc, such that all the particles of fluid crossing it may be considered as moving perpendicular to it.

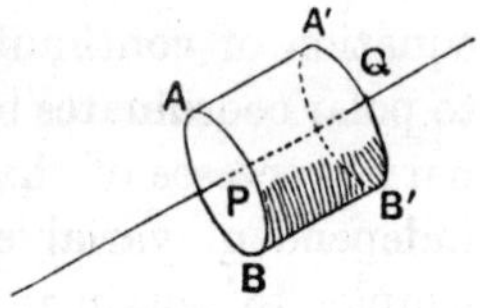

Let AA', BB', etc. be small arcs of the lines of motion through the bounding points of AB, and $A'B'$ the normal section through Q of the surface formed by these lines of motion.

Take ρ as the density of the fluid in PQ at the time t, κ the area of AB, and v the velocity at P; then the quantity of fluid which enters at AB during the time δt

$$= \kappa\rho v\, \delta t,$$

and that which flows out at $A'B'$

$$= \kappa\rho v\, \delta t + \frac{\partial}{\partial s}(\kappa\rho v\, \delta t)\, \delta s.$$

The excess of the former over the latter of these two expressions is the whole increase of the fluid in PQ during the time δt, and is

$$-\frac{\partial}{\partial s}(\kappa\rho v)\, \delta t\, \delta s:$$

but the mass of fluid at the time t being $\kappa\rho\,\delta s$, the increase in the time δt is also expressed by

$$\frac{\partial}{\partial t}(\kappa\rho\,\delta s)\,\delta t, \quad \text{or} \quad \frac{\partial}{\partial t}(\kappa\rho)\,\delta s\,\delta t,$$

and therefore

$$\frac{\partial}{\partial t}(\kappa\rho)+\frac{\partial}{\partial s}(\kappa\rho v)=0.$$

From the way in which this equation has been obtained, it will be seen that allowance is made for the expansion of the element which may in certain cases take place, and it is only in this way that κ can be an explicit function of the time. The small section AB may be taken arbitrarily, but the section $A'B'$ will depend, not only on the arc PQ, but also on the directions of the lines of motion passing through the bounding curve of AB; the variation of κ may therefore depend on the time explicitly, since these lines of motion may vary with the time.

1·52. Acceleration in Polar and Cylindrical Coordinates. Referring to the figure of **1·5**, if we take a right handed system of axes at the origin in the directions of the elements dr, $r\,d\theta$ and $r\sin\theta\,d\omega$, and let q_r, q_θ, q_ω denote the components of velocity in these directions, then the small displacements in time δt are

$$\delta r=q_r\,\delta t, \quad r\,\delta\theta=q_\theta\,\delta t, \quad r\sin\theta\,\delta\omega=q_\omega\,\delta t,$$

and the axes named above possess an angular velocity whose components are P, Q, R given by

$$P=\dot\omega\cos\theta \qquad Q=-\dot\omega\sin\theta \quad R=\dot\theta$$

$$=\frac{q_\omega}{r}\cot\theta, \qquad =-\frac{q_\omega}{r}, \qquad =\frac{q_\theta}{r}.$$

Also since the particle which is at (r,θ,ω) at time t is transferred in time δt to $\left(r+q_r\,\delta t,\ \theta+\frac{q_\theta}{r}\delta t,\ \omega+\frac{q_\omega}{r\sin\theta}\delta t\right)$, the rate of increase of a velocity component q_r following the motion of the fluid, as in **1·2**, is

$$\frac{Dq_r}{Dt}=\frac{\partial q_r}{\partial t}+q_r\frac{\partial q_r}{\partial r}+\frac{q_\theta}{r}\frac{\partial q_r}{\partial \theta}+\frac{q_\omega}{r\sin\theta}\frac{\partial q_r}{\partial \omega}.$$

But this and the corresponding rates of change of q_θ and q_ω are not the whole accelerations in the prescribed directions, because q_r, q_θ, q_ω are not velocities parallel to fixed axes, but parallel to axes rotating with angular velocities P, Q, R; so that, as in the general theory of moving axes, there are additional terms in the accelerations, viz.

$$-Rq_\theta+Qq_\omega, \quad -Pq_\omega+Rq_r \quad \text{and} \quad -Qq_r+Pq_\theta$$

or

$$-\frac{q_\theta^2+q_\omega^2}{r}, \quad -\frac{q_\omega^2\cot\theta}{r}+\frac{q_rq_\theta}{r} \quad \text{and} \quad \frac{q_rq_\omega}{r}+\frac{q_\theta q_\omega}{r}\cot\theta.$$

Hence the total components of acceleration in **polar coordinates** are

$$\left.\begin{aligned}&\frac{\partial q_r}{\partial t}+q_r\frac{\partial q_r}{\partial r}+\frac{q_\theta}{r}\frac{\partial q_r}{\partial\theta}+\frac{q_\omega}{r\sin\theta}\frac{\partial q_r}{\partial\omega}-\frac{q_\theta^2+q_\omega^2}{r}\\&\frac{\partial q_\theta}{\partial t}+q_r\frac{\partial q_\theta}{\partial r}+\frac{q_\theta}{r}\frac{\partial q_\theta}{\partial\theta}+\frac{q_\omega}{r\sin\theta}\frac{\partial q_\theta}{\partial\omega}-\frac{q_\omega^2\cot\theta}{r}+\frac{q_rq_\theta}{r}\\&\frac{\partial q_\omega}{\partial t}+q_r\frac{\partial q_\omega}{\partial r}+\frac{q_\theta}{r}\frac{\partial q_\omega}{\partial\theta}+\frac{q_\omega}{r\sin\theta}\frac{\partial q_\omega}{\partial\omega}+\frac{q_rq_\omega}{r}+\frac{q_\theta q_\omega}{r}\cot\theta\end{aligned}\right\}\quad\ldots(1).$$

By like arguments with **cylindrical coordinates** r, θ, z, if v_r, v_θ, v_z denote the velocities in the directions $dr, rd\theta, dz$, the components of acceleration are

$$\left.\begin{aligned}&\frac{\partial v_r}{\partial t}+v_r\frac{\partial v_r}{\partial r}+\frac{v_\theta}{r}\frac{\partial v_r}{\partial\theta}+v_z\frac{\partial v_r}{\partial z}-\frac{v_\theta^2}{r}\\&\frac{\partial v_\theta}{\partial t}+v_r\frac{\partial v_\theta}{\partial r}+\frac{v_\theta}{r}\frac{\partial v_\theta}{\partial\theta}+v_z\frac{\partial v_\theta}{\partial z}+\frac{v_rv_\theta}{r}\\&\frac{\partial v_z}{\partial t}+v_r\frac{\partial v_z}{\partial r}+\frac{v_\theta}{r}\frac{\partial v_z}{\partial\theta}+v_z\frac{\partial v_z}{\partial z}\end{aligned}\right\}\quad\ldots\ldots\ldots\ldots(2).$$

If in (2) we put $v_z=0$, we get the components of acceleration in polar coordinates in two dimensions.

From 1·5 (1) and (2) it is easy to see that the expression for the **dilatation in polar coordinates** is

$$\frac{\partial q_r}{\partial r}+\frac{2q_r}{r}+\frac{\partial q_\theta}{r\partial\theta}+\frac{q_\theta\cot\theta}{r}+\frac{1}{r\sin\theta}\frac{\partial q_\omega}{\partial\omega}\quad\ldots\ldots\ldots\ldots(3),$$

and in **cylindrical coordinates**

$$\frac{\partial v_r}{\partial r}+\frac{v_r}{r}+\frac{\partial v_\theta}{r\partial\theta}+\frac{\partial v_z}{\partial z}\quad\ldots\ldots\ldots\ldots\ldots\ldots\ldots(4).$$

1·6. The Boundary Surface. At any fixed boundary the velocity of the fluid normal to the surface must vanish, that is

$$lu+mv+nw=0$$

at every point of the boundary; l, m, n denoting the direction cosines of the normal.

At the surface of a solid moving in the fluid the normal velocity of the fluid must be equal to that of the solid. Also for any surface in the fluid composed of a given sheet of particles or, what is the same thing, for any surface which always contains the same fluid matter within it, we must have the normal velocity of the surface equal to the velocity in the same direction of a neighbouring particle of fluid. Thus if $\delta\nu$ is an element PP' of a normal to the surface

$$F(x, y, z, t) = 0 \qquad \text{.......................(1)},$$

and x, y, z are the coordinates of P, those of P' are $x+l\delta\nu$, $y+m\delta\nu$, $z+n\delta\nu$; where l, m, n are direction cosines of PP' and therefore proportional to $\partial F/\partial x$, $\partial F/\partial y$, $\partial F/\partial z$. But P' lies on the surface at time $t+\delta t$, therefore

$$F(x+l\delta\nu,\, y+m\delta\nu,\, z+n\delta\nu,\, t+\delta t) = 0 \qquad \text{.........(2)},$$

and from (1) and (2) we get, to the first power of $\delta\nu$ and δt,

$$\left(l\frac{\partial F}{\partial x} + m\frac{\partial F}{\partial y} + n\frac{\partial F}{\partial z}\right)\delta\nu + \frac{\partial F}{\partial t}\delta t = 0.$$

Again

$$\frac{u\dfrac{\partial F}{\partial x} + v\dfrac{\partial F}{\partial y} + w\dfrac{\partial F}{\partial z}}{\left\{\left(\dfrac{\partial F}{\partial x}\right)^2 + \left(\dfrac{\partial F}{\partial y}\right)^2 + \left(\dfrac{\partial F}{\partial z}\right)^2\right\}^{\frac{1}{2}}} = lu + mv + nw$$

$$= \text{normal velocity of particle of fluid}$$
$$= \text{normal velocity of surface}$$
$$= \dot{\nu}$$
$$= -\frac{\partial F}{\partial t}\bigg/\left(l\frac{\partial F}{\partial x} + m\frac{\partial F}{\partial y} + n\frac{\partial F}{\partial z}\right)$$
$$= -\frac{\partial F}{\partial t}\bigg/\left\{\left(\frac{\partial F}{\partial x}\right)^2 + \left(\frac{\partial F}{\partial y}\right)^2 + \left(\frac{\partial F}{\partial z}\right)^2\right\}^{\frac{1}{2}};$$

therefore
$$\frac{\partial F}{\partial t} + u\frac{\partial F}{\partial x} + v\frac{\partial F}{\partial y} + w\frac{\partial F}{\partial z} = 0 \qquad \text{...............(3)},$$

and the equation of every boundary surface must satisfy this differential equation.

1·61. Alternatively if we assume that a boundary surface always consists of the same particles of fluid, we may conclude at once that if $F(x, y, z, t) = 0$ be such a surface, then following the motion of the fluid

$$\frac{DF}{Dt} = 0, \quad \text{or} \quad \frac{\partial F}{\partial t} + u\frac{\partial F}{\partial x} + v\frac{\partial F}{\partial y} + w\frac{\partial F}{\partial z} = 0.$$

Though this hypothesis is generally true for continuous motion it may cease to hold in some cases of discontinuous or turbulent motion.

1·7. Stream Lines. A *stream line* or *line of flow* is a curve such that at any instant of time the tangent at any point of it is the direction of motion of the fluid at that point. A tubular space in the fluid bounded by lines of flow is called a *tube of flow*.

The direction of motion of the fluid particle at the point (x, y, z) is defined by the quantities u, v, w and therefore the differential equations of the stream lines are

$$\frac{dx}{u}=\frac{dy}{v}=\frac{dz}{w} \qquad \text{.......................(1).}$$

Except in the case of steady motion, u, v, w are always functions of the time and therefore the stream lines are continually changing with the time, and the actual path of any particle of the fluid will not in general coincide with a stream line. For if P, Q, R are consecutive points on a stream line at time t, a particle moving through P at this instant will move along PQ but when it arrives at Q at time $t+\delta t$, QR is no longer the direction of the velocity at Q and the particle will therefore cease to move along QR and move instead in the direction of the new velocity at Q. But if the motion be steady the stream lines remain unchanged as time progresses and they are also the paths of the particles of fluid.

The differential equations for the paths of the particles are

$$\dot{x}=u, \quad \dot{y}=v, \quad \dot{z}=w \qquad \text{.................(2),}$$

for when u, v, w are known functions of x, y, z, t these equations will determine x, y, z in terms of t and three arbitrary constants which might be taken to be a, b, c, the initial values of the coordinates of a particle, and hence the paths of the particles would be obtained.

1·71. The stream lines $dx/u=dy/v=dz/w$ are cut at right angles by the surfaces given by the differential equation

$$u\,dx+v\,dy+w\,dz=0 \qquad \text{.................(1);}$$

and the condition for the existence of such orthogonal surfaces is the condition that the last equation may admit of a solution of the form

$$\phi(x, y, z)=C \qquad \text{.......................(2),}$$

the analytical condition being

$$u\left(\frac{\partial w}{\partial y}-\frac{\partial v}{\partial z}\right)+v\left(\frac{\partial u}{\partial z}-\frac{\partial w}{\partial x}\right)+w\left(\frac{\partial v}{\partial x}-\frac{\partial u}{\partial y}\right)=0 \qquad \text{......(3).}$$

1·8. Velocity Potential. When the expression

$$udx+vdy+wdz$$

is an exact differential $-d\phi$, so that

$$u, v, w=-\frac{\partial\phi}{\partial x}, \quad -\frac{\partial\phi}{\partial y}, \quad -\frac{\partial\phi}{\partial z} \quad \text{............(1)},$$

then ϕ is called the *velocity potential* or *velocity function*.

It is clear that in this case

$$\frac{\partial w}{\partial y}=\frac{\partial v}{\partial z}, \quad \frac{\partial u}{\partial z}=\frac{\partial w}{\partial x}, \quad \frac{\partial v}{\partial x}=\frac{\partial u}{\partial y} \quad \text{...............(2)},$$

so that condition (3) of **1·71** is satisfied and surfaces exist which cut the stream lines orthogonally.

1·81. As an example consider the case in which

$$u=-c^2y/r^2, \quad v=c^2x/r^2, \quad w=0,$$

where r denotes distance from the z-axis, so that the velocity is wholly transversal and everywhere equal to c^2/r. These values satisfy the equation of continuity and therefore represent a possible motion.

The lines of flow are given by

$$\frac{dx}{-y}=\frac{dy}{x}=\frac{dz}{0}$$

or $$x^2+y^2=\text{const.}, \quad z=\text{const.}$$

In this case $$\frac{\partial v}{\partial x}=\frac{c^2(y^2-x^2)}{r^4}=\frac{\partial u}{\partial y},$$

so that conditions (2) of **1·8** are satisfied.

In fact $$udx+vdy+wdz=c^2d\left(\tan^{-1}\frac{y}{x}\right),$$

so that there is a velocity potential

$$\phi=-c^2\tan^{-1}\frac{y}{x},$$

and the planes $y=\kappa x$ cut the stream lines orthogonally.

1·82. It is possible however for the orthogonal surfaces to exist without a velocity potential. Take for instance the case

$$u=-\omega y, \quad v=\omega x, \quad w=0,$$

where again the velocity is transversal and varies as the distance from the z-axis, so that the whole mass rotates as if solid.

In this case we have the same lines of flow as in the last article, but $udx+vdy+wdz$ is not an exact differential, so there is no velocity potential though condition (3) of **1·71** is satisfied and

$$udx+vdy+wdz=0$$

leads to the family of planes $y=\kappa x$, which cut the stream lines orthogonally.

1·9. Irrotational and Rotational Motion. When the expressions

$$\frac{\partial w}{\partial y}-\frac{\partial v}{\partial z},\quad \frac{\partial u}{\partial z}-\frac{\partial w}{\partial x},\quad \frac{\partial v}{\partial x}-\frac{\partial u}{\partial y}$$

all vanish, the motion is said to be *irrotational*. When they do not all vanish the motion is said to be *rotational*.

The reason for this nomenclature will be given hereafter.

It will be noticed that, when a velocity potential exists, the motion is irrotational. Thus the motion of **1·81** is irrotational, and that of **1·82** is rotational.

EXAMPLES

1. A mass of fluid moves in such a way that each particle describes a circle in one plane about a fixed axis; shew that the equation of continuity is

$$\frac{\partial \rho}{\partial t}+\frac{\partial(\rho\omega)}{\partial\theta}=0,$$

where ω is the angular velocity of a particle whose azimuthal angle is θ at time t.

2. A mass of fluid is in motion so that the lines of motion lie on the surface of coaxial cylinders; shew that the equation of continuity is

$$\frac{\partial \rho}{\partial t}+\frac{1}{r}\frac{\partial(\rho v_\theta)}{\partial\theta}+\frac{\partial(\rho v_z)}{\partial z}=0,$$

where v_θ, v_z are the velocities perpendicular and parallel to z.

3. The particles of a fluid move symmetrically in space with regard to a fixed centre; prove that the equation of continuity is

$$\frac{\partial \rho}{\partial t}+u\frac{\partial \rho}{\partial r}+\frac{\rho}{r^2}\frac{\partial}{\partial r}(r^2 u)=0,$$

where u is the velocity at distance r.

4. Each particle of a mass of liquid moves in a plane through the axis of z; find the equation of continuity.

5. If the lines of motion are curves on the surfaces of cones having their vertices at the origin and the axis of z for common axis, prove that the equation of continuity is

$$\frac{\partial \rho}{\partial t}+\frac{\partial(\rho q_r)}{\partial r}+\frac{2\rho q_r}{r}+\frac{\operatorname{cosec}\theta}{r}\frac{\partial(\rho q_\omega)}{\partial\omega}=0.$$

6. If the lines of motion are curves on the surfaces of spheres all touching the plane of xy at the origin O, the equation of continuity is

$$r\sin\theta\frac{\partial \rho}{\partial t}+\frac{\partial(\rho v)}{\partial\phi}+\sin\theta\frac{\partial(\rho u)}{\partial\theta}+\rho u(1+2\cos\theta)=0,$$

where r is the radius CP of one of the spheres, θ the angle PCO, u the velocity in the plane PCO, v the perpendicular velocity, and ϕ the inclination of the plane PCO to a fixed plane through the axis of z.

7. If every particle moves on the surface of a sphere, prove that the equation of continuity is

$$\frac{\partial \rho}{\partial t}\cos\theta+\frac{\partial}{\partial\theta}(\rho\omega\cos\theta)+\frac{\partial}{\partial\phi}(\rho\omega'\cos\theta)=0,$$

ρ being the density, θ, ϕ the latitude and longitude of any element, and ω and ω' the angular velocities of the element in latitude and longitude respectively. (M.T. 1877.)

8. Shew that, if ξ, η, ζ be orthogonal coordinates and if U, V, W be the corresponding component velocities, the equation of continuity is

$$\frac{\partial \rho}{\partial t}+\rho(Us_1+Vs_2+Ws_3)+h_1\frac{\partial}{\partial\xi}(\rho U)+h_2\frac{\partial}{\partial\eta}(\rho V)+h_3\frac{\partial}{\partial\zeta}(\rho W)=0,$$

where $$\frac{1}{h_1^2}=\left(\frac{\partial x}{\partial\xi}\right)^2+\left(\frac{\partial y}{\partial\xi}\right)^2+\left(\frac{\partial z}{\partial\xi}\right)^2, \text{ etc. ......}$$

and s_1, s_2, s_3 are respectively the sums of the principal curvatures of the three orthogonal surfaces. (Coll. Exam. 1896.)

9. Shew that $$\frac{x^2}{a^2}\tan^2 t+\frac{y^2}{b^2}\cot^2 t=1$$

is a possible form for the bounding surface of a liquid, and find an expression for the normal velocity. (Coll. Exam. 1899.)

10. In the steady motion of homogeneous liquid if the surfaces $f_1=a_1$, $f_2=a_2$ define the stream lines, prove that the most general values of the velocity components u, v, w are

$$F(f_1,f_2)\frac{\partial(f_1,f_2)}{\partial(y,z)},\quad F(f_1,f_2)\frac{\partial(f_1,f_2)}{\partial(z,x)},\quad F(f_1,f_2)\frac{\partial(f_1,f_2)}{\partial(x,y)},$$

where F is any arbitrary function. (Coll. Exam. 1892.)

11. Shew that all necessary conditions can be satisfied by a velocity potential of the form $$\phi=\alpha x^2+\beta y^2+\gamma z^2,$$

and a bounding surface of the form

$$F\equiv ax^4+by^4+cz^4-\chi(t)=0,$$

where $\chi(t)$ is a given function of the time, and α, β, γ, a, b, c suitable functions of the time. (Trinity Coll. 1895.)

CHAPTER II

EQUATIONS OF MOTION

2·1. LET u, v, w be the components of velocity, ρ the density and p the pressure at the point (x, y, z) in a mass of fluid, and let X, Y, Z be the components of external force per unit mass at the same point.

As in **1·3**, consider a small rectangular parallelepiped hkl with its centre at (x, y, z), and resolve parallel to the x-axis; then we have

$$\rho hkl \frac{Du}{Dt} = \rho X hkl - \frac{\partial p}{\partial x} hkl,$$

for the last term can be shewn, as in **1·3**, to be the difference of the pressures on the two ends of area kl.

Hence
$$\frac{Du}{Dt} = X - \frac{1}{\rho}\frac{\partial p}{\partial x},$$

or

$$\left.\begin{aligned} \frac{\partial u}{\partial t} + u\frac{\partial u}{\partial x} + v\frac{\partial u}{\partial y} + w\frac{\partial u}{\partial z} &= X - \frac{1}{\rho}\frac{\partial p}{\partial x}. \\ \text{Similarly} \quad \frac{\partial v}{\partial t} + u\frac{\partial v}{\partial x} + v\frac{\partial v}{\partial y} + w\frac{\partial v}{\partial z} &= Y - \frac{1}{\rho}\frac{\partial p}{\partial y}, \\ \frac{\partial w}{\partial t} + u\frac{\partial w}{\partial x} + v\frac{\partial w}{\partial y} + w\frac{\partial w}{\partial z} &= Z - \frac{1}{\rho}\frac{\partial p}{\partial z}. \end{aligned}\right\} \quad \text{........(1)}$$

These are **Euler's Dynamical Equations.**

2·11. If the fluid be elastic we have to make use of the physical laws connecting pressure and density. Thus, if the temperature be constant, we have

$$p = \kappa\rho,$$

where κ is a constant. But if the changes that take place occur with such rapidity that there is not time for heat to enter or leave the fluid element, as is the case in the expansions and contractions of air that result in the propagation of sound waves, then the relation is the 'adiabatic' one,

$$p = \kappa\rho^\gamma,$$

where γ is a definite constant*.

* *Hydrostatics*, Art. 94.

2·12. In the case of a liquid, if Π be the external pressure upon its surface and p the pressure of the liquid at the surface, we shall have (neglecting surface tension)

$$p = \Pi,$$

and therefore at all points of the free surface

$$\frac{Dp}{Dt} = \frac{D\Pi}{Dt},$$

or if we suppose that Π depends only on the time

$$\frac{\partial p}{\partial t} + u\frac{\partial p}{\partial x} + v\frac{\partial p}{\partial y} + w\frac{\partial p}{\partial z} = \frac{\partial \Pi}{\partial t}.$$

2·2. Integration of the equations of motion. When a velocity potential exists and the external forces are derivable from a potential function, the equations of motion can always be integrated.

In this case $u, v, w = -\partial\phi/\partial x, \ -\partial\phi/\partial y, \ -\partial\phi/\partial z$;

and $\qquad X, Y, Z = -\partial V/\partial x, \ -\partial V/\partial y, \ -\partial V/\partial z$;

so that equations **2·1** (1) become

$$-\frac{\partial^2\phi}{\partial x\,\partial t} + \frac{\partial\phi}{\partial x}\frac{\partial^2\phi}{\partial x^2} + \frac{\partial\phi}{\partial y}\frac{\partial^2\phi}{\partial x\,\partial y} + \frac{\partial\phi}{\partial z}\frac{\partial^2\phi}{\partial x\,\partial z} = -\frac{\partial V}{\partial x} - \frac{1}{\rho}\frac{\partial p}{\partial x},$$

$$-\frac{\partial^2\phi}{\partial y\,\partial t} + \frac{\partial\phi}{\partial x}\frac{\partial^2\phi}{\partial x\,\partial y} + \frac{\partial\phi}{\partial y}\frac{\partial^2\phi}{\partial y^2} + \frac{\partial\phi}{\partial z}\frac{\partial^2\phi}{\partial y\,\partial z} = -\frac{\partial V}{\partial y} - \frac{1}{\rho}\frac{\partial p}{\partial y},$$

$$-\frac{\partial^2\phi}{\partial z\,\partial t} + \frac{\partial\phi}{\partial x}\frac{\partial^2\phi}{\partial x\,\partial z} + \frac{\partial\phi}{\partial y}\frac{\partial^2\phi}{\partial y\,\partial z} + \frac{\partial\phi}{\partial z}\frac{\partial^2\phi}{\partial z^2} = -\frac{\partial V}{\partial z} - \frac{1}{\rho}\frac{\partial p}{\partial z}.$$

Multiplying these equations by dx, dy, dz and adding we get

$$-d\frac{\partial\phi}{\partial t} + \tfrac{1}{2}d\left\{\left(\frac{\partial\phi}{\partial x}\right)^2 + \left(\frac{\partial\phi}{\partial y}\right)^2 + \left(\frac{\partial\phi}{\partial z}\right)^2\right\} + dV + \frac{1}{\rho}dp = 0,$$

or, if q denote the velocity,

$$-d\frac{\partial\phi}{\partial t} + \tfrac{1}{2}dq^2 + dV + \frac{1}{\rho}dp = 0 \quad \text{..............(1).}$$

Whence, assuming the existence of a functional relation between the pressure and the density, we get by integration

$$\int\frac{dp}{\rho} - \frac{\partial\phi}{\partial t} + \tfrac{1}{2}q^2 + V = C \quad \text{..................(2),}$$

where C is in general an arbitrary function of the time.

If the fluid be homogeneous and inelastic, the equation (2) becomes

$$\frac{p}{\rho} - \frac{\partial\phi}{\partial t} + \tfrac{1}{2}q^2 + V = C \quad \text{.....................(3).}$$

If the motion be steady $\partial\phi/\partial t=0$, and therefore

$$\frac{p}{\rho}+\tfrac{1}{2}q^2+V=C \text{ ....(4)},$$

where C is an absolute constant.

2·21. Steady Motion. Bernoulli's Theorem. Case of no Velocity Potential. We may obtain a similar equation when the motion is steady even though a velocity potential does not exist. Thus by considering the motion of a small cylinder of section κ with its axis of length δs along a stream line, if q be the velocity and S the component of external force per unit mass in direction of the stream line,

$$\rho\kappa\,\delta s\,\frac{Dq}{Dt}=\rho\kappa\,\delta s\,S-\kappa\frac{\partial p}{\partial s}\delta s,$$

and in this case

$$\frac{Dq}{Dt}=\frac{\partial q}{\partial t}+q\frac{\partial q}{\partial s},$$

so that

$$\frac{\partial q}{\partial t}+q\frac{\partial q}{\partial s}=S-\frac{1}{\rho}\frac{\partial p}{\partial s} \text{ ....(1)}.$$

If the motion be steady $\partial q/\partial t=0$, and if the external forces have a potential function such that $S=-\partial V/\partial s$, then by integrating along a stream line,

$$\int\frac{dp}{\rho}+\tfrac{1}{2}q^2+V=C \text{ ....(2)},$$

where C is a constant, whose value depends on the particular stream line chosen. This is *Bernoulli's Theorem.*

2·22. In general, when no velocity potential exists, we make use of equations **2·1** (1), in order to find the pressure at any point.

For instance, if a mass of liquid revolve uniformly without change of form or relative displacement about a fixed axis, there is no velocity potential, but taking the fixed axis as axis of z,

$$u=-\omega y,\quad v=\omega x,\quad w=0;$$

hence from equations **2·1** (1),

$$-\omega^2x=X-\frac{1}{\rho}\frac{\partial p}{\partial x},\quad -\omega^2y=Y-\frac{1}{\rho}\frac{\partial p}{\partial y},\quad 0=Z-\frac{1}{\rho}\frac{\partial p}{\partial z},$$

and therefore

$$\frac{1}{\rho}dp=X\,dx+Y\,dy+Z\,dz+\omega^2(x\,dx+y\,dy),$$

as in *Hydrostatics*, Art. 28.

For homogeneous liquid and conservative forces this becomes

$$\frac{p}{\rho}-\tfrac{1}{2}\omega^2(x^2+y^2)+V=\text{constant}.$$

At first sight this equation may appear to contradict **2·21** (2), but this is not so, for in that equation the constant C depends on the particular stream line; and in this particular case the velocity q is constant along a stream line, so that all the information we get from **2·21** (2) is that in this case

$$\int \frac{dp}{\rho} + V \text{ is constant along a stream line.}$$

2·23. When a velocity potential exists and the forces are conservative, the pressure is given by equation **2·2** (2) or (3).

Take, for instance, the case given in **1·81** in which there is a velocity potential $-c^2\theta$, while the velocity at distance r from the axis of z is c^2/r. Let z be measured vertically upwards and gravity be the only external force, then equation **2·2** (3) becomes

$$\frac{p}{\rho} + \frac{c^4}{2r^2} + gz = C.$$

If we take the pressure at the surface to be constant and assume that a is the value of z when r is infinite, we have for the equation of the surface

$$2g(x^2+y^2)(a-z) = c^4.$$

2·3. Equations of motion by the Flux Method. The equations of **2·1** can also be obtained by considering the changes of momentum that take place within a definite region of space due to the external forces acting throughout this region and to the fluid pressures on the boundary.

Thus if l, m, n are direction cosines of the outward-drawn normal to the element dS of any closed surface S drawn in the fluid and fixed in space, with the same notation the time-rate of increase of momentum parallel to the x-axis of the fluid inside S is $\frac{\partial}{\partial t}\iiint \rho u\,dx\,dy\,dz$, and this is composed of three parts:

(1) The rate of increase of x-momentum inside S due to the flow of momentum across the boundary, viz.

$$-\iint \rho u\,(lu+mv+nw)\,dS;$$

(2) The rate of increase of x-momentum due to the pressures on the boundary, viz. $-\iint lp\,dS$;

(3) The rate of increase of x-momentum due to the external forces acting throughout the region inside S, viz. $\iiint \rho X\,dx\,dy\,dz$.

Hence

$$\iiint \frac{\partial(\rho u)}{\partial t}\,dx\,dy\,dz = -\iint \rho u\,(lu+mv+nw)\,dS - \iint lp\,dS + \iiint \rho X\,dx\,dy\,dz;$$

and by transforming the surface integrals into volume integrals by Green's Theorem, we get

$$\iiint\left\{\frac{\partial}{\partial t}(\rho u)+\frac{\partial}{\partial x}(\rho u^2)+\frac{\partial}{\partial y}(\rho uv)+\frac{\partial}{\partial z}(\rho uw)+\frac{\partial p}{\partial x}-\rho X\right\}dx\,dy\,dz=0,$$

and since this must hold for all ranges of integration within the fluid, we must have at every point

$$\frac{\partial}{\partial t}(\rho u)+\frac{\partial}{\partial x}(\rho u^2)+\frac{\partial}{\partial y}(\rho uv)+\frac{\partial}{\partial z}(\rho uw)=\rho X-\frac{\partial p}{\partial x}.$$

But
$$\frac{\partial \rho}{\partial t}+\frac{\partial(\rho u)}{\partial x}+\frac{\partial(\rho v)}{\partial y}+\frac{\partial(\rho w)}{\partial z}=0,$$

and if we multiply this by u and subtract, we obtain

$$\frac{\partial u}{\partial t}+u\frac{\partial u}{\partial x}+v\frac{\partial u}{\partial y}+w\frac{\partial u}{\partial z}=X-\frac{1}{\rho}\frac{\partial p}{\partial x},$$

as before.

2·4. Equations referred to moving axes. Let U, V, W be the component velocities of the origin and P, Q, R the angular velocities of the frame of reference. Let u, v, w be the absolute velocities of the fluid at the point (x, y, z) rigidly connected to the frame and u', v', w' the velocities of the fluid at the same point relative to the frame.

We have

$$\left.\begin{aligned}u&=U+u'-yR+zQ\\ v&=V+v'-zP+xR\\ w&=W+w'-xQ+yP\end{aligned}\right\}\quad\text{.................(1).}$$

If we consider the increase in mass in a small rectangular element of volume attached to the frame of reference with its centre at (x, y, z) where ρ is the density of the fluid, we obtain as in 1·3

$$\frac{\partial \rho}{\partial t}+\frac{\partial \rho u'}{\partial x}+\frac{\partial \rho v'}{\partial y}+\frac{\partial \rho w'}{\partial z}=0\quad\text{.................(2)}$$

for the equation of continuity.

In the case of incompressible fluid this reverts to the standard form

$$\frac{\partial u}{\partial x}+\frac{\partial v}{\partial y}+\frac{\partial w}{\partial z}=0\quad\text{.....................(3).}$$

To obtain the equations of motion we proceed thus*:

Let k denote the component of the absolute velocity in a direction fixed in space whose direction cosines referred to the

* Greenhill, *Encyc. Brit.* 11th edition, Art. 'Hydromechanics'.

moving axes are l, m, n, i.e. $k = lu + mv + nw$. In time δt the coordinates of the fluid particle at (x, y, z) will have increased by $(u', v', w')\,\delta t$, so that lu will have become

$$(l+\delta l)\left\{u+\left(\frac{\partial u}{\partial t}+u'\frac{\partial u}{\partial x}+v'\frac{\partial u}{\partial y}+w'\frac{\partial u}{\partial z}\right)\delta t\right\}$$

+ terms containing higher powers of δt, as in **1·2**.

Whence we get

$$\begin{aligned}\frac{Dk}{Dt}=\frac{dl}{dt}u+\frac{dm}{dt}v+\frac{dn}{dt}w&+l\left(\frac{\partial u}{\partial t}+u'\frac{\partial u}{\partial x}+v'\frac{\partial u}{\partial y}+w'\frac{\partial u}{\partial z}\right)\\&+m\left(\frac{\partial v}{\partial t}+u'\frac{\partial v}{\partial x}+v'\frac{\partial v}{\partial y}+w'\frac{\partial v}{\partial z}\right)\\&+n\left(\frac{\partial w}{\partial t}+u'\frac{\partial w}{\partial x}+v'\frac{\partial w}{\partial y}+w'\frac{\partial w}{\partial z}\right)\end{aligned} \quad \ldots\ldots(4).$$

But since l, m, n are direction cosines referred to the moving axes of a line fixed in space, therefore

$$\frac{dl}{dt}-mR+nQ=0,\quad \frac{dm}{dt}-nP+lR=0,\quad \frac{dn}{dt}-lQ+mP=0 \ldots(5),$$

and

$$\frac{Dk}{Dt}=l\left(\frac{\partial u}{\partial t}-vR+wQ+u'\frac{\partial u}{\partial x}+v'\frac{\partial u}{\partial y}+w'\frac{\partial u}{\partial z}\right)+m(\ldots)+n(\ldots) \quad \ldots\ldots(6).$$

Again by resolving the external forces and the pressure in the direction (l, m, n) we obtain

$$\frac{Dk}{Dt}=l\left(X-\frac{1}{\rho}\frac{\partial p}{\partial x}\right)+m\left(Y-\frac{1}{\rho}\frac{\partial p}{\partial y}\right)+n\left(Z-\frac{1}{\rho}\frac{\partial p}{\partial z}\right) \ldots(7),$$

and since the choice of the direction (l, m, n) is arbitrary, a comparison of (6) and (7) gives the equations of motion in the form

$$\left.\begin{aligned}\frac{\partial u}{\partial t}-vR+wQ+u'\frac{\partial u}{\partial x}+v'\frac{\partial u}{\partial y}+w'\frac{\partial u}{\partial z}&=X-\frac{1}{\rho}\frac{\partial p}{\partial x}\\\frac{\partial v}{\partial t}-wP+uR+u'\frac{\partial v}{\partial x}+v'\frac{\partial v}{\partial y}+w'\frac{\partial v}{\partial z}&=Y-\frac{1}{\rho}\frac{\partial p}{\partial y}\\\frac{\partial w}{\partial t}-uQ+vP+u'\frac{\partial w}{\partial x}+v'\frac{\partial w}{\partial y}+w'\frac{\partial w}{\partial z}&=Z-\frac{1}{\rho}\frac{\partial p}{\partial z}\end{aligned}\right\} \ldots(8),$$

where the values of u', v', w' in terms of u, v, w and the velocities of the frame are given by (1).

2·41. The pressure equation. When a velocity potential exists, we substitute the values of u', v', w' from **2·4** (1) in (8), then write

$$u, v, w = -(\partial\phi/\partial x, \partial\phi/\partial y, \partial\phi/\partial z),$$

multiply the equations (8) by dx, dy, dz respectively, add and integrate. The result is

$$\int \frac{dp}{\rho} - \frac{\partial\phi}{\partial t} + \tfrac{1}{2}q^2 + P\left(y\frac{\partial\phi}{\partial z} - z\frac{\partial\phi}{\partial y}\right) + Q\left(z\frac{\partial\phi}{\partial x} - x\frac{\partial\phi}{\partial z}\right) + R\left(x\frac{\partial\phi}{\partial y} - y\frac{\partial\phi}{\partial x}\right) + V = F(t) \quad \text{......(1)},$$

where $$q^2 = (u-U)^2 + (v-V)^2 + (w-W)^2 \quad \text{............(2)}.$$

It should be noted that q^2 is what the square of the velocity of the fluid would become if a velocity equal and opposite to that of the origin were superposed on the fluid and the frame thus reducing the origin to rest.

Also that if Ω denotes the resultant angular velocity (P, Q, R) and K denotes the resultant moment of momentum per unit volume about fixed axes coinciding momentarily with the moving axes, then (1) may be written

$$\int \frac{dp}{\rho} - \frac{\partial\phi}{\partial t} + \tfrac{1}{2}q^2 - \frac{1}{\rho}(\Omega . K) + V = F(t) \quad \text{.........(3)},$$

where $(\Omega . K)$ denotes the scalar product of the two vectors.

2·5. Lagrange's Equations. Let a, b, c be the initial coordinates of a particle and x, y, z the coordinates of the same particle at time t, then a, b, c, t are the independent variables and our object is to determine x, y, z in terms of a, b, c, t and so investigate completely the motion. At time t the component accelerations of the fluid element $\delta x \delta y \delta z$ are $\partial^2 x/\partial t^2$, $\partial^2 y/\partial t^2$, $\partial^2 z/\partial t^2$, and if we assume the existence of a potential V for the external forces, we get as in **2·1**

$$\frac{\partial^2 x}{\partial t^2} = -\frac{\partial V}{\partial x} - \frac{1}{\rho}\frac{\partial p}{\partial x},$$

$$\frac{\partial^2 y}{\partial t^2} = -\frac{\partial V}{\partial y} - \frac{1}{\rho}\frac{\partial p}{\partial y},$$

$$\frac{\partial^2 z}{\partial t^2} = -\frac{\partial V}{\partial z} - \frac{1}{\rho}\frac{\partial p}{\partial z}.$$

To deduce equations containing only differentiations with regard to the independent variables a, b, c, t, we multiply these by $\partial x/\partial a$, $\partial y/\partial a$, $\partial z/\partial a$ and add, therefore

$$\frac{\partial^2 x}{\partial t^2}\frac{\partial x}{\partial a}+\frac{\partial^2 y}{\partial t^2}\frac{\partial y}{\partial a}+\frac{\partial^2 z}{\partial t^2}\frac{\partial z}{\partial a}=-\frac{\partial V}{\partial a}-\frac{1}{\rho}\frac{\partial p}{\partial a} \quad \text{.........(1).}$$

Similarly

$$\frac{\partial^2 x}{\partial t^2}\frac{\partial x}{\partial b}+\frac{\partial^2 y}{\partial t^2}\frac{\partial y}{\partial b}+\frac{\partial^2 z}{\partial t^2}\frac{\partial z}{\partial b}=-\frac{\partial V}{\partial b}-\frac{1}{\rho}\frac{\partial p}{\partial b} \quad \text{.........(2),}$$

and

$$\frac{\partial^2 x}{\partial t^2}\frac{\partial x}{\partial c}+\frac{\partial^2 y}{\partial t^2}\frac{\partial y}{\partial c}+\frac{\partial^2 z}{\partial t^2}\frac{\partial z}{\partial c}=-\frac{\partial V}{\partial c}-\frac{1}{\rho}\frac{\partial p}{\partial c} \quad \text{.........(3).}$$

These equations, together with the equation of continuity

$$\rho\frac{\partial(x, y, z)}{\partial(a, b, c)}=\rho_0,$$

constitute *Lagrange's Hydrodynamical Equations.*

2·51. Cauchy's Integrals. Assuming that ρ is a function of p, differentiate equations (2) and (3) of 2·5 with regard to c and b respectively and subtract, and we obtain after writing u, v, w for $\partial x/\partial t$, $\partial y/\partial t$, $\partial z/\partial t$,

$$\frac{\partial^2 u}{\partial t\partial b}\frac{\partial x}{\partial c}-\frac{\partial^2 u}{\partial t\partial c}\frac{\partial x}{\partial b}+\frac{\partial^2 v}{\partial t\partial b}\frac{\partial y}{\partial c}-\frac{\partial^2 v}{\partial t\partial c}\frac{\partial y}{\partial b}+\frac{\partial^2 w}{\partial t\partial b}\frac{\partial z}{\partial c}-\frac{\partial^2 w}{\partial t\partial c}\frac{\partial z}{\partial b}=0.$$

Integrate this equation with regard to t, and take u_0, v_0, w_0 as initial values; then

$$\frac{\partial u}{\partial b}\frac{\partial x}{\partial c}-\frac{\partial u}{\partial c}\frac{\partial x}{\partial b}+\frac{\partial v}{\partial b}\frac{\partial y}{\partial c}-\frac{\partial v}{\partial c}\frac{\partial y}{\partial b}+\frac{\partial w}{\partial b}\frac{\partial z}{\partial c}-\frac{\partial w}{\partial c}\frac{\partial z}{\partial b}=\frac{\partial w_0}{\partial b}-\frac{\partial v_0}{\partial c},$$

for initially $\partial x/\partial a=1$, $\partial x/\partial b=\partial x/\partial c=0$, etc., etc.

Now

$$\frac{\partial u}{\partial a}=\frac{\partial u}{\partial x}\frac{\partial x}{\partial a}+\frac{\partial u}{\partial y}\frac{\partial y}{\partial a}+\frac{\partial u}{\partial z}\frac{\partial z}{\partial a}, \text{ etc., etc.,}$$

and making these substitutions, the equation becomes

$$\left(\frac{\partial w}{\partial y}-\frac{\partial v}{\partial z}\right)\frac{\partial(y, z)}{\partial(b, c)}+\left(\frac{\partial u}{\partial z}-\frac{\partial w}{\partial x}\right)\frac{\partial(z, x)}{\partial(b, c)}+\left(\frac{\partial v}{\partial x}-\frac{\partial u}{\partial y}\right)\frac{\partial(x, y)}{\partial(b, c)}=\frac{\partial w_0}{\partial b}-\frac{\partial v_0}{\partial c}.$$

Writing

$$\frac{\partial w}{\partial y}-\frac{\partial v}{\partial z}=2\xi, \quad \frac{\partial u}{\partial z}-\frac{\partial w}{\partial x}=2\eta, \quad \frac{\partial v}{\partial x}-\frac{\partial u}{\partial y}=2\zeta,$$

we obtain the equations

$$\xi\frac{\partial(y,z)}{\partial(b,c)}+\eta\frac{\partial(z,x)}{\partial(b,c)}+\zeta\frac{\partial(x,y)}{\partial(b,c)}=\xi_0,$$

$$\xi\frac{\partial(y,z)}{\partial(c,a)}+\eta\frac{\partial(z,x)}{\partial(c,a)}+\zeta\frac{\partial(x,y)}{\partial(c,a)}=\eta_0,$$

$$\xi\frac{\partial(y,z)}{\partial(a,b)}+\eta\frac{\partial(z,x)}{\partial(a,b)}+\zeta\frac{\partial(x,y)}{\partial(a,b)}=\zeta_0.$$

Multiply these equations by $\partial x/\partial a$, $\partial x/\partial b$, $\partial x/\partial c$ respectively and add and take account of the equation of continuity

$$\rho\,\partial(x,y,z)/\partial(a,b,c)=\rho_0,$$

and we get
$$\frac{\xi}{\rho}=\frac{\xi_0}{\rho_0}\frac{\partial x}{\partial a}+\frac{\eta_0}{\rho_0}\frac{\partial x}{\partial b}+\frac{\zeta_0}{\rho_0}\frac{\partial x}{\partial c}.$$

Similarly
$$\frac{\eta}{\rho}=\frac{\xi_0}{\rho_0}\frac{\partial y}{\partial a}+\frac{\eta_0}{\rho_0}\frac{\partial y}{\partial b}+\frac{\zeta_0}{\rho_0}\frac{\partial y}{\partial c},$$

and
$$\frac{\zeta}{\rho}=\frac{\xi_0}{\rho_0}\frac{\partial z}{\partial a}+\frac{\eta_0}{\rho_0}\frac{\partial z}{\partial b}+\frac{\zeta_0}{\rho_0}\frac{\partial z}{\partial c}.$$

We notice that when a velocity potential exists $\xi=\eta=\zeta=0$, and from the foregoing equations it is evident that these quantities are always zero if their initial values are zero.

As we have already stated, when a velocity potential exists the motion is said to be irrotational and we therefore have the theorem that *the motion of a fluid under conservative forces, if once irrotational, is always irrotational*. This constitutes Cauchy's proof of this important theorem first enunciated by Lagrange.

When a velocity potential does not exist, the motion is called *rotational*. The reason for the phraseology employed to distinguish the two kinds of motion is given in the following article taken from a paper by Stokes*.

2·52. Physical Interpretation. Conceive an indefinitely small element of a fluid in motion to become solidified suddenly, and the fluid about it to be destroyed suddenly; let the form of the element be so taken that the resulting solid shall be that which is the simplest with respect to rotatory motion, namely, that which has its three principal moments about axes passing through the centre of gravity equal to each other, and therefore every axis passing through that point a principal axis, and consider the linear and angular motions of the element immediately after solidification.

By the instantaneous solidification velocities will be suddenly generated or destroyed in the different portions of the element, and a set of impulsive

* *Trans. Camb. Phil. Soc.* VIII. p. 287, or *Math. and Phys. Papers*, I, p. 112.

forces will be called into action. Let x, y, z be the coordinates of the centre of gravity G of the element at the instant of solidification, $x+x'$, $y+y'$, $z+z'$ those of any other point in it.

Let u, v, w be the velocities of G along the three axes just before solidification, u', v', w' the relative velocities of the point whose relative coordinates are x', y', z'.

Let $\bar{u}, \bar{v}, \bar{w}$ be the velocities of G, u_1, v_1, w_1 the relative velocities of the point (x', y', z'), and ξ, η, ζ the angular velocities just after solidification.

Since all the impulsive forces are internal,

$$\bar{u}=u, \quad \bar{v}=v, \quad \bar{w}=w.$$

Also, by the conservation of angular momentum,

$$\Sigma m\{y'(w_1-w')-z'(v_1-v')\}=0, \text{ etc.},$$

m denoting an element of the mass considered.

But

$$u_1=\eta z'-\zeta y',$$

$$u'=\frac{\partial u}{\partial x}x'+\frac{\partial u}{\partial y}y'+\frac{\partial u}{\partial z}z', \text{ ultimately},$$

and similar expressions hold good for the other quantities.

Substituting in the above equations, and observing that

$$\Sigma(my'z')=0, \quad \Sigma(mz'x')=0, \quad \Sigma(mx'y')=0,$$

and

$$\Sigma mx'^2=\Sigma my'^2=\Sigma mz'^2,$$

we have

$$2\xi=\frac{\partial w}{\partial y}-\frac{\partial v}{\partial z}, \quad 2\eta=\frac{\partial u}{\partial z}-\frac{\partial w}{\partial x}, \quad 2\zeta=\frac{\partial v}{\partial x}-\frac{\partial u}{\partial y}.$$

We see then that an indefinitely small element of the fluid of which the three principal moments about the centre of gravity are equal, if suddenly solidified and detached from the rest of the fluid, will begin to move with a motion of translation only if $u\,dx+v\,dy+w\,dz$ is an exact differential, but if this expression is not an exact differential the motion of the element will be rotational as well as translational; and this constitutes the reason for the nomenclature of **1·9**.

The quantities ξ, η, ζ are called the components of ***spin***. The term molecular rotation has been used in this sense, but there is no connection between the rotations and the molecules.

2·6. Assuming that the forces are conservative and ρ a function of p, we may write the equations of motion

$$\frac{Du}{Dt}=-\frac{\partial V}{\partial x}-\frac{1}{\rho}\frac{\partial p}{\partial x}=-\frac{\partial Q}{\partial x}, \text{ say},$$

$$\frac{Dv}{Dt}=-\frac{\partial V}{\partial y}-\frac{1}{\rho}\frac{\partial p}{\partial y}=-\frac{\partial Q}{\partial y},$$

$$\frac{Dw}{Dt}=-\frac{\partial V}{\partial z}-\frac{1}{\rho}\frac{\partial p}{\partial z}=-\frac{\partial Q}{\partial z};$$

so that

$$\frac{\partial}{\partial z}\frac{Dv}{Dt}=-\frac{\partial^2 Q}{\partial z\,\partial y}=\frac{\partial}{\partial y}\frac{Dw}{Dt},$$

therefore

$$\frac{D}{Dt}\left(\frac{\partial w}{\partial y}-\frac{\partial v}{\partial z}\right)+\frac{\partial u}{\partial y}\frac{\partial w}{\partial x}+\frac{\partial v}{\partial y}\frac{\partial w}{\partial y}+\frac{\partial w}{\partial y}\frac{\partial w}{\partial z}$$
$$-\frac{\partial u}{\partial z}\frac{\partial v}{\partial x}-\frac{\partial v}{\partial z}\frac{\partial v}{\partial y}-\frac{\partial w}{\partial z}\frac{\partial v}{\partial z}=0,$$

or by adding and subtracting $\frac{\partial v}{\partial x}\frac{\partial w}{\partial x}$, this equation may be written

$$\frac{D\xi}{Dt}=\xi\frac{\partial u}{\partial x}+\eta\frac{\partial v}{\partial x}+\zeta\frac{\partial w}{\partial x}-\xi\left(\frac{\partial u}{\partial x}+\frac{\partial v}{\partial y}+\frac{\partial w}{\partial z}\right)$$

or

$$\frac{D\xi}{Dt}=\xi\frac{\partial u}{\partial x}+\eta\frac{\partial v}{\partial x}+\zeta\frac{\partial w}{\partial x}-\xi\theta,$$

where $\theta=\frac{\partial u}{\partial x}+\frac{\partial v}{\partial y}+\frac{\partial w}{\partial z}$, but by the equation of continuity $\frac{1}{\rho}\frac{D\rho}{Dt}+\theta=0$

Hence we get

$$\frac{D}{Dt}\left(\frac{\xi}{\rho}\right)=\frac{\xi}{\rho}\frac{\partial u}{\partial x}+\frac{\eta}{\rho}\frac{\partial v}{\partial x}+\frac{\zeta}{\rho}\frac{\partial w}{\partial x},$$

$$\frac{D}{Dt}\left(\frac{\eta}{\rho}\right)=\frac{\xi}{\rho}\frac{\partial u}{\partial y}+\frac{\eta}{\rho}\frac{\partial v}{\partial y}+\frac{\zeta}{\rho}\frac{\partial w}{\partial y},$$

$$\frac{D}{Dt}\left(\frac{\zeta}{\rho}\right)=\frac{\xi}{\rho}\frac{\partial u}{\partial z}+\frac{\eta}{\rho}\frac{\partial v}{\partial z}+\frac{\zeta}{\rho}\frac{\partial w}{\partial z}.$$

Also observing that

$$\eta\frac{\partial v}{\partial x}+\zeta\frac{\partial w}{\partial x}=\eta\left(2\zeta+\frac{\partial u}{\partial y}\right)+\zeta\left(\frac{\partial u}{\partial z}-2\eta\right)=\eta\frac{\partial u}{\partial y}+\zeta\frac{\partial u}{\partial z},$$

the equations take the form

$$\left.\begin{aligned}\frac{D}{Dt}\left(\frac{\xi}{\rho}\right)&=\frac{\xi}{\rho}\frac{\partial u}{\partial x}+\frac{\eta}{\rho}\frac{\partial u}{\partial y}+\frac{\zeta}{\rho}\frac{\partial u}{\partial z}\\ \frac{D}{Dt}\left(\frac{\eta}{\rho}\right)&=\frac{\xi}{\rho}\frac{\partial v}{\partial x}+\frac{\eta}{\rho}\frac{\partial v}{\partial y}+\frac{\zeta}{\rho}\frac{\partial v}{\partial z}\\ \frac{D}{Dt}\left(\frac{\zeta}{\rho}\right)&=\frac{\xi}{\rho}\frac{\partial w}{\partial x}+\frac{\eta}{\rho}\frac{\partial w}{\partial y}+\frac{\zeta}{\rho}\frac{\partial w}{\partial z}\end{aligned}\right\}\quad\text{......................(1).}$$

These equations for the case of ρ constant were given by Stokes* and Helmholtz† and were extended to the form given above by Nanson‡.

From these equations Helmholtz concludes that if in a fluid element ξ, η, ζ are simultaneously zero, we also have

$$D\xi/Dt=D\eta/Dt=D\zeta/Dt=0.$$

Hence *those elements of fluid which at any instant have no rotation remain during the whole motion without rotation*. The justification for this conclusion is found in Stokes' paper already cited§. Thus in equations (1) we may assume that $\partial u/\partial x$, $\partial v/\partial x$, etc., are finite, and let L denote their superior limit, then ξ/ρ, η/ρ, ζ/ρ cannot increase faster than if they satisfied the equations

$$\frac{D}{Dt}\left(\frac{\xi}{\rho}\right)=\frac{D}{Dt}\left(\frac{\eta}{\rho}\right)=\frac{D}{Dt}\left(\frac{\zeta}{\rho}\right)=L(\xi+\eta+\zeta)/\rho \quad\text{...........(2);}$$

and if we put $\rho\Omega=\xi+\eta+\zeta$, we have

$$D\Omega/Dt=3L\Omega,$$

so that if Ω be not zero, by dividing by Ω and integrating we get

$$\Omega=Ce^{3Lt},$$

* *Loc. cit.* p. 23.

† *Crelle's Journal*, 1858; *Phil. Mag.* XXXIII, Fourth Series, 1867, p. 485.

‡ *Messenger of Math.* III, 1873, p. 120.

§ Also in *Math. and Phys. Papers*, II, p. 36, or *Camb. and Dub. Math. Journal*, III, p. 215.

and no value of C other than zero will allow Ω to vanish when $t=0$; but by hypothesis ξ, η, ζ, and therefore Ω also, are zero when $t=0$, therefore Ω is always zero. But Ω is the sum of three quantities which evidently cannot be negative, therefore each of them must be zero. And as ξ, η, ζ remain zero when they satisfy (2), still more will they do so when they satisfy (1).

2·7. Impulsive Action. If impulsive forces be made to act on a fluid, or if impulsive pressure be excited by a sudden change of motion of one of the boundaries, it can be shewn as in *Hydrostatics*, Art. 6, that the impulsive pressure at any point is the same in every direction and in the case of a liquid that the impulsive pressure is transmitted equally throughout the liquid. The incompressibility of the liquid implies infinitely rapid propagation of pressural effect, so that an impulsive pressure can be produced instantaneously throughout the liquid.

To find the relation between impulsive pressure and change of velocity.

Let ϖ denote the impulsive pressure and X', Y', Z' the extraneous impulses per unit mass of fluid at the point (x, y, z). Let u, v, w and u', v', w' denote the velocity components at this point just before and just after the impulsive action. Since impulses are measured by the change of momentum they produce, by considering a small parallelepiped $\delta x\,\delta y\,\delta z$ with centre at (x, y, z), we get

$$\rho\,(u'-u)\,\delta x\,\delta y\,\delta z=\rho X'\,\delta x\,\delta y\,\delta z-\frac{\partial\varpi}{\partial x}\,\delta x\,\delta y\,\delta z,$$

the last term representing the difference between the impulsive pressures on the two ends of area $\delta y\,\delta z$ found as in **1·3**.

Therefore

$$\left.\begin{aligned}\rho\,(u'-u)&=\rho X'-\frac{\partial\varpi}{\partial x}\\ \rho\,(v'-v)&=\rho Y'-\frac{\partial\varpi}{\partial y}\\ \rho\,(w'-w)&=\rho Z'-\frac{\partial\varpi}{\partial z}\end{aligned}\right\}\quad\text{..................(1).}$$

If there are no extraneous impulses the equations are equivalent to

$$d\varpi=-\rho\,(u'-u)\,dx-\rho\,(v'-v)\,dy-\rho\,(w'-w)\,dz,$$

or if ϕ, ϕ' denote the velocity potential just before and just after the impulsive action,

$$d\varpi=\rho\,(d\phi'-d\phi);$$

hence, by integration, when ρ is constant

$$\varpi = \rho\phi' - \rho\phi + C.$$

The constant C may be omitted, as an extra pressure, constant throughout the fluid, would not affect the motion.

2·71. Physical meaning of velocity potential. From the preceding article we see that any actual motion of a liquid, for which a single valued velocity potential exists, could be produced instantaneously from rest by a set of impulses properly applied, and if the liquid be regarded as of unit density the velocity potential is the impulsive pressure at any point.

We also conclude that when a state of rotational motion exists in a liquid, the motion could neither be created nor destroyed by impulsive pressures.

2·72. When there are no extraneous impulses and ρ is constant, by differentiating equations **2·7** (1), and making use of the equation of continuity, we obtain

$$\frac{\partial^2\varpi}{\partial x^2} + \frac{\partial^2\varpi}{\partial y^2} + \frac{\partial^2\varpi}{\partial z^2} = 0 \quad \text{..................(1)},$$

and the general problem of impulsive motion consists in obtaining a solution of this equation to satisfy the given boundary conditions.

2·73. It was pointed out by Stokes* that in selecting a solution to satisfy the given boundary conditions it is necessary also to note that the value of the fluid pressure, whether finite or impulsive pressure, cannot change abruptly from point to point in the fluid. He considers the following example: Suppose a mass of fluid to be at rest in a finite cylinder, whose axis coincides with the axis of z, the cylinder being entirely filled and closed at both ends. Suppose the cylinder to be moved by impact with initial velocity C in the direction of x; then the velocities are given by

$$u = C, \quad v = 0, \quad w = 0.$$

For these make $u\,dx + v\,dy + w\,dz$ an exact differential $-d\phi$, where ϕ satisfies (1) of **2·72**; they also make the normal velocity equal to that of the cylinder over the boundary, and give a value for the impulsive pressure, namely $C' - C\rho x$, which does not alter abruptly. But if we had supposed that ϕ was equal to $-Cx - C'\tan^{-1} y/x$ all the conditions would still have been satisfied, except that we should have obtained for the impulsive pressure a value $\varpi = C'' - \rho(Cx + C'\tan^{-1} y/x)$, in which the last term alters abruptly as $\tan^{-1} y/x$ passes through the value 2π. Hence the former was the correct solution of the problem.

* *Trans. Camb. Phil. Soc.* VIII, p. 105, or *Math. and Phys. Papers*, I, p. 23.

This is also an illustration of a theorem we shall have to discuss later, namely that cyclic irrotational motion cannot exist in simply connected space.

2·8. The following examples will serve to illustrate the application to particular cases of the principles of hydrodynamics.

(1) *A quantity of liquid occupies a length 2l of a straight tube of uniform small bore under the action of a force to a point in the tube varying as the distance from that point. It is required to determine the motion and the pressure.*

Let p be the pressure and u the velocity at a distance x from the fixed point O; and let z be the distance of the nearer free surface from O.

The equation of continuity is

$$\partial u/\partial x = 0.$$

The equation of motion is therefore

$$\frac{\partial u}{\partial t} = -\mu x - \frac{1}{\rho}\frac{\partial p}{\partial x}.$$

Integrate this equation with regard to x,

therefore $$x\frac{\partial u}{\partial t} = C - \tfrac{1}{2}\mu x^2 - \frac{p}{\rho},$$

and $p = 0$ when $x = z$ and when $x = z + 2l$,

therefore $$\frac{\partial u}{\partial t} = -\mu(z+l).$$

But clearly $$u = \dot{z},$$

therefore $$\ddot{z} + \mu(z+l) = 0,$$

hence $z + l = A\cos(\sqrt{\mu}t + \alpha)$, the constants being determined by the initial position and velocity.

Also $$p/\rho = -\tfrac{1}{2}\mu(x^2 - z^2) - (x-z)\frac{\partial u}{\partial t}$$

$$= -\tfrac{1}{2}\mu(x^2 - z^2) + \mu(x-z)(z+l),$$

and thus the pressure at any point is determined.

(2) *Oscillations of water in a bent uniform tube in a vertical plane.*

Let O be the lowest point of the tube, AB the equilibrium level of the water, h the height of AB above O, α, β the inclinations of the tube to the horizontal at A and B and θ its inclination at a distance s from O. Let a, b denote the lengths OA, OB and suppose that at time t the water is dis-

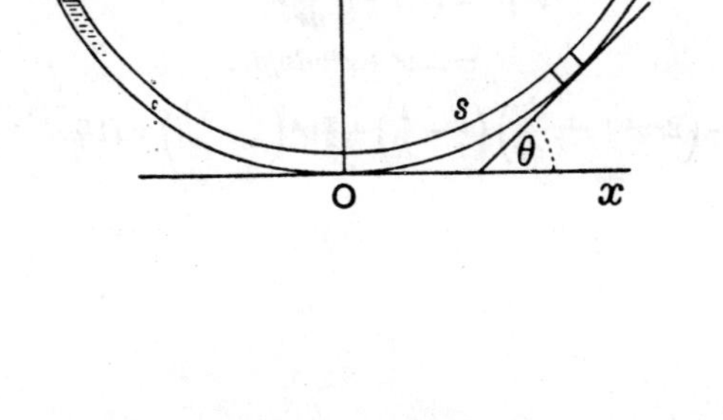

placed a small distance x along the tube from its equilibrium position. Then if u is the velocity the equation of continuity is $\partial u/\partial s=0$, so that the equation of motion is

$$\frac{\partial u}{\partial t}=-g\sin\theta-\frac{1}{\rho}\frac{\partial p}{\partial s},$$

or

$$\frac{\partial u}{\partial t}=-g\frac{dy}{ds}-\frac{1}{\rho}\frac{\partial p}{\partial s}.$$

Integrating this equation, we get

$$s\frac{\partial u}{\partial t}=-gy-\frac{p}{\rho}+F(t)$$

and by taking the values at the ends of the water,

$$(a+x)\frac{\partial u}{\partial t}=-g(h+x\sin\alpha)-\frac{\Pi}{\rho}+F(t),$$

and

$$-(b-x)\frac{\partial u}{\partial t}=-g(h-x\sin\beta)-\frac{\Pi}{\rho}+F(t),$$

where Π is the atmospheric pressure.

Therefore
$$(a+b)\frac{\partial u}{\partial t}=-gx(\sin\alpha+\sin\beta);$$

but $u=\dot{x}$, so that this equation represents oscillations of period

$$2\pi\sqrt{\{(a+b)/g(\sin\alpha+\sin\beta)\}}.$$

(3) *A mass of liquid surrounds a solid sphere of radius a, and its outer surface, which is a concentric sphere of radius b, is subject to a given constant pressure Π, no other forces being in action on the liquid. The solid sphere suddenly shrinks into a concentric sphere; it is required to determine the subsequent motion, and the impulsive action on the sphere.*

At time t, let p be the pressure and v' the velocity at distance r' from the centre; then the equation of motion is

$$\frac{\partial v'}{\partial t}+v'\frac{\partial v'}{\partial r'}=-\frac{1}{\rho}\frac{\partial p}{\partial r'};$$

and the equation of continuity is

$$r'^2v'=F(t),$$

therefore
$$\frac{F'(t)}{r'^2}+v'\frac{\partial v'}{\partial r'}=-\frac{1}{\rho}\frac{\partial p}{\partial r'}\quad\text{.......................(1).}$$

Let R, r be the radii of the external and internal boundaries at time t, and V, v their velocities; these quantities are functions of t only, and

$$V=\dot{R},\quad v=\dot{r}.$$

Integrating equation (1) with regard to r' from $r'=r$ to $r'=R$, we get

$$-F'(t)\left(\frac{1}{r}-\frac{1}{R}\right)+\tfrac{1}{2}(v^2-V^2)=\Pi/\rho.$$

But
$$F(t)=r^2v=R^2V,$$

therefore
$$F'(t)=2rv\dot{r}+r^2\frac{dv}{dr}\dot{r}$$
$$=2rv^2+r^2v\,dv/dr.$$

Hence
$$-\left(2rv^2+r^2v\frac{dv}{dr}\right)\left(\frac{1}{r}-\frac{1}{R}\right)+\tfrac{1}{2}v^2\left(1-\frac{r^4}{R^4}\right)=\Pi/\rho.$$

Putting $v^2=z$, and multiplying by $2r^2$, and observing that

$$R^3-r^3=b^3-a^3=c^3,$$

this becomes

$$2\Pi r^2/\rho=-\left\{\frac{1}{r}-\frac{1}{(r^3+c^3)^{\frac{1}{3}}}\right\}\frac{d}{dr}(zr^4)+zr^4\left\{\frac{1}{r^2}-\frac{r^2}{(r^3+c^3)^{\frac{4}{3}}}\right\}.$$

Integrating, we obtain

$$\frac{2}{3}\frac{\Pi}{\rho}\frac{a^3-r^3}{r^4}=v^2\left(\frac{1}{r}-\frac{1}{R}\right).$$

Take r for the radius of the solid sphere, and let ϖ denote the impulsive pressure at distance r'; then

$$d\varpi=-\rho v'dr'=-\rho\frac{r^2v\,dr'}{r'^2};$$

therefore, since $\varpi=0$ when $r'=R$,

$$\varpi/\rho=r^2v\left(\frac{1}{r}-\frac{1}{R}\right)$$

gives the impulsive pressure when $r'=r$.

The whole impulse on the sphere

$$=4\pi r^2\varpi=4\pi\rho r^3v(R-r)/R,$$

and the whole momentum destroyed

$$=\int_r^R 4\pi r'^2\rho v'dr'=4\pi\rho r^2v(R-r).$$

The velocity may also be obtained at once by help of the principle of energy.

For, the kinetic energy

$$=\tfrac{1}{2}\int_r^R 4\pi r'^2\rho v'^2dr'$$

$$=2\pi\rho\int_r^R\frac{r^4v^2}{r'^2}dr'=2\pi\rho r^4v^2\left(\frac{1}{r}-\frac{1}{R}\right),$$

and the work done by the outer pressure

$$=\int_b^R 4\pi R^2\Pi(-dR)$$

$$=\tfrac{4}{3}\pi\Pi(b^3-R^3)=\tfrac{4}{3}\pi\Pi(a^3-r^3).$$

Hence the equation of energy gives us at once the expression for the velocity.

EXAMPLES

1. If a bombshell explode at a great depth beneath the surface of the sea, prove that the impulsive pressure at any point varies inversely as the distance from the centre of the shell.

2. A straight tube of small bore, ABC, is bent so as to make the angle ABC a right angle, and AB equal to BC. The end C is closed; and the tube is placed with the end A upwards and AB vertical, and is filled with liquid. If the end C be opened, prove that the pressure at any point of the vertical tube is instantaneously diminished one-half; and find the instantaneous change of pressure at any point of the horizontal tube, the pressure of the atmosphere being neglected.

3. Steam is rushing from a boiler through a conical pipe, the diameters of the ends of which are D and d; if V and v be the corresponding velocities of the steam, and if the motion be supposed to be that of divergence from the vertex of the cone, prove that

$$\frac{v}{V}=\frac{D^2}{d^2}\cdot e^{\frac{v^2-V^2}{2k}},$$

where k is the pressure divided by the density, and supposed constant.

4. An elastic fluid, the weight of which is neglected, obeying Boyle's law, is in motion in a uniform straight tube; shew that on the hypothesis of parallel sections the velocity at any time t at a distance r from a fixed point in the tube is defined by the equation

$$\frac{\partial^2 v}{\partial t^2}+\frac{\partial}{\partial r}\left(2v\frac{\partial v}{\partial t}+v^2\frac{\partial v}{\partial r}\right)=k\frac{\partial^2 v}{\partial r^2}.$$

5. Air, obeying Boyle's law, is in motion in a uniform tube of small section; prove that if ρ be the density and v the velocity at a distance x from a fixed point at the time t,

$$\frac{\partial^2\rho}{\partial t^2}=\frac{\partial^2}{\partial x^2}\{(v^2+k)\,\rho\}.$$

6. Two equal closed cylinders, of height c, with their bases in the same horizontal plane, are filled, one with water, and the other with air of such a density as to support a column h of water, h being less than c. If a communication be opened between them at their bases, the height x, to which the water rises, is given by the equation

$$cx-x^2+ch\log\frac{c-x}{c}=0.$$

7. Water flows steadily along a pipe of variable cross section. If the pressure be 700 millimetres of mercury at a place where the velocity is 150 cms. per second, find the pressure at a place where the cross section of the pipe is twice as large. [Take the specific gravity of mercury as 13·6.]
(Univ. of London, 1907.)

8. A sphere of radius a is surrounded by infinite liquid of density ρ, the pressure at infinity being ϖ. The sphere is suddenly annihilated. Shew that the pressure at distance r from the centre immediately falls to

$$\varpi\left(1-\frac{a}{r}\right).$$

Shew further that if the liquid is brought to rest by impinging on a concentric sphere of radius $a/2$, the impulsive pressure sustained by the surface of this sphere is $\sqrt{7\varpi\rho a^2/6}$. (M.T. 1931.)

9. A spherical shell of homogeneous gravitating liquid, having no initial motion, is left to itself; find the pressure at any point during the collapse.

10. A mass of homogeneous liquid is moving so that the velocity at any point is proportional to the time, and that the pressure is given by

$$\frac{p}{\rho}=\mu xyz-\tfrac{1}{2}t^2(y^2z^2+z^2x^2+x^2y^2);$$

prove that this motion may have been generated from rest by finite natural forces independent of the time; and shew that, if the direction of motion at every point coincide with the direction of the acting force, each particle of the liquid describes a curve which is the intersection of two hyperbolic cylinders. (M.T. 1877.)

11. A given quantity of liquid moves, under no forces, in a smooth conical tube having a small vertical angle, and the distances of its nearer and farther extremities from the vertex at the time t are r and r'; shew that

$$2r\frac{d^2r}{dt^2}+\left(\frac{dr}{dt}\right)^2\left\{3-\frac{r}{r'}-\frac{r^2}{r'^2}-\frac{r^3}{r'^3}\right\}=0.$$

the pressures at the two surfaces being equal.

Shew also that the preceding equation results from supposing the vis viva of the mass of liquid to be constant; and that the velocity of the inner surface is given by the equations

$$V^2=Cr'/r^3(r'-r),\quad r'^3-r^3=c^3,$$

C and c being constants.

12. A portion of homogeneous fluid is confined between two concentric spheres radii A and a, and is attracted towards their centre by a force varying inversely as the square of the distance. The inner spherical surface is suddenly annihilated, and when the radii of the inner and outer surfaces of the fluid are r and R, the fluid impinges on a solid ball concentric with their surfaces; prove that the impulsive pressure at any point of the ball for different values of R and r varies as

$$\sqrt{\left\{(a^2-r^2-A^2+R^2)\left(\frac{1}{r}-\frac{1}{R}\right)\right\}}.$$

13. A fine tube whose section k is a function of its length s, in the form of a closed plane curve of area A, filled with ice, is moved in any manner. When the component angular velocity of the tube about a normal to its plane is Ω the ice melts without change of volume. Prove that the velocity of the fluid relatively to the tube at a point where the section is K at any subsequent time when ω is the angular velocity is

$$2A(\Omega-\omega)\div K\int\frac{ds}{k},$$

the integral being taken once round the tube. (M.T. 1873.)

14. A centre of force attracting inversely as the square of the distance is at the centre of a spherical cavity within an infinite mass of incompressible fluid, the pressure on which at an infinite distance is ϖ, and is such that the work done by this pressure on a unit of area through a unit of length is one-half the work done by the attractive force on a unit of volume of the fluid from infinity to the initial boundary of the cavity; prove that the time of filling up the cavity will be

$$\pi a\sqrt{\frac{\rho}{\varpi}}\{2-(\tfrac{3}{2})^{\frac{3}{2}}\};$$

a being the initial radius of the cavity, and ρ the density of the fluid. (M.T. 1874.)

15. A homogeneous liquid is contained between two concentric spherical surfaces, the radius of the inner being a and that of the outer indefinitely great. The fluid is attracted to the centre of these surfaces by a force $\phi(r)$, and a constant pressure Π is exerted at the outer surface.

Suppose $\int \phi(r)\,dr = \psi(r)$, and that $\psi(r)$ vanishes when r is infinite. Shew that if the inner surface is suddenly removed, the pressure at the distance r is suddenly diminished by

$$\Pi \frac{a}{r} - \frac{a\rho}{r}\psi(a).$$

Find $\phi(r)$ so that the pressure immediately after the inner surface is removed may be the same as it would be if no attractive force existed. Also with this value of $\phi(r)$, find the velocity of the inner boundary of the fluid at any period of the motion.

16. A stream in a horizontal pipe, after passing a contraction in the pipe at which its sectional area is A, is delivered at atmospheric pressure at a place where the sectional area is B. Shew that if a side tube is connected with the pipe at the former place, water will be sucked up through it into the pipe from a reservoir at a depth $\frac{s^2}{2g}\left(\frac{1}{A^2} - \frac{1}{B^2}\right)$ below the pipe; s being the delivery per second. (St John's Coll. 1896.)

17. A sphere whose radius at time t is $b + a\cos nt$ is surrounded by liquid extending to infinity under no forces. Prove that the pressure at distance r from the centre is less than the pressure at an infinite distance by

$$\rho \frac{n^2 a}{r}(b + a\cos nt)\left\{a(1 - 3\sin^2 nt) + b\cos nt + \frac{1}{2}\frac{a}{r^3}\sin^2 nt\,(b + a\cos nt)^3\right\}.$$

(Coll. Exam. 1913.)

18. A sphere of radius a is alone in an unbounded liquid, which is at rest at a great distance from the sphere and is subject to no external forces. The sphere is forced to vibrate radially keeping its spherical shape, the radius r at any time being given by $r = a + b\cos nt$. Shew that if Π is the pressure in the liquid at a great distance from the sphere the least pressure (assumed positive) at the surface of the sphere during the motion is $\Pi - n^2\rho b(a + b)$. (M.T. 1913.)

19. Shew that the rate per unit of time at which work is done by the internal pressures between the parts of a compressible fluid is

$$\iiint p\left(\frac{\partial u}{\partial x} + \frac{\partial v}{\partial y} + \frac{\partial w}{\partial z}\right) dx\,dy\,dz,$$

where p is the pressure, and (u, v, w) the velocity at any point, and the integration extends through the volume of the fluid. (St John's Coll. 1898.)

20. A sphere is at rest in an infinite mass of homogeneous liquid of density ρ, the pressure at infinity being ϖ. Shew that, if the radius R of the sphere varies in any manner, the pressure at the surface of the sphere at any time is

$$\varpi + \tfrac{1}{2}\rho\left\{\frac{d^2}{dt^2}(R^2) + \left(\frac{dR}{dt}\right)^2\right\}.$$

(Coll. Exam. 1900.)

21. An infinite mass of homogeneous incompressible fluid is at rest subject to a uniform pressure Π, and contains a spherical cavity of radius a, filled with gas at a pressure $m\Pi$; prove that, if the inertia of the gas be neglected, and Boyle's law be supposed to hold throughout the ensuing motion, the radius of the sphere will oscillate between the values a and na, where n is determined by the equation

$$1 + 3m \log n - n^3 = 0.$$

If m be nearly equal to 1, the time of an oscillation will be $2\pi\sqrt{\dfrac{a^2\rho}{3\Pi}}$, ρ being the density of the fluid. (M.T. 1869.)

22. A mass of liquid, of density ρ and volume $\frac{4}{3}\pi c^3$, is in the form of a spherical shell; a constant pressure Π is exerted on the external surface of the shell, there is no pressure on the internal surface, and no other forces act on the liquid; initially the liquid is at rest and the internal radius of the shell is $2c$, prove that the velocity of the internal surface, when its radius is c, is

$$\sqrt{\frac{14\Pi}{3\rho} \cdot \frac{2^{\frac{1}{3}}}{2^{\frac{1}{3}} - 1}}. \qquad \text{(Coll. Exam. 1904.)}$$

23. Investigate an expression for the change in an indefinitely short time in the mass of fluid contained within a spherical surface of small radius.

Prove that the momentum of the mass in the direction of the axis of x is greater than it would be if the whole were moving with the velocity at the centre by

$$\frac{1}{5}\frac{Ma^2}{\rho}\left\{\frac{\partial\rho}{\partial x}\frac{\partial u}{\partial x} + \frac{\partial\rho}{\partial y}\frac{\partial u}{\partial y} + \frac{\partial\rho}{\partial z}\frac{\partial u}{\partial z} + \frac{1}{2}\rho\left(\frac{\partial^2 u}{\partial x^2} + \frac{\partial^2 u}{\partial y^2} + \frac{\partial^2 u}{\partial z^2}\right)\right\}.$$

(M.T. 1876.)

24. An infinite fluid in which is a spherical hollow of radius a is initially at rest under the action of no forces. If a constant pressure Π is applied at infinity, shew that the time of filling up the cavity is

$$\pi^2 a (\rho/\Pi)^{\frac{1}{2}} \, 2^{\frac{5}{6}} \{\Gamma(\tfrac{1}{3})\}^{-3}. \qquad \text{(Trinity Coll. 1900.)}$$

25. A solid sphere of radius a is surrounded by a mass of liquid whose volume is $4\pi c^3/3$, and its centre is a centre of attractive force varying directly as the square of the distance. If the solid sphere be suddenly annihilated, shew that the velocity of the inner surface, when its radius is x, is given by

$$\dot{x}^2 x^3 \{(x^3 + c^3)^{\frac{1}{3}} - x\} = (2\Pi/3\rho + 2\mu c^3/9)(a^3 - x^3)(c^3 + x^3)^{\frac{1}{3}},$$

where ρ is the density, Π the external pressure and μ the absolute force. (M.T. 1881.)

26. A mass of gravitating fluid is at rest under its own attraction only; the free surface being a sphere of radius b and the inner surface a rigid concentric shell of radius a. Shew that if this shell suddenly disappear, the initial pressure at any point of the fluid at distance r from the centre is

$$\tfrac{2}{3}\pi\rho^2 (b - r)(r - a)\left(\frac{a + b}{r} + 1\right). \qquad \text{(Trinity Coll. 1902.)}$$

27. A spherical hollow of radius a initially exists in an infinite fluid, subject to constant pressure at infinity. Shew that the pressure at distance r from the centre when the radius of the cavity is x is to the pressure at infinity as

$$3x^2r^4 + (a^3 - 4x^3)\,r^3 - (a^3 - x^3)\,x^3 : 3x^2r^4.$$

(Trinity Coll. 1903.)

28. A spherical mass of liquid of radius b has a concentric spherical cavity of radius a, which contains gas at pressure p whose mass may be neglected: at every point of the external boundary of the liquid an impulsive pressure ϖ per unit area is applied. Assuming that the gas obeys Boyle's law, shew that when the liquid first comes to rest, the radius of the internal spherical surface will be

$$a \exp\{-\varpi^2 b/2p\rho a^2 (b-a)\},$$

where ρ is the density of the liquid. (M.T. 1900.)

29. A mass of homogeneous liquid, whose bounding surfaces are concentric spheres, is at rest under the action of no forces in a gas of uniform pressure. If the pressure of the external gas be suddenly increased, determine the instantaneous pressure in the liquid, and investigate the differential equation for the subsequent motion of the liquid and the pressure inside the shell at any time. (Coll. Exam. 1895.)

30. A volume $\frac{4}{3}\pi c^3$ of gravitating liquid, of density ρ, is initially in the form of a spherical shell of infinitely great radius. If the liquid shell contract under the influence of its own attraction, there being no external or internal pressure, shew that when the radius of the inner spherical surface is x, its velocity will be given by

$$V^2 = \frac{4\pi\gamma\rho z}{15x^3}(2z^4 + 2z^3x + 2z^2x^2 - 3zx^3 - 3x^4),$$

where γ is the constant of gravitation, and $z^3 = x^3 + c^3$. (M.T. 1899.)

31. A mass of uniform liquid is in the form of a thick spherical shell bounded by concentric spheres of radii a and b $(a<b)$. The cavity is filled with gas the pressure of which varies according to Boyle's law, and is initially equal to the atmospheric pressure Π, and the mass of which may be neglected. The outer surface of the shell is exposed to atmospheric pressure. Prove that if the system is symmetrically disturbed, so that each particle moves along the line joining it to the centre the time of a small oscillation is

$$2\pi a\{\rho(b-a)/3\Pi b\}^{\frac{1}{2}},$$

where ρ is the density of the liquid. (Coll. Exam. 1896.)

32. A mass of perfect incompressible fluid, of density ρ, is bounded by concentric spherical surfaces. The outer surface is contained by a flexible envelope which exerts continuously a uniform pressure Π and contracts from radius R_1 to radius R_2. The hollow is filled with a gas obeying Boyle's law, its radius contracts from c_1 to c_2, and the pressure of the gas is initially p_1. Initially the whole mass is at rest. Prove that, neglecting the mass of the gas, the velocity (v) of the inner surface when the configuration (R_2, c_2) is reached is given by

$$\frac{1}{2}v^2 = \frac{c_1^3}{c_2^3}\left\{\frac{1}{3}\left(1-\frac{c_2^3}{c_1^3}\right)\frac{\Pi}{\rho} - \frac{p_1}{\rho}\log\frac{c_1}{c_2}\right\}\Big/\left(1-\frac{c_2}{R_2}\right).$$

(Trinity Coll. 1908.)

33. An infinite mass of fluid is acted on by a force $\mu r^{-\frac{3}{2}}$ per unit mass directed to the origin. If initially the fluid is at rest and there is a cavity in the form of the sphere $r=c$ in it, shew that the cavity will be filled up after an interval of time $\left(\frac{2}{5\mu}\right)^{\frac{1}{2}} c^{\frac{5}{4}}$. (Trinity Coll. 1905.)

34. Explain on general grounds why two pulsating spheres in a liquid attract each other, if they are always in the same phase. (Coll. Exam. 1905.)

35. A mass of liquid of density ρ whose external surface is a long circular cylinder of radius a, which is subject to a constant pressure Π, surrounds a coaxial long circular cylinder of radius b. The internal cylinder is suddenly destroyed, shew that if v is the velocity at the internal surface when the radius is r, then

$$v^2 = \frac{2\Pi(b^2-r^2)}{\rho r^2 \log(r^2+a^2-b^2)/r^2}.$$ (Coll. Exam. 1894.)

36. Liquid is contained between two parallel planes; the free surface is a circular cylinder of radius a whose axis is perpendicular to the planes. All the liquid within a concentric circular cylinder of radius b is suddenly annihilated; prove that if ϖ be the pressure at the outer surface, the initial pressure at any point of the liquid distant r from the centre is

$$\varpi \frac{\log r - \log b}{\log a - \log b}.$$ (Coll. Exam. 1896.)

37. Prove that the differential equations of motion for a frictionless fluid are

$$\frac{1}{\rho}\frac{\partial p}{\partial x} - X + \frac{\partial u}{\partial t} - 2v\omega_3 + 2w\omega_2 + u\frac{\partial u}{\partial x} + v\frac{\partial u}{\partial y} + w\frac{\partial u}{\partial z}$$
$$-(\omega_2^2+\omega_3^2)x - \left(\frac{d\omega_3}{dt} - \omega_1\omega_2\right)y + \left(\frac{d\omega_2}{dt} + \omega_3\omega_1\right)z = 0,$$

and two similar equations; u, v, w being the components of the velocity at the time t at the point x, y, z *relative* to moving axes having component angular velocities $\omega_1, \omega_2, \omega_3$. (M.T. 1881.)

38. The motion of an incompressible fluid is referred to rectangular axes which are rotating with constant angular velocities $\theta_1, \theta_2, \theta_3$: prove that the equation of continuity is $\frac{\partial U}{\partial x} + \frac{\partial V}{\partial y} + \frac{\partial W}{\partial z} = 0$, and that the equations of motion are

$$\frac{\partial U}{\partial t} - 2V\zeta + 2W\eta = -\frac{\partial}{\partial x}\left(\frac{p}{\rho} + \tfrac{1}{2}Q^2\right) + X,$$

and two similar equations, where U, V, W are the velocities *relative* to the axes, and

$$Q^2 = U^2+V^2+W^2-(\theta_1^2+\theta_2^2+\theta_3^2)(x^2+y^2+z^2)+(\theta_1 x+\theta_2 y+\theta_3 z)^2.$$

(Trinity Coll. 1898.)

39. If the motion is irrotational and the axes to which the motion is referred rotate with angular velocities $\theta_1, \theta_2, \theta_3$, shew that

$$\frac{p}{\rho} + V + \tfrac{1}{2}q^2 + \theta_1(zv-yw) + \theta_2(xw-zu) + \theta_3(yu-xv) - \frac{\partial\phi}{\partial t}$$

is a function of the time. (M.T. 1898.)

CHAPTER III

PARTICULAR METHODS AND APPLICATIONS

3·1. Motion in Two Dimensions. The Current Function. When the motion is the same in all planes parallel to that of xy, and there is no velocity parallel to the z-axis, i.e. when u, v are functions of x, y only, and $w=0$, we may regard the motion as two-dimensional and consider only the circumstances in the plane xy; and when we speak of the flow across a curve in this plane we shall mean the flow across unit length of a cylinder whose trace on the plane xy is the curve in question, the generators of the cylinder being parallel to the z-axis.

The differential equation of the lines of flow in this case is

$$v\,dx-u\,dy=0 \quad \text{......................(1)},$$

and the equation of continuity is

$$\frac{\partial u}{\partial x}+\frac{\partial v}{\partial y}=0; \quad \text{or} \quad \frac{\partial u}{\partial x}=\frac{\partial(-v)}{\partial y};$$

which shews that the left-hand member of (1) is an exact differential, $d\psi$ say; i.e.

$$v\,dx-u\,dy=d\psi=\frac{\partial\psi}{\partial x}dx+\frac{\partial\psi}{\partial y}dy,$$

so that $$u=-\frac{\partial\psi}{\partial y} \text{ and } v=\frac{\partial\psi}{\partial x}.$$

This function ψ is called the **stream function** or the **current function**, and it is clear that the lines of flow are given by the equation

$$\psi=C,$$

where C is an arbitrary constant.

A property of the current function is that the difference of its values at two points represents the flow across any line joining the points.

For if ds be an element of a curve and θ the inclination of the tangent to the x-axis, the flow across the curve from right to left

$$=\int(v\cos\theta-u\sin\theta)\,ds=\int\left(\frac{\partial\psi}{\partial x}dx+\frac{\partial\psi}{\partial y}dy\right)$$

$$=\int d\psi=\psi_2-\psi_1;$$

where by 'from right to left' we mean in relation to an observer placed on the curve and looking in the direction in which s increases, the axes being so placed that rotation from x towards y is counterclockwise.

We might also define the value of the current function ψ at any point P as the amount of flow across a curve AP where A is some fixed point in the plane, for this makes

$$\psi = \int_A^P (v\cos\theta - u\sin\theta)\,ds$$

$$= \int_A^P (v\,dx - u\,dy).$$

And by varying the position of P, we get

$$v = \partial\psi/\partial x \quad \text{and} \quad u = -\partial\psi/\partial y,$$

in agreement with our former definition. Also it is easily seen that the velocity from right to left across any arc ds is $\partial\psi/\partial s$.

3·11. It is to be observed that the existence of the current function does not depend on whether the motion is irrotational or rotational. For the components of spin as defined in **2·52** we have

$$\xi = \frac{1}{2}\left(\frac{\partial w}{\partial y} - \frac{\partial v}{\partial z}\right) = 0; \quad \eta = \frac{1}{2}\left(\frac{\partial u}{\partial z} - \frac{\partial w}{\partial x}\right) = 0,$$

and

$$\zeta = \frac{1}{2}\left(\frac{\partial v}{\partial x} - \frac{\partial u}{\partial y}\right) = \frac{1}{2}\left(\frac{\partial^2\psi}{\partial x^2} + \frac{\partial^2\psi}{\partial y^2}\right).$$

Hence in irrotational motion the current function has to satisfy

$$\frac{\partial^2\psi}{\partial x^2} + \frac{\partial^2\psi}{\partial y^2} = 0.$$

3·2. Irrotational motion in two dimensions. When there is a velocity potential ϕ we have

$$\frac{\partial\phi}{\partial x} = -u = \frac{\partial\psi}{\partial y} \quad \text{and} \quad \frac{\partial\phi}{\partial y} = -v = -\frac{\partial\psi}{\partial x} \quad \text{.........(1)}$$

The equation of continuity is

$$\frac{\partial^2\phi}{\partial x^2} + \frac{\partial^2\phi}{\partial y^2} = 0,$$

and as we saw in **3·11**, ψ must satisfy the same equation.

The equations (1) shew that

$$\frac{\partial\phi}{\partial x}\frac{\partial\psi}{\partial x} + \frac{\partial\phi}{\partial y}\frac{\partial\psi}{\partial y} = 0,$$

so that the families of curves

$$\phi = \text{const.}, \quad \psi = \text{const.}$$

cut orthogonally at all their points of intersection.

These conditions are satisfied by taking $\phi+i\psi$ to be a function of the complex variable $x+iy$.

Thus if we write $\phi+i\psi=f(x+iy)$, we have

$$\frac{\partial\phi}{\partial x}+i\frac{\partial\psi}{\partial x}=f'(x+iy),$$

and
$$\frac{\partial\phi}{\partial y}+i\frac{\partial\psi}{\partial y}=if'(x+iy)=i\frac{\partial\phi}{\partial x}-\frac{\partial\psi}{\partial x},$$

so that
$$\frac{\partial\phi}{\partial x}=\frac{\partial\psi}{\partial y}, \text{ and } \frac{\partial\phi}{\partial y}=-\frac{\partial\psi}{\partial x}.$$

Such functions are called *conjugate functions*; and we see that if ϕ, ψ are two conjugate functions, a possible form of irrotational motion is obtained by taking the curves $\phi=$const. to be curves of equi-velocity potential, and the curves $\psi=$const. to be stream lines.

3·21. In the theory of functions of a complex variable, if z denote the complex variable $x+iy$, and w the function $\phi+i\psi$, the relation $w=f(z)$ implies that w has a definite differential coefficient with respect to z or that the limit of $\dfrac{f(z')-f(z)}{z'-z}$ as z' tends to z is independent of the path by which the point z' approaches z.

But
$$\frac{\delta w}{\delta z}=\frac{\delta(\phi+i\psi)}{\delta(x+iy)}=\frac{\left(\dfrac{\partial\phi}{\partial x}+i\dfrac{\partial\psi}{\partial x}\right)\delta x+\left(\dfrac{\partial\phi}{\partial y}+i\dfrac{\partial\psi}{\partial y}\right)\delta y}{\delta x+i\delta y},$$

and if this is to approach a definite limit as δx and δy tend to zero, independently of the ratio $\delta x:\delta y$, we must have

$$\frac{\partial\phi}{\partial y}+i\frac{\partial\psi}{\partial y}=i\left(\frac{\partial\phi}{\partial x}+i\frac{\partial\psi}{\partial x}\right).$$

Hence, as before,

$$\frac{\partial\phi}{\partial x}=\frac{\partial\psi}{\partial y} \text{ and } \frac{\partial\phi}{\partial y}=-\frac{\partial\psi}{\partial x};$$

and we have for the value of the differential coefficient

$$\frac{dw}{dz}=\frac{\partial\phi}{\partial x}+i\frac{\partial\psi}{\partial x}=\frac{\partial\phi}{\partial x}-i\frac{\partial\phi}{\partial y}.$$

It follows that any relation of the form $w=f(z)$, or

$$\phi+i\psi=f(x+iy),$$

represents a two-dimensional irrotational motion, in which the magnitude of the velocity at any point is given by $\left|\frac{dw}{dz}\right|$. For

$$\left|\frac{dw}{dz}\right| = \left\{\left(\frac{\partial\phi}{\partial x}\right)^2 + \left(\frac{\partial\phi}{\partial y}\right)^2\right\}^{\frac{1}{2}} = (u^2+v^2)^{\frac{1}{2}}$$

$$= \text{velocity}.$$

Also, if the curves $\phi=$const., $\psi=$const. are drawn, and δs_1 denotes the arc of the curve ψ intercepted between ϕ and $\phi+\delta\phi$, the velocity at P where ϕ and ψ intersect being normal to the curve ϕ is $-\frac{\partial\phi}{\partial s_1}$. Similarly if δs_2 be the arc of the curve ϕ intercepted between ψ and $\psi+\delta\psi$, the velocity at P as measured by the rate of flow across δs_2 is $-\frac{\partial\psi}{\partial s_2}$, where we adopt the convention of sign of **3·1**, so that with curves placed as in the figure $\partial\phi/\partial s_1 = \partial\psi/\partial s_2$, but if we interchange the relative positions of the ϕ and ψ curves we should obtain

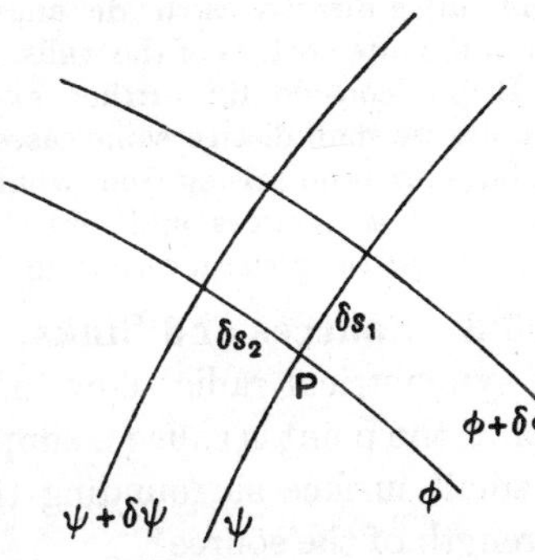

$$\partial\phi/\partial s_1 = -\partial\psi/\partial s_2.$$

3·22. Since the conditions $\frac{\partial\phi}{\partial x}=\frac{\partial\psi}{\partial y}$, $\frac{\partial\phi}{\partial y}=-\frac{\partial\psi}{\partial x}$ are also satisfied by the relation

$$-\psi+i\phi=f(x+iy),$$

it follows that from any given two-dimensional form of irrotational motion another may in general be deduced by interchanging the lines of equi-velocity potential and the stream lines.

If the motion be referred to polar coordinates, we have

$$\frac{\partial\phi}{\partial r}=\frac{\partial\phi}{\partial x}\cos\theta+\frac{\partial\phi}{\partial y}\sin\theta=\frac{\partial\psi}{\partial y}\cos\theta-\frac{\partial\psi}{\partial x}\sin\theta=\frac{1}{r}\left(\frac{\partial\psi}{\partial y}\frac{\partial y}{\partial\theta}+\frac{\partial\psi}{\partial x}\frac{\partial x}{\partial\theta}\right)=\frac{\partial\psi}{r\partial\theta};$$

and $$\frac{\partial\phi}{r\partial\theta}=-\frac{\partial\phi}{\partial x}\sin\theta+\frac{\partial\phi}{\partial y}\cos\theta=-\frac{\partial\psi}{\partial y}\sin\theta-\frac{\partial\psi}{\partial x}\cos\theta=-\frac{\partial\psi}{\partial r}.$$

3·23. As an example of the foregoing theory we might take

$$w=Az^2,$$

or $$\phi+i\psi=A(x+iy)^2;$$

giving $$\phi\equiv A(x^2-y^2)=\text{const.},$$

and $$\psi\equiv 2Axy=\text{const.},$$

for the lines of equi-velocity potential and the stream lines. These are two families of rectangular hyperbolas. Inasmuch as the axes $x=0$, $y=0$ are parts of the same stream line $\psi=0$, we may take the positive parts of the axes to be rigid boundaries and thus obtain a full representation of the steady motion of liquid in the angle made by two perpendicular walls.

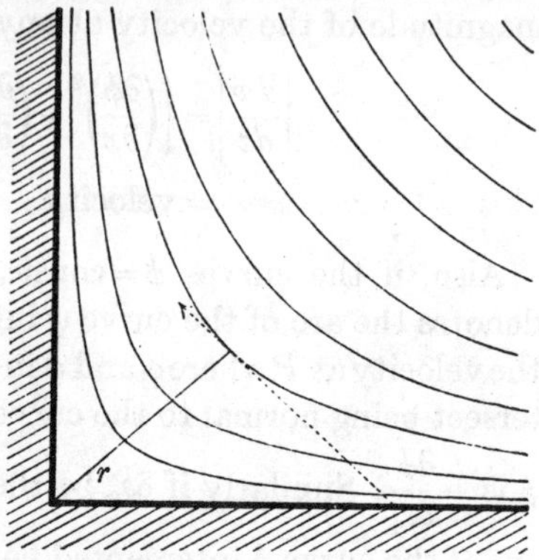

The velocity at any point

$$= |dw/dz| = |2Az| = 2Ar,$$

and varies directly as the distance from the intersection of the walls.

Before considering further examples we shall discuss some cases of liquid motion arising from what are known as 'sources' and 'sinks', taking first the general case of motion in three dimensions.

3·3. Sources and Sinks. If the motion of a liquid consists of symmetrical radial flow in all directions proceeding from a point, the point is called a simple source. If the total flow across a small surface surrounding the point is $4\pi m$, m is called the strength of the source*.

If ϕ be the velocity potential due to a simple source of strength m in liquid at rest at infinity, the velocity at distance r is $-\partial\phi/\partial r$, and the flow across a sphere of radius r is $-4\pi r^2 \partial\phi/\partial r$, therefore

$$-4\pi r^2 \frac{\partial \phi}{\partial r} = 4\pi m,$$

leading on integration to $\phi = m/r$.

A source of negative strength, or inward radial flow, is called a *sink*.

A source or sink implies the creation or annihilation of fluid at a point. Both are points at which the velocity potential and stream function become infinite, and they are to be regarded as due to the exigencies of analysis rather than as physical realities.

3·31. Doublets. A combination of a source of strength m and a sink of strength $-m$ at a small distance δs apart, where in the limit m is taken infinitely great and δs infinitely small but so that the product $m\,\delta s$ remains finite and equal to μ, is called a

* Some writers define the strength of the source to be the quantity of liquid produced in unit time, thus making the unit source 4π times as large as the one we have defined and introducing a symbol $m/4\pi$ instead of the m used in the text.

doublet of strength μ; and the line δs taken in the sense from $-m$ to $+m$ is called the axis of the doublet.

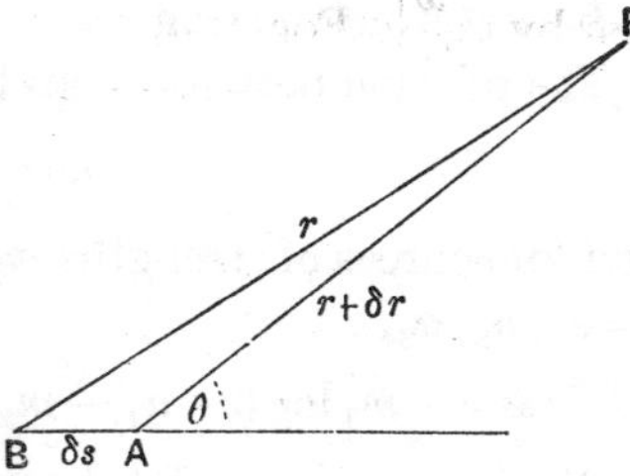

To find the velocity potential due to a doublet.

Let A, B denote the position of the source and sink and P be any point. Let $BP=r$, $AP=r+\delta r$, and suppose AP to make an angle θ with the axis of the doublet. Then by superposition, which is justified by the linearity of all equations that have to be satisfied,

$$\phi=-\frac{m}{r}+\frac{m}{r+\delta r}=-\frac{m\,\delta r}{r^2}$$

$$=\frac{m\,\delta s\cos\theta}{r^2}=\frac{\mu\cos\theta}{r^2}, \quad \text{or} \quad \mu\frac{\partial}{\partial s}\left(\frac{1}{r}\right);$$

so that the velocity potential due to a doublet may be obtained from the velocity potential due to a source by a differentiation in the direction of the axis of the doublet.

The components of velocity are

$$-\frac{\partial\phi}{\partial r}=\frac{2\mu\cos\theta}{r^3} \text{ along the radius vector,}$$

and $$-\frac{\partial\phi}{r\partial\theta}=\frac{\mu\sin\theta}{r^3} \text{ perpendicular to the radius vector, in the}$$

sense of θ increasing.

3·32. Sources and Sinks in Two Dimensions.

In two dimensions a source of strength m is such that the flow across any small curve surrounding it is $2\pi m$*.

If ϕ be the velocity potential due to such a source the flow across a circle of radius r is $-2\pi r\partial\phi/\partial r$, so that

$$-2\pi r\frac{\partial\phi}{\partial r}=2\pi m,$$

therefore $$\phi=-m\log r \quad \text{.......................(1).}$$

The curves of equi-velocity potential obviously are concentric circles. We may obtain the stream function from the consideration that $\phi+i\psi$ is a function of $x+iy$, or of $re^{i\theta}$, and since $\phi=-m\log r$, we must have

$$\psi=-m\theta \quad \text{..........................(2),}$$

* See footnote on p. 44.

and the stream lines are (as is otherwise obvious) straight lines radiating from the origin.

The relation between w and z for a single source is therefore

$$w = -m \log z,$$

and for sources of strengths $m_1, m_2, m_3, \ldots$ situated at the points $z = a_1, a_2, a_3, \ldots$

$$w = -m_1 \log (z - a_1) - m_2 \log (z - a_2) - m_3 \log (z - a_3) \ldots$$

leading to $\quad \phi = -\Sigma m \log r \quad$ and $\quad \psi = -\Sigma m\theta,$

where r denotes the length of the radius vector drawn from the source of strength m, and θ denotes the inclination of this radius vector to any fixed direction.

3·33. To take a simple case, let there be a source of strength m at the point $(a, 0)$ and a sink of strength $-m$ at the point $(-a, 0)$. Then

$$\phi = -m \log \frac{r}{r'},$$

and $$\psi = -m(\theta - \theta');$$

so that the stream lines are circles passing from source to sink, and the lines of equi-velocity potential are the orthogonal family of circles.

Since in this case

$$w = -m \log \frac{z-a}{z+a},$$

therefore

$$\frac{dw}{dz} = -\frac{2ma}{(z-a)(z+a)},$$

and the velocity $= \left| \frac{dw}{dz} \right| = \frac{2ma}{rr'}.$

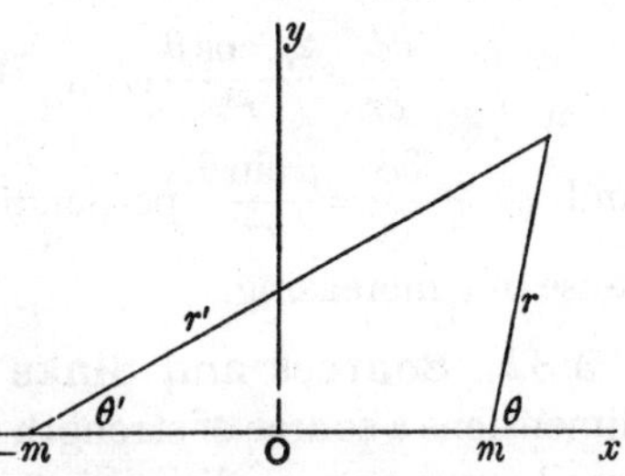

3·34. Doublets in Two Dimensions.

Referring to the figure of **3·31**, and with the same notation, we have

$$\begin{aligned} \phi &= m \log r - m \log (r + \delta r) \\ &= -m \log (1 + \delta r/r) \\ &= -m\,\delta r/r \\ &= m\,\delta s \cos \theta / r = \frac{\mu \cos \theta}{r}, \end{aligned}$$

where μ is the strength of the doublet.

The curves $\phi = \text{const.}$ in this case are clearly circles touching the y-axis at the origin.

We may obtain the stream function from the consideration

that $\phi + i\psi$ is a function of $x + iy$, or $re^{i\theta}$, and the form of ϕ suggests that

$$\phi + i\psi = \mu r^{-1} e^{-i\theta}$$
$$= \mu r^{-1} (\cos\theta - i\sin\theta),$$

so that
$$\psi = -\frac{\mu \sin\theta}{r}.$$

Hence the stream lines are circles touching the x-axis at the origin.

The relation between w and z for a single doublet of strength μ at the origin directed along the x-axis is therefore

$$w = \frac{\mu}{z};$$

and if the doublet makes an angle α with the x-axis, we have

$$\phi + i\psi = \mu r^{-1} e^{-i(\theta-\alpha)},$$

or
$$w = \frac{\mu e^{i\alpha}}{z}.$$

If the doublet be at the point $z = a$, the relation becomes

$$w = \frac{\mu e^{i\alpha}}{z - a};$$

and for any number of doublets of strengths $\mu_1, \mu_2, \mu_3, \ldots$ situated at $z = a_1, a_2, a_3, \ldots$ and making angles $\alpha_1, \alpha_2, \alpha_3, \ldots$ with the x-axis

$$w = \Sigma \frac{\mu e^{i\alpha}}{z - a}.$$

3·4. Images. If in a liquid a surface S can be drawn across which there is no flow, then any systems of sources, sinks and doublets on opposite sides of this surface may be said to be images of one another with regard to the surface. And if the surface S be regarded as a rigid boundary and the liquid removed from one side of it, the motion on the other side will remain unaltered.

3·41. *To find the image of a simple source with regard to a plane.*

If there are two equal sources of strength m at A and B on opposite sides of and equidistant from the plane OP, the normal velocity at P

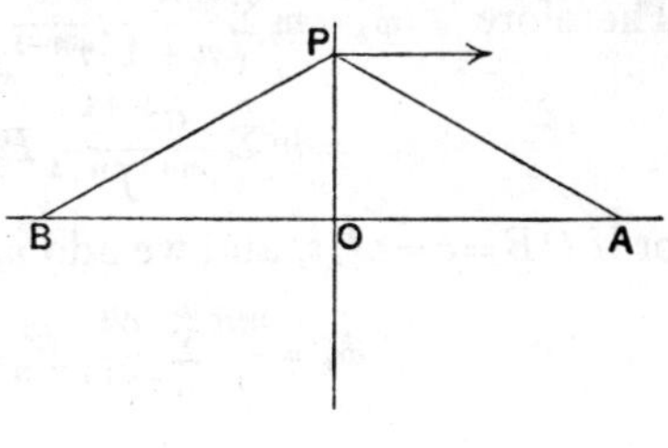

$$= -\frac{m}{AP^2}\cos OAP$$
$$+\frac{m}{BP^2}\cos OBP = 0;$$

that is, there is no flow across the plane. Therefore the image of a simple source with regard to a plane is an equal source equidistant from the plane.

Cor. The image of a doublet with regard to a plane is an equal doublet symmetrically placed.

3·42. *To find the image of a source with regard to a sphere.*

Let a be the radius of the sphere, $f\ (>a)$ the distance of the source A from the centre O, m the strength of the source and B the inverse point of A. We may regard the velocity potential as composed of two parts, viz. a part ϕ_1 due to the source alone when the sphere is not present, and a part ϕ_2 due to the presence of the sphere; this latter part will be the velocity potential of the required image system.

Taking O as origin and OA as axis, we have at any point $P(r, \theta)$

$$\phi_1 = m/AP = m(r^2+f^2-2rf\cos\theta)^{-\frac{1}{2}}$$

$$= \frac{m}{f}\left\{1+\sum_1^\infty \frac{r^n}{f^n} P_n(\mu)\right\},$$

where $\mu=\cos\theta$, and P_n is Legendre's coefficient of order n. This expression holds for all values of r less than f.

Since the motion is symmetrical about OA and the velocity potential has to satisfy Laplace's equation we may assume for ϕ_2 a series of the form

$$\phi_2 = \Sigma A_n \frac{a^n}{r^{n+1}} P_n.$$

We then have the condition that the velocity normal to the sphere is zero, i.e. $\frac{\partial}{\partial r}(\phi_1+\phi_2)=0$, when $r=a$.

Therefore $$\frac{m}{f}\sum_1^\infty \frac{na^{n-1}}{f^n} P_n - \sum_0^\infty (n+1)\frac{A_n}{a^2} P_n = 0,$$

for all values of θ, so that

$$A_0 = 0 \quad \text{and} \quad A_n = nma^{n+1}/(n+1)f^{n+1}.$$

Therefore $$\phi_2 = m\sum_1^\infty \frac{n}{n+1}\cdot\frac{a^{2n+1}}{r^{n+1}f^{n+1}} P_n$$

$$= m\sum_1^\infty \frac{a^{2n+1}}{r^{n+1}f^{n+1}} P_n - m\sum_1^\infty \frac{a^{2n+1}}{r^{n+1}f^{n+1}}\frac{P_n}{n+1},$$

or if $OB=c=a^2/f$, and we add and subtract a term,

$$\phi_2 = \frac{ma}{f}\sum_0^\infty \frac{c^n}{r^{n+1}} P_n - \frac{ma}{f}\sum_0^\infty \frac{c^n}{r^{n+1}}\frac{P_n}{n+1}.$$

The first term $=\dfrac{ma}{f\,(r^2+c^2-2rc\cos\theta)^{\frac{1}{2}}}$, and is therefore the velocity potential due to a source of strength ma/f at B.

Now for a source of unit strength at any point on OB at distance λ from O, we have a velocity potential

$$\chi=(r^2+\lambda^2-2r\lambda\cos\theta)^{-\frac{1}{2}}=\sum_0^\infty \frac{\lambda^n}{r^{n+1}}P_n,$$

so that
$$\int_0^\lambda \chi d\lambda=\sum_0^\infty \frac{\lambda^{n+1}}{r^{n+1}}\frac{P_n}{n+1}.$$

Therefore the second term in ϕ_2, viz.

$$-\frac{ma}{f}\sum_0^\infty \frac{c^n}{r^{n+1}}\frac{P_n}{n+1}=-\frac{ma}{cf}\int_0^c \chi d\lambda=-\frac{m}{a}\int_0^c \chi d\lambda,$$

and this is the velocity potential due to a continuous line distribution of sinks of strength $-m/a$ per unit length extending from O to B.

Hence the required image consists of a source of strength ma/f at the inverse point B, and a line sink of strength $-m/a$ per unit length extending from the centre to the inverse point*.

3·43. *To shew that the image with regard to a sphere of a doublet whose axis passes through the centre is a doublet at the inverse point.*

Regard the doublet as a source m at A and sink $-m$ at A', where $OA=f$, $AA'=\delta f$ and $m\delta f=\mu$.

The image of m at A is ma/f at B and a line sink of strength $-m/a$ per unit length from O to B.

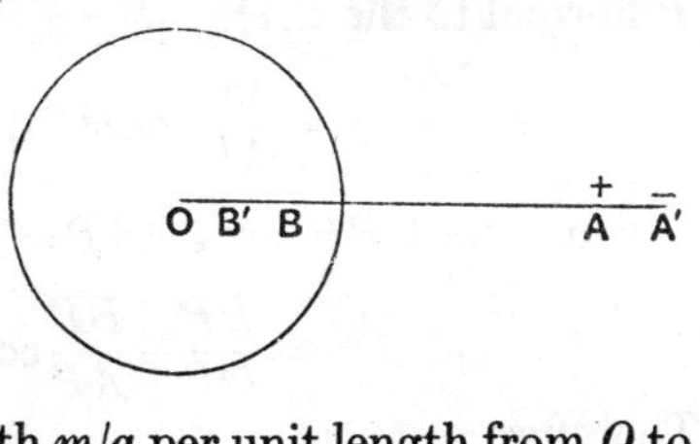

The image of $-m$ at A' is

$$-ma/(f+\delta f)$$

at B', that is

$$-ma/f+ma\,\delta f/f^2$$

at B'; and a line source of strength m/a per unit length from O to B'. Compounding these we get a doublet of strength $\dfrac{ma}{f}\,.\,BB'$, a source $ma\dfrac{\delta f}{f^2}$ and a sink $-\dfrac{m}{a}BB'$, all ultimately at the inverse point. But $OB=a^2/f$, therefore $BB'=\dfrac{a^2\delta f}{f^2}$, so that the source

* W. M. Hicks, *Phil. Trans.* 1880.

and sink cancel each other and there remains only the doublet of strength $\frac{ma^3}{f^3}\delta f$, or $\mu a^3/f^3$, at the inverse point in the opposite direction to the given doublet.

We might also obtain this result without assuming that of **3·42**, by supposing the image to be a doublet of strength m' at B and then determining the ratio of m' to m in order that the velocity normal to the sphere should be zero.

3·44. Images in Two Dimensions. It is easy to see that the image of a simple source with regard to a straight line in the plane of motion is an equal source equidistant from the line, and that the image of a doublet is an equal doublet symmetrically placed with regard to the line. But we must remember that as our two-dimensional motion is the motion of a liquid occupying three dimensions, what we call a simple source is a line source perpendicular to the plane of motion, and by the image of the simple source with regard to a line we mean the image of the line source with regard to a plane parallel to itself, the image being an equal line source equidistant from and parallel to the same plane.

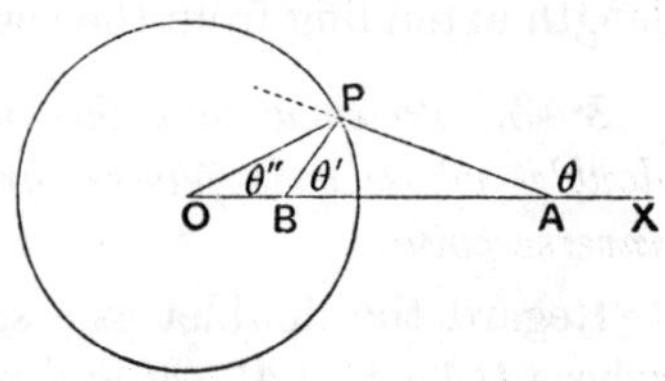

With regard to a circle, if we have a simple source m at A and place an equal source m at the inverse point B the velocity at P normal to the circle

$$=\frac{m}{AP}\cos OPA+\frac{m}{BP}\cos OPB.$$

But $\quad \cos OPB=\cos OAP=(AP+OP\cos PBA)/OA$

$$=\frac{BP}{OP}+\frac{BP}{AP}\cos PBA.$$

Therefore

$$\text{normal velocity}=\frac{m}{AP}\cos OPA+\frac{m}{OP}+\frac{m}{AP}\cos PBA=\frac{m}{OP}.$$

Hence if we place a sink $-m$ at O the normal velocity will be zero, so that the image system consists of an equal source at B and an equal sink at O*.

* Kirchhoff, *Ann. Phys. Chem.* 1845.

Alternatively if we place sources of strength m at A and B and an equal sink at O, the equations of the stream lines are

$$m\theta + m\theta' - m\theta'' = \text{constant},$$

where θ, θ', θ'' are vectorial angles at A, B, O.

For any point P on the circle we have

$$\begin{aligned}\theta + \theta' - \theta'' &= PAX + PBA - POA \\ &= OPA + POA + PBA - POA \\ &= \pi,\end{aligned}$$

so that the circle is a stream line and this verifies that for this arrangement of sources and sink there will be no flow across the boundary.

Cor. In like manner the image of a two-dimensional doublet at A with regard to a circle is another doublet at the inverse point B, the axes of the doublets making supplementary angles with

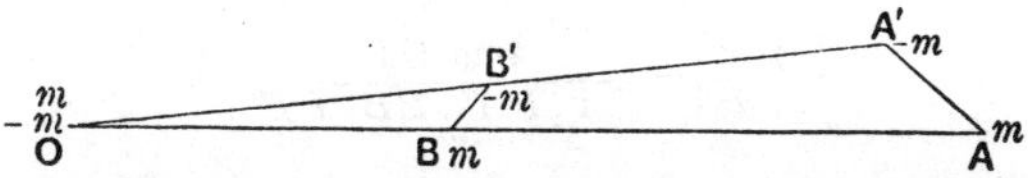

the radius OBA. This is clear from the figure and it is also seen that the moments of the doublets at B and A are in the ratio $BB' : AA'$, or $a^2 : f^2$, if a is the radius of the circle and $OA = f$.

3·5. Conjugate Functions. As a further example of the use of conjugate functions let us consider the relation

$$w = -m \log \frac{z^2 - a^2}{z^2 + a^2}.$$

This may also be written

$$w = -m \log \frac{(z-a)(z+a)}{(z-ia)(z+ia)};$$

so that $\phi = -m \log \dfrac{r_1 r_2}{r_3 r_4}$,

and $\psi = -m(\theta_1 + \theta_2 - \theta_3 - \theta_4)$,

where the symbols are used as in the figure and A, A', B, B' are the points $(a, 0)$, $(-a, 0)$, $(0, a)$, $(0, -a)$.

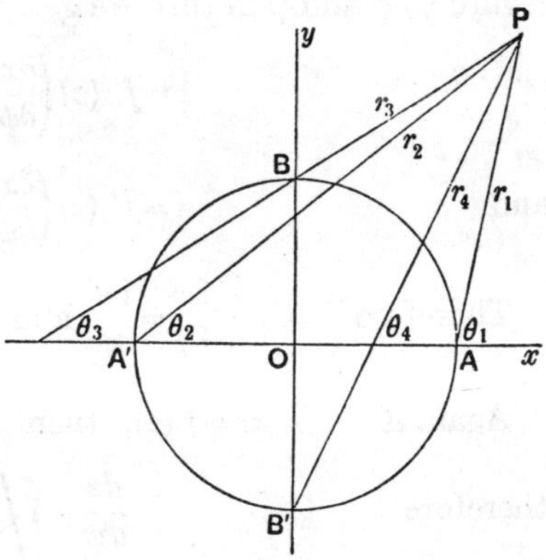

The circle $ABA'B'$ is the stream line $\psi = -m\pi/2$, as can be seen by taking P on the circle, and the axes are the stream line $\psi = 0$.

From **3·32** we see that the motion could be produced by equal sources at A, A' and equal sinks at B, B' all of strength m. And it is clear that the axes or the circle or both might be taken as fixed boundaries, and we have thus solved the problem of the motion in the quadrant, inside or outside the circle, due to an equal source and sink at the ends of the radii.

The velocity at any point may be found by compounding the components due to each source and sink, or more simply as the value of $\left|\frac{dw}{dz}\right|$.

Thus we have after a little reduction

$$\frac{dw}{dz} = -\frac{4mza^2}{z^4 - a^4}$$

$$= -\frac{4ma^2 r(\cos\theta + i\sin\theta)}{r^4\cos 4\theta - a^4 + ir^4\sin 4\theta};$$

so that the velocity $= \left|\frac{dw}{dz}\right| = \frac{4ma^2 r}{(r^8 + a^8 - 2r^4a^4\cos 4\theta)^{\frac{1}{2}}}$.

We may also observe that

$$\frac{dw}{dz} = -\frac{4mza^2}{(z-a)(z+a)(z-ia)(z+ia)},$$

so that we also have the velocity

$$\left|\frac{dw}{dz}\right| = \frac{4ma^2 OP}{PA.PA'.PB.PB'}.$$

3·51. It is sometimes convenient to use relations of the form $z = F(w)$ instead of $w = F(z)$.

If $\phi + i\psi$ is a function of $x + iy$ it follows that $x + iy$ is a function of $\phi + i\psi$.

Thus if $\phi + i\psi = f(z) = f(x+iy)$, then by differentiating with regard to ϕ and ψ in turn we get

$$1 = f'(z)\left(\frac{\partial x}{\partial\phi} + i\frac{\partial y}{\partial\phi}\right),$$

and

$$i = f'(z)\left(\frac{\partial x}{\partial\psi} + i\frac{\partial y}{\partial\psi}\right).$$

Therefore $\qquad \frac{\partial x}{\partial\phi} = \frac{\partial y}{\partial\psi}$ and $-\frac{\partial x}{\partial\psi} = \frac{\partial y}{\partial\phi}$.

Again if $\qquad w = f(z)$, then $1 = f'(z)\frac{dz}{dw}$,

therefore $\qquad \frac{dz}{dw} = 1\Big/\frac{dw}{dz}$.

But if q denote the velocity

$$q = \left|\frac{dw}{dz}\right|, \text{ so that } \frac{1}{q} = \left|\frac{dz}{dw}\right|.$$

Also, from above,

$$\frac{dz}{dw} = \frac{1}{f'(z)} = \frac{\partial x}{\partial\phi} + i\frac{\partial y}{\partial\phi}.$$

Therefore

$$\frac{1}{q^2}=\left(\frac{\partial x}{\partial \phi}\right)^2+\left(\frac{\partial y}{\partial \phi}\right)^2, \text{ similarly } =\left(\frac{\partial x}{\partial \psi}\right)^2+\left(\frac{\partial y}{\partial \psi}\right)^2,$$

$$=\frac{\partial x}{\partial \phi}\frac{\partial y}{\partial \psi}-\frac{\partial x}{\partial \psi}\frac{\partial y}{\partial \phi},$$

or

$$=\frac{\partial (x, y)}{\partial (\phi, \psi)}.$$

We also notice that

$$\frac{dz}{dw}=\frac{1}{\frac{dw}{dz}}=\frac{1}{\frac{\partial \phi}{\partial x}+i\frac{\partial \psi}{\partial x}}=\frac{1}{-u+iv}=-\frac{u+iv}{u^2+v^2};$$

so that $-\dfrac{dz}{dw}$ is a vector in the direction of the velocity whose modulus is the reciprocal of the velocity*.

3·52. Now consider $z=c\cosh w$,

or $x+iy=c\cosh(\phi+i\psi)$,

so that $x=c\cosh\phi\cos\psi, \quad y=c\sinh\phi\sin\psi.$

By eliminating ψ and ϕ in turn we get

$$\frac{x^2}{\cosh^2\phi}+\frac{y^2}{\sinh^2\phi}=c^2,$$

and

$$\frac{x^2}{\cos^2\psi}-\frac{y^2}{\sin^2\psi}=c^2;$$

equations which define ϕ and ψ respectively as functions of x and y, and by giving different values to ϕ and to ψ in these equations we get the curves of equi-velocity potential and the stream lines.

These are confocal ellipses and hyperbolas. The foci $(\pm c, 0)$ correspond to the values $\phi=0$, $\psi=n\pi$, and the velocity q is given by

$$\frac{1}{q}=\left|\frac{dz}{dw}\right|=c\sinh w=c\sinh(\phi+i\psi),$$

and at the foci this is zero, so that the velocity in the corresponding motion would be infinite at the foci.

If we take the hyperbolas $\psi=$ const. as the stream lines, the stream line $\psi=n\pi$ will be the part of the x-axis outside the foci and this might be made a rigid boundary, so that we should then have the case of liquid streaming through a slit of breadth $2c$ in an infinite plane, but the results of the theory could not be realized in practice because the theory makes the velocity infinite at the edges of the slit. We shall return to this point later.

* Kirchhoff, *Mechanik*, p. 291.

3·6. Steady motion—Efflux of Liquid. We shall now consider some further application of the equations of motion, particularly cases of *steady motion*, that is motion in which the velocity components at any point are independent of the time. As we have seen in **2·21**, in this case, for a liquid, we have the equation

$$\frac{p}{\rho}+\tfrac{1}{2}q^2+V=C,$$

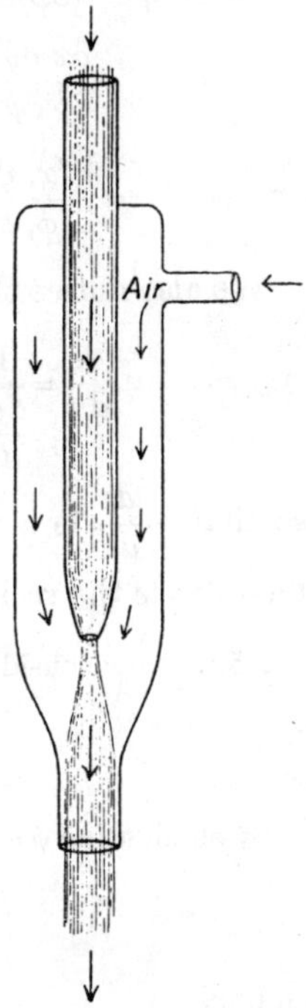

where C may be an absolute constant, or a constant depending on a particular stream line. This equation shews that neglecting the external forces *the smaller the pressure the greater the velocity* and vice versa. Thus in the case of water flowing through a pipe if the pipe is narrowed the velocity is increased and the pressure is consequently diminished. This is an important principle. A practical application of it is seen in jet exhaust pumps, one of which is shewn in the figure, the air being sucked in at the narrow portion of the jet.

3·61. Consider the case of a vessel kept constantly full of water and having a horizontal orifice in its base from which the water issues at a uniform rate. Let A, a be the areas of the free surface and the orifice, U, u the velocities at the free surface and the orifice, and h the depth of the orifice below the free surface.

If z be measured downwards from the free surface $V=-gz$, so that

$$\frac{p}{\rho}+\tfrac{1}{2}q^2-gz=C;$$

and if Π denotes the atmospheric pressure, at the free surface

$$\frac{\Pi}{\rho}+\tfrac{1}{2}U^2=C,$$

and at the orifice
$$\frac{\Pi}{\rho}+\tfrac{1}{2}u^2-gh=C,$$

so that
$$u^2=U^2+2gh;$$

but the condition of continuity of the water requires that

$$AU = au,$$

therefore $$u^2 = 2ghA^2/(A^2 - a^2),$$

and if the orifice be small, the ratio a/A may be neglected, and $u^2 = 2gh$ approximately.

This is **Torricelli's Theorem.**

If the vessel be not kept constantly full, the motion will not be steady, but when the orifice is small compared to the area of the free surface of water the motion may be taken as being approximately steady, and the expression $\sqrt{(2gh)}$ may be employed as the velocity of the issuing liquid.

3·62. The Clepsydra. On this hypothesis we can find the form of a vessel of revolution with a small aperture at its lowest point so that the surface of the water in it may descend uniformly.

At time t let x be the height of the free surface above the orifice, πy^2 its area, and σ the area of the orifice. Then, approximately,

$$\text{velocity at the orifice} = \sqrt{(2gx)};$$

but if U is the uniform velocity at the free surface

$$\pi y^2 U = \sigma \sqrt{2gx},$$

therefore $$y^4 \propto x \quad \text{or} \quad y^4 = a^3 x$$

gives the form of the vessel required.

This is the theory of the Clepsydra or ancient water clock.

3·63. The Contracted Vein. When liquid issues through a small orifice in the thin base of a vessel, it is observed that the issuing stream is not cylindrical, but, near the orifice, is contracted so that its sectional area is less than the area of the orifice; and afterwards the stream expands. The ratio of the area of the section of the 'contracted vein' to the area of the orifice is called the *coefficient of contraction* and it can be shewn that this coefficient is greater than ·5 and less than unity.

Neglecting external forces suppose liquid of density ρ to be escaping through an orifice of section σ in the bottom of a vessel in which the pressure is p_1 to a region in which the pressure is p_0. Theoretically the velocity acquired in passing from pressure p_1 to pressure p_0 is given by

$$\tfrac{1}{2}\rho q^2 = p_1 - p_0 \quad \text{.....................(1)}.$$

At the edge of the orifice σ the pressure is p_0, but in the interior of the area of the orifice the pressure is somewhat higher.

The actual velocity of the liquid in the plane of the orifice is therefore q at the edge, but falls off somewhat towards the interior. It follows that the actual rate of discharge is less than σq and this for two reasons. First because the velocity at the edge is not perpendicular to the plane of the orifice, and it is the resolved velocity that determines the discharge, and secondly because the mean actual velocity itself falls short of q.

If σ' be the area of the section of the jet at a place where the velocity at every point of the section is parallel and uniform, and therefore by equation (1) equal to q, the discharge is $\sigma' q$; and since this is less than σq it follows that σ' is less than σ, or the coefficient of contraction is less than unity.

The quantity of momentum carried away by the jet in unit time is $\rho\sigma' q^2$ and the force generating this momentum is the force necessary to hold the vessel at rest. If the whole interior surface of the vessel were subject to the pressure $p_1 - p_0$ this force would have no existence.

But on account of the orifice the equilibrium of pressures is disturbed and a force $(p_1 - p_0)\sigma$ is uncompensated. But this assumes that the internal pressure would be uniform and equal to p_1 over the whole of the bottom of the vessel, whereas at the edge of the orifice itself it is p_0 and for a sensible distance will vary between p_0 and p_1, we may therefore call the force that produces momentum $(p_1 - p_0)(\sigma + d\sigma)$, where $d\sigma$ is a small positive quantity.

Hence $$\rho\sigma' q^2 = (p_1 - p_0)(\sigma + d\sigma),$$

but $$\tfrac{1}{2}\rho q^2 = p_1 - p_0,$$

therefore $$\sigma' = \tfrac{1}{2}(\sigma + d\sigma),$$

or the coefficient of contraction is greater than ·5.

This discussion is based on that given by Lord Rayleigh*. If the hole in the vessel be replaced by a thin tube projecting into the interior of the vessel and the tube be long enough for the sides of the vessel to be sufficiently removed from the region of rapid flow to allow the pressure on them to be treated as constant, $d\sigma$ is evanescent and $\sigma' = \frac{1}{2}\sigma$. This form of opening is known as Borda's mouthpiece†.

* *Phil. Mag.* II, 1876, p. 441, or *Scientific Papers*, I, p. 297, and a letter to *Engineering*, April 10, 1876.

† *Mémoires de l'Acad. des Sciences*, 1766.

An exact method of treating the problem regarded as a problem in two dimensions was developed by Kirchhoff* and discussed in detail with numerical results by Lord Rayleigh†. We shall have more to say on this subject in Chapter VI.

3·64. Efflux of Gases. For a gas the steady motion equation is

$$\int \frac{dp}{\rho} + \tfrac{1}{2}q^2 + V = C.$$

Consider the efflux of gas from a vessel in which the pressure is p_1 and density ρ_1 to an atmosphere of density ρ_0 at pressure p_0. In practice the adiabatic law will hold true approximately, so that $p = \kappa\rho^\gamma$. Neglecting external forces the velocity acquired is given by

$$\tfrac{1}{2}q^2 = -\int_{\rho_1}^{\rho_0} \frac{dp}{\rho} = -\kappa \int_{\rho_1}^{\rho_0} \gamma\rho^{\gamma-2}\,d\rho$$

$$= \frac{\kappa\gamma}{\gamma-1}(\rho_1{}^{\gamma-1} - \rho_0{}^{\gamma-1})$$

$$= \frac{\gamma}{\gamma-1}\left(\frac{p_1}{\rho_1} - \frac{p_0}{\rho_0}\right);$$

or

$$q^2 = \frac{2\gamma}{\gamma-1}\frac{p_1}{\rho_1}\left\{1 - \left(\frac{p_0}{p_1}\right)^{\frac{\gamma-1}{\gamma}}\right\},$$

which is the usual formula for the efflux of gases.

It follows that a diminution of pressure p_0 accompanies an increase of velocity and vice versa, and this is the explanation of a common experiment which is performed as follows: One end of a tube is fitted into a hole in a disc of cardboard, the end of the tube being flush with the surface of the cardboard; if a piece of paper is placed over this end of the tube, blowing through the tube will cause the paper to remain in contact with the card: but as soon as the current of air ceases the paper falls off.

3·7. Steady Motion. Transverse acceleration. *In steady irrotational motion, or in steady motion in which the Bernoulli constant* (**2·21**) *has the same value along neighbouring stream lines,*

$$\frac{q}{r} = \frac{\partial q}{\partial n},$$

where r is the radius of curvature of a stream line and $\partial/\partial n$ *is a differentiation along the principal normal towards the centre of curvature.*

* *Mechanik*, chap. XXII. † *Loc. cit.*

This result was given as an example in earlier editions of this book; we now state it as a theorem because of the important place which it has taken in the theory of discontinuous motions*.

There are several methods of proof: thus, we may take q^2/r as the normal acceleration of an element of the fluid and resolve along the normal to a stream line inwards, getting

$$\frac{q^2}{r}=-\frac{\partial V}{\partial n}-\frac{1}{\rho}\frac{\partial p}{\partial n} \quad \text{.....................(1)}.$$

But by Bernoulli's Theorem

$$\frac{p}{\rho}+V+\tfrac{1}{2}q^2=C \quad \text{.......................(2)},$$

and if the motion is irrotational or in any case in which C has the same value for neighbouring stream lines, by differentiating along the normal

$$\frac{1}{\rho}\frac{\partial p}{\partial n}+\frac{\partial V}{\partial n}+q\frac{\partial q}{\partial n}=0 \quad \text{.....................(3)}.$$

Thence, by comparing (1) and (3)

$$\frac{q}{r}=\frac{\partial q}{\partial n} \quad \text{..............................(4)}.$$

Or, as a two-dimensional problem in steady irrotational motion the theorem is a purely kinematical one.

Thus let PP', QQ' be elements of stream lines and PQ, $P'Q'$ lines of equivelocity potential ϕ, $\phi+\delta\phi$. Then PQ, $P'Q'$ are normals to the stream lines and meet in the centre of curvature O. Also

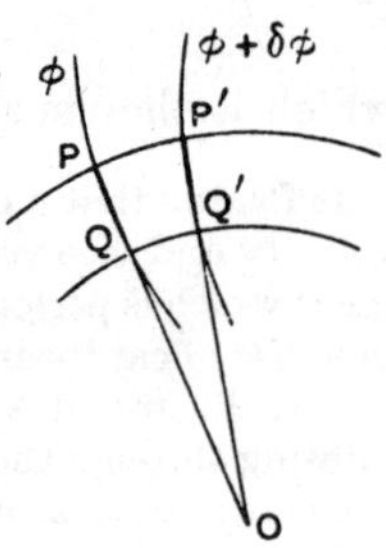

$$PO=r, \quad PQ=\delta n.$$

Then $q=$ vel. at P along PP'

$$=-\delta\phi/PP',$$

and $q+\frac{\partial q}{\partial n}\delta n=$ vel. at Q along QQ'

$$=-\delta\phi/QQ';$$

so that
$$1+\frac{1}{q}\frac{\partial q}{\partial n}\delta n=\frac{PP'}{QQ'}=\frac{PO}{QO}=\frac{r}{r-\delta n}=1+\frac{\delta n}{r},$$

whence we get $\frac{q}{r}=\frac{\partial q}{\partial n}$, as before.

3·71. An immediate corollary to the last theorem is that in steady irrotational motion if the stream lines are straight the velocity has the same value at all points of a cross section of the stream. It also follows that when the stream lines are curved the

* *V. The Physics of Solids and Fluids*, by P. P. Ewald, Th. Pöschl and L. Prandtl, 1930, p. 225.

velocity increases as we cross the stream lines towards the centre of curvature, and from Bernoulli's Theorem, neglecting external forces, it therefore follows that the pressure decreases as we cross

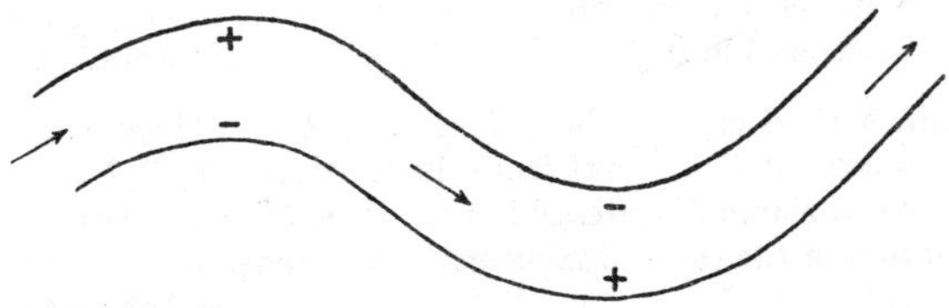

the stream lines in the same sense. Thus when a stream flows between curved banks the pressures will on this account have greatest values (+) on the outsides and least values (−) on the insides of the curves, but they may also be affected by a transverse circulation of the liquid.

3·72. Now suppose that the thick line in the figure represents the common surface of two streams flowing with different velocities one over the other: the dotted lines representing lines of flow. If we mark the positions of the excesses and defects of pressure in the two streams above and below the common surface

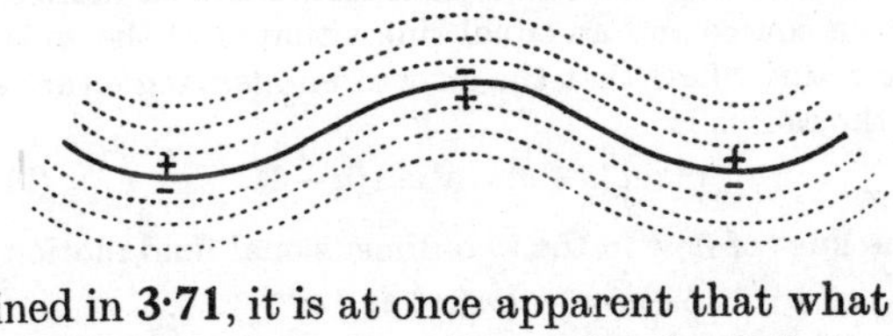

as explained in **3·71**, it is at once apparent that what the figure depicts cannot be a permanent state, for the defects of pressure on one side of the common surface are all opposite to excesses of pressure on the other side, so that any slight unevenness (departure from the plane) in the surface of separation will tend immediately to become exaggerated. As will be seen later the surface is a vortex sheet and the effect of disturbance is that it rolls up on itself into more or less concentrated vortices*. This is what actually happens when streams from different sources converge; but when a stream flows over a sharp edge and the fluid behind the edge does not possess the general velocity of the stream the phenomenon is rather different. A vortex sheet begins to be formed but is not fully developed. It curls round on itself and something in the nature of a concentrated vortex is formed.

* *V.* L. Rosenhead, *Proc. Roy. Soc.* A, CXXXIV, 1931, p. 187.

EXAMPLES

1. Liquid is streaming steadily and irrotationally in two dimensions in the region bounded by one branch of a hyperbola and its minor axis: determine the stream lines. (St John's Coll. 1901.)

2. Within a rigid circular boundary of radius a there is a source of strength m at a point P distant b from the centre; at X, Y, the extremities of the diameter through P, are equal sinks. Find the velocity potential and stream function of the (two-dimensional) fluid motion. (St John's Coll. 1900.)

3. In the case of two-dimensional fluid motion due to a simple source A outside a circular disc, prove that the asymptotes of the stream lines all pass through the same point and are parallel to the tangents to them at the point A. (Coll. Exam. 1905.)

4. Find the Cartesian equation of the lines of plane flow, when fluid is streaming from three equal sources situated at the corners of an equilateral triangle; and make a rough sketch of their configuration. (St John's Coll. 1896.)

5. Find the stream function of the two-dimensional motion due to two equal sources and an equal sink midway between them; sketch the stream lines and find the velocity at any point.

In a region bounded by a fixed quadrantal arc and its radii, deduce the motion due to a source and an equal sink situated at the ends of one of the bounding radii. Shew that the stream line leaving either end at an angle α with the radius is

$$r^2 \sin(\alpha+\theta) = a^2 \sin(\alpha-\theta).$$ (M.T. 1911.)

6. Find the lines of flow in the two-dimensional fluid motion given by

$$\phi + i\psi = -\tfrac{1}{2} n (x+iy)^2 e^{2int}.$$

Prove or verify that the paths of the particles of the fluid (in polar coordinates) may be obtained by eliminating t from the equations

$$r\cos(nt+\theta) - x_0 = r\sin(nt+\theta) - y_0 = nt(x_0 - y_0).$$

(Coll. Exam. 1908.)

7. λ denoting a variable parameter, and f a given function, find the condition that $f(x, y, \lambda) = 0$ should be a possible system of stream lines for steady irrotational motion in two dimensions. (Coll. Exam. 1893.)

8. If a homogeneous liquid is acted on by a repulsive force from the origin, the magnitude of which at distance r from the origin is μr per unit mass, shew that it is possible for the liquid to move steadily, without being constrained by any boundaries, in the space between one branch of the hyperbola $x^2 - y^2 = a^2$ and the asymptotes; and find the velocity potential. (Coll. Exam. 1902.)

9. In the case of the two-dimensional fluid motion produced by a source of strength m placed at a point S outside a rigid circular disc of radius a whose centre is O, shew that the velocity of slip of the fluid in

contact with the disc is greatest at the points where the lines joining S to the ends of the diameter at right angles to OS cut the circle; and prove that its magnitude at these points is

$$2m \, . \, OS/(OS^2 - a^2). \qquad \text{(Coll. Exam. 1908.)}$$

10. A source of fluid situated in space of two dimensions, is of such strength that $2\pi\rho\mu$ represents the mass of fluid of density ρ emitted per unit of time. Shew that the force necessary to hold a circular disc at rest in the plane of the source is $2\pi\rho\mu^2 a^2/r\,(r^2-a^2)$, where a is the radius of the disc and r the distance of the source from its centre. In what direction is the disc urged by the pressure? (M.T. 1893.)

11. Between the fixed boundaries $\theta = \frac{1}{4}\pi$ and $\theta = -\frac{1}{4}\pi$ there is a two-dimensional liquid motion due to a source of strength m at the point $(r=a,\ \theta=0)$, and an equal sink at the point $(r=b,\ \theta=0)$. Shew that the stream function is

$$-m \tan^{-1}\left\{\frac{r^4(a^4-b^4)\sin 4\theta}{r^8 - r^4(a^4+b^4)\cos 4\theta + a^4b^4}\right\}.$$

(Coll. Exam. 1901.)

12. A two-dimensional liquid motion is due to a source of strength m at the point whose polar coordinates are $(a, 0)$ and a sink of equal strength at the point $(b, 0)$, between the fixed boundaries $\theta = \frac{1}{4}\pi$ and $\theta = -\frac{1}{4}\pi$. Shew that the velocity at (r, θ) is

$$\frac{4m(a^4-b^4)r^3}{(r^8 - 2a^4r^4\cos 4\theta + a^8)^{\frac{1}{2}}(r^8 - 2b^4r^4\cos 4\theta + b^8)^{\frac{1}{2}}}.$$

(Trinity Coll. 1905.)

13. Prove that for liquid circulating irrotationally in the part of the plane between two non-intersecting circles the curves of constant velocity are Cassini's ovals. (St John's Coll. 1898.)

14. Between the fixed boundaries $\theta = \frac{1}{6}\pi$ and $\theta = -\frac{1}{6}\pi$ there is a two-dimensional liquid motion due to a source at the point $(r=c,\ \theta=\alpha)$, and a sink at the origin, absorbing water at the same rate as the source produces it. Find the stream function, and shew that one of the stream lines is a part of the curve

$$r^3 \sin 3\alpha = c^3 \sin 3\theta. \qquad \text{(M.T. 1901.)}$$

15. What arrangement of sources and sinks will give rise to the function $w = \log(z - a^2/z)$?

Draw a rough sketch of the stream lines in this case, and prove that two of them subdivide into the circle $r=a$, and the axis of y.

(St John's Coll. 1911.)

16. An area A is bounded by that part of the x-axis for which $x > a$ and by that branch of $x^2 - y^2 = a^2$ which is in the positive quadrant. There is a two-dimensional unit source at $(a, 0)$ which sends out liquid uniformly in all directions. Shew by means of the transformation $w = \log(z^2 - a^2)$ that in steady motion the stream lines of the liquid within the area A are portions of rectangular hyperbolas. Draw the stream lines corresponding to $\psi = 0,\ \frac{1}{4}\pi$ and $\frac{1}{2}\pi$. If ρ_1 and ρ_2 are the distances of a point P within the fluid from the points $(\pm a, 0)$, shew that the velocity of the fluid at P is measured by $2OP/\rho_1\rho_2$, O being the origin. (M.T. 1904.)

17. Find the velocity potential when there is a source and an equal sink inside a circular cavity and shew that one of the stream lines is an arc of the circle which passes through the source and sink and cuts orthogonally the boundary of the cavity. (Coll. Exam. 1894.)

18. Prove that, in the two-dimensional liquid motion due to any number of sources at points on a circle, the circle is a stream line provided that there is no boundary and that the algebraic sum of the strengths of the sources is zero.

Shew that the same is true if the region of flow is bounded by a circle which cuts orthogonally the circle in question.

(St John's Coll. 1908.)

19. In the part of an infinite plane bounded by a circular quadrant AB and the productions of the radii OA, OB, there is a two-dimensional motion due to the production of liquid at A, and its absorption at B, at the uniform rate m. Find the velocity potential of the motion; and shew that the fluid which issues from A in the direction making an angle μ with OA follows the path whose polar equation is

$$r = a\sin^{\frac{1}{2}} 2\theta\,[\cot\mu + \sqrt{(\cot^2\mu + \operatorname{cosec}^2 2\theta)}]^{\frac{1}{2}},$$

the positive sign being taken for all the square roots. (M.T. 1902.)

20. In the case of the motion of liquid in a part of a plane bounded by a straight line due to a source in the plane, prove that if $m\rho$ is the mass of fluid (of density ρ) generated at the source per unit of time the pressure on the length $2l$ of the boundary immediately opposite to the source is less than that on an equal length at a great distance by

$$\frac{1}{2}\frac{m^2\rho}{\pi^2}\left\{\frac{1}{c}\tan^{-1}\frac{l}{c} - \frac{l}{l^2+c^2}\right\},$$

where c is the distance of the source from the boundary.

(St John's Coll. 1898.)

21. Within a circular boundary of radius a there is a two-dimensional liquid motion due to a source producing liquid at the rate m, at a distance f from the centre, and an equal sink at the centre. Find the velocity potential, and shew that the resultant of the pressure on the boundary is

$$\rho m^2 f^3/\{2a^2\pi(a^2-f^2)\},$$

where ρ is the density.

Deduce, as a limit, the velocity potential due to a doublet at the centre.

(St John's Coll. 1905.)

22. Use the method of images to prove that if there be a source m at the point (z_0) in a fluid bounded by the lines $\theta=0$ and $\theta=\pi/3$, the solution is

$$\phi + i\psi = -m\log\{(z^3-z_0^3)(z^3-z_0'^3)\},$$

where $z_0 = x_0 + iy_0$ and $z_0' = x_0 - iy_0$. (Coll. Exam. 1906.)

23. A source S and a sink T of equal strengths m are situated within the space bounded by a circle whose centre is O. If S and T are at equal distances from O on opposite sides of it and on the same diameter AOB, shew that the velocity of the liquid at any point P is

$$2m.\frac{OS^2+OA^2}{OS}.\frac{PA.PB}{PS.PS'.PT.PT'},$$

where S' and T' are the inverse points of S and T with respect to the circle. (Coll. Exam. 1901.)

24. Within a rigid boundary in the form of the circle

$$(x+\alpha)^2+(y-4\alpha)^2=8\alpha^2$$

there is liquid motion due to a doublet of strength μ at the point $(0, 3\alpha)$, with its axis along the axis of y. Shew that the velocity potential is

$$\mu\left[4\frac{x-3\alpha}{(x-3\alpha)^2+y^2}+\frac{y-3\alpha}{x^2+(y-3\alpha)^2}\right].$$

(Coll. Exam. 1903.)

25. The internal boundary of a liquid is composed of the two orthogonal circles $x^2+y^2+2y=1$ and $x^2+y^2-2y=1$. A source producing liquid at the rate m is placed at one of the points of intersection $(z=1)$; shew that the complex of the fluid motion is $\frac{m}{2\pi}\log\{z(z^2+1)/(z-1)^4\}$, and that the two circles are the only stream line possessing double points. (Coll. Exam. 1910.)

26. In two-dimensional irrotational fluid motion shew that, if the stream lines are confocal ellipses

$$x^2/(a^2+\lambda)+y^2/(b^2+\lambda)=1,$$

$$\psi=A\log(\sqrt{a^2+\lambda}+\sqrt{b^2+\lambda})+B,$$

and the velocity at any point is inversely proportional to the square root of the rectangle under the focal radii of the point. (Coll. Exam. 1894.)

27. Liquid flows steadily and irrotationally in two dimensions in a space with fixed boundaries the cross section of which consists of the two lines $\theta=\pm\frac{1}{10}\pi$ and the curve $r^5\cos 5\theta=\kappa^5$; prove that, if V is the velocity of the liquid in contact with one of the plane boundaries at unit distance from their intersection, the volume of liquid which passes per unit time through a circular ring in the plane $\theta=0$ is $\frac{1}{8}\pi Va^2(a^4+12a^2c^2+8c^4)$, where a is the radius of the ring, and c the distance of its centre from the intersection of the plane boundaries. (Coll. Exam. 1896.)

28. Shew that any two-dimensional irrotational motion of a liquid may be transformed into any other by multiplying the velocity of each particle of the fluid by e^P and turning its direction round through an angle Q, where P, $-Q$ are suitably chosen conjugate functions of x, y. (Coll. Exam. 1906.)

29. In a two-dimensional liquid motion ϕ and ψ are the velocity potential and current function; shew that a second fluid motion exists in which ψ is the velocity potential and $-\phi$ the current function; and prove that if the first motion be due to sources and sinks, the second motion can be built up by replacing a source and an equal sink by a line of doublets uniformly distributed along any curve joining them. (Coll. Exam. 1899.)

30. A line source is in the presence of an infinite plane on which is placed a semi-circular cylindrical boss; the direction of the source is parallel to the axis of the boss, the source is at distance c from the plane and the axis of the boss, whose radius is a. Shew that the radius to the point on the boss at which the velocity is a maximum makes an angle θ with the radius to the source, where

$$\theta = \cos^{-1}\frac{a^2+c^2}{\sqrt{2(a^4+c^4)}}. \qquad \text{(Coll. Exam. 1907.)}$$

31. A source and a sink, each of strength μ, exist in an infinite liquid on opposite sides of, and at equal distances c from, the centre of a rigid sphere of radius a. Shew that the velocity potential V may be expressed in the form

$$V = \frac{2\mu}{c}\sum_{n=0}^{n=\infty}\left\{\left(\frac{r}{c}\right)^{2n+1} + \frac{2n+1}{2n+2}\cdot\frac{c}{a}\cdot\left(\frac{a^2}{rc}\right)^{2n+2}\right\} P_{2n+1}(\cos\theta),$$

θ being the vectorial angle measured from the diameter of the sphere on which the source and sink lie, and $r<c$; and find an expression for V when $r>c$. (M.T. 1900.)

32. If a fluid be in motion with a velocity potential $\phi = z\log r$, and if the density at a point fixed in space be independent of the time, shew that the surfaces of equal density are of the form $r^2(\log r - \frac{1}{2}) - z^2 = f(\theta, \rho)$; where ρ is the density and z, r, θ the cylindrical coordinates. (Coll. Exam. 1897.)

33. A single source is placed in an infinite perfectly elastic fluid, which is also a perfect conductor of heat; shew that if the motion be steady, the velocity V at a distance r from the source satisfies the equation

$$\left(V - \frac{\kappa}{V}\right)\frac{\partial V}{\partial r} = \frac{2\kappa}{r};$$

and hence that

$$r = \frac{1}{\sqrt{V}}e^{\frac{V^2}{4\kappa}}. \qquad \text{(Coll. Exam. 1905.)}$$

34. If fluid fill the region of space on the positive side of the x-axis, which is a rigid boundary, and if there be a source m at the point $(0, a)$ and an equal sink at $(0, b)$, and if the pressure on the negative side of the boundary be the same as the pressure of the fluid at infinity, shew that the resultant pressure on the boundary is $\pi\rho m^2(a-b)^2/ab(a+b)$, where ρ is the density of the fluid. (Coll. Exam. 1906.)

35. In a steady two-dimensional motion of an incompressible liquid the stream lines are given by $x = f_1(k, c)$, $y = f_2(k, c)$, where c is a parameter defining a stream line and k is a parameter defining a point on a stream line. Shew that the particle at the point given by k_0, c_0 at time t_0 will be at the point given by k, c_0 at time t, where

$$t - t_0 = \left[C\int_{k_0}^{k}\frac{\partial(x, y)}{\partial(k, c)}dk\right]_{c=c_0},$$

and C is a function of c. (M.T. 1920.)

36. An infinite mass of liquid is moving irrotationally and steadily under the influence of a source of strength μ and an equal sink at a distance

$2a$ from it. Prove that the kinetic energy of the liquid which passes in unit time across the plane which bisects at right angles the line joining the source and sink is $\frac{8}{7}\pi\rho\mu^3/a^4$, ρ being the density of the liquid.

(Coll. Exam. 1896.)

37. Draw the stream lines $\psi=0$, $\psi=\pi$ and some of the intermediate stream lines for the motion given by the equation

$$z=w+e^w. \qquad \text{(Trinity Coll. 1895.)}$$

38. Trace the stream lines along which $\psi=0$ and ϕ diminishes from $+\infty$ to $-\infty$ in the two cases

$$(1)\ x+iy=2(\phi+i\psi)^{\frac{3}{2}},$$

$$(2)\ x+iy=(\phi+i\psi-1)^{\frac{3}{2}}+(\phi+i\psi+1)^{\frac{3}{2}},$$

and indicate roughly the form of the stream lines for which ψ has a positive value. (Univ. of London, 1909.)

39. The space on one side of an infinite plane wall, $y=0$, is filled with inviscid, incompressible fluid, moving at infinity with velocity U in the direction of the axis of x. The motion of the fluid is wholly two-dimensional, in the (x, y) plane. A doublet of strength μ is at a distance a from the wall, and points in the negative direction of the axis of x. Shew that if μ is less than $4a^2U$, the pressure of the fluid on the wall is a maximum at points distant $\sqrt{3}a$ from O, the foot of the perpendicular from the doublet on the wall, and is a minimum at O.

If μ is equal to $4a^2U$, find the points where the velocity of the fluid is zero, and shew that the stream lines include the circle

$$x^2+(y-a)^2=4a^2,$$

where the origin is taken at O. (M.T. 1934.)

CHAPTER IV

GENERAL THEORY OF IRROTATIONAL MOTION

4·1. IN this chapter we shall examine in general terms the nature of irrotational motion and the circumstances under which it is produced. In the first place let us analyse the most general type of displacement of an element of fluid.

Let u, v, w be the components of velocity of the particle at the point P whose coordinates are x, y, z. Then the relative velocities of the particle at P' whose coordinates are $x+\mathbf{x}$, $y+\mathbf{y}$, $z+\mathbf{z}$ at the instant considered will be

$$\left.\begin{aligned} \mathbf{u} &= \frac{\partial u}{\partial x}\mathbf{x}+\frac{\partial u}{\partial y}\mathbf{y}+\frac{\partial u}{\partial z}\mathbf{z} \\ \mathbf{v} &= \frac{\partial v}{\partial x}\mathbf{x}+\frac{\partial v}{\partial y}\mathbf{y}+\frac{\partial v}{\partial z}\mathbf{z} \\ \mathbf{w} &= \frac{\partial w}{\partial x}\mathbf{x}+\frac{\partial w}{\partial y}\mathbf{y}+\frac{\partial w}{\partial z}\mathbf{z} \end{aligned}\right\} \quad \ldots\ldots\ldots\ldots\ldots\ldots(1),$$

neglecting squares and products of $\mathbf{x}$, $\mathbf{y}$, $\mathbf{z}$.

If we put

$$\left.\begin{aligned} & a=\frac{\partial u}{\partial x}, \quad b=\frac{\partial v}{\partial y}, \quad c=\frac{\partial w}{\partial z} \\ & f=\frac{1}{2}\left(\frac{\partial w}{\partial y}+\frac{\partial v}{\partial z}\right), \quad g=\frac{1}{2}\left(\frac{\partial u}{\partial z}+\frac{\partial w}{\partial x}\right), \quad h=\frac{1}{2}\left(\frac{\partial v}{\partial x}+\frac{\partial u}{\partial y}\right) \\ & \xi=\frac{1}{2}\left(\frac{\partial w}{\partial y}-\frac{\partial v}{\partial z}\right), \quad \eta=\frac{1}{2}\left(\frac{\partial u}{\partial z}-\frac{\partial w}{\partial x}\right), \quad \zeta=\frac{1}{2}\left(\frac{\partial v}{\partial x}-\frac{\partial u}{\partial y}\right) \end{aligned}\right\} \ldots(2),$$

equations (1) may be written

$$\left.\begin{aligned} \mathbf{u} &= a\mathbf{x}+h\mathbf{y}+g\mathbf{z}+\eta\mathbf{z}-\zeta\mathbf{y} \\ \mathbf{v} &= h\mathbf{x}+b\mathbf{y}+f\mathbf{z}+\zeta\mathbf{x}-\xi\mathbf{z} \\ \mathbf{w} &= g\mathbf{x}+f\mathbf{y}+c\mathbf{z}+\xi\mathbf{y}-\eta\mathbf{x} \end{aligned}\right\} \quad \ldots\ldots\ldots\ldots\ldots(3).$$

Thus the relative motion in the most general case consists of two parts: a motion in the direction of the normal to the surface

$$a\mathbf{x}^2+b\mathbf{y}^2+c\mathbf{z}^2+2f\mathbf{y}\mathbf{z}+2g\mathbf{z}\mathbf{x}+2h\mathbf{x}\mathbf{y}=\text{const.} \quad \ldots(4),$$

and a rotation of which the component angular velocities are ξ, η, ζ. The former motion is called a *pure strain**, it is such that lines drawn parallel to any one of three mutually perpendicular

* For a fuller discussion of this subject see Kelvin and Tait, *Natural Philosophy*, Arts. 155–185, or Love, *Mathematical Theory of Elasticity*, chap. I.

directions (the axes of the quadric (4)) undergo elongation at a uniform rate. Thus if the equation of the quadric referred to its principal axes be

$$a'\mathbf{x}'^2+b'\mathbf{y}'^2+c'\mathbf{z}'^2=\text{const.},$$

the velocities due to the pure strain, parallel to the axes, are

$$\mathbf{u}'=a'\mathbf{x}', \quad \mathbf{v}'=b'\mathbf{y}', \quad \mathbf{w}'=c'\mathbf{z}',$$

so that a', b', c' are the time-rates of elongation of lines parallel to the axes of x', y', z'. If there is no change of volume, as in the case of a liquid, it is clear that a', b', c' cannot be independent; in fact we have

$$a'+b'+c'=a+b+c$$
$$=\frac{\partial u}{\partial x}+\frac{\partial v}{\partial y}+\frac{\partial w}{\partial z}=0.$$

Hence the most general displacement of a fluid element consists of a pure strain compounded with a rotation; and this analysis of the motion is unique, for if we were to compound together a pure strain and a rotation both arbitrarily assumed and endeavour to adapt them so as to result in a given displacement of a fluid element, the equations to determine the axes of the strain-quadric and the components of spin would be exactly those we have used above.

In accordance with **2·52**, ξ, η, ζ are the components of spin, and if they are all zero the motion is *irrotational*, and in this case the relative displacement of a fluid element consists of a pure strain only.

4·11. Flow and Circulation. If A, P be any two points in a fluid the value of the integral

$$\int_A^P (u\,dx+v\,dy+w\,dz),$$

or

$$\int_A^P \left(u\frac{dx}{ds}+v\frac{dy}{ds}+w\frac{dz}{ds}\right)ds,$$

taken along any path from A to P, is called *the flow along that path from A to P*.

When a velocity potential exists, the flow from A to P is equal to

$$-\int_A^P \left(\frac{\partial\phi}{\partial x}dx+\frac{\partial\phi}{\partial y}dy+\frac{\partial\phi}{\partial z}dz\right)$$
$$=\phi_A-\phi_P.$$

The flow round a closed curve is called the **circulation** round the curve. If a *single-valued* velocity potential exists the circulation round any closed curve is clearly zero; and we shall see presently that if the velocity potential is many-valued there are closed curves for which the circulation is zero, though it is not zero for all such paths.

4·2. Stokes's Theorem. We shall now shew that the circulation round any closed curve drawn in a fluid is equal to twice the surface integral of the normal component of spin taken over any surface having the curve for boundary, provided the surface lies wholly in the fluid: i.e. we shall prove that

$$\int u\,dx + v\,dy + w\,dz = 2\iint (l\xi + m\eta + n\zeta)\,dS,$$

where l, m, n are direction cosines of the normal to the element dS of the surface and the other symbols have the usual meanings; and throughout this theorem sense of circulation on the surface is to be associated with the positive direction of the normal to the surface by the right-handed or the left-handed screw convention according as the axes of coordinates are right-handed or left-handed.

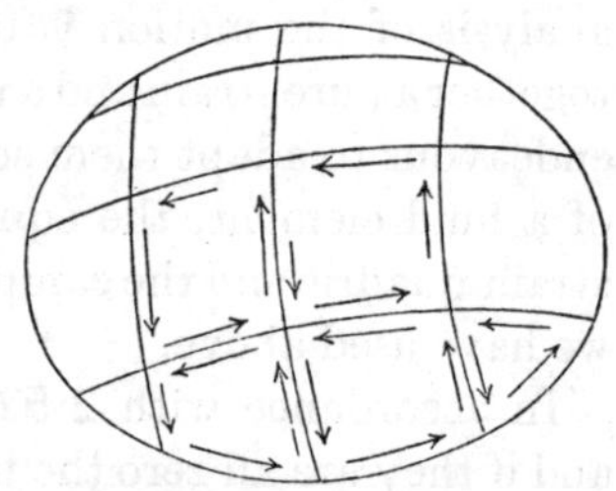

In the first place we observe that any surface can be divided up into small areas by drawing a net-work of lines across it as in the figure; and if we take the sum of the circulations round each mesh of the surface, the flow along all lines common to two meshes will be taken twice in opposite directions, so that the result will be the circulation round the boundary.

Now with the notation of **4·1**, let the point (x, y, z) be a point P within a mesh and let $(x+\mathbf{x}, y+\mathbf{y}, z+\mathbf{z})$ and $(x+\mathbf{x}+d\mathbf{x}, \ldots\ldots)$ be points P', P'' on its boundary. The circulation round the mesh is then

$$\int\{(u+\mathbf{u})\,d\mathbf{x} + (v+\mathbf{v})\,d\mathbf{y} + (w+\mathbf{w})\,d\mathbf{z}\},$$

and substituting from **4·1** (3), this becomes

$$\int\{(u + a\mathbf{x} + h\mathbf{y} + g\mathbf{z} + \eta\mathbf{z} - \zeta\mathbf{y})\,d\mathbf{x} + \ldots + \ldots\}$$

or

$$\int d\{u\mathbf{x} + v\mathbf{y} + w\mathbf{z} + \tfrac{1}{2}(a, b, c, f, g, h)(\mathbf{x}, \mathbf{y}, \mathbf{z})^2\}$$
$$+\int\{\xi(\mathbf{y}\,d\mathbf{z} - \mathbf{z}\,d\mathbf{y}) + \eta(\mathbf{z}\,d\mathbf{x} - \mathbf{x}\,d\mathbf{z}) + \zeta(\mathbf{x}\,d\mathbf{y} - \mathbf{y}\,d\mathbf{x})\}.$$

The former of these integrals taken round the mesh is clearly zero, and in the latter ξ, η, ζ are constants for the mesh, being values at a definite point P, and their coefficients are twice the projections on the coordinate planes of the area $PP'P''$, hence if dS denotes the area of the mesh the circulation round it is

$$2(l\xi + m\eta + n\zeta)\,dS.$$

By summation we get the circulation round any closed curve

$$= 2\iint (l\xi + m\eta + n\zeta)\,dS.$$

Hence the theorem follows as stated.

The proof that we have given above is stated in terms of hydrodynamical ideas, but the theorem is one of pure analysis and is true for any functions u, v, w which are continuous and differentiable throughout a region including the ranges of integration*.

In the language of vectors the theorem is expressed by saying that 2ξ, 2η, 2ζ are components of a vector 2ω which is the 'curl' of the vector q whose components are u, v, w. Thus 2ω is the curl of q, when the surface integral of the normal component of 2ω over any surface is equal to the line integral of the component of q round the boundary; and the result may be written

$$2(\xi, \eta, \zeta) = \operatorname{curl}(u, v, w).$$

4·21. The foregoing theorem will still be true for a surface which is bounded by more than one closed curve; as for example the shaded area in the accompanying figure, provided the circulations round the boundary curves are taken with proper signs. We can see this by regarding the boundary as a continuous curve $ABCADEFDA$ and observing that the total flow along AD and DA is zero.

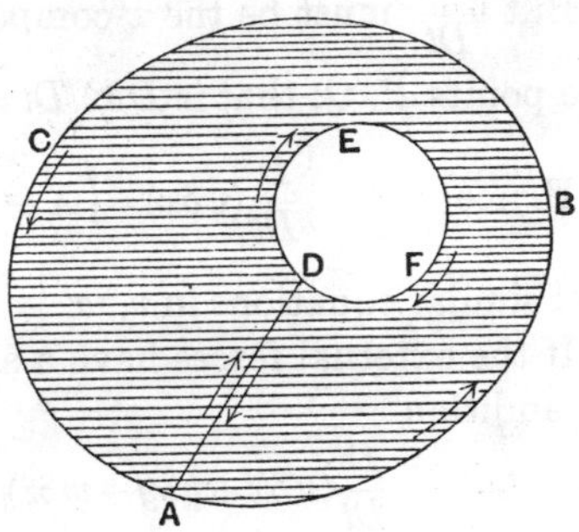

* This theorem, generally known as Stokes's Theorem, first appeared in print as a question set by Stokes in the Smith's Prize Examination in 1854, but it occurs in a letter from Kelvin to Stokes dated July 2, 1850. See Stokes, *Math. and Phys. Papers*, v, p. 321 footnote. Stokes however appears to have priority in the use of the vector which is the subject of the surface integral.

4·22. Irrotational Motion. If ξ, η, ζ are all zero, that is, in the case of irrotational motion, the circulation round any closed curve is zero, provided that the closed curve can be regarded as the boundary of a surface every part of which lies within the fluid. When this is the case the curve or circuit is said to be *reducible*; that is, it can be contracted to a point without passing out of the fluid. If the circuit be *irreducible* we cannot conclude that the circulation is zero. Thus if the last figure represents fluid filling the space between two infinite cylinders, the circuit ABC is irreducible, but it will still be true, as in **4·21**, that the circulations round ABC and DEF are together zero if the motion is irrotational, so that the circulations in the same sense round the circuits ABC and DFE are equal, whence it follows that the circulation in all circuits going once round the inner cylinder in the same sense is constant and the same for all. We shall have more to say on this point later under the heading of multiply-connected space.

4·23. Constancy of Circulation. Let AB be any line of particles in the fluid and moving with it.

Let P, Q be two consecutive points on the line; (x, y, z), $(x+\delta x,\ y+\delta y,\ z+\delta z)$ their coordinates; u, v, w the velocity components at P and $u+\delta u$, $v+\delta v$, $w+\delta w$ those at Q. Then

$$\frac{D}{Dt}(u\,\delta x)=\frac{Du}{Dt}\delta x+u\frac{D\delta x}{Dt}.$$

But $\dfrac{D\delta x}{Dt}$ must be the x-component of the relative velocity of the points P, Q; that is $D\delta x/Dt=\delta u$.

Hence
$$\frac{D}{Dt}(u\,\delta x)=\left(X-\frac{1}{\rho}\frac{\partial p}{\partial x}\right)\delta x+u\,\delta u;$$

and similar equations in v, w.

If the external forces have a single-valued potential Ω we get by addition

$$\frac{D}{Dt}(u\,\delta x+v\,\delta y+w\,\delta z)=-\delta\Omega-\frac{\delta p}{\rho}+\tfrac{1}{2}\delta q^2,$$

where $q^2=u^2+v^2+w^2$.

And by integration along the line from A to B

$$\frac{D}{Dt}\left\{\int_A^B(u\,dx+v\,dy+w\,dz)\right\}=\left[\tfrac{1}{2}q^2-\Omega-\int\frac{dp}{\rho}\right]_A^B.$$

This gives the rate of change of flow along any line moving with the fluid.

If there be any integrable functional relation between the pressure and density and we make the line a closed circuit the right-hand side of the last equation vanishes. Whence it follows that *the circulation in any closed path moving with the fluid is constant for all time*. This is true whether the motion be rotational or irrotational, the only assumptions being that the external forces are conservative and that there is a relation between the pressure and the density.

The foregoing proof is due to Kelvin*.

4·24. From the theorem of **4·23** it is easy to deduce the theorem of the **Permanence of Irrotational Motion** proved in **2·51**. For at any instant at which the motion of a fluid is irrotational the circulation in all reducible circuits in the fluid vanishes, but the circulation in any such circuit is constant for all time and therefore remains zero. Hence, at any subsequent time, by **4·2**,

$$\iint (l\xi + m\eta + n\zeta)\, dS = 0,$$

where the integration may be taken over any surface lying wholly in the fluid, and this requires that

$$\xi = \eta = \zeta = 0$$

at every point in the fluid, so that the motion is always irrotational.

4·25. Components of spin in Cylindrical and in Polar Coordinates. Using **cylindrical coordinates**, let (r, θ, z) be the centre of an element of volume whose diameters are of lengths $\delta r, r\,\delta\theta, \delta z$, let v_r, v_θ, v_z be the components of velocity in these directions, and ξ, η, ζ corresponding components of spin.

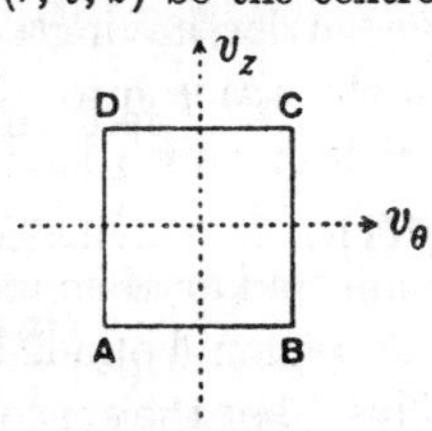

Let $ABCD$ be a central section of the element with diameters $r\,\delta\theta$ and δz.

Then

$$2\xi\,.\,r\,\delta\theta\,\delta z = \text{circulation round } ABCD \quad \ldots(1).$$

The contributions to the circulation from the sides AB and CD are

$$v_\theta r\,\delta\theta - \frac{1}{2}\frac{\partial}{\partial z}(v_\theta r\,\delta\theta)\,\delta z \quad \text{and} \quad -\left\{v_\theta r\,\delta\theta + \frac{1}{2}\frac{\partial}{\partial z}(v_\theta r\,\delta\theta)\,\delta z\right\}$$

making up $$-\frac{\partial v_\theta}{\partial z}\,.\,r\,\delta\theta\,\delta z.$$

* 'On Vortex Motion,' *Trans. Roy. Soc. Edin.* xxv, 1869; also *Math. and Phys. Papers*, iv, p. 49.

Similarly the contributions of the sides BC and DA are

$$v_z\,\delta z+\frac{1}{2}\frac{\partial}{r\,\partial\theta}(v_z\,\delta z)\,r\,\delta\theta \quad\text{and}\quad -\left\{v_z\,\delta z-\frac{1}{2}\frac{\partial}{r\,\partial\theta}(v_z\,\delta z)\,r\,\delta\theta\right\}$$

making up $\dfrac{\partial v_z}{r\,\partial\theta}.r\,\delta\theta\,\delta z.$

Hence from (1) we get

$$\left.\begin{aligned} 2\xi&=\frac{\partial v_z}{r\,\partial\theta}-\frac{\partial v_\theta}{\partial z}.\\ \text{Similarly}\qquad 2\eta&=\frac{\partial v_r}{\partial z}-\frac{\partial v_z}{\partial r}\\ \text{and}\qquad 2\zeta&=\frac{1}{r}\frac{\partial(v_\theta r)}{\partial r}-\frac{\partial v_r}{r\,\partial\theta}\end{aligned}\right\}\quad\ldots\ldots\ldots\ldots\ldots\ldots(2).$$

In like manner using **polar coordinates,** let (r,θ,ω) be the centre of an element of volume whose diameters are of lengths δr, $r\,\delta\theta$, $r\sin\theta\,\delta\omega$, and let q_r, q_θ, q_ω be the components of velocity in these directions and ξ, η, ζ corresponding components of spin. Then by taking the circulation round central sections of this element of volume as above, we can shew that

$$2\xi.r^2\sin\theta\,\delta\theta\,\delta\omega=\frac{\partial(q_\omega r\sin\theta\,\delta\omega)}{r\,\partial\theta}\,r\,\delta\theta-\frac{\partial(q_\theta r\,\delta\theta)}{r\sin\theta\,\partial\omega}.r\sin\theta\,\delta\omega,$$

$$\left.\begin{aligned} \text{or}\qquad 2\xi&=\frac{1}{r\sin\theta}\left(\frac{\partial(q_\omega\sin\theta)}{\partial\theta}-\frac{\partial q_\theta}{\partial\omega}\right);\\ \text{similarly}\qquad 2\eta&=\frac{1}{r\sin\theta}\frac{\partial q_r}{\partial\omega}-\frac{1}{r}\frac{\partial(q_\omega r)}{\partial r}\\ \text{and}\qquad 2\zeta&=\frac{1}{r}\frac{\partial(q_\theta r)}{\partial r}-\frac{\partial q_r}{r\,\partial\theta}.\end{aligned}\right\}\quad\ldots\ldots\ldots\ldots\ldots(3).$$

4·3. Classification of Regions of Space. A region in which every closed curve can be contracted to a point without passing out of the region is called a *simply-connected region.* Otherwise the space is *multiply-connected.* In any multiply-connected space it is possible to draw at least one section of the region, or insert one barrier, having a closed curve for boundary, without breaking up the space into disconnected regions. A region of space for which one such barrier can be drawn is said to be doubly-connected. If $n-1$ such barriers can be drawn, the region is n-ply connected or of connectivity n.

A region bounded by a single surface such as a sphere or ellipsoid or the space between two closed surfaces one within the other such as concentric spheres is simply-connected, for every closed curve within it is reducible and no barrier can be drawn across it without dividing it into two disconnected regions as is seen in Fig. 1. But the space inside an anchor ring is doubly-connected for one barrier can be drawn without dividing the space into disconnected regions (Fig. 2).

Fig. 3 represents an anchor ring and another tubular region communicating with it, forming a triply-connected region; and in like manner Fig. 4 shews a quadruply-connected region. It will be seen that in each of Figs. 2–4 the maximum number of barriers have been inserted without dividing the region into disconnected parts.

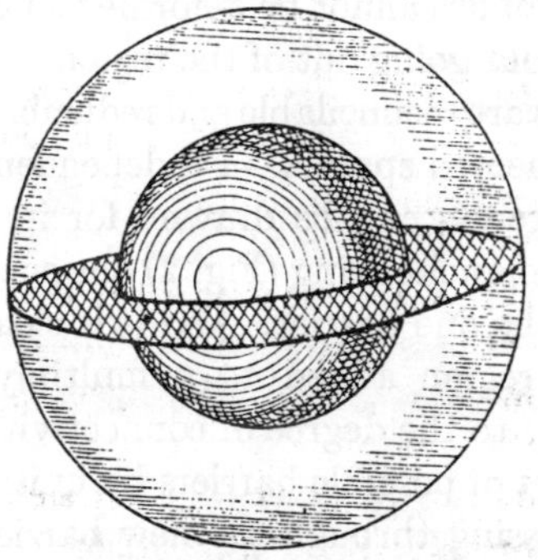

Fig. 1

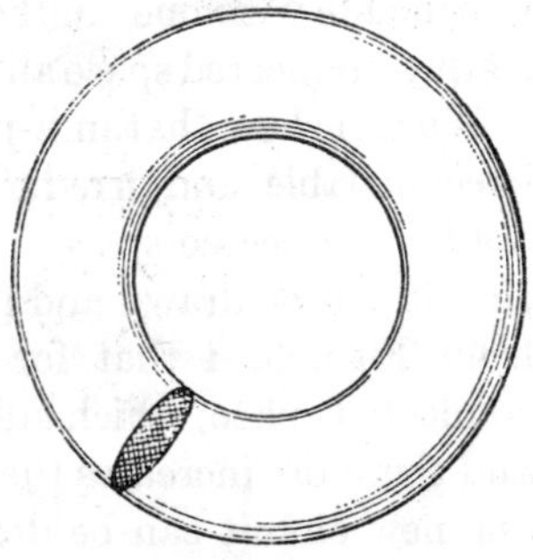

Fig. 2

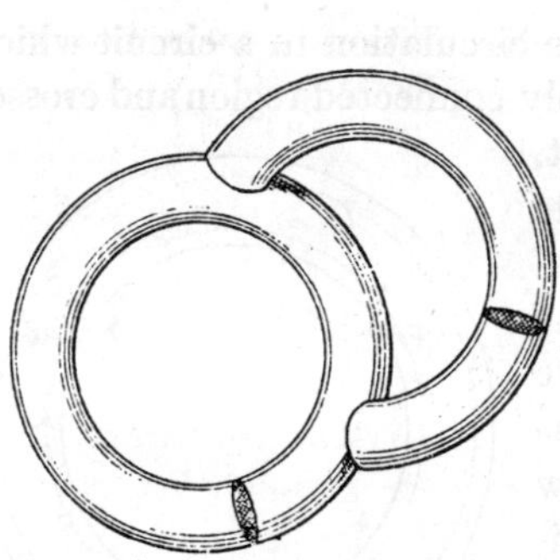

Fig. 3

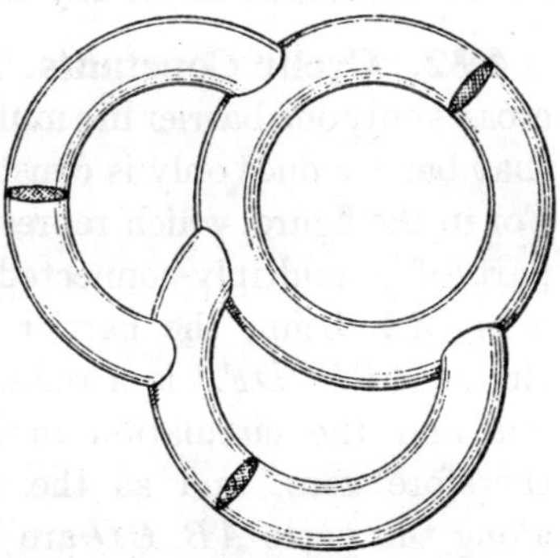

Fig. 4

In the same way the space outside the regions shewn in Figs. 2, 3, 4 are respectively doubly-, triply- and quadruply-connected, thus for the space outside the anchor ring a barrier might be drawn filling the opening of the ring, for such a barrier would be bounded by a closed curve and would not divide the external space into disconnected portions; and similarly for the other figures.

When in a multiply-connected region all barriers have been inserted that can be inserted without dividing the region into disconnected parts, if these barriers are regarded as temporary boundaries the region will have been reduced to a simply-connected one. This will be obvious from a study of the figures.

4·31. Circuits in a given region may be called *reconcilable* or *irreconcilable*, according as they can or cannot be deformed so as to coincide with one another without going out of the region. In simply-connected space all circuits are reconcilable and reducible.

We can shew that in n-ply connected space $n-1$ independent irreconcilable and irreducible circuits can be drawn; for in a doubly-connected space such as an anchor ring (Fig. 2) one such circuit can be drawn and it cuts the one barrier. And it is clear from Figs. 3, 4 that for every region added to a multiply-connected space, which adds unity to the degree of connectivity and therefore increases the number of possible barriers by unity, one new circuit can be drawn passing through the new barrier and not reconcilable with any existing circuit. Thus in Fig. 3, which represents a triply-connected region, two such circuits can be drawn, and so on for any degree of connectivity.

4·32. Cyclic Constants. The circulation in a circuit which crosses only one barrier in a multiply-connected region and crosses that barrier once only is constant. For in the figure, which represents part of a multiply-connected region, XY being the barrier, the circuit $ABECDFA$ is a reducible one and the circulation in it is therefore zero, and as the flow along the parts AB, CD are ultimately equal and opposite when A coincides with D and B with C, therefore the circulations in closed circuits $BECB$, $DFAD$ are equal and opposite; or the circulations in any two such circuits taken in the same sense are equal to a constant κ, and if the circuit crosses the barrier p times in the same sense the circulation will be $p\kappa$. κ is called the *cyclic constant* of the circuit.

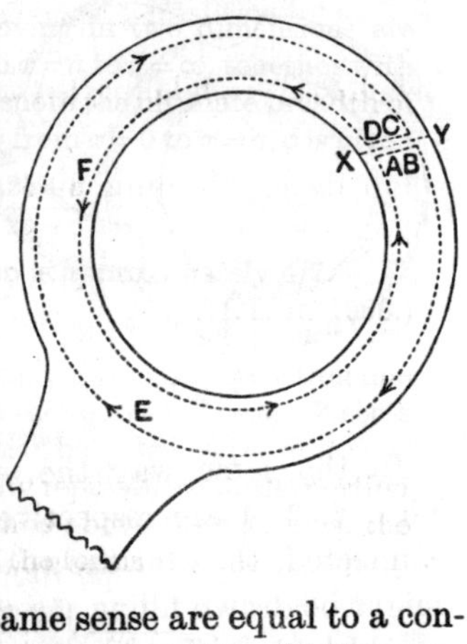

In the same way if $\kappa_1, \kappa_2, \ldots \kappa_{n-1}$ be the cyclic constants of the $n-1$ irreducible circuits of an n-ply connected space, the circulation in any compound circuit will be $p_1\kappa_1+p_2\kappa_2+\ldots+p_{n-1}\kappa_{n-1}$, where p_r denotes the excess of the number of crossings of the rth barrier in the positive sense over the number of crossings in the negative sense. Motion in which the circulation in every circuit does not vanish is called cyclic motion.

4·4. Nature of the Problems to be discussed. The types of irrotational fluid motion with which we shall be chiefly concerned, in what follows, may be classified thus:

(i) A finite mass of liquid is enclosed within a given boundary and possibly limited internally by other boundaries. Liquid motion is set up by giving a definite motion to one or more of the boundaries, or by applying given impulses to one or more of the boundaries.

(ii) An infinite mass of liquid is limited internally by the surfaces of one or more bodies, and either

(*a*) the liquid is at rest at infinity and the bodies are in motion; or

(*b*) the liquid has a uniform constant velocity at infinity, and the bodies are at rest or in motion.

We propose to prove the determinateness of these problems; i.e. that a definite liquid motion will result from definite motions of the boundaries, or from the application of definite impulses to the boundaries.

As we have seen already, irrotational motion implies the existence of a velocity potential ϕ which satisfies Laplace's equation

$$\frac{\partial^2\phi}{\partial x^2}+\frac{\partial^2\phi}{\partial y^2}+\frac{\partial^2\phi}{\partial z^2}=0, \text{ or } \nabla^2\phi=0;$$

and the solution of any problem in irrotational motion depends on finding a solution of the equation $\nabla^2\phi=0$ that will give the correct values to the normal velocity $\partial\phi/\partial n$, or to ϕ which may be taken as a measure of the impulse, over the boundaries. In this respect the problem is akin to the general problem of electrostatics.

We do not propose to prove the existence of a potential function which will satisfy given boundary conditions, but we shall prove that if the problem has a solution it is a definite one; so that, in any particular case in which we have found a solution that fits the circumstances of the case, we shall know that since only one solution is possible our solution is the right one.

We shall begin by proving a theorem of Green's which is of fundamental importance in physical investigations.

4·5. Green's Theorem*. Let ϕ, ϕ' be two functions of x, y, z which with their first and second derivatives are finite and single-valued throughout the region considered; and let S denote a closed surface bounding any singly-connected region of space and ∂n an element of the normal at a point on this boundary drawn *into* the region considered, then

$$\iiint\left(\frac{\partial\phi}{\partial x}\frac{\partial\phi'}{\partial x}+\frac{\partial\phi}{\partial y}\frac{\partial\phi'}{\partial y}+\frac{\partial\phi}{\partial z}\frac{\partial\phi'}{\partial z}\right)dx\,dy\,dz$$

$$=-\iint\phi\frac{\partial\phi'}{\partial n}dS-\iiint\phi\nabla^2\phi'\,dx\,dy\,dz$$

$$=-\iint\phi'\frac{\partial\phi}{\partial n}dS-\iiint\phi'\nabla^2\phi\,dx\,dy\,dz \quad \text{......(1)},$$

where the surface integrals are taken over the closed surface S and the volume integrals throughout the space enclosed.

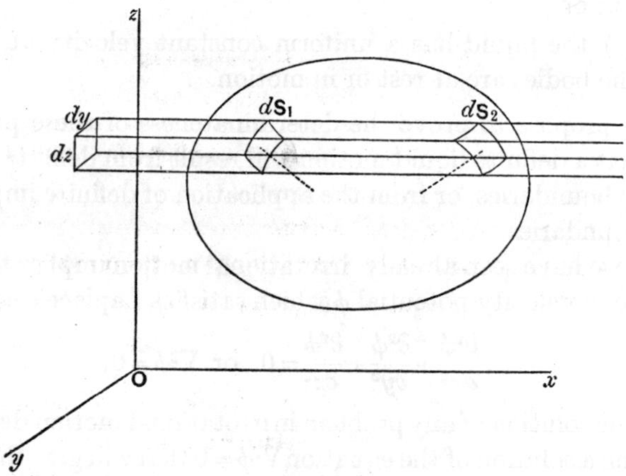

To prove this, integrate $\iiint\frac{\partial\phi}{\partial x}\frac{\partial\phi'}{\partial x}dx\,dy\,dz$ by parts, integrating along a prism of section $dy\,dz$ which intersects the surface in elements dS_1, dS_2 where the inward-drawn normals have x-direction cosines l_1, l_2.

* G. Green, *Essay on Electricity and Magnetism*, 1828, or *Math. Papers* (ed. Ferrers), p. 23.

The result is

$$\iint\left[\phi\frac{\partial\phi'}{\partial x}\right]dy\,dz-\iiint\phi\frac{\partial^2\phi'}{\partial x^2}\,dx\,dy\,dz;$$

where
$$\left[\phi\frac{\partial\phi'}{\partial x}\right]dy\,dz=-\phi_2\frac{\partial\phi_2'}{\partial x}l_2dS_2-\phi_1\frac{\partial\phi_1'}{\partial x}l_1dS_1$$
$$=-\phi\frac{\partial\phi'}{\partial x}\,l\,dS,$$

where in this expression dS is taken to include the two elements of area at the ends of the prism.

Hence

$$\iiint\frac{\partial\phi}{\partial x}\frac{\partial\phi'}{\partial x}\,dx\,dy\,dz=-\iint\phi l\frac{\partial\phi'}{\partial x}\,dS-\iiint\phi\frac{\partial^2\phi'}{\partial x^2}\,dx\,dy\,dz,$$

and by similar treatment of the remaining terms of the first expression in (1), and remembering that

$$l\frac{\partial\phi'}{\partial x}+m\frac{\partial\phi'}{\partial y}+n\frac{\partial\phi'}{\partial z}=\frac{\partial\phi'}{\partial n},$$

we prove the first expression equal to the second; and by interchanging ϕ and ϕ' it becomes equal to the third.

4·51. The statement of the theorem needs modification if the given region includes discontinuities in the values of ϕ, ϕ' or their first derivatives. But the theorem is still true if we surround the point or surface of discontinuity by a closed surface and exclude the enclosed space from the region of integration, provided that the remaining space is singly-connected and we include in the surface integrals integration over the extra surface or surfaces that we have introduced.

4·52. Deductions from Green's Theorem. We shall now make some deductions from Green's Theorem, but we remark at the outset that many of these are capable of very simple independent proof.

(i) Put $\phi'=$ constant. Then

$$-\iint\frac{\partial\phi}{\partial n}dS=\iiint\nabla^2\phi\,dx\,dy\,dz;$$

and if ϕ satisfies Laplace's equation, we also have

$$\iint\frac{\partial\phi}{\partial n}dS=0.$$

If ϕ denotes a velocity potential this result means that the *total* flow of liquid into any closed region at any instant is zero.

(ii) If ϕ, ϕ' are both velocity potentials,

$$\iint \phi \frac{\partial \phi'}{\partial n} dS = \iint \phi' \frac{\partial \phi}{\partial n} dS;$$

a reciprocal theorem which has a physical meaning if we bear in mind that, if ρ denotes density, $\rho\phi$, $\rho\phi'$ denote impulsive pressures that would produce the motions instantaneously and $\partial\phi/\partial n$, $\partial\phi'/\partial n$ are the velocities of the boundaries at which these pressures may be supposed to be applied.

(iii) Put $\phi' = \phi$. Then, if ϕ is a velocity potential,

$$\iiint \left\{ \left(\frac{\partial \phi}{\partial x}\right)^2 + \left(\frac{\partial \phi}{\partial y}\right)^2 + \left(\frac{\partial \phi}{\partial z}\right)^2 \right\} dx\,dy\,dz = -\iint \phi \frac{\partial \phi}{\partial n} dS.$$

Hence if q be the velocity and ρ the density of the liquid, we have for the kinetic energy of the liquid within S

$$\tfrac{1}{2}\rho \iiint q^2\,dx\,dy\,dz = -\tfrac{1}{2}\rho \iint \phi \frac{\partial \phi}{\partial n} dS.$$

Since $\rho\phi$ is the impulsive pressure that would set up the motion instantaneously from rest, and $-\partial\phi/\partial n$ is the inward normal velocity at the surface, therefore the last result is an example of the theorem that the kinetic energy set up by impulses, in a system starting from rest, is the sum of the products of each impulse and half the velocity of its point of application. The result also shews that the kinetic energy of a given mass of liquid moving irrotationally in simply-connected space depends only on the motion of its boundaries.

4·53. For the present we shall consider that ϕ is the velocity potential of a liquid in singly-connected space. From **4·52** (iii) we see that, if the boundaries are at rest or if $\phi = 0$ over the boundaries, we must have

$$\iiint q^2\,dx\,dy\,dz = 0,$$

so that $q = 0$ at every point. Hence irrotational motion is impossible in a closed singly-connected region with fixed boundaries. Also if a closed vessel full of liquid which moves irrotationally is suddenly brought to rest the liquid is also brought to rest.

4·54. Uniqueness Theorem. *There cannot be two different forms of irrotational motion for a given confined mass of liquid whose boundaries have prescribed velocities or are subject to given impulses.* For if two such motions are possible let ϕ_1, ϕ_2 denote their velocity potentials, then at all points of the boundaries either $\partial\phi_1/\partial n = \partial\phi_2/\partial n$, or else $\phi_1 = \phi_2$. But $\phi_1 - \phi_2$ will also satisfy Laplace's equation and represent an irrotational motion in which either the boundary velocity $\partial(\phi_1 - \phi_2)/\partial n$ is zero or $\phi_1 - \phi_2$ is zero over the boundary. Hence in this case, by **4·53**, the liquid is at rest, or $\phi_1 - \phi_2$ is constant everywhere. Therefore the two motions are the same.

4·55. Mean Potential over Spherical Surface. If a region lying wholly in the liquid be bounded by a spherical surface the mean value of the velocity potential over the surface is equal to its value at the centre of the sphere.

For if ϕ_r denote the mean value of ϕ over a sphere of radius r, we have

$$\phi_r = \frac{1}{4\pi r^2}\iint \phi\, dS = \frac{1}{4\pi}\iint \phi\, d\omega,$$

where $d\omega$ is the solid angle which the element dS subtends at the centre of the sphere.

Therefore
$$\frac{\partial \phi_r}{\partial r} = \frac{1}{4\pi}\iint \frac{\partial \phi}{\partial r}\, d\omega = \frac{1}{4\pi r^2}\iint \frac{\partial \phi}{\partial r}\, dS;$$

and the last integral is zero by **4·52** (i), so that ϕ_r is independent of the radius r; consequently the mean value of ϕ is the same over all spheres having the same centre, and by continually diminishing the radius we get that this mean value is the same as the value of ϕ at the centre. This theorem is due to Gauss.

4·56. We shall now extend the last theorem to the case where the region in which the motion takes place is *periphractic*, that is bounded internally by one or more surfaces.

Suppose that a sphere of radius r in the liquid encloses one or more closed surfaces and that the total flow across these surfaces into the given region is $4\pi M$. There must be accordingly an equal flow outwards across the sphere so that

$$\iint \frac{\partial \phi}{\partial r}\, dS = -4\pi M,$$

or
$$\iint \frac{\partial \phi}{\partial r}\, d\omega = -\frac{4\pi M}{r^2},$$

where $d\omega$ has the same meaning as before.

This may also be written

$$\frac{1}{4\pi}\frac{\partial}{\partial r}\iint \phi\, d\omega = -\frac{M}{r^2},$$

and by integrating with respect to r, we get

$$\frac{1}{4\pi}\iint \phi\, d\omega = \frac{M}{r} + C,$$

or
$$\phi_r = \frac{1}{4\pi r^2}\iint \phi\, dS = \frac{M}{r} + C \quad \text{.................(1)},$$

where C is constant with respect to r, but has yet to be proved independent of the position of the sphere.

Supposing the liquid to extend to infinity and to be at rest there, let the sphere be displaced a small distance δx in any direction without altering its radius, then the consequent change in ϕ_r is

$$\frac{\partial \phi_r}{\partial x}\delta x = \frac{1}{4\pi r^2}\iint \frac{\partial \phi}{\partial x}\delta x\, dS = \frac{\partial C}{\partial x}\delta x.$$

Hence $\partial C/\partial x$ is equal to the mean value of $\partial\phi/\partial x$ taken over the sphere. But $\partial\phi/\partial x$ vanishes at infinity and so does its mean value over an infinite sphere; therefore $\partial C/\partial x$ is zero when the sphere has a very large radius. But C is the same for all spheres having the same centre, therefore C is not altered by displacing the sphere, and the result (1) is true for all spheres provided they lie within the liquid and enclose the same internal boundaries*.

4·57. From the previous two articles it follows that the velocity potential ϕ cannot have a maximum value at a point within the liquid, for if there were such a point and a sphere were described with this point as centre the mean velocity potential over this sphere would be less than at its centre. Similarly there cannot be a point at which ϕ has a minimum value.

By a similar argument† the square of the velocity cannot have a maximum within the liquid. For when ϕ satisfies Laplace's equation so does $\partial\phi/\partial x$, therefore the theorem of **4·55** is true when we write $\partial\phi/\partial x$ for ϕ, so that $\partial\phi/\partial x$ cannot have a maximum or minimum at a point in the liquid. Now take the axis of x in the direction of the velocity at a point P, so that $(\partial\phi/\partial x)^2$ is the square of the velocity at P. Then since $(\partial\phi/\partial x)^2$ has no maximum there

* Kirchhoff, *Mechanik*, p. 191.

† *Ibid.* p. 186.

must be points Q in the vicinity of P at which $(\partial\phi/\partial x)^2$ is greater than the square of the velocity at P, and much more then is $(\partial\phi/\partial x)^2+(\partial\phi/\partial y)^2+(\partial\phi/\partial z)^2$, or the square of the velocity at Q, greater than the square of the velocity at P. Hence the square of the velocity cannot be a maximum at P. It will be apparent in what follows that it may have a zero minimum value.

4·6. Liquid extending to Infinity. When the liquid extends to infinity the arguments of **4·53, 4·54** cannot be applied directly without examining the value of $\iint \phi \dfrac{\partial\phi}{\partial n} dS$ over an infinite boundary surface; for, though the velocity may vanish at infinity, it does not necessarily follow that this integral vanishes when taken over an infinite area.

As a first step in this discussion we shall make a further deduction from Green's Theorem.

If ϕ, ϕ' both satisfy Laplace's equation, within a region bounded by a surface S, we have

$$\iint \phi \frac{\partial\phi'}{\partial n} dS = \iint \phi' \frac{\partial\phi}{\partial n} dS \quad \text{.................(1).}$$

Let P be any point within the region, and put $\phi'=1/r$, where r is the distance from P. Since ϕ' becomes infinite at P we must exclude P from the region to which the theorem (1) is applied by surrounding it by a surface, say a sphere of small radius ϵ and surface Σ. This surface must be added to the range of integration, and we get

$$\iint \phi \frac{\partial \frac{1}{r}}{\partial n} dS + \iint \phi\left(-\frac{1}{\epsilon^2}\right) d\Sigma = \iint \frac{1}{r}\frac{\partial\phi}{\partial n} dS + \iint \frac{1}{\epsilon}\frac{\partial\phi}{\partial n} d\Sigma.$$

Since $d\Sigma=\epsilon^2 d\omega$, where $d\omega$ is the solid angle subtended at P by $d\Sigma$, therefore the second integral tends to $-4\pi\phi_P$ as ϵ tends to zero, where ϕ_P denotes the value of ϕ at P. For the same reason the fourth integral tends to zero with ϵ. Hence we have

$$\phi_P = \frac{1}{4\pi}\iint \phi \frac{\partial \frac{1}{r}}{\partial n} dS - \frac{1}{4\pi}\iint \frac{1}{r}\frac{\partial\phi}{\partial n} dS \quad \text{.........(2).}$$

Now consider an infinite mass of liquid bounded internally by certain finite surfaces S and let us apply the last result, taking for external boundary a sphere Σ of large radius R with its centre at P. We have for any point P in the liquid

$$\phi_P = \frac{1}{4\pi}\iint \phi \frac{\partial \frac{1}{r}}{\partial n} dS - \frac{1}{4\pi}\iint \frac{1}{r}\frac{\partial \phi}{\partial n} dS + \frac{1}{4\pi}\iint \phi \frac{\partial \frac{1}{r}}{\partial n} d\Sigma - \frac{1}{4\pi}\iint \frac{1}{r}\frac{\partial \phi}{\partial n} d\Sigma$$

$$= \frac{1}{4\pi}\iint \phi \frac{\partial \frac{1}{r}}{\partial n} dS - \frac{1}{4\pi}\iint \frac{1}{r}\frac{\partial \phi}{\partial n} dS + \frac{1}{4\pi R^2}\iint \phi\, d\Sigma - \frac{1}{4\pi R}\iint \frac{\partial \phi}{\partial n} d\Sigma.$$

Assuming that the total flow of liquid across the internal boundaries is zero and that the velocity vanishes at infinity, by **4·56** the third integral is a definite constant C. And the total flow across the sphere Σ is also zero, so that the fourth integral is zero. Therefore

$$\phi_P = C + \frac{1}{4\pi}\iint \phi \frac{\partial \frac{1}{r}}{\partial n} dS - \frac{1}{4\pi}\iint \frac{1}{r}\frac{\partial \phi}{\partial n} dS \quad \text{.........(3).}$$

Now let P move to an infinite distance from the inner boundaries S, the integrands then tend to zero and the range of integration is finite, so that the integrals vanish and we see that the velocity potential ϕ tends to a definite constant limit at infinity, when the velocity vanishes at infinity*.

Now apply **4·52** (iii) to the space between the inner boundaries S and a sphere Σ of large radius R and we get

$$\iiint q^2\,dx\,dy\,dz = -\iint \phi \frac{\partial \phi}{\partial n} dS - \iint \phi \frac{\partial \phi}{\partial n} d\Sigma.$$

Also because of the constancy of the whole mass of liquid

$$\iint \frac{\partial \phi}{\partial n} dS + \iint \frac{\partial \phi}{\partial n} d\Sigma = 0;$$

and on the sphere Σ as its radius increases ϕ tends to a constant limit C, therefore

$$\iiint q^2\,dx\,dy\,dz = -\iint (\phi - C)\frac{\partial \phi}{\partial n} dS,$$

where the surface integral extends to the inner boundaries only.

* It cannot be assumed that ϕ must be constant at infinity if its space-derivatives all vanish there. For example, if $\phi = \log r$ then $\partial\phi/\partial r = 1/r$ and vanishes as $r \to \infty$, but ϕ becomes infinite.

Hence if the inner boundaries are at rest, or if $\phi - C = 0$ over the boundaries, we get

$$\iiint q^2\,dx\,dy\,dz = 0,$$

so that $q = 0$ everywhere. That is, irrotational motion is impossible in a liquid at rest at infinity unless its inner boundaries are in motion.

4·61. Further, if the value of $\partial\phi/\partial n$, or of ϕ, is prescribed over the inner boundaries there is only one motion possible. For if two different motions of the liquid were possible having equal values of $\partial\phi/\partial n$ or of ϕ at each point of the boundaries, let ϕ_1, ϕ_2 denote their velocity potentials; then $\phi_1 - \phi_2$ satisfies $\nabla^2\phi = 0$, and is also the velocity potential of a motion giving zero velocity or making $\phi - C$ zero over the boundaries. Hence as in the last article the velocity in this case is zero everywhere, that is the two motions are the same.

4·62. Referring to **4·4** we have now only to consider the case in which the liquid has uniform constant velocity at infinity; and the determinateness of the problem in this case follows from the consideration that the problem of the relative motion is not affected by imposing on the whole mass of liquid and its boundaries a velocity equal and opposite to the velocity at infinity. The liquid is then at rest at infinity and it follows from **4·61** that if the velocities of the boundaries are prescribed or if given impulses are applied to them there is only one possible motion of the liquid.

4·7. Minimum Kinetic Energy. If a mass of liquid be set in motion by giving prescribed velocities to its boundaries, the Kinetic Energy in the actual motion is less than that in any other motion consistent with the same motion of the boundaries.

Let T be the kinetic energy of the motion of which ϕ is the velocity potential, and T_1 the kinetic energy of any other possible state of motion in which the velocity components at (x, y, z) are u_1, v_1, w_1. These components must satisfy the equation of continuity

$$\frac{\partial u_1}{\partial x} + \frac{\partial v_1}{\partial y} + \frac{\partial w_1}{\partial z} = 0 \quad \text{.....................(1)},$$

and give the same normal boundary velocity as in the other motion, which condition is expressed by a relation

$$lu_1 + mv_1 + nw_1 = lu + mv + nw \quad \text{...........(2)}.$$

Now

$$T_1 - T$$
$$= \tfrac{1}{2}\rho \iiint (u_1{}^2 + v_1{}^2 + w_1{}^2)\, dx\,dy\,dz - \tfrac{1}{2}\rho \iiint (u^2 + v^2 + w^2)\, dx\,dy\,dz$$
$$= \tfrac{1}{2}\rho \iiint \{2u(u_1 - u) + \ldots + \ldots + (u_1 - u)^2 + \ldots + \ldots\}\, dx\,dy\,dz.$$

But, by an integration similar to that used in the proof of Green's Theorem,

$$\iiint \{u(u_1 - u) + v(v_1 - v) + w(w_1 - w)\}\, dx\,dy\,dz$$
$$= -\iiint \left\{\frac{\partial\phi}{\partial x}(u_1 - u) + \ldots + \ldots\right\} dx\,dy\,dz$$
$$= \iint \phi\{l(u_1 - u) + m(v_1 - v) + n(w_1 - w)\}\, dS$$
$$+ \iiint \phi\left\{\frac{\partial}{\partial x}(u_1 - u) + \frac{\partial}{\partial y}(v_1 - v) + \frac{\partial}{\partial z}(w_1 - w)\right\} dx\,dy\,dz$$

$= 0$, from (1) and (2).

Hence

$$T_1 - T = \tfrac{1}{2}\rho \iiint \{(u_1 - u)^2 + (v_1 - v)^2 + (w_1 - w)^2\}\, dx\,dy\,dz$$
$$= \text{a positive quantity.}$$

Hence the theorem follows. This theorem is due to Lord Kelvin*, and was subsequently generalized by him so as to apply to all dynamical systems started impulsively from rest†.

4·71. Kinetic Energy of an Infinite Mass of Liquid moving irrotationally.

We have, as in **4·6**,

$$\iiint q^2\, dx\,dy\,dz = -\iint (\phi - C)\frac{\partial\phi}{\partial n}\, dS,$$

where C is a constant and the surface integral extends to the inner boundaries of the liquid; and, if the total flow across the inner boundaries is zero,

$$\iint \frac{\partial\phi}{\partial n}\, dS = 0,$$

so that the kinetic energy is

$$-\tfrac{1}{2}\rho \iint \phi\frac{\partial\phi}{\partial n}\, dS.$$

* *Camb. and Dub. Math. Journal*, 1849, p. 92, or *Math. and Phys. Papers*, I, p. 107.
† Kelvin and Tait, *Natural Philosophy*, § 312.

4·8. Irrotational Motion in multiply-connected Space. We have seen in **4·32** that the circulation in any circuit in an $(n+1)$ply-connected region is of the form

$$p_1\kappa_1 + p_2\kappa_2 + \dots + p_n\kappa_n \quad \dots\dots\dots\dots\dots\dots(1),$$

where the κ's are the cyclic constants of the n irreducible circuits, and the p's are integers.

If
$$\phi = -\int_A^P (u\,dx + v\,dy + w\,dz) \quad \dots\dots\dots\dots\dots(2),$$

be the flow along a path from a fixed point A to a variable point P, the value of ϕ depends on the particular path; because, if ABP and ACP are two paths, the circulation round $ABPCA$ is not generally zero. Hence ϕ is indeterminate or many-valued to the extent of the addition of an expression of the form (1).

By displacing P parallel to the axes in turn we obtain from (2)

$$u = -\partial\phi/\partial x, \quad v = -\partial\phi/\partial y, \quad w = -\partial\phi/\partial z;$$

and these are single-valued expressions whether ϕ be multiple-valued or not.

4·81. Kelvin's Modification of Green's Theorem. In our proof of Green's Theorem in **4·5** we assumed that ϕ, ϕ' were single-valued functions in the region considered, but if either be a many-valued or cyclic function the formula needs modification. Thus, if we suppose ϕ to be cyclic, the second expression in **4·5** (1) must be corrected so as to take account of the indeterminateness of ϕ. We can do this by supposing all the barriers that are necessary to reduce the region under consideration to a simply-connected space to be inserted: then we may regard ϕ as single-valued throughout this region and the correction to be made consists therefore in including in the range of the surface integral both sides of each of the barriers.

If $d\sigma_r$ be an element of area of one of the barriers and κ_r the corresponding cyclic constant, we have to take $\iint \phi \dfrac{\partial\phi'}{\partial n} d\sigma_r$ over both sides of the barrier. The values of $\dfrac{\partial\phi'}{\partial n}$, being taken in opposite directions on opposite sides of the barrier, are equal in magnitude but opposite in sign at corresponding points; while the value of ϕ on the positive side of the barrier exceeds the value on the negative side by the cyclic constant κ_r, so that the contribution of this barrier to the surface integral is $\kappa_r \iint \frac{\partial \phi'}{\partial n} d\sigma$ taken once over the barrier.

Hence the theorem becomes

$$\iiint \left(\frac{\partial \phi}{\partial x}\frac{\partial \phi'}{\partial x} + \frac{\partial \phi}{\partial y}\frac{\partial \phi'}{\partial y} + \frac{\partial \phi}{\partial z}\frac{\partial \phi'}{\partial z} \right) dx\,dy\,dz$$

$$= -\iint \phi \frac{\partial \phi'}{\partial n} dS - \sum_{r=1}^{n} \kappa_r \iint \frac{\partial \phi'}{\partial n} d\sigma_r - \iiint \phi \nabla^2 \phi' \,dx\,dy\,dz \quad \ldots (1).$$

No extra terms arise because of the indeterminateness of ϕ in the last integral, if we suppose that $\nabla^2\phi' = 0$, for the indeterminate part of ϕ is a constant.

It is clear that the coefficient of each κ is the total flow in the positive direction across each barrier due to a velocity potential ϕ'.

If we assume ϕ' to be cyclic with cyclic constants κ_1', κ_2', etc., we get another relation similar to (1) in which ϕ, ϕ' are interchanged and κ_r' is written for κ_r.

4·82. Kinetic Energy of Cyclic Irrotational Motion. If we put $\phi' = \phi$ in **4·81**, and take ϕ to be a velocity potential, we get for the kinetic energy of the motion

$$T = \tfrac{1}{2}\rho \iiint \left\{ \left(\frac{\partial \phi}{\partial x}\right)^2 + \left(\frac{\partial \phi}{\partial y}\right)^2 + \left(\frac{\partial \phi}{\partial z}\right)^2 \right\} dx\,dy\,dz$$

$$= -\tfrac{1}{2}\rho \iint \phi \frac{\partial \phi}{\partial n} dS - \tfrac{1}{2}\rho \sum_{r=1}^{n} \kappa_r \iint \frac{\partial \phi}{\partial n} d\sigma_r \quad \ldots\ldots\ldots (1).$$

This assumes, of course, that the barriers do not obstruct the motion of the liquid, but move along with it.

If the liquid extend to infinity as in **4·71**, we must replace the first term on the right by

$$-\tfrac{1}{2}\rho \iint (\phi - C) \frac{\partial \phi}{\partial n} dS \quad \ldots\ldots\ldots\ldots\ldots\ldots (2),$$

where C is a constant and the integral extends to the internal boundaries of the liquid, the C term being omitted if the *total* flow across these inner boundaries is zero.

4·83. Determinateness of Irrotational Motion in multiply-connected Space. If the cyclic constants $\kappa_1, \kappa_2, \ldots \kappa_n$ are given and the boundary velocities, we can shew that the motion is determinate. For supposing the space to be rendered simply-connected by the introduction of suitable barriers, let there be two possible motions represented by velocity potentials ϕ, ϕ' which both have the same cyclic constants. Then $\phi - \phi'$ will be a velocity potential having no cyclic constants, i.e. the velocity potential of an acyclic motion, in which, in addition, the velocity is zero at all boundaries. Hence by **4·54** and **4·61** the two motions are identical.

4·831. Example. Let us take, as an example, two-dimensional irrotational motion in the space between two coaxial circular cylinders; and suppose that the velocity at distance r from the axis is c^2/r at right angles to the radius vector.

We have seen in **1·81** that the velocity potential is given by

$$\phi = -c^2 \tan^{-1}\frac{y}{x}.$$

This is a many-valued function, the region being doubly-connected, and the

$$\begin{aligned} \text{cyclic constant } \kappa &= \text{circulation} \\ &= 2\pi r \times c^2/r \\ &= 2\pi c^2, \end{aligned}$$

so that the circulation in any closed path is $n\kappa$ or $2\pi n c^2$, where n is the number of times the path embraces the cylinder.

To find the kinetic energy of the liquid contained between unit lengths of the cylinders we may proceed directly taking

$$T = \tfrac{1}{2}\rho \int_a^b \frac{c^4}{r^2} 2\pi r\, dr = \pi\rho c^4 \log b/a,$$

where a and b are the radii of the inner and outer cylinders; or we may shew that we get the same result from the expression (1) of **4·82**. The first integral in that expression is zero because $\partial\phi/\partial n$ vanishes over the fixed boundaries.

For the second integral, $-\tfrac{1}{2}\rho\kappa \iint \frac{\partial\phi}{\partial n}\, d\sigma$, we may take as barrier a plane through the axis of the cylinders; $-\partial\phi/\partial n$, the velocity perpendicular to the barrier, is then the whole velocity c^2/r, and the integral becomes

$$\tfrac{1}{2}\rho\kappa \int_a^b \frac{c^2}{r}\, dr = \tfrac{1}{2}\rho\kappa c^2 \log b/a = \pi\rho c^4 \log b/a.$$

4·9. Motion regarded as due to Sources and Doublets.

Referring to the theorem represented by **4·6** (2), viz.

$$\phi_P = \frac{1}{4\pi}\iint \phi \frac{\partial \frac{1}{r}}{\partial n}\, dS - \frac{1}{4\pi}\iint \frac{1}{r}\frac{\partial\phi}{\partial n}\, dS,$$

it follows from **3·3** and **3·31** that the velocity potential at P is the same as if the motion in the region bounded by the surface S were due to a distribution over S of simple sources with a density $-\frac{1}{4\pi}\frac{\partial\phi}{\partial n}$ per unit area, together with a distribution of doublets with axes pointing inwards along the normals to the surface of density $\phi/4\pi$ per unit area.

Now let a closed surface S be drawn in a liquid and let ϕ, ϕ' denote the velocity potentials of possible motions inside and outside S respectively, with the condition that ϕ' vanishes at infinity. If P is any point inside S, we have

$$\phi_P = \frac{1}{4\pi}\iint \phi \frac{\partial \frac{1}{r}}{\partial n} dS - \frac{1}{4\pi}\iint \frac{1}{r}\frac{\partial \phi}{\partial n} dS.$$

Also since P is not within the region of velocity potential ϕ'

$$0 = \frac{1}{4\pi}\iint \phi' \frac{\partial \frac{1}{r}}{\partial n'} dS - \frac{1}{4\pi}\iint \frac{1}{r}\frac{\partial \phi'}{\partial n'} dS,$$

where ∂n, $\partial n'$ are drawn inwards and outwards from the surface S, so that $\partial/\partial n = -\partial/\partial n'$. Then by addition

$$\phi_P = \frac{1}{4\pi}\iint (\phi - \phi')\frac{\partial \frac{1}{r}}{\partial n} dS - \frac{1}{4\pi}\iint \frac{1}{r}\left(\frac{\partial \phi}{\partial n} + \frac{\partial \phi'}{\partial n'}\right) dS \quad \ldots(1).$$

If we take $\phi' = \phi$ at the surface S, we have

$$\phi_P = -\frac{1}{4\pi}\iint \frac{1}{r}\left(\frac{\partial \phi}{\partial n} + \frac{\partial \phi'}{\partial n'}\right) dS \quad \ldots\ldots\ldots\ldots\ldots(2);$$

and, if we take $\dfrac{\partial \phi}{\partial n} = \dfrac{\partial \phi'}{\partial n}$, we get

$$\phi_P = \frac{1}{4\pi}\iint (\phi - \phi')\frac{\partial \frac{1}{r}}{\partial n} dS \quad \ldots\ldots\ldots\ldots\ldots\ldots(3).$$

Equation (2) shews that when the velocity potential is continuous but the normal flow across S is discontinuous the motion inside S might be produced by a distribution over the surface of simple sources of density $-\dfrac{1}{4\pi}\left(\dfrac{\partial \phi}{\partial n} + \dfrac{\partial \phi'}{\partial n'}\right)$ per unit area.

Equation (3) shews that when the normal velocity across the surface is continuous, but the velocity potential discontinuous, the motion inside S might be produced by a distribution over the surface of doublets with axes along the normals inwards of density $(\phi - \phi')/4\pi$ per unit area. Such a distribution might be called a *double sheet*.

MISCELLANEOUS EXAMPLES

1. Explain the meaning of the term *rotational* as applied to fluid motion; and determine the character of the circulatory motion of fluid, round a straight axis, which is not rotational.

Shew that, in such a case, minute bubbles of air in the circulating fluid will be sucked in towards the axis. (St John's Coll. 1896.)

2. When a body immersed in a fluid executes periodic vibrations it appears to exert an attraction on other bodies at rest in the fluid. Give a general explanation of this phenomenon. (Coll. Exam. 1903.)

3. Prove that if the velocity potential at any instant be λxyz, the velocity at any point $(x+\xi, y+\eta, z+\zeta)$ relative to the fluid at the point (x, y, z), where ξ, η, ζ are small, is normal to the quadric $x\eta\zeta+y\zeta\xi+z\xi\eta=$ constant, with centre at (x, y, z). (Trinity Coll. 1897.)

4. Prove that if $\lambda=\frac{\partial u}{\partial t}-v\left(\frac{\partial v}{\partial x}-\frac{\partial u}{\partial y}\right)+w\left(\frac{\partial u}{\partial z}-\frac{\partial w}{\partial x}\right)$

and μ, ν are two similar expressions, then $\lambda dx+\mu dy+\nu dz$ is a perfect differential, if the forces are conservative and the density is constant. (Coll. Exam. 1902.)

5. Shew that, if a heterogeneous incompressible liquid moves irrotationally under the action of conservative forces, the surfaces of equal pressure and equal density coincide; and that a homogeneous liquid cannot move irrotationally under the action of non-conservative forces. (Coll. Exam. 1901.)

6. Shew that the theorem, that under certain conditions, the motion of a frictionless fluid, if once irrotational, will always be so, is true also when each particle is acted on by a frictional resistance varying as its velocity. (Coll. Exam. 1895.)

7. If p denote the pressure, V the potential of the external forces and q the velocity of a homogeneous liquid moving irrotationally, shew that $\nabla^2 q^2$ is positive; and $\nabla^2 p$ is negative provided that $\nabla^2 V=0$. Hence prove that the velocity cannot have a maximum value and the pressure cannot have a minimum value at a point in the interior of the liquid. (Coll. Exam. 1900.)

8. Shew that in the motion of a fluid in two dimensions if the coordinates (x, y) of an element at any time be expressed in terms of the initial coordinates (a, b) and the time, the motion is irrotational if

$$\frac{\partial(\dot{x}, x)}{\partial(a, b)}+\frac{\partial(\dot{y}, y)}{\partial(a, b)}=0.$$ (Coll. Exam. 1903.)

9. Prove that, if

$$\phi=-\tfrac{1}{2}(ax^2+by^2+cz^2),\quad V=\tfrac{1}{2}(lx^2+my^2+nz^2),$$

where a, b, c, l, m, n are functions of the time and $a+b+c=0$, irrotational motion is possible with a free surface of equi-pressure if

$$(l+a^2+\dot{a})\,e^{2\int a\,dt},\quad (m+b^2+\dot{b})\,e^{2\int b\,dt},\quad (n+c^2+\dot{c})\,e^{2\int c\,dt}$$

are constants. (Coll. Exam. 1903.)

10. Shew that if the velocity potential of an irrotational fluid motion is equal to

$$A(x^2+y^2+z^2)^{-\frac{3}{2}}z\tan^{-1}\frac{y}{x},$$

the lines of flow lie on the series of surfaces

$$x^2+y^2+z^2=\kappa^{\frac{2}{3}}(x^2+y^2)^{\frac{2}{3}}. \qquad \text{(Coll. Exam. 1899.)}$$

11. A thin stratum of incompressible fluid is contained between two concentric spheres; shew that the velocity at any point is equivalent to the components

$$-\frac{1}{\sin\theta}\frac{\partial\psi}{\partial\omega},\ \frac{\partial\psi}{\partial\theta}$$

along the meridian and parallel respectively. Also if the fluid be homogeneous and the motion irrotational, prove that

$$\frac{\partial\phi}{\partial\theta}=\frac{1}{\sin\theta}\frac{\partial\psi}{\partial\omega},\quad -\frac{1}{\sin\theta}\frac{\partial\phi}{\partial\omega}=\frac{\partial\psi}{\partial\theta},$$

and deduce that $\phi+i\psi=F(e^{i\omega}\tan\frac{1}{2}\theta)$. (St John's Coll. 1906.)

12. In the case of irrotational motion in two dimensions, on the surface of a sphere, shew that the velocity potential is of the form

$$f\left(\frac{x+iy}{r+z}\right)+f\left(\frac{x-iy}{r+z}\right),$$

r being the radius of the sphere and x, y, z the coordinates of a point referred to rectangular axes through the centre of the sphere.

(Coll. Exam. 1893.)

13. A rigid envelope is filled with homogeneous frictionless liquid; shew that it is not possible, by any movements applied to the envelope, to set its contents into motion which will persist after the envelope has come to rest.

(St John's Coll. 1898.)

14. A space is bounded by an ideal fixed surface S drawn in a homogeneous incompressible fluid satisfying the conditions for the continued existence of a velocity potential ϕ under conservative forces. Prove that the rate per unit time at which energy flows across S into the space bounded by S is

$$-\rho\iint\frac{d\phi}{dt}\frac{\partial\phi}{\partial n}dS,$$

where ρ is the density and ∂n an element of the normal to dS drawn into the space considered. (M.T. 1908.)

15. Deduce from the principle that the kinetic energy set up is a minimum that, if a mass of incompressible liquid be given at rest, completely filling a closed vessel of any shape and if any motion of the liquid be produced suddenly by giving arbitrarily prescribed normal velocities to all the points of its bounding surface subject to the condition of constant volume, the motion produced is irrotational. (Thomson and Tait.)

16. If q is the resultant velocity at any point of a fluid which is moving irrotationally in two dimensions, prove that

$$\left(\frac{\partial q}{\partial x}\right)^2+\left(\frac{\partial q}{\partial y}\right)^2=q\nabla^2 q. \qquad \text{(Univ. of London, 1911.)}$$

17. Shew that the curvature of a stream line in steady motion is $\frac{1}{q^2}\frac{\partial}{\partial\nu}\left(\frac{p}{\rho}+V\right)$, where p, ρ, q are the pressure, density and velocity of the liquid, V the potential of the external forces, and $\partial\nu$ is an element of the principal normal to the stream line, and hence obtain the velocity potential of the two-dimensional irrotational motion for which the stream lines are confocal ellipses. (Coll. Exam. 1900.)

18. Prove that in acyclic irrotational motion of a homogeneous fluid the total momentum of the fluid contained within a sphere of any radius is equivalent to a single vector through the centre of the sphere. (Univ. of London, 1915.)

19. Incompressible fluid of density ρ is contained between two coaxial circular cylinders, of radii a and b $(a<b)$, and between two rigid planes perpendicular to the axis at a distance l apart. The cylinders are at rest and the fluid is circulating in irrotational motion, its velocity being V at the surface of the inner cylinder. Prove that the kinetic energy is $\pi\rho a^2 lV^2 \log b/a$. (Trinity Coll. 1896.)

20. Liquid of density ρ is flowing in two dimensions between the oval curves $r_1 r_2 = a^2$, $r_1 r_2 = b^2$, where r_1, r_2 are the distances measured from two fixed points: if the motion is irrotational and quantity q per unit time crosses any line joining the bounding curves, then the kinetic energy is

$$\pi\rho q^2/\log b/a. \qquad \text{(Trinity Coll. 1895.)}$$

21. A thin sheet of incompressible fluid moves on the surface of a sphere of unit radius. Shew that the velocity potential and stream function are conjugate functions of the Cartesian coordinates of the stereographic projection of any point; and that if the boundary move as a rigid curve on the sphere and its axis of instantaneous rotation cut the sphere in O, the stream function at any point P of the boundary differs from $\omega \cos OP$ by a constant, where ω is the instantaneous angular velocity of the boundary. (M.T. 1896.)

CHAPTER V

SPECIAL PROBLEMS OF IRROTATIONAL MOTION IN TWO DIMENSIONS

5·1. In Chapter III we introduced the stream function ψ for motion in two dimensions and found expressions for it in certain cases. We propose now to make use of it for the determination of two-dimensional irrotational motion produced by the motion of a cylinder in an infinite mass of liquid at rest at infinity, or for the disturbance produced in a steady stream by the presence of a fixed cylinder. For the sake of simplicity we shall suppose the cylinder to be of unit length, and the liquid and the cylinder to be confined between two smooth parallel planes at right angles to the axis of the cylinder.

The stream function ψ must satisfy Laplace's equation $\nabla^2\psi=0$ at all points of the liquid and must also satisfy the boundary conditions as follows:

(1) When the liquid is at rest at infinity then at infinity $\partial\psi/\partial x=0$ and $\partial\psi/\partial y=0$.

(2) At any fixed boundary the normal velocity must be zero, or the boundary must coincide with a stream line $\psi=\text{const.}$

(3) At the boundary of the moving cylinder, the normal component of the velocity of the liquid must be equal to the normal component of the velocity of the cylinder.

Condition (3) may be expressed by a formula for ψ as follows: let a point of the cross section of the cylinder chosen as origin have velocities U, V parallel to the axes of x and y and let the cylinder turn with angular velocity ω, so that the velocity of a point whose coordinates are x, y has components

$$U-\omega y, \quad V+\omega x.$$

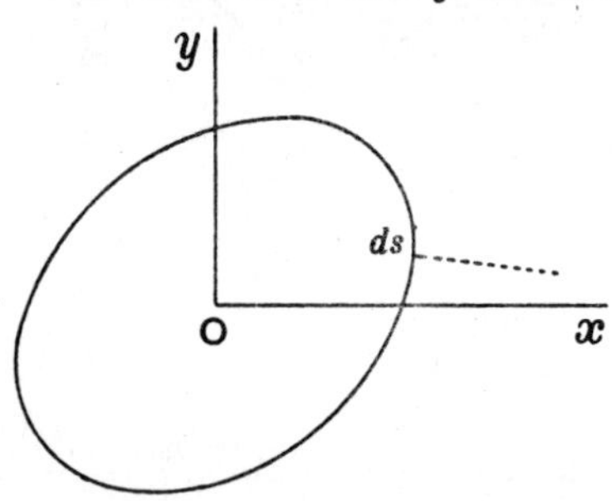

Let ds be an element of arc of the cross section of the cylinder. The velocity of the liquid in the direction of the outward normal is $-\partial\psi/\partial s$, and the cosines of the angles which this normal makes with the axes are dy/ds and $-dx/ds$, so that

$$-\frac{\partial\psi}{\partial s}=(U-\omega y)\frac{dy}{ds}-(V+\omega x)\frac{dx}{ds}.$$

Whence, by integrating along the arc, we get

$$\psi=Vx-Uy+\tfrac{1}{2}\omega(x^2+y^2)+C \quad \text{.........(1).}$$

This is the condition for the most general type of motion of the cylinder and of course includes a simple translation $\omega=0$ and say $V=0$ so that

$$\psi=-Uy+C \quad \text{.....................(2),}$$

or a simple rotation $U=V=0$ and

$$\psi=\tfrac{1}{2}\omega(x^2+y^2)+C \quad \text{....................(3).}$$

5·2. Circular Cylinder. The solution of the problem indicated in **5·1**, viz. to determine a two-dimensional irrotational motion satisfying given boundary conditions, has been effected in a limited number of cases; and the method of solution has frequently been an inverse one. That is to say, instead of a direct investigation of a solution of $\nabla^2\psi=0$ which will satisfy given boundary conditions, known solutions have been studied to see what kind of boundary conditions each will satisfy and the problems have not been formulated until their solutions have been obtained. As an example let us consider the motion represented by the functional relation

$$w=Az^{-1}, \quad \text{.........................(1),}$$

or

$$\phi+i\psi=\frac{A}{r}(\cos\theta-i\sin\theta).$$

This gives $\psi=-\dfrac{A\sin\theta}{r}$, and if we take this value for ψ in the boundary equation **5·1** (2) we have

$$\frac{A\sin\theta}{r}=-Ur\sin\theta+C.$$

This equation represents a family of curves, and if we put $C=0$ and $A=Ua^2$, the family includes a circle of radius a. Hence

$$\psi=-\frac{Ua^2}{r}\sin\theta,\quad \phi=\frac{Ua^2}{r}\cos\theta$$

are the stream function and velocity potential due to the motion of a circular cylinder of radius a moving with velocity U parallel to the x-axis; the origin being always on the axis of the cylinder.

We observe that the velocity potential and stream function are the same as for a two-dimensional doublet of strength Ua^2 on the axis of the cylinder in an infinite mass of liquid.

The case of liquid streaming with general velocity U past a fixed cylinder of radius a may be deduced from the foregoing case by imposing a velocity $-U$ parallel to the x-axis on both the cylinder and the liquid. The cylinder is then reduced to rest and we have to add to the velocity potential a term Ux to correspond to the additional velocity, that is $Ur\cos\theta$; hence a term $Ur\sin\theta$ must be added to ψ, so that

$$\phi = U\left(r + \frac{a^2}{r}\right)\cos\theta, \quad \psi = U\left(r - \frac{a^2}{r}\right)\sin\theta.$$

Hence the equation $(r - a^2/r)\sin\theta = \text{const.}$ represents the stream lines *relative to the cylinder*, and this is true whether the cylinder be moving or at rest.

5·21. Another method of solving problems of the same class is to find a velocity potential that will satisfy the given boundary conditions, i.e. to find a ϕ that will satisfy $\nabla^2\phi = 0$ at every point of the liquid, and make the normal velocity $-\partial\phi/\partial n$ assume the proper values at the boundaries.

In this connection it is useful to remember that in polar coordinates in two dimensions Laplace's equation takes the form

$$\frac{\partial^2\phi}{\partial r^2} + \frac{1}{r}\frac{\partial\phi}{\partial r} + \frac{1}{r^2}\frac{\partial^2\phi}{\partial\theta^2} = 0,$$

and that it has solutions of the form

$$r^n\cos n\theta, \quad r^n\sin n\theta,$$

where n is any integer, positive or negative. Hence the sum of any number of terms of the form

$$A_n r^n\cos n\theta, \quad B_n r^n\sin n\theta$$

is also a solution.

Reverting to the problem of liquid streaming past a fixed circular cylinder, with the notation of **5·2**, the uniform stream in the negative direction of the x-axis is represented by

$$\phi = Ux = Ur\cos\theta,$$

and we have to add a term or terms to represent the disturbance due to the cylinder. Since the disturbance vanishes at infinity these terms can only involve negative powers of r.

The boundary condition is $\frac{\partial \phi}{\partial r}=0$, when $r=a$, and if we assume that

$$\phi = Ur\cos\theta + \frac{A}{r}\cos\theta,$$

this leads to $\qquad U - A/a^2 = 0, \text{ or } A = a^2U,$

whence as before $\qquad \phi = U\left(r+\frac{a^2}{r}\right)\cos\theta,$

and the conjugate function is

$$\psi = U\left(r-\frac{a^2}{r}\right)\sin\theta.$$

5·22. Two Coaxial Cylinders. As a further example let us consider a problem of *initial* motion. Let a cylinder of radius a be surrounded by a coaxial cylinder of radius b, the space between the cylinders being filled with liquid. Suppose the cylinders to be moved suddenly parallel to themselves in directions *at right angles* with velocities U, V respectively.

The boundary conditions for the velocity potential ϕ are:

$$\text{(i) when } r=a,\ \frac{\partial\phi}{\partial r} = -U\cos\theta,$$

$$\text{(ii) when } r=b,\ \frac{\partial\phi}{\partial r} = -V\sin\theta.$$

To satisfy these assume that

$$\phi = \left(Ar+\frac{B}{r}\right)\cos\theta + \left(Cr+\frac{D}{r}\right)\sin\theta;$$

then $\qquad -U\cos\theta = \left(A-\frac{B}{a^2}\right)\cos\theta + \left(C-\frac{D}{a^2}\right)\sin\theta,$

and $\qquad -V\sin\theta = \left(A-\frac{B}{b^2}\right)\cos\theta + \left(C-\frac{D}{b^2}\right)\sin\theta,$

for all values of θ. Hence

$$A-\frac{B}{a^2} = -U, \quad C-\frac{D}{a^2}=0,$$

$$A-\frac{B}{b^2}=0, \quad C-\frac{D}{b^2} = -V;$$

from which we get

$$\phi = -\frac{a^2U}{a^2-b^2}\left(r+\frac{b^2}{r}\right)\cos\theta + \frac{b^2V}{a^2-b^2}\left(r+\frac{a^2}{r}\right)\sin\theta,$$

and the conjugate function

$$\psi = -\frac{a^2U}{a^2-b^2}\left(r-\frac{b^2}{r}\right)\sin\theta - \frac{b^2V}{a^2-b^2}\left(r-\frac{a^2}{r}\right)\cos\theta.$$

It must be remembered however that these equations only represent the motion at the instant when the cylinders are coaxial.

5·23. Equations of Motion of a Circular Cylinder. Reverting to the case of 5·2—a cylinder moving in a liquid at rest at infinity—we have to calculate the forces acting on the cylinder owing to the presence of the liquid. If the extraneous forces have a potential Ω and act on the cylinder and liquid alike their resultant effect is, from Hydrostatical considerations, a force equal to the difference between the forces exerted on the cylinder and the liquid displaced, i.e. if σ, ρ are the densities of the cylinder and liquid the resultant extraneous force is $(\sigma-\rho)/\sigma$ times what it would be if the liquid were not present. Omitting the extraneous forces, the part of the pressure due to the motion is to be found from the equation

$$\frac{p}{\rho}=F(t)+\frac{\partial\phi}{\partial t}-\tfrac{1}{2}q^2 \quad\text{.....................(1)}$$

of 2·2 (3).

Let the centre of the section be the point $z_0\equiv x_0+iy_0$, so that if U, V denote the components of velocity of the cylinder, $U=\dot{x}_0$, $V=\dot{y}_0$.

Let

$$w=a^2\frac{(U+iV)}{z-z_0} \quad\text{.....................(2),}$$

this being the same type of relation as 5·2 (1) with the constants adjusted to give the correct liquid velocity normal to the surface of the cylinder. For if we put $z-z_0=re^{i\theta}$, so that r denotes distance from the axis of the cylinder, we have

$$\phi+i\psi=\frac{a^2}{r}(U+iV)(\cos\theta-i\sin\theta)$$

and

$$\phi=\frac{a^2}{r}(U\cos\theta+V\sin\theta) \quad\text{..................(3),}$$

making the normal velocity on $r=a$

$$-\frac{\partial\phi}{\partial r}=U\cos\theta+V\sin\theta.$$

Again since

$$\dot{z}_0=\dot{x}_0+i\dot{y}_0=U+iV,$$

therefore

$$\frac{\partial w}{\partial t}=\frac{a^2(\dot{U}+i\dot{V})}{z-z_0}+\frac{a^2(U+iV)^2}{(z-z_0)^2},$$

or

$$\frac{\partial\phi}{\partial t}+i\frac{\partial\psi}{\partial t}=\frac{a^2}{r}(\dot{U}+i\dot{V})(\cos\theta-i\sin\theta)$$
$$+\frac{a^2}{r^2}(U+iV)^2(\cos 2\theta-i\sin 2\theta).$$

Hence on $r=a$

$$\frac{\partial\phi}{\partial t}=a\,(\dot{U}\cos\theta+\dot{V}\sin\theta)+(U^2-V^2)\cos 2\theta+2UV\sin 2\theta \quad \dots(4).$$

Also $$q^2=\left|\frac{dw}{dz}\right|^2=\left|-a^2\frac{(U+iV)}{(z-z_0)^2}\right|^2=\frac{a^4(U^2+V^2)}{r^4} \quad \dots\dots\dots(5).$$

Now the components of force on the cylinder are

$$\mathbf{X}=-\int_0^{2\pi} ap\cos\theta\, d\theta \text{ and } \mathbf{Y}=-\int_0^{2\pi} ap\sin\theta\, d\theta.$$

Putting $r=a$ in (5), substituting from (4) and (5) in (1) and performing the integrations, we find that

$$\left.\begin{aligned}\mathbf{X}&=-\pi\rho a^2\dot{U}=-M'\dot{U}\\ \text{and}\quad \mathbf{Y}&=-\pi\rho a^2\dot{V}=-M'\dot{V},\end{aligned}\right\} \quad \dots\dots\dots\dots\dots\dots(6)$$

where M' is the mass of liquid displaced by the cylinder (of unit length).

Hence if M denotes the mass of the cylinder and X', Y' the components of what the extraneous force on the cylinder would be if no liquid were present, the equations of motion are of the form

$$M\dot{U}=-M'\dot{U}+\frac{\sigma-\rho}{\sigma}X',$$

or $$M\dot{U}=\frac{M}{M+M'}\cdot\frac{\sigma-\rho}{\sigma}X',$$

or $M\dot{U}=\dfrac{\sigma-\rho}{\sigma+\rho}X'$ and a similar equation in $\dot{V}$ and Y'.

Hence the effect of the presence of the liquid is to reduce the extraneous forces in the ratio $\sigma-\rho:\sigma+\rho$.

Result (6) implies that if the cylinder were to move with uniform velocity the resultant pressure set up by the motion or the resistance to motion would be zero. This is of course contrary to experience. It will be seen later that a small amount of friction in the liquid alters very considerably the character of the motion in the immediate neighbourhood of the cylinder so that the result obtained above does not apply in the case of a real liquid.

5·231. We may also obtain result (6) of **5·23** from the principle of energy. By **4·71** the kinetic energy of the liquid is given by

$$T=-\tfrac{1}{2}\rho\int\phi\frac{\partial\phi}{\partial r}ds$$

integrated round the boundary of the cylinder. And since, if U denotes the velocity of the cylinder, $\phi = U\frac{a^2}{r}\cos\theta$, therefore

$$T = \tfrac{1}{2}\rho a^2 U^2 \int_0^{2\pi} \cos^2\theta\, d\theta$$
$$= \tfrac{1}{2}\pi\rho a^2 U^2 = \tfrac{1}{2}M'U^2.$$

Hence the presence of the liquid may be considered to increase the effective inertia of the cylinder by an amount M'. And if X denote the force parallel to the axis of x,

$$\frac{d}{dt}(\tfrac{1}{2}MU^2 + \tfrac{1}{2}M'U^2) = \text{rate at which work is being done}$$
$$= XU;$$

so that
$$(M+M')\frac{dU}{dt} = X,$$

or
$$M\frac{dU}{dt} = X - M'\frac{dU}{dt};$$

so that the pressure of the liquid, apart from any extraneous force acting on it, is equivalent to a force $-M'dU/dt$ opposing the motion.

5·24. Circulation about a moving Cylinder. To complete the discussion of irrotational motion of a liquid about a moving cylinder, we must include the possibility of cyclic motion, since the liquid occupies a doubly-connected region. The solution is completed by adding to the velocity potential and stream functions terms that will correspond to a constant circulation κ about the cylinder.

The appropriate form may be found thus: by taking the circulation round a circle of radius r whose centre is at the origin, we get

$$-\frac{\partial\phi}{r\partial\theta}\,.\,2\pi r = \kappa,$$

so that $\phi = -\frac{\kappa\theta}{2\pi}$ and the conjugate function is $\psi = \frac{\kappa}{2\pi}\log r$.

Hence
$$\phi + i\psi = \frac{i\kappa}{2\pi}(\log r + i\theta) \text{ or } w = \frac{i\kappa}{2\pi}\log z \quad \text{.........(1).}$$

Hence, with the notation of **5·23**, we may put for the whole motion

$$w = a^2\frac{(U+iV)}{z-z_0} + \frac{i\kappa}{2\pi}\log(z-z_0) \quad \text{..................(2).}$$

This gives
$$\frac{dw}{dz} = -a^2\frac{(U+iV)}{(z-z_0)^2} + \frac{i\kappa}{2\pi(z-z_0)},$$

so that
$$q^2 = \left|\frac{dw}{dz}\right|^2 = \left|-\frac{a^2}{r^2}(U+iV)e^{-2i\theta} + \frac{i\kappa e^{-i\theta}}{2\pi r}\right|$$
$$\times\left|-\frac{a^2}{r^2}(U-iV)e^{2i\theta} - \frac{i\kappa e^{i\theta}}{2\pi r}\right|$$
$$= \frac{a^4}{r^4}(U^2+V^2) + \frac{\kappa^2}{4\pi^2 r^2} + \frac{\kappa a^2}{\pi r^3}(U\sin\theta - V\cos\theta) \quad \text{...(3).}$$

Again $\partial\phi/\partial t$ has the same value as in **5·23** (4), plus a term arising from the circulation, viz. the real part of $-i\kappa(U+iV)/2\pi(z-z_0)$, or

$$\kappa(V\cos\theta - U\sin\theta)/2\pi r.$$

Whence by substituting in **5·23** (1) and integrating we get

$$\left.\begin{aligned} \mathbf{X} &= -\int_0^{2\pi} ap\cos\theta\,d\theta = -\pi\rho a^2\dot{U} - \kappa\rho V \\ \text{and}\qquad \mathbf{Y} &= -\int_0^{2\pi} ap\sin\theta\,d\theta = -\pi\rho a^2\dot{V} + \kappa\rho U \end{aligned}\right\} \qquad\ldots\ldots\ldots\ldots(4).$$

Hence if, as before, M denotes the mass of unit length of the cylinder and $M' = \pi\rho a^2$ and there are no extraneous forces, the equations of motion are

$$\left.\begin{aligned} (M+M')\,\dot{U} &= -\kappa\rho V \\ (M+M')\,\dot{V} &= \kappa\rho U \end{aligned}\right\} \qquad\ldots\ldots\ldots\ldots(5).$$

These equations give $U\dot{U} + V\dot{V} = 0$, or

$$U^2 + V^2 = \text{const.} \qquad\ldots\ldots\ldots\ldots(6),$$

and

$$\frac{U\dot{V} - V\dot{U}}{U^2+V^2} = \frac{\kappa\rho}{M+M'},$$

or

$$\dot{\epsilon} = \kappa\rho/(M+M') \qquad\ldots\ldots\ldots\ldots(7),$$

where $\epsilon = \tan^{-1}(V/U)$ is the inclination of the direction of motion to the axis of x.

Equations (5) shew that the cylinder is acted on by a force $\kappa\rho$ (velocity) at right angles to the path. We shall see subsequently that this force is independent of the cross section of the cylinder.

Equations (6) and (7) shew that the cylinder describes a circle of radius $(M+M')(U^2+V^2)^{\frac{1}{2}}/\kappa\rho$ with constant velocity $(U^2+V^2)^{\frac{1}{2}}$ in the sense of the cyclic motion.

Suppose now that the liquid and the cylinder are subject to a field of force of the nature of gravity in the negative direction of the axis of y. Then if σ be the density of the cylinder, the equations of motion are

$$\pi\sigma a^2\dot{U} = -\pi\rho a^2\dot{U} - \kappa\rho V,$$

and

$$\pi\sigma a^2\dot{V} = -\pi\rho a^2\dot{V} + \kappa\rho U - \pi(\sigma-\rho)a^2 g,$$

or say

$$\dot{U} + nV = 0$$

and

$$\dot{V} - nU = -g'.$$

The solutions of which are

$$U = g'/n - c\sin(nt+\alpha)$$

and

$$V = c\cos(nt+\alpha);$$

so that, if x, y denote the coordinates of the centre of the cylinder referred to fixed axes, by another integration

$$x = x_0 + g'\frac{t}{n} + \frac{c}{n}\cos(nt+\alpha)$$

and

$$y = y_0 + \frac{c}{n}\sin(nt+\alpha),$$

so that the path is a trochoid.

The existence of the transverse force due to circulation was first investigated by Lord Rayleigh* as the explanation of the swerve of a ball in tennis, golf, cricket or baseball, the circulation of the air being due through friction to the spin of the ball. The same force is the basis of modern Aerodynamics. Since the force clearly only depends on the relative motion of the cylinder and the liquid, it will be unaltered if we superpose on the whole mass a velocity equal and opposite to that of the cylinder, so that the cylinder will then be at rest in a stream of liquid circulating about it.

5·25. In the case of a fixed circular cylinder in a steady stream with a circulation κ superposed, we have

$$\phi = U\left(r + \frac{a^2}{r}\right)\cos\theta - \frac{\kappa\theta}{2\pi} \quad \text{.........................(1),}$$

where the velocity of the stream at infinity is $-U$ parallel to Ox.

The velocity on the cylinder $r=a$ is therefore

$$-\frac{\partial\phi}{r\,\partial\theta} = 2U\sin\theta + \frac{\kappa}{2\pi a} \quad \text{.........................(2).}$$

If there were no circulation there would be points of zero velocity on the cylinder at $\theta=0$ and $\theta=\pi$, the former being the point at which the oncoming stream divides. But when there is circulation the positions on the cylinder of these critical points is given by

$$\sin\theta = -\kappa/4\pi a U \quad \text{.............................(3),}$$

and they only exist when $$|\kappa| < 4\pi U a \quad \text{..............................(4).}$$

The lines of flow are then as indicated in the figure, N, N' being points of zero velocity. It is clear that any point on the circumference might be

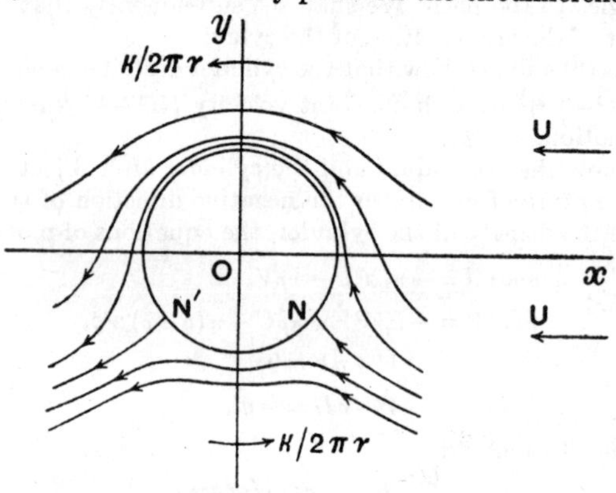

made a critical point by a suitable choice of the ratio κ/U; and we shall see later that this fact has an important bearing in the theory of aerofoils.

When (4) is not satisfied because the circulation is relatively too large there are no points of zero velocity on the cylinder but there is such a point below the cylinder on the axis of y in the figure. At this point a stream line crosses itself and the liquid between this stream line and the cylinder circulates continually round it and is not carried onwards by the stream.

* See Lord Rayleigh, 'On the Irregular Flight of a tennis ball', *Mess. of Math.* 1877, or *Sci. Papers*, I, p. 344. Also Greenhill, *Mess. of Math.* 1880.

5·3. Conjugate Functions. Elliptic Cylinders. Suppose that we have a relation

$$w=f(z), \text{ or } \phi+i\psi=f(x+iy),$$

and that in addition

$$z=F(\zeta), \text{ or } x+iy=F(\xi+i\eta),$$

so that x, y are conjugate functions of ξ, η. Then ϕ, ψ are also conjugate functions of ξ, η. For, the elimination of z gives a functional relation

$$w=\chi(\zeta), \text{ or } \phi+i\psi=\chi(\xi+i\eta),$$

from which we obtain

$$\frac{\partial\phi}{\partial\xi}=\frac{\partial\psi}{\partial\eta},\ \frac{\partial\phi}{\partial\eta}=-\frac{\partial\psi}{\partial\xi}.$$

Since
$$\frac{\partial\phi}{\partial x}=\frac{\partial\phi}{\partial\xi}\frac{\partial\xi}{\partial x}+\frac{\partial\phi}{\partial\eta}\frac{\partial\eta}{\partial x},$$

and
$$\frac{\partial\phi}{\partial y}=\frac{\partial\phi}{\partial\xi}\frac{\partial\xi}{\partial y}+\frac{\partial\phi}{\partial\eta}\frac{\partial\eta}{\partial y},$$

therefore, by squaring and adding and remembering that

$$\frac{\partial\xi}{\partial x}=\frac{\partial\eta}{\partial y} \text{ and } \frac{\partial\xi}{\partial y}=-\frac{\partial\eta}{\partial x},$$

we get
$$\left(\frac{\partial\phi}{\partial x}\right)^2+\left(\frac{\partial\phi}{\partial y}\right)^2=\left\{\left(\frac{\partial\phi}{\partial\xi}\right)^2+\left(\frac{\partial\phi}{\partial\eta}\right)^2\right\}\left\{\left(\frac{\partial\xi}{\partial x}\right)^2+\left(\frac{\partial\xi}{\partial y}\right)^2\right\},$$

or
$$\left|\frac{dw}{dz}\right|=\left|\frac{dw}{d\zeta}\right|\left|\frac{d\zeta}{dz}\right|.$$

Similarly we can prove that

$$\frac{\partial^2\phi}{\partial x^2}+\frac{\partial^2\phi}{\partial y^2}=h^2\left\{\frac{\partial^2\phi}{\partial\xi^2}+\frac{\partial^2\phi}{\partial\eta^2}\right\}, \text{ where } h=\left|\frac{d\zeta}{dz}\right|.$$

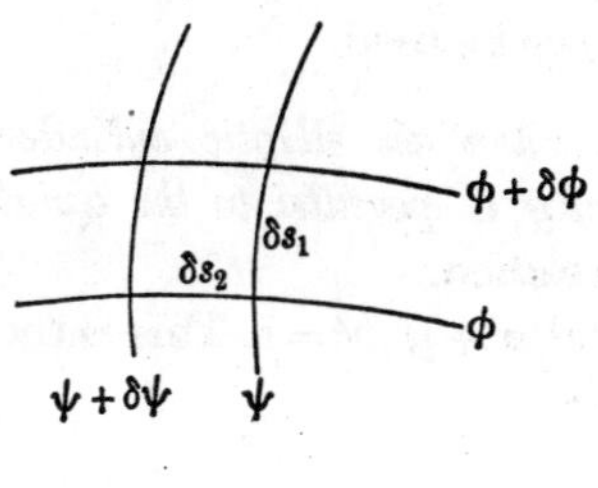

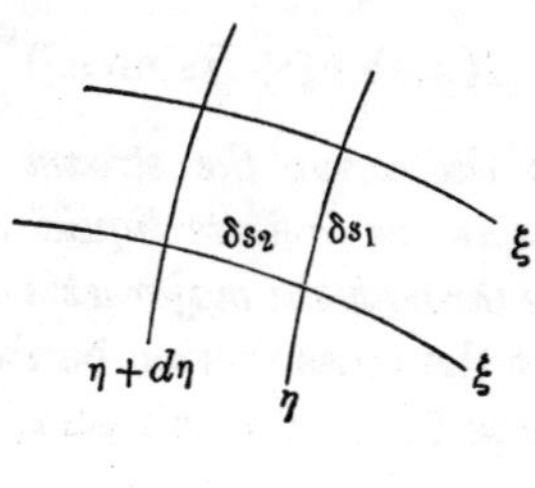

Geometrically, if we draw the curves $\phi=\text{const.}$, $\psi=\text{const.}$ and δs_1, δs_2 denote elements of ψ intercepted between ϕ and $\phi+\delta\phi$, and of ϕ intercepted between ψ and $\psi+\delta\psi$, we have

$$\left(\frac{\partial\phi}{\partial s_1}\right)^2=\left(\frac{\partial\phi}{\partial x}\right)^2+\left(\frac{\partial\phi}{\partial y}\right)^2=h^2\left\{\left(\frac{\partial\phi}{\partial\xi}\right)^2+\left(\frac{\partial\phi}{\partial\eta}\right)^2\right\}$$

and $$\left(\frac{\partial\psi}{\partial s_2}\right)^2=\text{the same expressions.}$$

Though for the curves $\xi=\text{const.}$, $\eta=\text{const.}$ the corresponding relations are of course

$$\frac{\partial\xi}{\partial s_1}=h=\frac{\partial\eta}{\partial s_2}.$$

5·31. Elliptic Cylinders.

The relation $z=c\cosh\zeta$ or $x+iy=c\cosh(\xi+i\eta)$

gives $$x=c\cosh\xi\cos\eta \quad\text{and}\quad y=c\sinh\xi\sin\eta.$$

Let ξ have all values from zero to infinity and η all values from 0 to 2π; then $$\xi=\text{const.}\quad\text{and}\quad \eta=\text{const.}$$

represent confocal ellipses and hyperbolas respectively, viz.

$$\frac{x^2}{c^2\cosh^2\xi}+\frac{y^2}{c^2\sinh^2\xi}=1 \quad\text{and}\quad \frac{x^2}{c^2\cos^2\eta}-\frac{y^2}{c^2\sin^2\eta}=1,$$

the distance between the foci being $2c$, and in any particular ellipse η denotes the eccentric angle.

In dealing with elliptic cylinders, it is useful to observe that the equation

$$\frac{\partial^2\psi}{\partial\xi^2}+\frac{\partial^2\psi}{\partial\eta^2}=0$$

has solutions of the type

$$\left.\begin{matrix}\cosh\\ \sinh\\ \exp\end{matrix}\right\}(n\xi)\left.\begin{matrix}\cos\\ \sin\end{matrix}\right\}(n\eta);$$

and that $e^{-n\xi}$ must be used when vanishing at infinity is required, i.e. when the liquid extends to infinity. For confocal ellipses the form $(A\cosh n\xi+B\sinh n\xi)\begin{matrix}\cos\\ \sin\end{matrix}(n\eta)$ may be used.

To determine the stream function when an elliptic cylinder moves in an infinite liquid with velocity U parallel to the axial plane through the major axis of a cross section.

Let the cross section be the ellipse $x^2/a^2+y^2/b^2=1$. This is the same as $\xi=\alpha$, if $a=c\cosh\alpha$, $b=c\sinh\alpha$.

The boundary condition is $\psi = -Uy + \text{constant}$, where $\xi = \alpha$, i.e. where $y = c \sinh \alpha \sin \eta$.

Since the effect is to vanish at infinity and $\sin \eta$ is the only variable factor in the boundary condition we must therefore assume a complex relation which gives to ψ the form $e^{-\xi} \sin \eta$.

Assume therefore that

$$\phi + i\psi = Ae^{-(\xi + i\eta)},$$

so that

$$\psi = -Ae^{-\xi} \sin \eta.$$

Then at the boundary $\xi = \alpha$, we must have

$$-Ae^{-\alpha} \sin \eta = -Uc \sinh \alpha \sin \eta + B$$

for all values of η. This requires that $B = 0$, and $A = Uce^{\alpha} \sinh \alpha$.

Hence

$$\psi = -Uce^{\alpha - \xi} \sinh \alpha \sin \eta$$

is a stream function which will make the boundary of the ellipse a stream line, when the cylinder moves with velocity U.

Also

$$ce^{\alpha} \sinh \alpha = be^{\alpha} = b(a+b)/c = b\sqrt{\frac{a+b}{a-b}};$$

therefore

$$\left.\begin{aligned} \psi &= -Ub\sqrt{\frac{a+b}{a-b}}\, e^{-\xi} \sin \eta \\ \phi &= Ub\sqrt{\frac{a+b}{a-b}}\, e^{-\xi} \cos \eta \end{aligned}\right\} \quad \text{..............(1).}$$

and so (the second line).

To examine whether this is a correct solution it is easy to verify that it makes the velocity vanish at infinity.

If the cylinder moves parallel to the axial plane through the minor axis of its cross section with velocity V, we get in like manner

$$\left.\begin{aligned} \psi &= Va\sqrt{\frac{a+b}{a-b}}\, e^{-\xi} \cos \eta \\ \phi &= Va\sqrt{\frac{a+b}{a-b}}\, e^{-\xi} \sin \eta \end{aligned}\right\} \quad \text{.................(2).}$$

and (the second line).

The forms of these results are the same for all confocal ellipses and therefore this last result includes the case of a plane lamina of breadth $2c$ moving at right angles to itself in the liquid; the ellipse in this case reducing to the straight line joining the foci and the formulae becoming

$$\psi = Vce^{-\xi} \cos \eta,$$

$$\phi = Vce^{-\xi} \sin \eta.$$

But these equations would make the velocity infinite at the edges ($\xi=0, \eta=0$), and therefore cannot represent real conditions. In reality there is a region of 'dead water' behind the body, separated by surfaces of discontinuity from the moving liquid. The foregoing analysis assumes continuous motion.

5·32. Liquid streaming past a fixed Elliptic Cylinder. This case may be deduced from **5·31** by superposing on the liquid and cylinder a velocity equal and opposite to that of the cylinder. Thus when the general velocity of the stream is $-U$ parallel to the major axis, we must add Ux to the value of ϕ, and Uy to the value of ψ; so that

$$\phi = Ub\sqrt{\frac{a+b}{a-b}}\,e^{-\xi}\cos\eta + U\sqrt{a^2-b^2}\cosh\xi\cos\eta,$$

and
$$\psi = -Ub\sqrt{\frac{a+b}{a-b}}\,e^{-\xi}\sin\eta + U\sqrt{a^2-b^2}\sinh\xi\sin\eta.$$

5·33. Elliptic Cylinder rotating in an infinite Mass of Liquid at rest at Infinity. If ω be the angular velocity the boundary condition is

$$\psi = \tfrac{1}{2}\omega(x^2+y^2) + C; \quad \textbf{(5.1 (3))}$$

or, putting $x = c\cosh\xi\cos\eta$ and $y = c\sinh\xi\sin\eta$,

$$\psi = \tfrac{1}{4}\omega c^2(\cosh 2\xi + \cos 2\eta) + C, \text{ where } \xi = \alpha.$$

Since the effect is to vanish at infinity and the only variable term in the boundary condition is $\cos 2\eta$, therefore we must assume a complex relation which gives to ψ the form $e^{-2\xi}\cos 2\eta$.

Assume therefore that

$$\phi + i\psi = Aie^{-2(\xi+i\eta)},$$

so that
$$\psi = Ae^{-2\xi}\cos 2\eta.$$

Hence at the boundary $\xi=\alpha$, we must have

$$Ae^{-2\alpha}\cos 2\eta = \tfrac{1}{4}\omega c^2(\cosh 2\alpha + \cos 2\eta) + C$$

for all values of η. And this is the case, provided

$$A = \tfrac{1}{4}\omega c^2 e^{2\alpha} \quad\text{and}\quad C = -\tfrac{1}{4}\omega c^2\cosh 2\alpha.$$

Therefore $\psi = \frac{1}{4}\omega c^2 e^{2\alpha-2\xi}\cos 2\eta$ gives a stream function which makes the boundary of the ellipse a stream line, when the cylinder rotates with angular velocity ω.

Since $c^2 e^{2\alpha} = (a+b)^2$, we may write the results

$$\psi = \tfrac{1}{4}\omega (a+b)^2 e^{-2\xi} \cos 2\eta,$$

and

$$\phi = \tfrac{1}{4}\omega (a+b)^2 e^{-2\xi} \sin 2\eta.$$

It is easy to verify that the velocity vanishes at infinity.

5·34. Any of the previous motions may be superposed. Thus if the elliptic cylinder be moving parallel to itself with velocity v in a direction making an angle θ with the major axis of the cross section, we have from **5·31**

$$\phi = v\sqrt{\frac{a+b}{a-b}}\, e^{-\xi} (b\cos\eta\cos\theta + a\sin\eta\sin\theta),$$

and

$$\psi = -v\sqrt{\frac{a+b}{a-b}}\, e^{-\xi} (b\sin\eta\cos\theta - a\cos\eta\sin\theta).$$

5·35. Circulation about an Elliptic Cylinder. If in **5·34** the irrotational motion is cyclic, with circulation κ round the cylinder, we can take this into account by means of the function

$$\phi + i\psi = \frac{i\kappa}{2\pi}(\xi + i\eta).$$

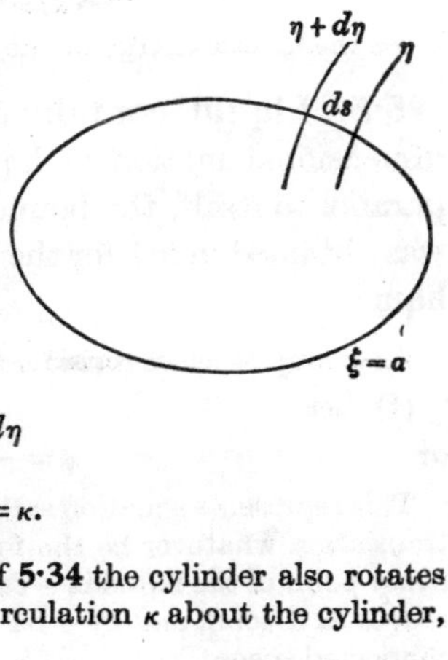

To verify that this gives the correct value to the circulation, we have that the circulation

$$= \int -\frac{\partial\phi}{\partial s}\, ds$$

taken round the cylinder,

$$= \int_0^{2\pi} -\frac{\partial\phi}{\partial\eta}\, d\eta$$

$$= \int_0^{2\pi} \frac{\kappa}{2\pi}\, d\eta = \kappa.$$

Hence if in addition to the velocity v of **5·34** the cylinder also rotates with angular velocity ω, and there is a circulation κ about the cylinder, we have

$$\phi = v\sqrt{\frac{a+b}{a-b}}\, e^{-\xi} (b\cos\eta\cos\theta + a\sin\eta\sin\theta) + \tfrac{1}{4}\omega (a+b)^2 e^{-2\xi} \sin 2\eta - \frac{\kappa\eta}{2\pi},$$

and

$$\psi = -v\sqrt{\frac{a+b}{a-b}}\, e^{-\xi} (b\sin\eta\cos\theta - a\cos\eta\sin\theta) + \tfrac{1}{4}\omega (a+b)^2 e^{-2\xi} \cos 2\eta + \frac{\kappa\xi}{2\pi}.$$

5·4. Kinetic Energy.

In any of these cases of a cylinder moving in liquid at rest at infinity, the expression for the kinetic energy is, as in **4·71**,

$$T = -\tfrac{1}{2}\rho \int \phi \frac{\partial\phi}{\partial n}\, ds,$$

where the integration is now round the perimeter of the cylinder, and we are supposing as before that the liquid is confined between two smooth planes at unit distance apart. But $-\partial\phi/\partial n$ is the normal velocity outwards, and $\partial\psi/\partial s$ is the normal velocity inwards, so that

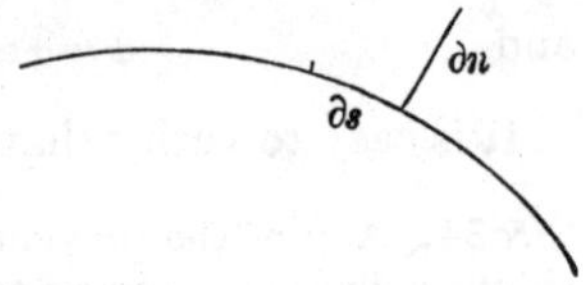

$$\partial\phi/\partial n = \partial\psi/\partial s,$$

and therefore

$$T = -\tfrac{1}{2}\rho\int \phi\, d\psi.$$

As an example consider the rotating elliptic cylinder of **5·33**, bounded by the ellipse $\xi = \alpha$. Here we have on the boundary

$$\psi = \tfrac{1}{4}\omega(a+b)^2 e^{-2\alpha}\cos 2\eta,$$

and
$$\phi = \tfrac{1}{4}\omega(a+b)^2 e^{-2\alpha}\sin 2\eta,$$

so that
$$T = \tfrac{1}{16}\rho\omega^2(a+b)^4 e^{-4\alpha}\int_0^{2\pi}\sin^2 2\eta\, d\eta = \tfrac{1}{16}\pi\rho\omega^2(a^2-b^2)^2$$

gives the kinetic energy of the liquid.

5·5. Liquid contained in Cylinders.

In cases of two-dimensional motion of liquid contained in a cylinder moving parallel to itself, the boundary condition is clearly the same as was obtained in **5·1** for the motion of a cylinder surrounded by a liquid.

As examples let us consider the following:

(1) Let
$$w = -Uz,$$

or
$$\phi = -Ux, \quad \psi = -Uy.$$

This represents a motion satisfying the boundary condition for uniform translation whatever be the form of the boundary; and the velocity at every point of the liquid is $-\partial\phi/\partial x$ or U, so that the liquid in the cylinder moves as if solid, and by **4·54** this is the only motion possible in simply connected space.

(2) Let
$$w = -iAz^2,$$

or
$$\phi = Ar^2\sin 2\theta \quad \psi = -Ar^2\cos 2\theta$$
$$= 2Axy, \qquad = -A(x^2-y^2).$$

Let us adapt these forms to the boundary condition for uniform rotation assuming the liquid to be contained in a rotating cylinder. From **5·1**, at the boundary we must have

$$\tfrac{1}{2}\omega(x^2+y^2) - B = -A(x^2-y^2),$$

or
$$(\tfrac{1}{2}\omega + A)x^2 + (\tfrac{1}{2}\omega - A)y^2 = B.$$

Hence the boundary of the section may be an ellipse

$$x^2/a^2 + y^2/b^2 = 1,$$

provided
$$a^2(\tfrac{1}{2}\omega + A) = b^2(\tfrac{1}{2}\omega - A)$$

or
$$A = -\tfrac{1}{2}\omega\frac{a^2-b^2}{a^2+b^2}.$$

Therefore

$$\phi = -\omega\frac{a^2-b^2}{a^2+b^2}xy \quad \text{and} \quad \psi = \tfrac{1}{2}\omega\frac{a^2-b^2}{a^2+b^2}(x^2-y^2)$$

determine the motion of the liquid in the rotating elliptic cylinder referred to fixed axes momentarily coinciding with the axes of the cross section.

If q denote the velocity,

$$q^2 = \omega^2\left(\frac{a^2-b^2}{a^2+b^2}\right)^2(x^2+y^2).$$

Hence the kinetic energy T of unit length is given by

$$2T = \tfrac{1}{4}\omega^2\frac{(a^2-b^2)^2}{a^2+b^2}\pi\rho ab.$$

If we require the motion of the liquid *relative to the cylinder*, we may proceed thus: The velocities in space of the particle, whose coordinates are (x, y) referred to the moving axes of the cross section, are $\dot{x}-\omega y$ and $\dot{y}+\omega x$; therefore

$$\dot{x}-\omega y = -\frac{\partial\phi}{\partial x} = \omega\frac{a^2-b^2}{a^2+b^2}y,$$

and

$$\dot{y}+\omega x = -\frac{\partial\phi}{\partial y} = \omega\frac{a^2-b^2}{a^2+b^2}x;$$

so that

$$\dot{x} = \frac{2a^2}{a^2+b^2}\omega y,$$

and

$$\dot{y} = -\frac{2b^2}{a^2+b^2}\omega x.$$

Hence

$$\ddot{x}+\frac{4a^2b^2}{(a^2+b^2)^2}\omega^2 x = 0,$$

which leads on integration to

$$x = B\cos\left(\frac{2ab}{a^2+b^2}\omega t+\alpha\right),$$

and therefore

$$y = -\frac{b}{a}B\sin\left(\frac{2ab}{a^2+b^2}\omega t+\alpha\right).$$

It follows that the motion is simple harmonic motion; the paths of the particles being ellipses similar to the boundary ellipse, described in time $\pi(a^2+b^2)/ab\omega$.

Or, to get the relative motion, we may impose on the whole system the angular velocity ω reversed. That is, we must increase ψ by $-\tfrac{1}{2}\omega(x^2+y^2)$. This makes

$$\psi = \tfrac{1}{2}\omega\frac{a^2-b^2}{a^2+b^2}(x^2-y^2)-\tfrac{1}{2}\omega(x^2+y^2)$$

$$= -\frac{\omega a^2b^2}{a^2+b^2}\left(\frac{x^2}{a^2}+\frac{y^2}{b^2}\right),$$

shewing that the stream lines are similar ellipses.

(3) Another simple case is that of a rotating prism whose section is an equilateral triangle. For this we take

$$w = iAz^3,$$

or

$$\phi = -Ar^3\sin 3\theta, \quad \psi = Ar^3\cos 3\theta = A(x^3-3xy^2).$$

The boundary condition for rotation gives, in this case,

$$A(x^3-3xy^2)=\tfrac{1}{2}\omega(x^2+y^2)+B,$$

to be satisfied at all points of the boundary.

To include the line $x=a$ in the boundary, we must take

$$Aa^3=\tfrac{1}{2}\omega a^2+B,$$

and

$$-3Aa=\tfrac{1}{2}\omega;$$

so that the equation becomes

$$x^3-3xy^2+3a(x^2+y^2)=4a^3,$$

or

$$(x-a)(x-\sqrt{3}y+2a)(x+\sqrt{3}y+2a)=0.$$

These three lines form an equilateral triangle with its centre at the origin; and the motion of liquid in a prism having this triangle for section and rotating with angular velocity ω is given by

$$\phi=\frac{\omega}{6a}r^3\sin 3\theta,\quad \psi=-\frac{\omega}{6a}r^3\cos 3\theta.$$

5·51. The stream function has been determined for the motion of liquid produced by moving cylinders of a great variety of forms. We have discussed some of the simplest cases very fully and append here a list of other cases with references to shew where the investigations may be found.

1. Rotating rectangular prism or box. Stokes, *Trans. Camb. Phil. Soc.* VIII, or *Math. and Sci. Papers*, I, p. 60. Ferrers, *Quart. Journal*, XV, p. 83. Greenhill, *ibid.* p. 144. Basset, *Hydrodynamics*, I, p. 96.

2. Rotating semicircle. Hicks, *Mess. of Math.* VIII, p. 42.

3. Rotating quadrantal sector of a circle. Greenhill, *ibid.* p. 89.

4. Rotating sector of a circle. Stokes, *Trans. Camb. Phil. Soc.* VIII, or *Math. and Sci. Papers*, I, p. 305. Greenhill, *Mess. of Math.* X, p. 83. Basset, *Hydrodynamics*, I, p. 98. Lamb, *Hydrodynamics*, 1932, p. 89.

5. Rotating rectangle bounded by two concentric circular arcs and two radii. Greenhill, *Mess. of Math.* IX, p. 35.

6. Rotating arcs of confocal ellipse and hyperbola. Ferrers, *Quart. Journal*, XVII, p. 227.

7. Rotating arcs of two confocal parabolas. *Ibid.*

8. Confocal elliptic cylinders—translation and rotation. Greenhill, *Quart. Journal*, XVI, p. 227, and *Encyc. Brit.* 11th edition, 'Hydromechanics'.

9. Rotation and translation of inverse of an ellipse. Basset, *Quart. Journal*, XIX, p. 190, XXI, p. 336, and *Hydrodynamics*, I, p. 102.

10. Rotation and translation of a lemniscate. Basset, *Quart. Journal*, XX, p. 234, and *Hydrodynamics*, I, p. 106.

5·6. Applications of the Theory of Functions of a Complex Variable. Some well-known theorems in the theory of functions of a complex variable have direct applications to the kind of hydrodynamical problems considered in this chapter. In particular **Cauchy's Theorem**, that *if C is a closed curve in a region within which $f(z)$ is a regular function of z then* $\int_C f(z)\,dz=0$; with its immediate corollary that *if C' is another closed curve inside C or surrounding C and $f(z)$ is regular in the region formed of C, C' and the part of the plane between them then*

$$\int_C f(z)\,dz=\int_{C'} f(z)\,dz.$$

Also the integral theorem in the **theory of residues** that *if $f(z)$ is regular on a closed curve C and at all points within it save at a number of 'poles' then*

$$\int_C f(z)\,dz=2\pi i\,(\textit{sum of residues of } f(z) \textit{ at its poles inside } C);$$

where, if in the neighbourhood of a point $z=a$, $f(z)$ can be expressed in the form

$$g(z)+\frac{b_1}{z-a}+\frac{b_2}{(z-a)^2}+\ldots+\frac{b_n}{(z-a)^n},$$

and $g(z)$ is regular at a, then $f(z)$ is said to have a *pole* of order n at a, and the coefficient of $(z-a)^{-1}$ viz. b_1 is called the *residue* of $f(z)$ at a.

5·61. Theorem of Blasius. *In a steady two-dimensional irrotational motion given by the relation $w=f(z)$, or $\phi+i\psi=f(x+iy)$, if the hydrodynamical pressures on the contour of a fixed cylinder are represented by a force* (**X**, **Y**) *and a couple* **N** *about the origin of coordinates, then*

$$\mathbf{X}-i\mathbf{Y}=\tfrac{1}{2}i\rho\int_C\left(\frac{dw}{dz}\right)^2 dz$$

and

$$\mathbf{N}=\textit{real part of } -\tfrac{1}{2}\rho\int_C\left(\frac{dw}{dz}\right)^2 z\,dz,$$

where the integrations are round any contour which surrounds the cylinder.*

Let the normal to the cylinder at the point (x, y) make an angle θ with the axis of x, then

$$\mathbf{X}=-\int p\cos\theta\,ds \quad\text{and}\quad \mathbf{Y}=-\int p\sin\theta\,ds \quad \ldots\ldots(1);$$

where in steady motion

$$p=A-\tfrac{1}{2}\rho q^2=A-\tfrac{1}{2}\rho(u^2+v^2).$$

* Blasius, *Zeits. f. Math. u. Phys.* LVIII, 1910. The proof given above was outlined in a Tripos question in 1933.

Therefore

$$\mathbf{X} = \tfrac{1}{2}\rho\int_{C'}(u^2+v^2)\,dy \quad\text{and}\quad \mathbf{Y} = -\tfrac{1}{2}\rho\int_{C'}(u^2+v^2)\,dx \ldots(2),$$

where the integrals are round the contour C' of the cylinder.

Now the contour of the cylinder is a stream line and on every stream line $dx/u = dy/v$, so that

$$\left.\begin{aligned}\mathbf{X} &= \tfrac{1}{2}\rho\int_{C'}\{2uv\,dx-(u^2-v^2)\,dy\}\\ \text{and}\quad \mathbf{Y} &= -\tfrac{1}{2}\rho\int_{C'}\{(u^2-v^2)\,dx+2uv\,dy\}\end{aligned}\right\}\ldots\ldots\ldots\ldots(3).$$

Again since $\dfrac{dw}{dz} = \dfrac{\partial\phi}{\partial x} + i\dfrac{\partial\psi}{\partial x} = -u+iv,$

therefore

$$\tfrac{1}{2}i\rho\int_{C'}\left(\frac{dw}{dz}\right)^2 dz = \tfrac{1}{2}i\rho\int_{C'}(u^2-v^2-2iuv)(dx+i\,dy)$$

$$= \mathbf{X} - i\mathbf{Y} \ldots\ldots\ldots\ldots\ldots\ldots\ldots\ldots\ldots\ldots\ldots(4).$$

Now in the plane outside the cylinder a singularity in the function $(dw/dz)^2$ would only be occasioned by a physical singularity in the fluid, such as a 'source' or a 'vortex'. It follows that if we take a larger contour C surrounding C' and such that between C' and C there are no such singularities, or, more generally, such that when such singularities exist the sum of the residues of $(dw/dz)^2$ at all poles between C' and C is zero, then the integrals of this function have the same value for all such contours and

$$\mathbf{X} - i\mathbf{Y} = \tfrac{1}{2}i\rho\int_{C}\left(\frac{dw}{dz}\right)^2 dz \ldots\ldots\ldots\ldots\ldots\ldots(5).$$

Again, with the same notation, the total moment about the origin of the pressure on the cylinder is

$$\begin{aligned}\mathbf{N} &= \int_{C'}(y\cos\theta - x\sin\theta)\,p\,ds\\ &= \int_{C'}p\,(x\,dx+y\,dy)\\ &= -\tfrac{1}{2}\rho\int_{C'}(u^2+v^2)(x\,dx+y\,dy) \quad\ldots\ldots\ldots\ldots(6).\end{aligned}$$

Making the same use as before of the stream line relation $dx/u = dy/v$, it is easy to see that

$$\mathbf{N} = \tfrac{1}{2}\rho\int_{C'}[\{(u^2-v^2)\,y-2uv\,x\}\,dy - \{(u^2-v^2)\,x+2uv\,y\}\,dx]$$

and that this is the real part of

$$-\tfrac{1}{2}\rho\int_{C'}(u-iv)^2(x+iy)(dx+idy) \quad\text{or}\quad -\tfrac{1}{2}\rho\int_{C'}\left(\frac{dw}{dz}\right)^2 z\,dz \;\ldots(7),$$

and subject to the same limitation as before regarding singularities in the liquid the integral may be taken round any contour which surrounds the cylinder.

The advantage of being able to use any such contour will become evident later. It lies in the fact that if all parts of the contour lie at a great distance from the cylinder it is sufficient to use an approximation to the expression for w as a function of z.

5·7. Steady Streaming with Circulation. Theorem of Kutta and Joukowski. The relation

$$w=\frac{i\kappa}{2\pi}\log z \;\ldots\ldots\ldots\ldots\ldots\ldots(1),$$

or

$$\phi=\frac{-\kappa\theta}{2\pi},\quad \psi=\frac{\kappa}{2\pi}\log r$$

represents fluid motion in which ϕ decreases by κ in making a circuit of any contour which encloses the origin; i.e. motion with circulation κ.

Let this circulation be superposed upon a steady stream $w=Uz$, in which the velocity in the direction of the axis of x is $-U$, and let there be a fixed cylinder of some form in the finite region of the plane, its cross section containing the origin. The disturbance of the stream caused by the cylinder can be represented at a great distance by terms of the form

$$w=\frac{A}{z}+\frac{B}{z^2}+\ldots$$

where $A, B, \ldots$ depend on U and κ, so that at a great distance from the origin

$$w=Uz+\frac{i\kappa}{2\pi}\log z+\frac{A}{z}+O\left(\frac{1}{z^2}\right)\;\ldots\ldots\ldots\ldots(2).$$

Then by **5·61** the force exerted on the cylinder is given by

$$\mathbf{X}-i\mathbf{Y}=\tfrac{1}{2}i\rho\int_C\left(\frac{dw}{dz}\right)^2 dz$$

$$=\tfrac{1}{2}i\rho\int_C\left(U+\frac{i\kappa}{2\pi z}-\frac{A}{z^2}+\ldots\right)^2 dz$$

taken round any contour at a great distance from the origin. Expanding the integrand in the form $\frac{1}{2}i\rho\left\{U^2+\frac{i\kappa U}{\pi z}+\ldots\right\}$, the function is seen to have a pole at the origin with residue $-\rho\kappa U/2\pi$, so that the integral $= -i\rho\kappa U$ (**5·6**).

The value of the integral might also be obtained directly by taking for contour a circle of large radius R; i.e. by writing $z = Re^{i\theta}$. It appears that when we make $R\to\infty$ the only term which contributes to the result is the term

$$-\tfrac{1}{2}\rho\int_C \frac{U\kappa}{\pi z}\,dz \text{ or } -\tfrac{1}{2}\rho\int_0^{2\pi}\frac{U\kappa Rie^{i\theta}}{\pi Re^{i\theta}}\,d\theta,$$

which is equal to $-i\rho\kappa U$.

Hence we have $\mathbf{X} - i\mathbf{Y} = -i\rho\kappa U$, giving

$$\mathbf{X}=0 \text{ and } \mathbf{Y}=\rho\kappa U \quad \text{..............(3)*.}$$

There is thus a transverse force on the cylinder at right angles to the stream.

The couple on the cylinder might be calculated as the real part of $-\frac{1}{2}\rho\int_C\left(\frac{dw}{dz}\right)^2 z\,dz$ (**5·61**), but when we substitute for w from (2) above, it will be found that the terms which contribute to the result contain A, so that the couple depends on the form of the cylinder.

It will be observed that the velocity $\kappa/2\pi r$ due to circulation increases the general velocity of the stream on one side of the cylinder and decreases it on the other; that the pressure is

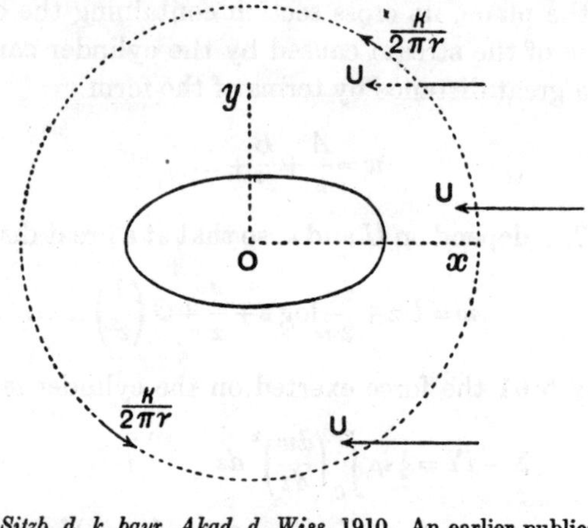

* Kutta, *Sitzb. d. k. bayr. Akad. d. Wiss.* 1910. An earlier publication is attributed to Joukowski, 1906; see also Joukowski, *Aérodynamique*, 1916, p. 139.

greater on the side of less velocity causing a resultant force on the cylinder across the stream towards the side of greater velocity.

Since the hydrodynamical pressure on the cylinder is only due to the relative motion of the fluid and the cylinder, any common velocity may be superposed on the cylinder and fluid without affecting the result, so that the same formulae will give the resultant pressure when the fluid is at rest at infinity and the cylinder is in motion.

5·71. Example. Consider the relation

$$w=\frac{\kappa}{\pi}\tan^{-1}\frac{z}{c} \quad\text{......(1),}$$

or

$$\tan\frac{\pi}{\kappa}(\phi+i\psi)=\frac{x+iy}{c}.$$

Since $\tan\frac{\pi}{\kappa}(\phi-i\psi)=\frac{x-iy}{c}$, it is easy to eliminate ϕ and ψ in turn and obtain the equations

$$x^2+\left(y-c\coth\frac{2\pi\psi}{\kappa}\right)^2=c^2\operatorname{cosech}^2\frac{2\pi\psi}{\kappa} \quad\text{......(2),}$$

and

$$\left(x+c\cot\frac{2\pi\phi}{\kappa}\right)^2+y^2=c^2\operatorname{cosec}^2\frac{2\pi\phi}{\kappa} \quad\text{......(3),}$$

so that the curves $\phi=$const. and $\psi=$const. are orthogonal families of coaxial circles, with $z=\pm ic$ as the limiting points C, C'.

Consider a two-dimensional flow in which the ψ-circles are stream lines and let $\psi=\alpha$ be a fixed circular boundary (cross section of a cylinder) of radius $a=c\operatorname{cosech}\frac{2\pi\alpha}{\kappa}$, with its centre A at the point $\left(0, c\coth\frac{2\pi\alpha}{\kappa}\right)$.

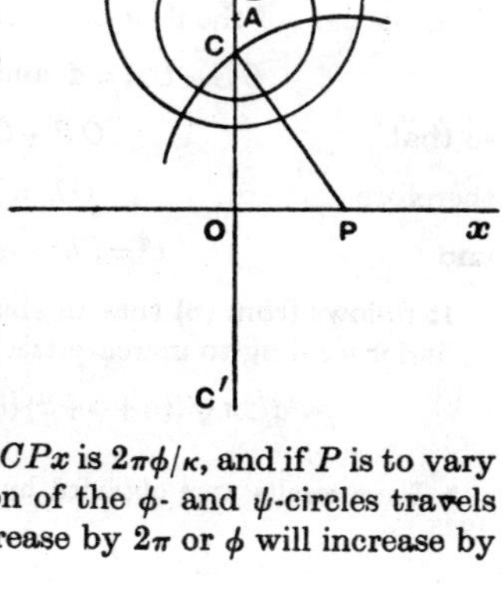

Though (1) represents w as a many-valued function, yet the velocity components being given by

$$-u+iv=\frac{dw}{dz}=\frac{\kappa}{\pi}\frac{c}{z^2+c^2} \quad\text{...(4)}$$

are single-valued and define a definite motion.

Again the circulation round the cylinder is the decrease in ϕ in going round any ψ-circle. But if P is the centre of a ϕ-circle, it is clear from (3) that the angle CPx is $2\pi\phi/\kappa$, and if P is to vary its position so that a point of intersection of the ϕ- and ψ-circles travels round the latter, the angle CPx will increase by 2π or ϕ will increase by

κ, so that the circulation round the cylinder is $-\kappa$ in the positive sense or κ in the clockwise sense.

We can also verify the sense of the motion by considering the velocity on the axis Oy; putting $x=0$ in (4) gives

$$u=-\frac{\kappa}{\pi}\frac{c}{c^2-y^2},\quad v=0 \quad \text{.......................(5)}$$

making u positive above the cylinder and negative below it.

Now apply the theorem of Blasius to find the resultant fluid pressure on the cylinder.

We have
$$\frac{dw}{dz}=\frac{\kappa}{\pi}\frac{c}{z^2+c^2},$$

and
$$X-iY=\tfrac{1}{2}i\rho\frac{\kappa^2c^2}{\pi^2}\int_C\frac{dz}{(z^2+c^2)^2}$$

integrated round any contour between which and the given circle there is no singularity in the integrand. The integrand has a pole at $z=ic$, and to find its residue there we write $z=ic+\zeta$ where ζ is small. Then

$$\frac{1}{(z^2+c^2)^2}=\frac{1}{(2ic\zeta+\zeta^2)^2}=-\frac{1}{4c^2\zeta^2}\left(1-\frac{i\zeta}{2c}\right)^{-2}$$
$$=-\frac{1}{4c^2\zeta^2}-\frac{i}{4c^3\zeta}+\dots$$

so that the residue is $-i/4c^3$, and the value of the integral is $\pi/2c^3$.

Hence
$$X-iY=\frac{i\rho\kappa^2}{4\pi c},$$

or
$$X=0 \text{ and } Y=-\rho\kappa^2/4\pi c \quad \text{.....................(6).}$$

Now the fluid might have as an external boundary any other stream line, e.g. the x-axis, giving flow round a cylinder parallel to a wall; or a larger ψ-circle, say a circle of radius b with its centre B at a distance d from that of the given circle A.

In the former case, when the x-axis is a rigid boundary, if the cylinder of radius a has its axis at a distance a' from the boundary, then $c^2=a'^2-a^2$, and the force per unit length on the cylinder towards the wall is from (6) $\rho\kappa^2/4\pi\sqrt{(a'^2-a^2)}$*.

In the case of the flow between two cylinders, we have

$$OB-OA=d \text{ and } OB^2-b^2=c^2=OA^2-a^2,$$

so that
$$OB+OA=(b^2-a^2)/d;$$

therefore
$$OB=(b^2-a^2+d^2)/2d$$

and
$$c^2=\{(b^2-a^2+d^2)^2-4b^2d^2\}/4d^2.$$

It follows from (6) that in this case the force per unit length on either cylinder tending to increase the distance between their axes is

$$\rho\kappa^2 d/2\pi\sqrt{\{(a+b+d)(b+d-a)(b+a-d)(b-d+a)\}}^*.$$

* These results were obtained by Cisotti, *Atti della R. Acad. dei Lincei*, 6 A, I, 1925.

5·72. Reaction on a Moving Cylinder. Other Formulae. Let the motion of the cylinder be defined as in **5·1** by the velocity components U, V of a point of its cross section and by an angular velocity ω. Let the reaction on the cylinder per unit length be represented by a force **X**, **Y** and a couple **N**; we shall obtain expressions for **X**, **Y** and **N** as products of U, V, ω and certain line integrals.

In steady motion the pressure in the liquid surrounding the cylinder is given, as in **2·41**, by

$$p/\rho = \text{const.} - \tfrac{1}{2}\{(u-U)^2+(v-V)^2\} + \omega(xv-yu) \quad \text{......(1)},$$

and

$$\mathbf{X} = -\int_C lp\,ds,\ \mathbf{Y} = -\int_C mp\,ds \text{ and } \mathbf{N} = -\int_C (mx-ly)p\,ds \quad \text{...(2)},$$

where l, m are direction cosines of the outward normal to the element ds of the contour C of the cylinder.

The method is now to substitute for p from (1) in (2) and transform the integrals by making use of Green's Theorem in the form

$$\int_C (lP+mQ)\,ds = \iint_A \left(\frac{\partial P}{\partial x}+\frac{\partial Q}{\partial y}\right) dx\,dy,$$

where the contour C in the first integral is the complete boundary of the area A of integration on the right, the contour is described in such a sense as to have the area on the left and here (l, m) represents the normal drawn outwards from A.

We must therefore assume an outer boundary for the liquid and take account of contour integrals on this part of the boundary. It is convenient to take a fixed circle of large radius as the outer boundary. Since no tubes of flow can end in the liquid or on a fixed boundary, therefore all such tubes start from the moving cylinder and return to it. The motion is therefore of the general type which would be produced by a doublet or doublets, so that (in two dimensions) the velocity potential at a great distance is of order $1/r$ and the velocity is of order $1/r^2$. On the large circle we may also assume that $l/x = m/y$, so that the factor $mx - ly$ is zero.

Taking the variable terms in the pressure we have

$$\mathbf{X} = +\tfrac{1}{2}\rho\int_C l(u^2+v^2)\,ds - \rho\int_C l\{u(U-\omega y)+v(V+\omega x)\}\,ds \quad \text{...(3)}.$$

Considering the first integral, since on an infinite circle it clearly vanishes, we have by Green's Theorem

$$\tfrac{1}{2}\int_C l(u^2+v^2)\,ds = -\iint\left(u\frac{\partial u}{\partial x}+v\frac{\partial v}{\partial x}\right)dx\,dy$$

$$= -\iint\left(u\frac{\partial u}{\partial x}+v\frac{\partial u}{\partial y}\right)dx\,dy$$

$$= -\iint\left\{\frac{\partial u^2}{\partial x}+\frac{\partial uv}{\partial y}-u\left(\frac{\partial u}{\partial x}+\frac{\partial v}{\partial y}\right)\right\}dx\,dy$$

$$= \int_C (lu^2+muv)\,ds \;\ldots\ldots\ldots\ldots\ldots\ldots\ldots(4),$$

where we have made use of the equation of continuity, and again note that the integral round the infinite circle vanishes.

Hence (3) becomes

$$\mathbf{X} = \rho\int_C u\,(lu+mv)\,ds - \rho\int_C l\{u\,(U-\omega y)+v\,(V+\omega x)\}\,ds \ldots(5)$$

where the integration is round the contour of the cylinder. But on this contour the normal velocity of the liquid is equal to that of the cylinder, so that

$$lu+mv = l\,(U-\omega y)+m\,(V+\omega x) \;\ldots\ldots\ldots\ldots(6).$$

Therefore $\quad \mathbf{X} = \rho\int_C (mu-lv)\,(V+\omega x)\,ds;$

or, since $m=-dx/ds$ and $l=dy/ds$,

$$\left.\begin{aligned}\mathbf{X} &= \rho\int_C (V+\omega x)\frac{\partial\phi}{\partial s}\,ds\\ \text{and similarly}\quad \mathbf{Y} &= -\rho\int_C (U-\omega y)\frac{\partial\phi}{\partial s}\,ds\end{aligned}\right\}\ldots\ldots\ldots\ldots(7)^*.$$

This gives

$$\mathbf{X}+i\mathbf{Y} = -i\rho\,(U+iV)\int_C\frac{\partial\phi}{\partial s}\,ds + \rho\omega\int_C z\frac{\partial\phi}{\partial s}\,ds.$$

If there is a circulation κ round the cylinder then $-\int_C\frac{\partial\phi}{\partial s}\,ds=\kappa$, and

$$\mathbf{X}+i\mathbf{Y} = i\kappa\rho\,(U+iV)+\rho\omega\int z\frac{\partial\phi}{\partial s}\,ds \;\ldots\ldots\ldots\ldots(8),$$

including as a special case the theorem of Kutta and Joukowski.

* This discussion is based on a paper by Lamb, *Reports and Memoranda of the Aeronautical Research Committee*, 1218, 1929, also in *Hydrodynamics*, 1932, p. 184, which includes the effects of acceleration and contains results (7) and (12), and (8) in slightly different form.

If there is no circulation, then by an integration by parts we get

$$\mathbf{X}+i\mathbf{Y}=-\rho\omega\int_C \phi\,dz \quad\text{.................(9).}$$

Again, from (1) and (2)

$$\mathbf{N}=\tfrac{1}{2}\rho\int_C (u^2+v^2)(mx-ly)\,ds$$
$$-\rho\int_C \{u(U-\omega y)+v(V+\omega x)\}(mx-ly)\,ds \text{...(10),}$$

and since the first integral would vanish when taken round the infinite circle its value is, by Green's theorem,

$$-\rho\iint\left\{x\left(u\frac{\partial u}{\partial y}+v\frac{\partial v}{\partial y}\right)-y\left(u\frac{\partial u}{\partial x}+v\frac{\partial v}{\partial x}\right)\right\}dx\,dy;$$

and, by making use of the relations $\dfrac{\partial u}{\partial y}=\dfrac{\partial v}{\partial x}$ and $\dfrac{\partial u}{\partial x}+\dfrac{\partial v}{\partial y}=0$, this is seen to be equivalent to

$$-\rho\iint\left\{\frac{\partial xuv}{\partial x}+\frac{\partial xv^2}{\partial y}-\frac{\partial yu^2}{\partial x}-\frac{\partial yuv}{\partial y}\right\}dx\,dy,$$

so that $\mathbf{N}=\rho\displaystyle\int_C (lu+mv)(xv-yu)\,ds$

$$-\rho\int_C \{u(U-\omega y)+v(V+\omega x)\}(mx-ly)\,ds \text{ ...(11);}$$

whence by using (6) we find that

$$\mathbf{N}=-\rho\int_C (mu-lv)(xU+yV)\,ds$$
$$=-\rho\int_C (Ux+Vy)\frac{\partial\phi}{\partial s}\,ds \quad\text{.......................(12).}$$

When there is no circulation, an integration by parts gives

$$\mathbf{N}=\rho\int_C \phi\,(U\,dx+V\,dy),$$

or $\mathbf{N}=\text{real part of } \rho(U-iV)\displaystyle\int_C \phi\,dz$ ....(13).

When there is no circulation the formulae (9) and (13) may be further modified thus: taking the formula **5·1** (1) for ψ on the contour C, we have

$$\int_C \psi\,dz=\int_C \{Vx-Uy+\tfrac{1}{2}\omega(x^2+y^2)\}(dx+i\,dy);$$

and *provided that the origin is the centroid of the cross section of the cylinder*

$$\int_C (x^2+y^2)\,dx = -\iint \frac{\partial}{\partial y}(x^2+y^2)\,dx\,dy = 0,$$

and

$$\int_C (x^2+y^2)\,dy = \iint \frac{\partial}{\partial x}(x^2+y^2)\,dx\,dy = 0,$$

so that

$$\int_C \psi\,dz = A\,(U+iV) \quad \text{..................(14)},$$

where A is the area enclosed by the contour.

It follows that (9) may be written

$$\mathbf{X}+i\mathbf{Y} = -\rho\omega \int_C w\,dz + i\rho A\omega\,(U+iV) \quad \text{......(9')};$$

and we may write (13), adding a purely imaginary term,

$$\mathbf{N} = \text{real part of} \left\{\rho\,(U-iV)\int_C \phi\,dz + i\rho A\,(U^2+V^2)\right\},$$

or, from (14)

$$\mathbf{N} = \text{real part of} \left\{\rho\,(U-iV)\int_C (\phi+i\psi)\,dz\right\}$$

i.e.

$$\mathbf{N} = \text{real part of } \rho\,(U-iV)\int_C w\,dz \quad \text{...............(13')}.$$

5·8. Formulae for Momentum. Consider the case of a two-dimensional motion represented by the relation $w=f(z)$. The components H_x, H_y of the momentum of the liquid bounded by a contour C are given by

$$H_x+iH_y = \rho\iint (u+iv)\,dx\,dy$$

$$= \rho\iint\left(-\frac{\partial\psi}{\partial y} + i\frac{\partial\psi}{\partial x}\right)dx\,dy \quad \text{...........(1)}$$

integrated over the area bounded by the contour C. It follows that

$$H_x+iH_y = \rho\int_C \psi\,(dx+i\,dy) = \rho\int_C \psi\,dz \quad \text{.........(2)}.$$

Alternatively, instead of (1), we have

$$H_x+iH_y = -\rho\iint\left(\frac{\partial\phi}{\partial x} + i\frac{\partial\phi}{\partial y}\right)dx\,dy$$

$$= \rho\int_C \phi\,(i\,dx - dy) = i\rho\int_C \phi\,dz \quad \text{......(3)}.$$

Also, by adding (2) and (3), we get

$$H_x + iH_y = \tfrac{1}{2} i\rho \int_C \overline{w}\,dz, \quad (\overline{w} = \phi - i\psi) \quad \text{.........(4).}$$

It follows in the same way, that, if the liquid is contained between two contours C, C' of which C' is the outer, then

$$H_x + iH_y = \rho \int_{C'} \psi\,dz - \rho \int_C \psi\,dz \quad \text{...........(5),}$$

and similar formulae corresponding to (3) and (4).

These expressions for the linear momentum may lead to results independent of the shapes of the contours. For example, consider the momentum produced in liquid contained between two long cylinders, set in motion impulsively, so that their velocity components are U, V and U', V' at right angles to their lengths.

Then, on C,

$$\psi = Vx - Uy \quad \textbf{(5·1)}$$

and

$$\int_C \psi\,dz = \int_C (Vx - Uy)(dx + i\,dy)$$

$$= A\,(U + iV).$$

Similarly

$$\int_{C'} \psi\,dz = A'\,(U' + iV'),$$

where A, A' are the areas of the cross sections of the cylinders.

Hence, from (5)

$$H_x + iH_y = \rho A'\,(U' + iV') - \rho A\,(U + iV)$$

or

$$H_x = M'U' - MU \quad \text{and} \quad H_y = M'V' - MV,$$

where H_x, H_y are momenta per unit length of cylinder, and M, M' denote the masses of liquid which unit lengths of the cylinders would contain.

5·9. Example. *An elliptic cylinder, semi-axes a and b, is held with its length perpendicular to, and its major axis making an angle θ with, the direction of a stream of velocity V. Prove that the magnitude of the couple per unit length on the cylinder due to the fluid pressure is $\pi\rho\,(a^2 - b^2)\,V^2 \sin\theta\cos\theta$, and determine its sense.* (M.T. 1903.)

Let

$$w = A\cosh(\zeta - \gamma) \quad \text{..........................(1),}$$

where A is real, $\zeta = \xi + i\eta$, $\gamma = \alpha + i\beta$ and $z = c\cosh\zeta$.

This makes $\psi = 0$ on $\xi = \alpha$, which we take to be the boundary of the cylinder.

Then

$$-u + iv = \frac{dw}{dz} = \frac{A\sinh(\zeta - \gamma)}{c\sinh\zeta} \quad \text{....................(2).}$$

At a great distance from the cylinder ξ is large, and $u=-V\cos\theta$, $v=-V\sin\theta$, so that (2) takes the form

$$Ve^{-i\theta}=\frac{A}{c}\frac{e^{\zeta-\gamma}}{e^{\zeta}}=\frac{Ae^{-(\alpha+i\beta)}}{c},$$

giving $A=cVe^{\alpha}$ and $\beta=\theta$.

Therefore $$\frac{dw}{dz}=Ve^{\alpha}(\cosh\gamma-\sinh\gamma\coth\zeta)$$
$$=Ve^{\alpha}\left(\cosh\gamma-\sinh\gamma\frac{z}{\sqrt{(z^2-c^2)}}\right).$$

For large values of z this gives

$$\frac{dw}{dz}=Ve^{\alpha}\left\{\cosh\gamma-\sinh\gamma\left(1+\frac{c^2}{2z^2}\right)\right\}$$
$$=Ve^{\alpha}\left(e^{-\gamma}-\frac{c^2\sinh\gamma}{2z^2}\right).$$

Hence $$z\left(\frac{dw}{dz}\right)^2=V^2e^{2\alpha}z\left(e^{-\gamma}-\frac{c^2\sinh\gamma}{2z^2}\right)^2.$$

This function has a pole at the origin with residue

$$-V^2c^2e^{2\alpha-\gamma}\sinh\gamma=-\tfrac{1}{2}V^2c^2(e^{2\alpha}-e^{-2i\beta}),$$

so that $\int_C\left(\frac{dw}{dz}\right)^2 z\,dz$ taken round a large contour surrounding the cylinder has the value

$$-\pi iV^2c^2(e^{2\alpha}-\cos 2\theta+i\sin 2\theta).$$

By the theorem of Blasius the couple on the cylinder is $-\frac{1}{2}\rho$ times the real part of this integral, i.e. $-\pi\rho V^2(a^2-b^2)\sin\theta\cos\theta$. The $-$ sign in relation to the direction assumed for V above indicates that the couple tends to set the cylinder broadside to the stream.

The result may also be obtained simply from **5·72** (13′).

EXAMPLES

1. An infinite circular cylinder of radius a is in motion in homogeneous fluid which extends to infinity and is at rest there. Shew that at any moment the pressure at a point of the fluid at distance r from the axis of the cylinder exceeds the hydrostatic pressure by

$$\rho\left[\frac{a^2}{r}f_1+\frac{a^2}{r^2}\left\{\left(1-\frac{a^2}{2r^2}\right)u_1{}^2-\left(1+\frac{a^2}{2r^2}\right)v_1{}^2\right\}\right],$$

where f_1 is the component acceleration of the centre of the cylinder in the direction of r, u_1 and v_1 are the component velocities in and perpendicular to that direction. (Trinity Coll. 1904.)

2. In the case of the two-dimensional motion of a liquid streaming past a fixed circular disc, the velocity at infinity is u in a fixed direction where u is variable. Shew that the maximum value of the velocity at any point of the fluid is $2u$. Prove that the force necessary to hold the disc at rest is $2m\dot{u}$, where m is the mass of liquid displaced by the disc.

(Coll. Exam. 1907.)

3. Shew that when a cylinder moves uniformly in a given straight line in an infinite liquid, the path of any point in the fluid is given by the equations

$$\frac{dz}{dt}=\frac{Va^2}{(z'-Vt)^2};\quad \frac{dz'}{dt}=\frac{Va^2}{(z-Vt)^2},$$

where V = velocity of cylinder, a its radius, and z, z' are $x+iy, x-iy$ where x, y are the coordinates measured from the starting point of the axis, along and perpendicular to its direction of motion. (Coll. Exam. 1897.)

4. The space between two fixed coaxial circular cylinders of radii a and b, and between two planes perpendicular to the axis and distant c apart, is occupied by liquid of density ρ. Shew that the velocity potential of a motion whose kinetic energy shall equal a given quantity T is given by $A\theta$, where

$$\pi\rho A^2 c\log b/a=T.$$

Work out the same problem for the space between two confocal elliptic cylinders. (St John's Coll. 1903.)

5. A circular cylinder of radius a is moving with velocity U along the axis of x; shew that the motion produced by the cylinder in a mass of fluid at rest is given by the complex function

$$w=\phi+i\psi=a^2U/(z-Ut),$$

where

$$z=x+iy.$$

Find the magnitude and direction of the velocity in the fluid; and deduce that for a marked particle of the fluid, whose polar coordinates are r, θ referred to the centre of the cylinder as origin,

$$\frac{1}{r}\frac{dr}{dt}+i\frac{d\theta}{dt}=\frac{U}{r}\left(\frac{a^2}{r^2}e^{i\theta}-e^{-i\theta}\right)\ \text{and}\ \left(r-\frac{a^2}{r}\right)\sin\theta=b.$$

Hence prove that the path of such a particle is the elastic curve given by

$$\rho\,(y-\tfrac{1}{2}b)=\tfrac{1}{4}a^2,$$

where ρ is the radius of curvature of the path. (St John's Coll. 1911.)

6. An infinite cylinder of radius a and density σ is surrounded by a fixed concentric cylinder of radius b, and the intervening space is filled with liquid of density ρ. Prove that the impulse per unit length necessary to start the inner cylinder with velocity V is

$$\frac{\pi a^2}{b^2-a^2}\{(\sigma+\rho)\,b^2-(\sigma-\rho)\,a^2\}\,V.\quad \text{(Trinity Coll. 1912.)}$$

7. A stream of water of great depth is flowing with uniform velocity V over a plane level bottom. An infinite cylinder, of which the cross section is a semicircle of radius a, lies on its flat side with its generating lines making an angle α with the undisturbed stream lines. Prove that the resultant fluid pressure per unit length on the curved surface is

$$2a\Pi-\tfrac{5}{3}\rho a V^2\sin^2\alpha,$$

where Π is the fluid pressure at a great distance from the cylinder. (Trinity Coll. 1896.)

8. The space between two infinitely long coaxial cylinders of radii a and b respectively is filled with homogeneous liquid of density ρ and the inner cylinder is suddenly moved with velocity U perpendicular to the axis, the

outer one being kept fixed. Shew that the resultant impulsive pressure on a length l of the inner cylinder is

$$\pi\rho a^2 l \frac{b^2+a^2}{b^2-a^2} U. \qquad \text{(M.T. 1882.)}$$

9. Verify that the stream functions for uniform streaming parallel to the axes past a solid, bounded by those parts of the circles

$$(x+1)^2+y^2=2, \quad (x-1)^2+y^2=2$$

which are external to each other, are

$$\psi = y\left[1+\frac{1}{x^2+y^2}-\frac{2}{(x+1)^2+y^2}-\frac{2}{(x-1)^2+y^2}\right]$$

and
$$\psi = -x+\frac{x}{x^2+y^2}+\frac{2(x+1)}{(x+1)^2+y^2}+\frac{2(x-1)}{(x-1)^2+y^2};$$

and, when the stream is inclined at an angle α to the line of centres, find the equation to the stream line that divides on the solid. (M.T. 1894.)

10. If a long circular cylinder of radius a moves in a straight line at right angles to its length in liquid at rest at infinity, shew that when a particle of liquid in the plane of symmetry, initially at distance b in advance of the axis of the cylinder, has moved through a distance c, then the cylinder has moved through a distance

$$c+\frac{b^2-a^2}{b+a\coth\dfrac{c}{a}}. \qquad \text{(M.T. 1931.)}$$

11. A circular cylinder is fixed across a stream of velocity U with circulation κ round the cylinder. Shew that the maximum velocity in the liquid is $2U+\frac{\kappa}{2\pi a}$, where a is the radius of the cylinder. (M.T. 1927.)

12. An elliptic cylinder, the semi-axes of whose cross section are a and b, is moving with velocity U parallel to the major axis of its cross section, through an infinite liquid of density ρ which is at rest at infinity, the pressure there being Π. Prove that in order that the pressure may everywhere be positive

$$\rho U^2 < 2a^2\Pi/(2ab+b^2). \qquad \text{(M.T. 1906.)}$$

13. In the two-dimensional irrotational motion of a liquid streaming past a fixed elliptic disc $x^2/a^2+y^2/b^2=1$, the velocity at infinity being parallel to the major axis and equal to V, prove that if

$$x+iy=c\cosh(\xi+i\eta),$$

$$a^2-b^2=c^2 \text{ and } a=c\cosh\alpha,\ b=c\sinh\alpha,$$

the velocity at any point is given by

$$q^2=V^2\frac{a+b}{a-b}\cdot\frac{\sinh^2(\xi-\alpha)+\sin^2\eta}{\sinh^2\xi+\sin^2\eta},$$

and that it has its maximum value $V(a+b)/a$ at the end of the minor axis. (Coll. Exam. 1899.)

14. An infinite two-dimensional stream whose velocity potential is $\sum_1^\infty A_n r^n \cos n\theta$, is disturbed by the insertion of a stationary cylindrical

obstacle $r=c$. Shew that the pressure on the cylinder is in the direction $\theta=0$ and of amount $\sum_1^\infty \lambda_n A_n A_{n+1}$, where the λ's are independent of the A's. (M.T. 1921.)

15. Shew that with proper choice of units the motion of an infinite liquid produced by the motion of an elliptic cylinder parallel to one of its principal axes is given by the complex function

$$w=e^{-\zeta}, \text{ where } z=2\cosh\zeta.$$

Deduce the formulae

$$x=\phi\left(1+\frac{1}{\phi^2+\psi^2}\right), \quad y=\psi\left(1-\frac{1}{\phi^2+\psi^2}\right)$$

and trace the curves $\phi=\text{const.}$, $\psi=\text{const.}$, indicating which parts are of physical interest. (St John's Coll. 1909.)

16. Prove that the relative stream lines of the liquid bounded by the hyperbolic cylinders

$$x(x-y)-a^2=0, \quad y(x+y)-b^2=0$$

are the quartic curves

$$\{x(x-y)-a^2\}\{y(x+y)-b^2\}=\text{const.} \qquad \text{(M.T. 1881.)}$$

17. If liquid be contained between two confocal elliptic cylinders, and two planes perpendicular to the axes, prove that if the outer cylinder be made to rotate about its axis, the inner will begin to rotate with $\operatorname{sech} 2(\beta-\alpha)$ times the angular velocity of the outer cylinder, supposing $c\cosh\alpha$, $c\sinh\alpha$ the semi-axes of the inner cylinder, and $c\cosh\beta$, $c\sinh\beta$ of the outer; neglecting the inertia of the cylinder. (M.T. 1881.)

18. An elliptic cylinder is placed in a steady stream which at infinity makes an angle α with the major axis of the cylinder. Shew that on the ellipse the pressure is greatest at the points where the stream divides, and least at the points where the fluid is moving parallel to the stream as it meets the ellipse. (Trinity Coll. 1906.)

19. Prove that when an infinitely long cylinder of density σ whose cross section is an ellipse of semi-axes a, b is immersed in an infinite liquid of density ρ the square of its radius of gyration about its axis is effectively increased by the quantity

$$\frac{\rho}{8\sigma}\frac{(a^2-b^2)^2}{ab}. \qquad \text{(Univ. of London, 1907.)}$$

20. Determine the character of the two-dimensional fluid motion inside the ellipse (a, b), for which the stream function is $k\left(\dfrac{x^2}{a^2}+\dfrac{y^2}{b^2}\right)$; and find the pressure at each point in the cross section when there is no field of force. (St John's Coll. 1901.)

21. An infinite elliptic cylinder with semi-axes a, b is rotating round its axis with angular velocity ω, in an infinite liquid of density ρ which is at rest at infinity. Shew that if the fluid is under the action of no forces the moment of the fluid pressure on the cylinder round the centre is $\frac{1}{8}\pi\rho c^4\dfrac{d\omega}{dt}$, where $c^2=a^2-b^2$. (Coll. Exam. 1902.)

22. The space between two confocal elliptic cylinders (a_0, b_0) and (a_1, b_1) and two planes perpendicular to their axis is filled with liquid. If both cylinders be made to rotate about their common axis with angular velocity ω, the kinetic energy of the motion set up is

$$\tfrac{1}{8} M\omega^2 c^4 (b_1 a_0 - b_0 a_1)/(a_1 a_0 - b_1 b_0)(a_1 b_1 - a_0 b_0),$$

M being the mass of the liquid, and $2c$ the distance between the foci. (St John's Coll. 1900.)

23. An elliptic cylinder whose semi-axes are $c\cosh\alpha$, $c\sinh\alpha$ is divided in two by a plane through the axis of the cylinder and the major axis of its cross section. An infinite liquid of density ρ streams past the cylinder, its velocity U at infinity being uniform and parallel to the major axis of the cross section of the cylinder. Shew that in consequence of the motion of the liquid the pressure between the two portions of the cylinder is diminished by

$$\rho c U^2 e^{\alpha} \sinh\alpha \{2\cosh\alpha + e^{\alpha}\sinh\alpha \log\tanh\tfrac{1}{2}\alpha\}$$

per unit length of the cylinder. (M.T. 1899.)

24. A fixed elliptic cylinder whose principal axes are $c\cosh\beta$, $c\sinh\beta$ is surrounded by infinite liquid in which there is a source of strength m at the point $c\cosh\gamma$, 0; prove that if β is very small the stream function of the motion is

$$\psi = m\tan^{-1}\frac{\sin\xi\sinh\eta}{\cos\xi\cosh\eta - \cosh\gamma} + \frac{m\beta\sin\xi}{\cosh(\gamma+\eta) - \cos\xi},$$

where $x + iy = c\cos(\xi - i\eta)$. (Coll. Exam. 1900.)

25. A thin shell in the form of an infinitely long elliptic cylinder, semi-axes a and b, is rotating about its axis in an infinite liquid otherwise at rest. It is filled with the same liquid. Prove that the ratio of the kinetic energy of the liquid inside to that of the liquid outside is $2ab : a^2 + b^2$. (M.T. 1926.)

26. A long circular cylinder moves through an infinite liquid, which is at rest at infinity, with a velocity u at right angles to the axis. If the cross section is not quite circular but has for equation

$$r = a(1 + \epsilon\cos n\theta),$$

where ϵ is small, shew that when the motion is parallel to the axis of x, the approximate value of the velocity potential is

$$ua\left\{\frac{a}{r}\cos\theta + \epsilon\frac{a^{n+1}}{r^{n+1}}\cos(n+1)\theta - \epsilon\frac{a^{n-1}}{r^{n-1}}\cos(n-1)\theta\right\}.$$

(Coll. Exam. 1901.)

27. Liquid of density ρ is circulating irrotationally between two confocal elliptic cylinders $\xi = \alpha$, $\xi = \beta$, where

$$x + iy = c\cosh(\xi + i\eta).$$

Prove that, if κ is the circulation, the kinetic energy per unit length of cylinder is

$$\tfrac{1}{4}\rho\kappa^2(\beta - \alpha)/\pi.$$

(M.T. 1925.)

28. If ξ, η be conjugate functions of x and y, such that the curves for which ξ is constant are closed ovals surrounding the origin, then the kinetic energy and moment of momentum of homogeneous fluid of density

ρ contained between two curves ξ_1 and ξ_2, which are rotating with unit angular velocity about the origin, can be expressed in the form $\frac{1}{2}Mk^2$ and Mk^2 respectively, where

$$Mk^2 = \tfrac{1}{2}\rho \int (x^2+y^2)\frac{\partial\phi}{\partial\eta}\,d\eta$$

taken round the boundaries. (M.T. 1895.)

29. Shew that the angular momentum, of a two-dimensional motion of a homogeneous fluid, about an axis perpendicular to the plane of the motion, is $\rho \int \varpi\phi\,ds$, the integral being taken round a cross section of the containing vessel, where ϖ is the perpendicular from the axis to the normal of the cross section, ρ is the density and ϕ the velocity potential.

If the vessel be rotating with angular velocity ω, and $I\omega$, $I_0\omega$ are the angular momenta about the axis of rotation, and the line of centroids of the cross sections respectively, find an expression for $I-I_0$ in a form which does not depend on the shape of the vessel. (M.T. 1897.)

30. Prove that, if $2a$, $2b$ are the axes of the cross section of an elliptic cylinder placed across a stream in which the velocity at infinity is U parallel to the major axis of the cross section, the velocity at a point $(a\cos\eta,\ b\sin\eta)$ on the surface is

$$U(a+b)\sin\eta\,(b^2\cos^2\eta + a^2\sin^2\eta)^{-\frac{1}{2}};$$

and that, in consequence of the motion of the liquid, the resultant thrust (per unit length) on that half cylinder on which the stream impinges is diminished by

$$\frac{2b^2\rho U^2}{a-b}\left\{1-\left(\frac{a+b}{a-b}\right)^{\frac{1}{2}}\tan^{-1}\left(\frac{a-b}{a+b}\right)^{\frac{1}{2}}\right\},$$

where ρ is the density of the liquid. (M.T. 1924.)

31. An infinite cylinder contains fluid and is rotating with angular velocity ω about its axis Oz. Shew that the two-dimensional irrotational motion of the fluid may be determined by use of the *relative* stream function χ, where χ is constant on the boundary, and satisfies the equation $\dfrac{\partial^2\chi}{\partial x^2}+\dfrac{\partial^2\chi}{\partial y^2}=-2\omega$ at internal points.

Shew that the kinetic energy of the fluid is less than its kinetic energy when it is rotating as a rigid body with the same angular velocity by

$$\tfrac{1}{2}\rho\iint\left\{\left(\frac{\partial\chi}{\partial x}\right)^2+\left(\frac{\partial\chi}{\partial y}\right)^2\right\}dx\,dy.$$

(Univ. of London, 1915.

32. A circular cylinder of radius a and infinite length lies on a plane in an infinite depth of liquid. The velocity of the liquid at a great distance from the cylinder is U perpendicular to the generators, and the motion is irrotational and two-dimensional. Verify that the stream function is the imaginary part of

$$w = \pi a U \coth(\pi a/z),$$

where z is a complex variable zero on the line of contact and real on the plane. Prove also that the pressures at the two ends of the diameter of the cylinder normal to the plane differ by $\pi^4\rho U^2/32$. (M.T. 1929.)

33. A hollow vessel of the form of an equilateral triangular prism, filled with liquid, is struck excentrically by a given blow in a plane perpendicular to the axis and bisecting the three edges; find the initial motion of the vessel. (M.T. 1887.)

34. What is the nature of the motion in the neighbourhood of the origin, when, $f(z)$ being continuous finite and one-valued in that neighbourhood,

$$(1)\quad \frac{dw}{dz}=\frac{m}{z}+f(z),$$

$$(2)\quad \frac{dw}{dz}=\frac{im}{z}+f(z),$$

$$(3)\quad \frac{dw}{dz}=\frac{M}{z^2}+f(z),$$

m and M being real? (Univ. of London, 1911.)

35. Find the steady motion in two dimensions of an incompressible liquid, such that the stream lines are all ellipses similar to

$$x^2/a^2+y^2/b^2=1,$$

which is possible under the action of external forces whose components at the point xy are $X=Axy^2$, $Y=Bx^2y$, where A and B are constants. (Dublin Univ. 1911.)

36. In a two-dimensional irrotational motion of an incompressible fluid, the space between two cylinders whose cross sections are the curves C_1 and C_2 is completely filled with fluid, and C_1 is wholly inside C_2. If the velocity components are $-\partial\phi/\partial x$ and $-\partial\phi/\partial y$, and ϕ is single-valued, shew that

$$\int_{C_1} l\phi\,ds-\int_{C_2} l\phi\,ds=\int_{C_1} x\frac{\partial\phi}{\partial n}\,ds-\int_{C_2} x\frac{\partial\phi}{\partial n}\,ds,$$

where l is the cosine of the angle between the outward normal and the axis of x, and the differentiation is along the outward normal.

An infinite solid cylinder, whose section is the curve C, moves with velocity U along the axis of x in an infinite expanse of inviscid, incompressible fluid, of constant density ρ, and ϕ is the (single-valued) velocity potential of the fluid motion, defined as above. Shew that T, the kinetic energy of the fluid per unit length, is equal to $\frac{1}{2}\rho U\int_C l\phi\,ds$.

If for large values of z

$$\phi+i\psi=\frac{U(\lambda+i\mu)}{z}+O\left(\frac{1}{|z|^2}\right),$$

use the equality above to prove that

$$T=\tfrac{1}{2}\rho U^2(2\pi\lambda-A),$$

where A is the area enclosed by C. (M.T. 1934.)

CHAPTER VI

THE USE OF CONFORMAL REPRESENTATION. DISCONTINUOUS MOTION. FREE STREAM LINES. AEROFOILS

6·1. Conformal Representation. If

$$\xi + i\eta = f(x+iy), \quad \text{or} \quad t = f(z),$$

and we take (ξ, η) and (x, y) to be rectangular coordinates of points in two planes which we may call the t plane and the z plane, then the point (ξ, η) in the t plane corresponds to the point (x, y) in the z plane and the functional relation between t and z implies (**3·21**) that at an ordinary point the ratio $\delta t/\delta z$ of small corresponding elements tends to a limit which is independent of the direction of δz. Thus let P, P_1, P_2 be near points z, z_1, z_2 and Q, Q_1, Q_2 the corresponding points t, t_1, t_2. Then we may write

$$z_1 - z = r_1 e^{i\theta_1}, \quad z_2 - z = r_2 e^{i\theta_2},$$
$$t_1 - t = \rho_1 e^{i\phi_1}, \quad t_2 - t = \rho_2 e^{i\phi_2};$$

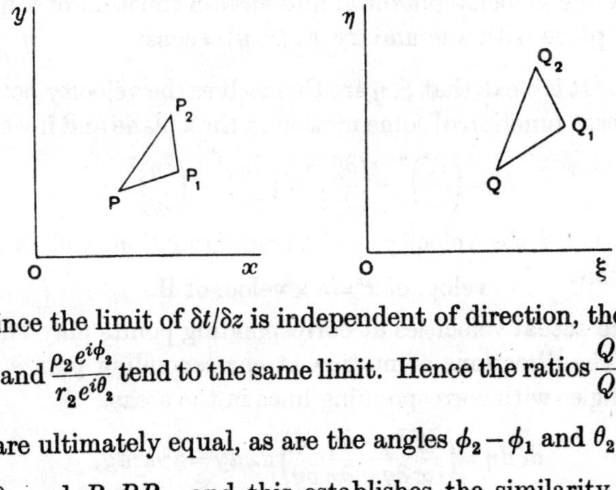

and since the limit of $\delta t/\delta z$ is independent of direction, therefore $\dfrac{\rho_1 e^{i\phi_1}}{r_1 e^{i\theta_1}}$ and $\dfrac{\rho_2 e^{i\phi_2}}{r_2 e^{i\theta_2}}$ tend to the same limit. Hence the ratios $\dfrac{QQ_1}{QQ_2}$ and $\dfrac{PP_1}{PP_2}$ are ultimately equal, as are the angles $\phi_2 - \phi_1$ and $\theta_2 - \theta_1$ or Q_1QQ_2 and P_1PP_2, and this establishes the similarity of the corresponding infinitesimal elements of the two planes, though corresponding finite areas of the two planes are not similar. Such a relation between the two planes is called the *conformal representation* of either plane on the other.

It is to be observed that the similarity of infinitesimal elements of the two planes will not hold good at points at which dt/dz is zero or infinite, as for example at a branch point of a multiple-valued function. Thus, the origin is a branch point in the t plane of the function

$$t=z^{\frac{1}{2}},$$

and as z describes a circular arc of angle α round the origin, t describes an arc of angle $\frac{1}{2}\alpha$ so that corresponding elements of the planes are not similar.

Now let there be two areas occupied by a fluid in motion. Let ξ, η be the coordinates of a point Π in one, and x, y the coordinates of a corresponding point P in the other. Let ϕ, ψ be the velocity potential and current function of any motion within the chosen area in the t plane given by

$$\phi+i\psi=\chi_1(\xi+i\eta),$$

and let the boundary be $\psi\equiv F_1(\xi,\eta)=\text{const.}$ If we substitute for ξ, η their values in terms of x, y, we get a relation

$$\phi+i\psi=\chi_2(x+iy),$$

and, if $F_1(\xi,\eta)=F_2(x,y)$, the corresponding boundary in the z plane is $\psi\equiv F_2(x,y)=\text{const.}$ Hence the same functions ϕ and ψ are now the velocity potential and stream function of a motion in the z plane with a boundary $F_2(x,y)=\text{const.}$

6·11. It is clear that ξ, η are themselves the velocity potential and stream function of some motion in the z plane and if we write

$$h^2=\left(\frac{\partial\xi}{\partial x}\right)^2+\left(\frac{\partial\xi}{\partial y}\right)^2=\left(\frac{\partial\eta}{\partial x}\right)^2+\left(\frac{\partial\eta}{\partial y}\right)^2,$$

we may call h the velocity of the transformation, and as in **5·3** we see that

$$\text{veloc. of } P=h\times\text{veloc. of }\Pi.$$

Thus the actual velocities at corresponding points may be compared. The directions of motion at corresponding points make equal angles with corresponding lines in the areas.

Since
$$d\xi\, d\eta=\left(\frac{\partial\xi}{\partial x}\frac{\partial\eta}{\partial y}-\frac{\partial\eta}{\partial x}\frac{\partial\xi}{\partial y}\right)dx\,dy=h^2dx\,dy,$$

corresponding elementary areas in the t and z planes are in the ratio $h^2:1$. Hence the kinetic energies of the two fluids that occupy corresponding areas are equal. Thus the whole kinetic energies of the two motions are equal, but differently distributed over the areas of motion.

6·12. *If a source exist in one fluid there will be a source at the corresponding point of the other fluid.* This follows at once from the fact that ψ is the same at corresponding points in the two fluids, so that $\int d\psi$ taken along corresponding arcs of curves must have the same value. That is, the flow across corresponding arcs is the same. At a pair of corresponding points at which t and z possess no singularities a small curve surrounding one corresponds to a small curve surrounding the other, and $\int d\psi$ round either curve represents the flow across it. Hence to a source at one such point must correspond a source of equal strength at the other. But care must be taken at a zero, infinity or branch point of the function that t is of z or that z is of t. A source will always correspond to a source but the strengths may differ; thus in the case $t=z^{\frac{1}{2}}$, since a semicircle round $t=0$ corresponds to a circle round $z=0$ and the flow across both is the same, if there be a source of strength m at $z=0$ the corresponding source at $t=0$ must be of strength $2m$.

If a doublet of strength m exists in the z plane at a point which occasions no singularity in t there will clearly be a doublet at the corresponding point in the t plane, the axes of the doublets will be in corresponding directions, i.e. they will make equal angles with any two corresponding lines through the points, and the strength m' of the doublet in the t plane will be given by

$$m'/m=|dt/dz|=h,$$

for the strength of a doublet is the product of the strength of a source and an infinitesimal length.

6·121. Example. Consider the transformation

$$t=z^{\kappa}, \quad 0<\kappa<1.$$

If we use polar coordinates r, θ in the z plane and ρ, χ in the t plane, this relation may be written

$$\rho e^{i\chi}=r^{\kappa}e^{i\kappa\theta},$$

so that

$$\chi=\kappa\theta, \text{ and } \rho=r^{\kappa}.$$

Suppose there to be liquid in the z plane bounded by the real axis, i.e. from $\theta=0$ to $\theta=\pi$. The corresponding boundaries in the t plane are $\theta=0$ and $\theta=\kappa\pi$.

First let the motion in the z plane be due to a source of strength m at the origin, then

$$\phi+i\psi=-m\log z.$$

The corresponding motion in the t plane is therefore given by

$$\phi+i\psi=-m\log t^{\frac{1}{\kappa}}=-\frac{m}{\kappa}\log t,$$

and this represents motion due to a source of strength m/κ at the origin in an area of the t plane bounded by $\theta=0$ and $\theta=\kappa\pi$.

Secondly if the motion in the z plane is due to a source m at $z=a$, we must introduce an equal source at the image point a' with regard to the real axis in order to make the real axis a stream line. Then we have

$$\phi+i\psi=-m\log(z-a)(z-a').$$

If $b=a^\kappa$ and $b'=a'^\kappa$ be the points in the t plane corresponding to a and a' the motion in the t plane is given by

$$\phi+i\psi=-m\log(t^{\frac{1}{\kappa}}-b^{\frac{1}{\kappa}})(t^{\frac{1}{\kappa}}-b'^{\frac{1}{\kappa}}).$$

To investigate the form of this expression in the neighbourhood of the point b, we write $t=b+\delta t$, and it is easily seen that the variable part of $\phi+i\psi$ reduces to $-m\log\delta t$ or $-m\log(t-b)$. Hence it follows that in this case the motion in the t plane is due to a source of strength m at b.

6·2. We may use this method, by proper choice of formulae of transformation, to deduce the motion with a complicated boundary from that with a simpler boundary. Thus to find the motion of a fluid with sources or doublets $P_1, P_2, \ldots$ within an infinite area on the z plane with a boundary $F_2(x,y)=0$. First suppose the sources and doublets removed and try to find a steady acyclic motion of fluid with the same boundary. If this can be done, let ξ, η be the velocity potential and stream function, so that η is constant along the boundary F_2, say $\eta=k$. Then use ξ, η as the formulae of transformation; the boundary F_2 transforms into the straight line $\eta=k$ and the area of motion transforms into the infinite area on one side of this line. Now replace the sources and doublets $P_1, P_2, \ldots$ by corresponding sources and doublets $\Pi_1, \Pi_2, \ldots$ in the t plane. The motion in the t plane due to the sources and doublets $\Pi_1, \Pi_2, \ldots$ can generally be inferred by placing single images for each on the other side of the line $\eta=k$, and so we obtain $\phi+i\psi$ in terms of $\xi+i\eta$ for the motion in the t plane, and substituting for ξ, η in terms of x and y we get $\phi+i\psi$ in terms of $x+iy$ giving the motion in the z plane due to the sources and doublets $P_1, P_2, \ldots$.

6·21. Examples. 1. *To find the motion in the space bounded by $x=0$, $y=0$, $y=b$ due to a source at the origin.*

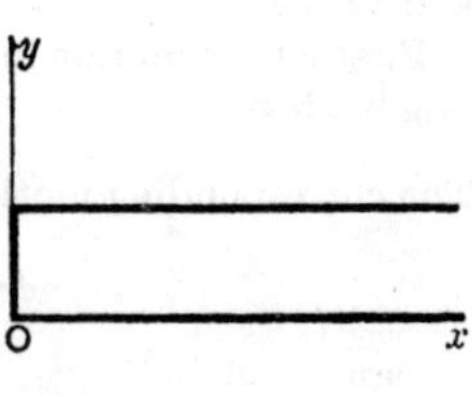

We want a solution of

$$\frac{\partial^2\eta}{\partial x^2}+\frac{\partial^2\eta}{\partial y^2}=0$$

that will make η constant when $x=0$, $y=0$, or $y=b$.

If we put $\eta = f(x)\sin\frac{\pi y}{b}$, we get

$$\frac{\partial^2 f}{\partial x^2} - \frac{\pi^2}{b^2} f = 0,$$

so that
$$f(x) = A\sinh\frac{\pi x}{b} + B\cosh\frac{\pi x}{b},$$

and we shall have $\eta = 0$ when $x = 0$ if $B = 0$.

Hence
$$\eta = A\sinh\frac{\pi x}{b}\sin\frac{\pi y}{b},$$

and the conjugate function is

$$\xi = A\cosh\frac{\pi x}{b}\cos\frac{\pi y}{b};$$

so that
$$t = \xi + i\eta = A\cosh\frac{\pi}{b}(x+iy) = A\cosh\frac{\pi z}{b}$$

transforms the given boundary into the straight line $\eta = 0$, and the point $\xi = A$, $\eta = 0$ corresponds to $x = 0$, $y = 0$.

If we place a source of strength m at this point, we have for the motion in the t plane
$$\phi + i\psi = -m\log(t - A).$$

Therefore the motion in the z plane is given by

$$\phi + i\psi = -m\log A\left(\cosh\frac{\pi z}{b} - 1\right),$$

or omitting an additive constant

$$\phi + i\psi = -2m\log\sinh\frac{\pi z}{2b};$$

and it is to be observed that since the straight boundary in the t plane corresponds to a right angle at O in the z plane, the motion in the z plane is due to a source of strength $2m$.

2. *Verify that, if r, s be real positive constants,*

$$z = x + iy, \quad a = \rho e^{i\beta}, \quad c^{-1} = r^{-1} + s^{-1},$$

the steady motion outside both the circles $x^2 + y^2 + 2sx = 0$, $x^2 + y^2 - 2rx = 0$, due to a doublet at the point $z = a$, outside both the circles, of strength μ and inclination α to the axis of x, is given by putting $\phi + i\psi$ equal to

$$\frac{c\mu\pi}{\rho^2}\left[e^{i(\alpha-2\beta)}\cot c\pi\left(\frac{1}{z} - \frac{1}{a}\right) - e^{-i(\alpha-2\beta)}\cot c\pi\left(\frac{1}{z} - \frac{1}{a_0}\right)\right],$$

where $z = a_0$ is the inverse point to $z = a$ with regard to either one of the circles.

(M.T. 1896.)

The transformation $t = \dfrac{i}{z}$, or $\xi + i\eta = \dfrac{y + ix}{x^2 + y^2}$,

$$\text{i.e. } \xi = \frac{y}{x^2 + y^2}, \quad \eta = \frac{x}{x^2 + y^2},$$

makes the given circles correspond to straight lines $\eta = -1/2s$, $\eta = 1/2r$ in the t plane, and the space between these lines clearly corresponds to the space outside the circles.

To correspond to the doublet of strength μ at $z = a$ we must take one of strength $\mu\left|\dfrac{dt}{dz}\right|_{z=a} = \dfrac{\mu}{\rho^2}$ at the point $\xi + i\eta = t = \dfrac{i}{a} = \dfrac{\sin\beta + i\cos\beta}{\rho}$.

To get the direction of this doublet in the t plane, we observe that the doublet μ at a makes an angle α with Ox and therefore makes an angle $\frac{1}{2}\pi-(2\beta-\alpha)$ with the circle of the same coaxial family that passes through a; and this circle transforms into a parallel to $O\xi$ in the t plane so that the doublet in the t plane makes an angle $\frac{1}{2}\pi-(2\beta-\alpha)$ with $\eta=(\cos\beta)/\rho$.

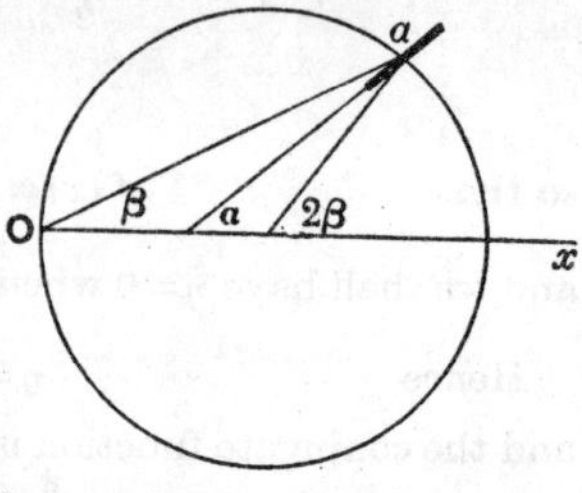

Instead of taking an infinite series of images of this doublet in the parallel boundaries $\eta=-1/2s$, $\eta=1/2r$ in the t plane, we make another substitution which transforms the strip of the t plane into the upper half of a plane z'; viz.

$$z'=e^{2\pi c\left(t+\frac{i}{2s}\right)},\quad\text{or}\quad r'e^{i\theta'}=x'+iy'=e^{2\pi c\left(\xi+i\eta+\frac{i}{2s}\right)},$$

for this makes $\theta'=2\pi c\left(\eta+\dfrac{1}{2s}\right)$, so that, as η increases from $-1/2s$ to $1/2r$, θ' increases from 0 to π. And we may now omit the constant factor $e^{\pi ci/s}$ from z' as it only implies a rotation of the axes about the origin.

There is now therefore a single boundary $y'=0$; and the last doublet $\dfrac{\mu}{\rho^2}$ at $t=\dfrac{i}{a}$ becomes a doublet $\dfrac{\mu}{\rho^2}\left|\dfrac{dz'}{dt}\right|_{t=i/a}$ at $z'=e^{2\pi ci/a}$; i.e. a doublet

$$\mu'=\frac{\mu}{\rho^2}\left|2\pi c e^{2\pi ci/a}\right|=\frac{2\pi\mu c}{\rho^2}e^{2\pi c(\sin\beta)/\rho}.$$

Also since the doublet in the t plane makes an angle $\frac{1}{2}\pi+\alpha-2\beta$ with $\eta=(\cos\beta)/\rho$, therefore the doublet μ' makes the same angle in the z' plane with the radius $\theta'=(2\pi c\cos\beta)/\rho$, or an angle

$$\gamma=\tfrac{1}{2}\pi+\alpha-2\beta+(2\pi c\cos\beta)/\rho,$$

with Ox'.

This doublet gives rise to a motion represented by

$$w=\frac{\mu' e^{i\gamma}}{z'-e^{2\pi ci/a}}=\frac{2\pi\mu ci}{\rho^2}e^{i(\alpha-2\beta)}\frac{e^{2\pi ci/a}}{e^{2\pi ci/z}-e^{2\pi ci/a}}$$

$$=\frac{2\pi\mu ci}{\rho^2}e^{i(\alpha-2\beta)}\frac{e^{-\pi ci\left(\frac{1}{z}-\frac{1}{a}\right)}}{e^{\pi ci\left(\frac{1}{z}-\frac{1}{a}\right)}-e^{-\pi ci\left(\frac{1}{z}-\frac{1}{a}\right)}}$$

$$=\frac{\pi\mu c}{\rho^2}e^{i(\alpha-2\beta)}\frac{\cos\pi c\left(\dfrac{1}{z}-\dfrac{1}{a}\right)-i\sin\pi c\left(\dfrac{1}{z}-\dfrac{1}{a}\right)}{\sin\pi c\left(\dfrac{1}{z}-\dfrac{1}{a}\right)}$$

$$=\frac{\pi\mu c}{\rho^2}e^{i(\alpha-2\beta)}\cot\pi c\left(\frac{1}{z}-\frac{1}{a}\right)+\text{const.}$$

But this doublet will require an equal doublet as an image symmetrically placed with regard to the line $y'=0$, and this can easily be shewn to give rise to the other term in w.

6·3. Discontinuous Motion. We have now arrived at a point in the theory of motion of a perfect fluid at which it is again necessary to emphasize the difference between this and real fluids. In any steady motion, apart from external causes, we have

$$p/\rho = C - \tfrac{1}{2}q^2,$$

where C is a constant, so that for large velocities the pressure will be negative. This is physically possible, and in the case of a liquid there is then the question whether 'cavitation' accompanied by evaporation takes place or not. In many cases single liquid elements would be in the region of low pressure for too short a time for the necessary transfer of heat to cause evaporation, but if the liquid contained dissolved gas a separation might take place. In the flow of a gas there is no question of cavitation because indefinite expansion is possible. So far as we are concerned with ideal liquid, cavitation would prevent the establishment of the large negative pressure and infinite velocities which the theory of continuous flow sometimes requires, e.g. in **5·31**. But in *real* fluids viscosity is the important consideration, and the slightest amount of viscosity is effective in so modifying the motion, before any cavitation takes place, that no large negative pressures are brought into being. In the perfect fluid theory it is assumed that when an obstacle hinders the flow of a stream there is generally a region of 'dead-water' behind the obstacle; this region is separated from the rest by a surface of stream lines and there is a discontinuity in the tangential velocity as we cross the surface*. We have seen in **3·72** reasons why such a surface should be unstable, but in the analysis which follows in the examples considered in the next few articles we shall proceed on the hypothesis of the existence of a steady state.

6·31. We propose now to consider some cases of discontinuous two-dimensional motion, such as the flow of liquid through an aperture, and the impact of a stream on a plane lamina. The earliest solutions of problems of this nature were by Helmholtz†,

* This idea of discontinuity was enunciated by Stokes, 'On the Critical Values of the Sums of Periodic Series', *Trans. Camb. Phil. Soc.* VIII, or *Math. and Phys. Papers*, I, p. 310.

† 'Ueber discontinuirliche Flüssigkeitsbewegungen', *Berlin. Monatsberichte*, 1868.

and Kirchhoff* who developed a general method of treatment applicable to cases in which the fixed boundaries are rectilinear, and where there may also be surfaces of constant pressure which may be free surfaces of the liquid or surfaces separating a portion of liquid at rest from the rest of the liquid. The given fixed boundaries are portions of stream lines, the other boundaries may be regarded as free stream lines and the solution of the problem will determine their form and position. Along the fixed boundaries the direction of the velocity is known but not its magnitude, and along the free stream lines, the pressure being constant, the velocity is constant in magnitude though its direction is not known. We must point out however that the problems of flow of a jet and flow past an obstacle are different in character, and that *in the latter case the results of the free stream line theory are of little practical importance because they are widely at variance with reality.*

6·32. In any particular case it is our object to find a suitable relation between w and z, i.e. to express ϕ and ψ in terms of x and y. When we have found the equation of the stream lines, $\psi=\text{const.}$, it will of course include the equations of the fixed boundaries.

For this purpose Kirchhoff introduced the intermediate function

$$\zeta = -\frac{dz}{dw} = \frac{u+iv}{q^2}, \quad (3\cdot51)$$

$$= e^{i\theta}/q,$$

where θ is the inclination to the x axis of the velocity q, so that θ is constant along a fixed boundary and q is constant along a free stream line. Kirchhoff then shewed how, by conformal representation, to obtain a relation between w and this function ζ, and the elimination of ζ between this relation and $dz/dw = -\zeta$ gives on integration a relation between w and z.

6·33. In our two-dimensional problem we have a certain region on the z plane bounded by stream lines, that is, lines for which ψ is constant, so that the corresponding region on the w plane will be bounded by straight lines parallel to the ϕ axis. The method that we shall use for obtaining the relation between w and

* *Crelle*, 1869. See also *Mechanik*, chaps. XXI, XXII.

z consists in making two intermediate transformations*. Thus consider the function

$$\Omega = \log \zeta = \log q^{-1} + i\theta.$$

Since the figure in the z plane is bounded by lines for which either θ is constant or q is constant, and we may by suitable choice of units take unity to be the constant value of q along the free stream lines, hence if the z plane is conformally represented on the Ω plane the fixed boundaries (θ = constant) on the z plane will correspond to lines parallel to the real axis on the Ω plane, and the free stream lines ($q = 1$) on the z plane will correspond to portions of the imaginary axis on the Ω plane. Thus the figure on the Ω plane is rectangular and bounded by straight lines.

We next make use of a theorem due to Schwarz† and Christoffel‡ by which a rectilinear polygon in one plane can be transformed into the real axis in another plane, which we will call the t plane. This theorem enables us to determine the relations between Ω and t and between w and t that will transform our figures in both the Ω and w planes into the real axis in the t plane, so that points which ought to correspond in the Ω and w planes both correspond to the same point on the real axis in the t plane. The elimination of t then gives w in terms of Ω or $\log(-dz/dw)$ and hence we get the required relation between w and z, though it is sometimes more convenient to retain t as a variable parameter.

6·4. Theorem of Schwarz and Christoffel.

If $z = x + iy$ and $t = \xi + i\eta$ then any polygon bounded by straight lines in the z plane can be transformed into the axis of ξ, points inside the polygon corresponding to points on one side of the axis of ξ; and the relation that effects this transformation is

$$\frac{dz}{dt} = A\,(t-\xi_1)^{\frac{\alpha_1}{\pi}-1}\,(t-\xi_2)^{\frac{\alpha_2}{\pi}-1}\ldots(t-\xi_n)^{\frac{\alpha_n}{\pi}-1},$$

where $\alpha_1, \alpha_2, \ldots \alpha_n$ are the internal angles of the polygon in the z plane, and $\xi_1, \xi_2, \ldots \xi_n$ are the points on the axis of ξ that correspond to the angular points of the polygon in the z plane.

* See Love, 'On the Theory of Discontinuous Fluid motions in two dimensions', *Proc. Camb. Phil. Soc.* VII, p. 175.

† 'Ueber einige Abbildungsaufgaben', *Crelle*, LXX, 1869, p. 105.

‡ 'Sul problema delle temperature stazionare', *Annali di Matematica*, I, 1867, p. 89.

To verify this, we observe that dz/dt is never zero or infinite except at the points $\xi_1, \xi_2, \ldots \xi_n$ on the real axis of ξ. Also if $dz/dt = Re^{i\theta}$, where R is real, the argument θ remains unchanged so long as t is real and does not pass through any of the values $\xi_1, \xi_2, \ldots \xi_n$; hence the argument of dz is constant so long as t lies between any two of the values $\xi_1, \xi_2, \ldots \xi_n$, and all points z which correspond to points between ξ_r and ξ_{r+1}, say, on the axis of ξ, lie on a straight line in the z plane.

Hence it appears that points on one side of the axis of ξ in the t plane correspond to points within a polygon on the z plane and that the points $\xi_1, \xi_2, \ldots \xi_n$ correspond to the corners.

Now consider the change in the argument of dz/dt as t, moving along the ξ axis, passes through the point ξ_r. It is clear that the only factor that will give rise to any change is $(t-\xi_r)^{\frac{\alpha_r}{\pi}-1}$, and we

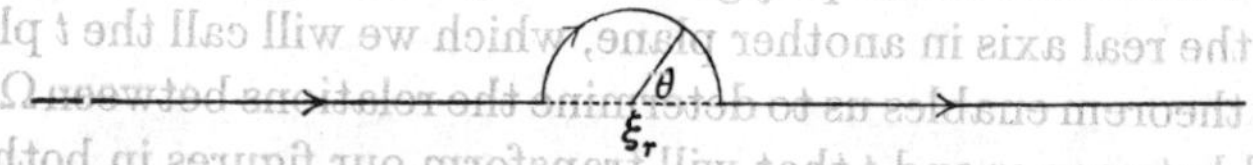

can make the passage by making the path near ξ_r a semicircle of small radius ϵ with centre at ξ_r as in the figure. On this semicircle $t-\xi_r = \epsilon e^{i\theta}$, so that

$$(t-\xi_r)^{\frac{\alpha_r}{\pi}-1} = \epsilon^{\frac{\alpha_r}{\pi}-1} e^{i\left(\frac{\alpha_r}{\pi}-1\right)\theta};$$

and as the semicircle is described θ changes from π to zero, hence the argument of dz/dt increases by $\pi-\alpha_r$. There is, therefore, a change of argument in the z plane amounting to $\pi-\alpha_r$, so that the lines in the z plane corresponding to $\xi_{r-1}\xi_r$ and $\xi_r\xi_{r+1}$ make an angle $\pi-\alpha_r$ with one another and the internal angle of the polygon corresponding to the corner ξ_r is α_r.

6·41. When we wish to transform a given polygon in the z plane into the axis of ξ in the t plane, the values of $\alpha_1, \alpha_2, \ldots \alpha_n$ are known, and as regards the values of $\xi_1, \xi_2, \ldots \xi_n$ three of them may be chosen arbitrarily and the others then depend on the dimensions of the polygon. For in order to construct a polygon similar to a given polygon of n sides we must have $n-3$ relations between the lengths of the sides. Any arbitrary distribution of the points $\xi_1, \xi_2, \ldots \xi_n$, provided they are taken in the proper order, will correspond to a polygon whose sides are in the right directions but, if the polygon is to have definite shape, only three of the points $\xi_1, \xi_2, \ldots \xi_n$ can be chosen arbitrarily.

By consideration of the function

$$\frac{d}{dt}\left\{\log\frac{dz}{dt}\right\}=\Sigma\frac{\alpha_r/\pi-1}{t-\xi_r},$$

it can be shewn that if the point $\xi=\infty$ be taken to correspond with one corner of the polygon the corresponding factor in the expression for dz/dt is omitted*.

6·42. As indicated in **6·33** the cases with which we are concerned will be those in which the polygon is rectangular. For a rectangle

$$\alpha_1=\alpha_2=\alpha_3=\alpha_4=\frac{\pi}{2},$$

and if the corners correspond to the points $\xi_1, \xi_2, \xi_3, \xi_4$ on the ξ axis we have

$$\frac{dz}{dt}=\frac{A}{\sqrt{\{(t-\xi_1)(t-\xi_2)(t-\xi_3)(t-\xi_4)\}}}.$$

(i) If we take $\xi_1=-1$, $\xi_2=1$, $\xi_3=\infty$ it is clear that two sides of the rectangle are infinite, so that we must also have $\xi_4=-\infty$, and the relation is, in this case,

$$\frac{dz}{dt}=\frac{A}{\sqrt{t^2-1}}.$$

This gives $z=A\cosh^{-1}t+B$; and if we take $B=0$, which only means moving the origin in the z plane, we have

$$t=\cosh z/A,$$

and the following values correspond:

$$t=1,\ -1,\ \infty,\ -\infty;\quad z=0,\ i\pi A,\ \infty,\ \infty+i\pi A.$$

The area in the z plane is then a strip of breadth πA parallel to the real axis and extending from $x=0$ to $x=\infty$,

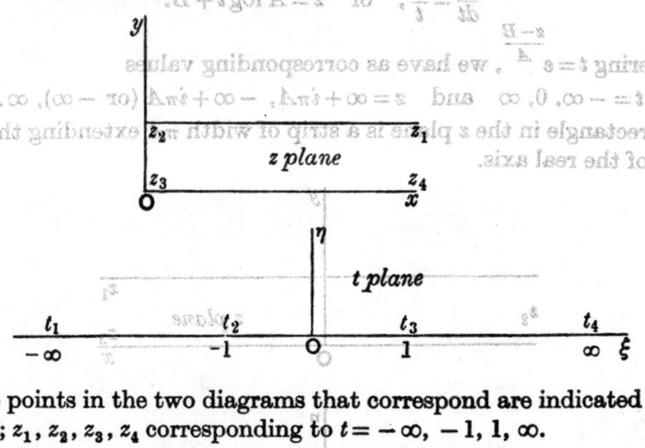

and the points in the two diagrams that correspond are indicated by like suffixes; z_1, z_2, z_3, z_4 corresponding to $t=-\infty, -1, 1, \infty$.

(ii) Another method of representing on the t axis the corners of the same strip of the z plane is to regard the strip as a triangle of zero angle in the

* See Forsyth's *Theory of Functions*, Art. 268.

direction $x=\infty$, we may then take any three points on the ξ axis in the t plane to correspond to the corners, say the points $t=-1, a, 1$, as shewn in the figure.

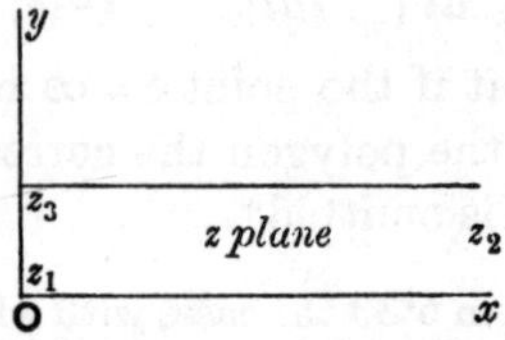

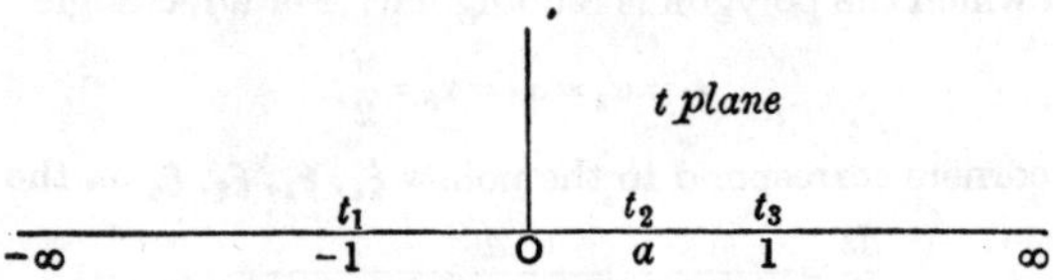

The relation connecting z and t is then

$$\frac{dz}{dt}=\frac{A}{(t-a)\sqrt{t^2-1}},$$

which gives on integration

$$z=\frac{iA}{\sqrt{1-a^2}}\cosh^{-1}\frac{at-1}{t-a}+B,$$

or

$$z=C\cosh^{-1}\frac{at-1}{t-a}+B.$$

If we choose the constant B so that $z=0$ when $t=-1$ we find $B=0$; then $t=1$ makes $z=i\pi C$ so that the width of the strip is πC, and $t=a$ makes $z=\infty$ as it ought.

(iii) As another case let us consider what sort of rectangle will correspond to the four points $t=-\infty, 0, 0, \infty$. The relation between t and z is

$$\frac{dz}{dt}=\frac{A}{t}, \quad \text{or} \quad z=A\log t+B.$$

Considering $t=e^{\frac{z-B}{A}}$, we have as corresponding values

$$t=-\infty, 0, \infty \quad \text{and} \quad z=\infty+i\pi A, \ -\infty+i\pi A \ (\text{or } -\infty), \ \infty.$$

So the rectangle in the z plane is a strip of width πA extending the whole length of the real axis.

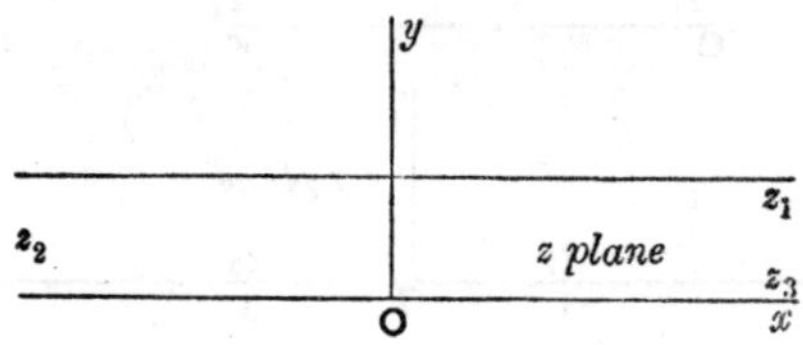

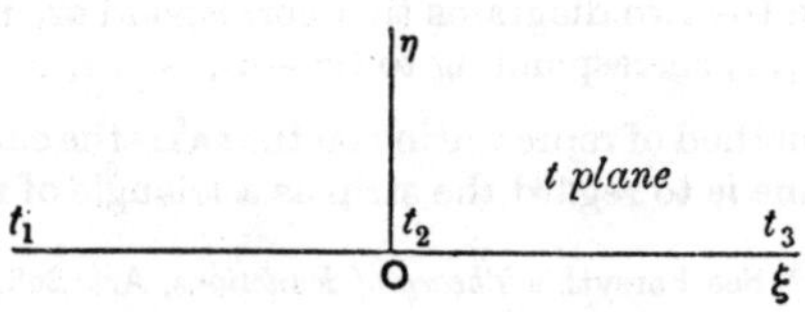

6·5. We shall now apply the foregoing theory to some Examples, in every case assuming the velocity to be unity along the free stream lines and neglecting all external forces.

Jet of Liquid through a Slit in a Plane Barrier. We assume that the sides of the vessel containing the liquid are infinitely distant from the slit compared to its breadth. In the diagrams fixed boundaries and lines that correspond to them are indicated by thick lines, free stream lines by thin lines, and the arrows indicate the direction of flow. Remembering that velocity is in the direction in which velocity potential decreases ($q = -\partial\phi/\partial s$), we may place $\phi = \infty$, $\phi = -\infty$ at opposite ends of the stream. For convenience we suppose the boundary stream lines to be $\psi = 0$, $\psi = \pi$. The region on the w plane which is to correspond to the given region on the

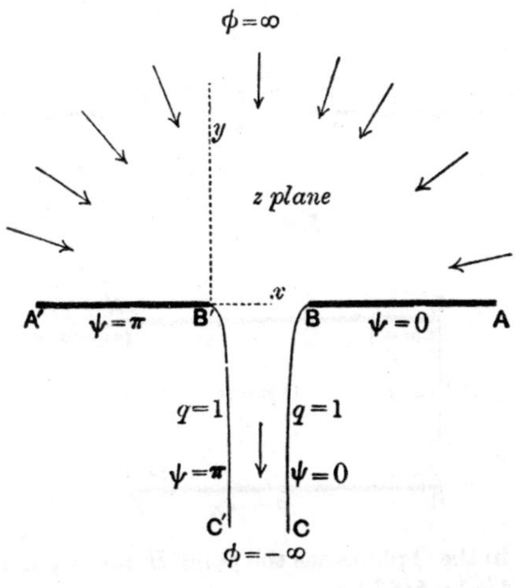

z plane is therefore seen to be a strip of width π extending along and above the axis of ϕ from $\phi = -\infty$ to $\phi = \infty$.

We have now to transform the z plane on to the Ω plane, where $\Omega = \log q^{-1} + i\theta$. In the z plane we take the origin at B', then for the velocity along $A'B'$ we have $\theta = 0$ and along AB $\theta = -\pi$. Hence in the Ω plane the lines $A'B'$, AB are $\theta = 0$ and $\theta = -\pi$, and the lines corresponding to the free stream lines BC, $B'C'$ for which $q = 1$ are parts of the imaginary axis.

We have now to transform the areas in the w plane and in the Ω plane into the upper half of the t plane so that corresponding corners in the w and Ω planes are represented by the same point on the real t axis.

Before lettering our w and Ω diagrams it will be convenient to choose particular points on the real t axis to correspond to them, since as we saw in **6·41** three such points may be chosen arbitrarily. Thus we may take the edges of the slit B, B' to correspond to $t = 1$, $t = -1$ and let A correspond to $t = \infty$. The w diagram is then as indicated, where we may take the line BB' to be $\phi = 0$ so that B is the origin in this diagram.

The relation between w and t is as in **6·42** (iii)

$$w = A\log t + B,$$

and $\qquad w=0$ when $t=1$, so that $B=0$;

also $\qquad w=i\pi$ when $t=-1$. But $\log(-1)=i\pi$,

therefore $\qquad A=1$ and $w=\log t$.

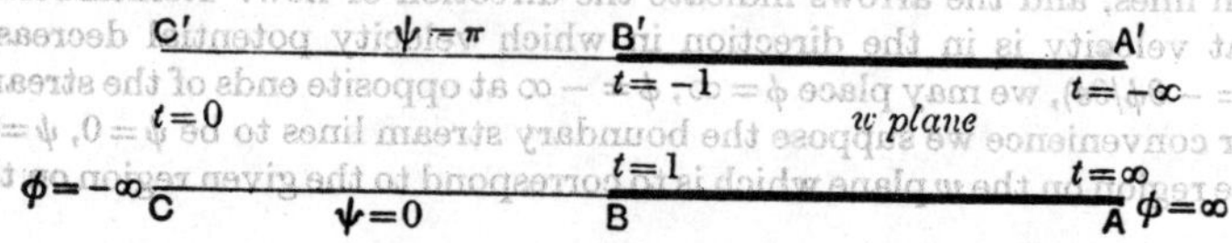

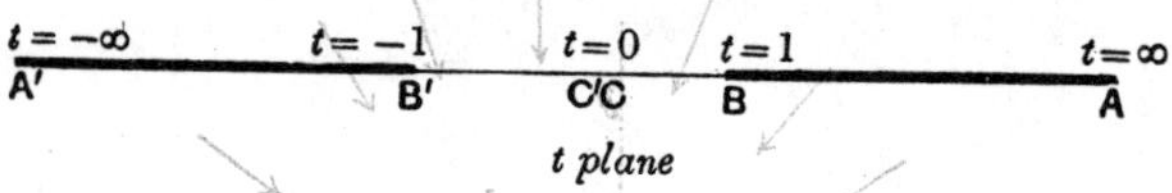

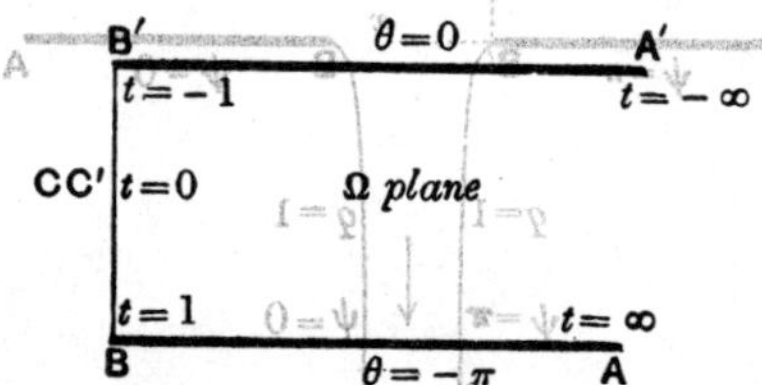

The diagram in the Ω plane has the point B' for origin, and the relation between Ω and t is by **6·42** (i)

$$\Omega = C\cosh^{-1}t + D,$$

and $\qquad \Omega=-i\pi$ when $t=1$, so that $D=-i\pi$;

also $\qquad \Omega=0$ when $t=-1$. But $\cosh^{-1}(-1)=i\pi$,

therefore $\qquad \Omega=\cosh^{-1}t-i\pi$, or $t=-\cosh\Omega$.

But $\qquad \Omega=\log\zeta$ or $\log\left(-\dfrac{dz}{dw}\right)$,

hence we have $\qquad \cosh\log\zeta=-t=-e^{w}$,

or $\qquad \zeta+\zeta^{-1}=-2e^{w}$.

From which we deduce

$$-\frac{dz}{dw}=\zeta=-e^{w}\pm\sqrt{e^{2w}-1},$$

and the fact that ζ or $e^{i\theta}/q$ is infinite when $\psi=0$ and $\phi=\infty$ determines that the lower sign must be taken.

Hence we get $\qquad \dfrac{dz}{dw}=e^{w}+\sqrt{e^{2w}-1},$

and the integral of this is

$$z = e^w + \sqrt{e^{2w}-1} - \tan^{-1}\sqrt{e^{2w}-1} - 1,$$

adjusting the constant so that $z=0$ when $w=0$.

To find the equation of a free stream line, we have along the stream line $B'C'$

$$-\frac{\partial\phi}{\partial s} = q = 1,$$

so that $\phi = -s$ measuring s from the origin B'. Hence on this stream line $\psi = \pi$ we have

$$s = -\phi = -\text{real part of } w = -\text{real part of } \log t,$$

where t is real and lies between -1 and 0; also $q=1$ so that

$$i\theta = \Omega = \cosh^{-1} t - i\pi \quad \text{or} \quad t = -\cos\theta,$$

where θ varies from 0 to $-\frac{1}{2}\pi$.

Hence on the stream line $B'C'$

$$s = \log(-\sec\theta).$$

But $$dx/ds = \cos\theta,$$

since θ gives the direction of the curve, therefore

$$dx = \sin\theta\, d\theta,$$

and $$x = 1 - \cos\theta,$$

the constant being determined by the consideration that $\theta = 0$ when $x = 0$.

Similarly $$y = \log(\tan\theta + \sec\theta) - \sin\theta.$$

Since the ultimate breadth of the jet when the free stream lines become parallel is π, and this is attained when $\theta = -\frac{\pi}{2}$, for which the value of x is unity, it follows that the breadth of the slit is $\pi + 2$ and the coefficient of contraction $\pi/(\pi+2)$.

6·51. Borda's Mouthpiece. We shall now consider the efflux of liquid through a pipe projecting into the containing vessel, being the case to which reference was made in **3·63**, but restricted to two dimensions and assuming that the sides of the vessel are so far away as not to affect the problem.

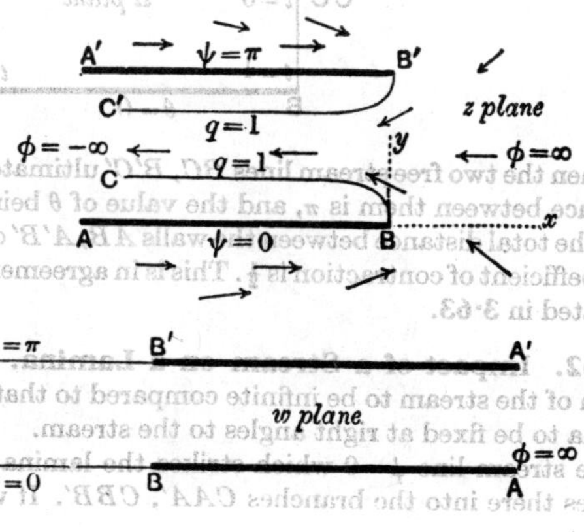

We shall adopt so far as possible the same notation and lettering as in 6·5. The boundary stream lines ABC, $A'B'C'$ are $\psi=0$ and $\psi=\pi$, so that the diagram in the w plane is the same as in the last case. If we take the same set of corresponding points on the real axis in the t plane as before, we have the same diagram in the t plane, and the relation between w and t is still

$$w=\log t.$$

The diagram in the Ω plane is also the same as before but now the line AB is $\theta=0$ and $A'B'$ is $\theta=2\pi$ so in the relation

$$\Omega=C\cosh^{-1}t+D$$

we have $\Omega=0$ when $t=1$, so that $D=0$,

and $\Omega=2i\pi$ when $t=-1$, so that, since $\cosh^{-1}(-1)=i\pi$,

we have $C=2$ and $\Omega=2\cosh^{-1}t$.

With the origin at B in the z plane (also in the w and Ω planes) we get along the free stream line BC, or $\psi=0$,

$$-s=\phi=w=\log t,$$

where t ranges from 1 to 0 and

since $q=1$, $i\theta=\Omega=2\cosh^{-1}t$, so that $t=\cos\frac{1}{2}\theta$,

and $s=\log\sec\frac{1}{2}\theta$.

Then $dx/ds=\cos\theta$ and $dy/ds=\sin\theta$

give $x=\sin^2\frac{1}{2}\theta-\log\sec\frac{1}{2}\theta$ and $y=\frac{1}{2}(\theta-\sin\theta)$

as the equations for the free stream line BC.

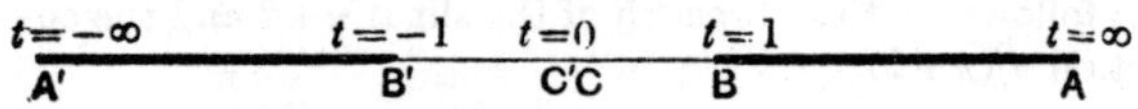

t plane

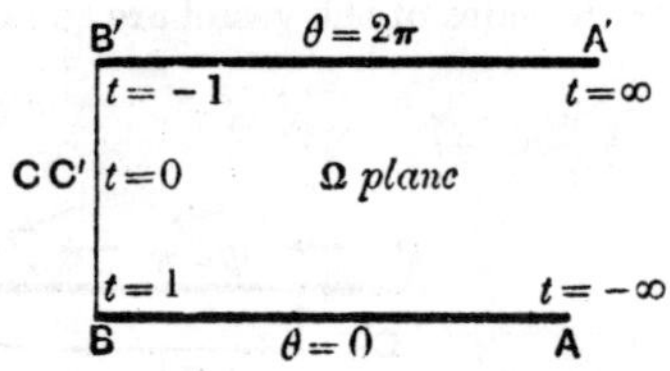

When the two free stream lines BC, $B'C'$ ultimately become parallel the distance between them is π, and the value of θ being π, we get $y=\frac{1}{2}\pi$, so that the total distance between the walls AB, $A'B'$ of the opening is 2π and the coefficient of contraction is $\frac{1}{2}$. This is in agreement with Borda's theory as stated in 3·63.

6·52. Impact of a Stream on a Lamina. We shall suppose the width of the stream to be infinite compared to that of the lamina and the lamina to be fixed at right angles to the stream.

The stream line $\psi=0$ which strikes the lamina at its middle point C divides there into the branches CAA', CBB'. If we take $\phi=0$ at C, the region on the z plane occupied by liquid corresponds to the whole w plane regarded as bounded by the double line from the origin to $\phi = -\infty$, $\psi = 0$.

We may clearly choose a transformation on to the t plane so that the points A', A, C, B, B' correspond to $t = \infty$, 1, 0, -1, $-\infty$. The relation between w and t is then

$$\frac{dw}{dt} = At,$$

for the interior angle of the w polygon is 2π. This gives

$$w = \tfrac{1}{2}At^2, \text{ since } w = 0 \text{ when } t = 0 \quad \ldots\ldots\ldots\ldots\ldots(1).$$

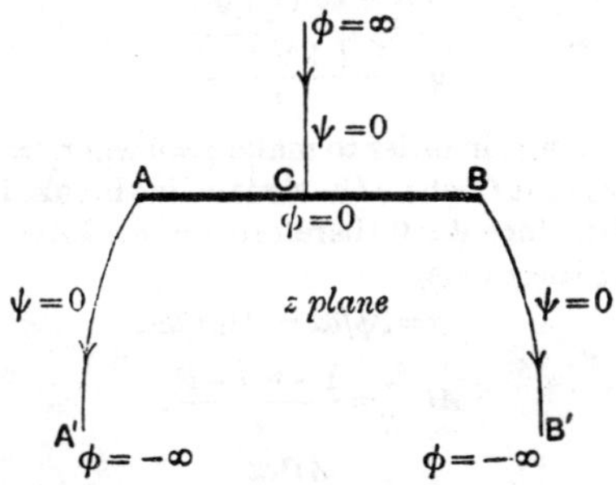

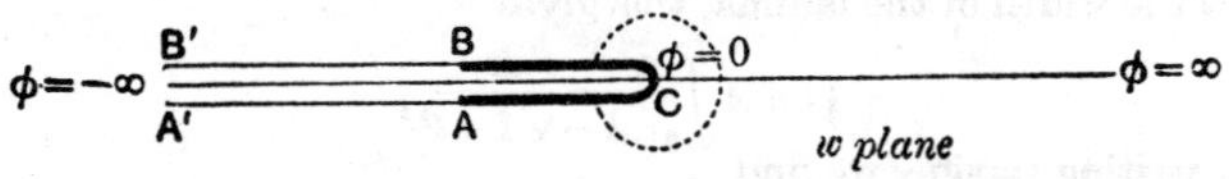

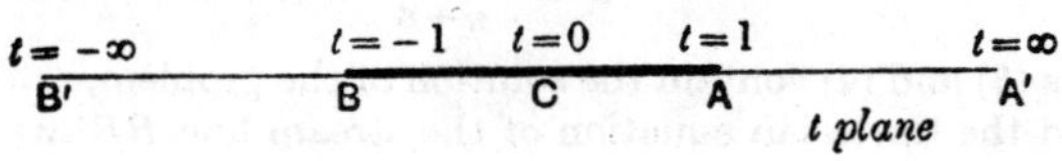

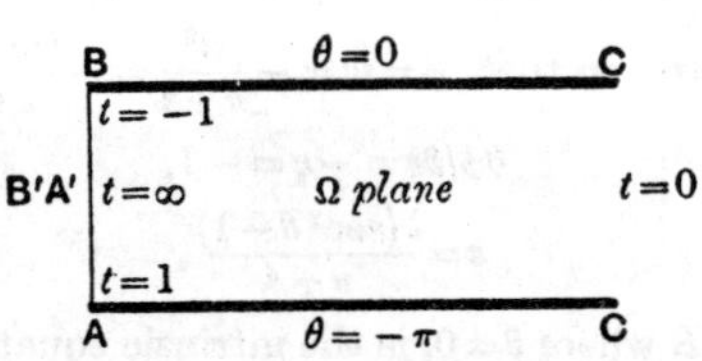

To get the diagram on the Ω plane we have $\theta = 0$ along CB, and $\theta = -\pi$ along CA and $q = 1$ along BB' and AA'. Hence the diagram must be as indicated and the relation between Ω and t is by **6·42** (ii)

$$\frac{d\Omega}{dt} = \frac{B}{t\sqrt{t^2-1}},$$

or

$$\Omega = C\cosh^{-1}\left(-\frac{1}{t}\right) + D.$$

But when $t=-1$ the diagram shews that $\Omega=0$, therefore $D=0$; and when $t=1$, we have $\Omega=-i\pi$, but $\cosh^{-1}(-1)=i\pi$, therefore $C=-1$.

Hence $$\Omega=-\cosh^{-1}(-1/t) \quad \text{or} \quad \frac{1}{t}=-\cosh\Omega \quad \ldots\ldots\ldots\ldots(2),$$

but $$\Omega=\log\zeta, \quad \text{therefore} \quad t=\frac{-2\zeta}{1+\zeta^2} \quad \ldots\ldots\ldots\ldots(3).$$

We have now to determine the constant A in equation (1), and its value must depend on the width of the lamina.

Along the stream line CB, since $\theta=0$ therefore $\zeta=1/q$ and

$$t=-2q/(1+q^2),$$

which gives $$q=\frac{-1+\sqrt{1-t^2}}{t}.$$

We take the positive sign in order to make $q=0$ when $t=0$, for the velocity must be zero at the point C where the stream line breaks into two branches.

Again, along CB, since $\psi=0$ therefore $\phi=w=\frac{1}{2}At^2$, and, the velocity being wholly along the x axis,

$$-q=\partial\phi/\partial x=A\,t\,dt/dx.$$

Therefore $$At\frac{dt}{dx}=\frac{1-\sqrt{1-t^2}}{t},$$

or $$dx=\frac{At^2dt}{1-\sqrt{1-t^2}}.$$

If l is the width of the lamina, this gives

$$\tfrac{1}{2}l=A\int_0^{-1}\frac{t^2dt}{1-\sqrt{1-t^2}},$$

and by writing $t=\sin\chi$ we find

$$\tfrac{1}{2}l=-A(1+\tfrac{1}{4}\pi), \quad \text{so that} \quad A=-\frac{2l}{\pi+4},$$

and $$w=-\frac{lt^2}{\pi+4} \quad \ldots\ldots\ldots\ldots(4).$$

Relations (3) and (4) contain the solution of the problem.

To find the Cartesian equation of the stream line BB' we have $q=1$, so that
$$i\theta=\Omega=-\cosh^{-1}(-1/t), \quad \text{or} \quad \cos\theta=-1/t.$$

Also $$\psi=0, \quad \text{so that} \quad \phi=w=-\frac{lt^2}{\pi+4}=-\frac{l\sec^2\theta}{\pi+4}.$$

Again $$\partial\phi/\partial s=-q=-1,$$

therefore $$s=\frac{l(\sec^2\theta-1)}{\pi+4},$$

measuring s from B where $\theta=0$, is the intrinsic equation.

Then $$dx=\cos\theta\,ds=2l\sec\theta\tan\theta\,d\theta/(\pi+4),$$
so that, taking the origin at C,

$$x=\frac{2l}{\pi+4}\left(\sec\theta+\frac{\pi}{4}\right),$$

and $$dy=\sin\theta\,ds=2l\sec\theta\tan^2\theta\,d\theta/(\pi+4);$$

whence $$y=\frac{l}{\pi+4}\{\sec\theta\tan\theta-\log(\sec\theta+\tan\theta)\}.$$

6·53. The same problem with oblique impact. We may proceed in the same way, but the stream line that divides is not in this case the one that strikes the barrier at its middle point.

We get a similar set of diagrams (see next page) wherein, in this case, the points A', A, C, B, B' correspond to $t=\infty$, 1, a, -1, $-\infty$, and the relation between w and t is

$$\frac{dw}{dt}=A(t-a),$$

or

$$w=\tfrac{1}{2}A(t-a)^2 \quad \text{.......................(1)},$$

since $w=0$ when $t=a$.

Also for the relation between Ω and t we have by **6·42** (ii)

$$\frac{d\Omega}{dt}=\frac{C}{(t-a)\sqrt{t^2-1}},$$

or

$$\Omega=C\cosh^{-1}\frac{at-1}{t-a}+D.$$

But when $t=-1$, the diagram shews that $\Omega=0$, therefore $D=0$; and when $t=1$ we have $\Omega=-i\pi$; but $\cosh^{-1}(-1)=i\pi$, therefore $C=-1$.

Hence

$$\Omega=-\cosh^{-1}\frac{at-1}{t-a},$$

or

$$\frac{at-1}{t-a}=\cosh\Omega=\tfrac{1}{2}(\zeta+\zeta^{-1}) \quad \text{...................(2)}.$$

Also, if the stream makes an acute angle α with the barrier, the final direction of AA' and BB' is given by $\theta=-(\pi-\alpha)$ when $t=\infty$. Hence

$$i(\pi-\alpha)=\cosh^{-1}a, \quad \text{or} \quad a=-\cos\alpha.$$

Therefore (2) may be written

$$-\frac{t\cos\alpha+1}{t+\cos\alpha}=\cosh\Omega=\tfrac{1}{2}(\zeta+\zeta^{-1}) \quad \text{.................(3)}.$$

On the barrier from A to C

$$\theta=-\pi \quad \text{and} \quad \zeta=-1/q,$$

and from C to B

$$\theta=0 \quad \text{and} \quad \zeta=1/q;$$

therefore

$$\frac{q^2+1}{2q}=\pm\frac{t\cos\alpha+1}{t+\cos\alpha},$$

the upper or lower sign according as t lies between 1 and $-\cos\alpha$ or between $-\cos\alpha$ and -1.

This makes

$$q=\pm\frac{t\cos\alpha+1-\sin\alpha\sqrt{1-t^2}}{t+\cos\alpha},$$

the signs being adjusted so that q shall not become infinite when $t=-\cos\alpha$.

Also along the barrier

$$\psi=0 \text{ and } \phi=w=\tfrac{1}{2}A(t-a)^2,$$

so that

$$q=\mp\frac{\partial\phi}{\partial x}=\mp A(t-a)\frac{dt}{dx},$$

the upper or lower sign according as we are on CB or CA, since these are the directions of q.

Hence

$$A(t+\cos\alpha)\frac{dt}{dx}=-\frac{t\cos\alpha+1-\sin\alpha\sqrt{1-t^2}}{t+\cos\alpha},$$

and

$$dx=-A(t\cos\alpha+1+\sin\alpha\sqrt{1-t^2})\,dt.$$

Integrating this and taking the origin at the middle point of AB so that $t=1$, $t=-1$ give equal and opposite values for x, we obtain

$$x=-\tfrac{1}{2}A\{(t^2-1)\cos\alpha+2t+\sin\alpha\,(t\sqrt{1-t^2}+\sin^{-1}t)\}\;\ldots\ldots(4).$$

If we put $t=-1$ we get half the width l of the barrier, so that

$$l=\tfrac{1}{2}A\,(4+\pi\sin\alpha)\;\ldots\ldots\ldots\ldots\ldots\ldots\ldots\ldots(5).$$

Hence $$w=l\,(t+a\cos\alpha)^2/(4+\pi\sin\alpha).$$

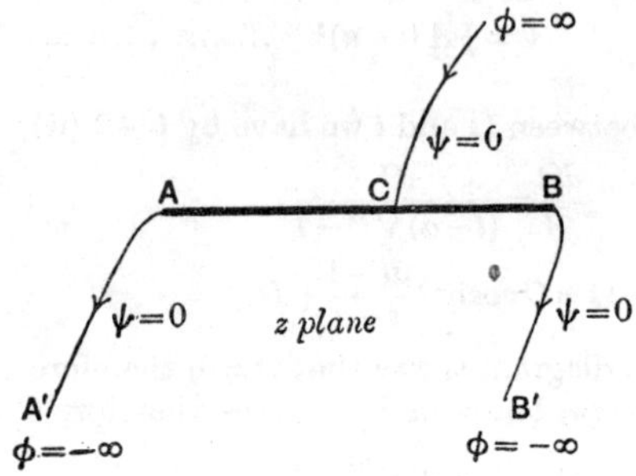

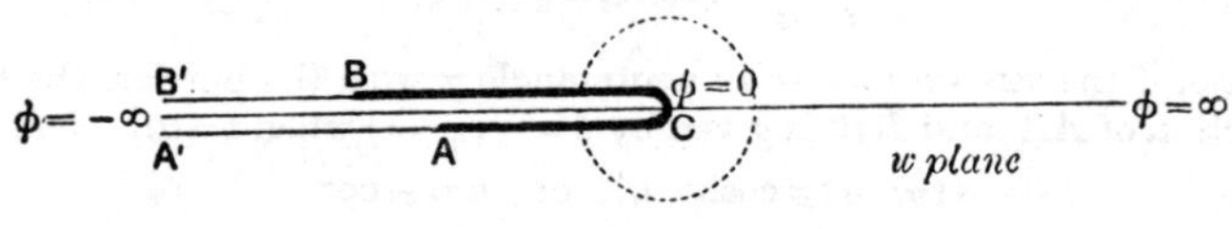

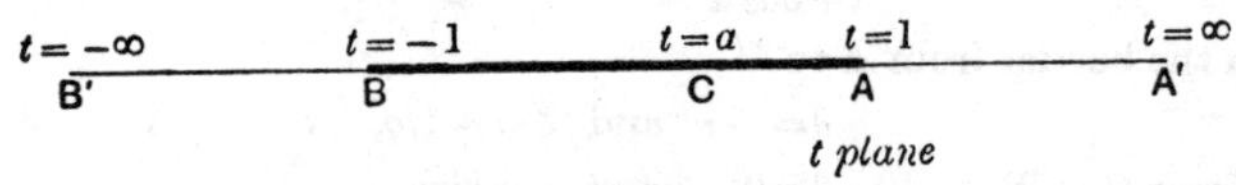

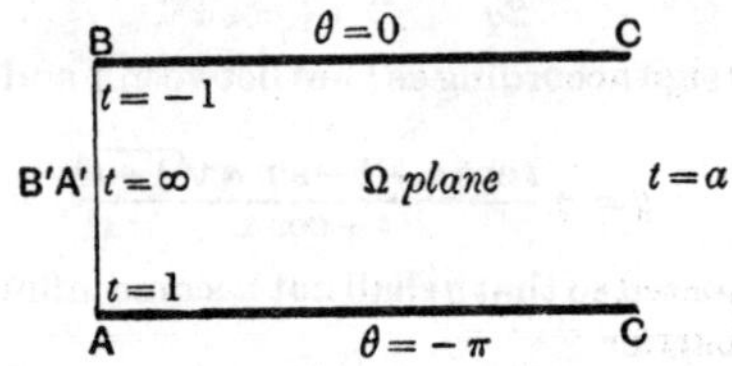

If in (4) we put $t=-\cos\alpha$ we get for the distance from the middle of the barrier to the point where the stream divided

$$x=\tfrac{1}{2}A\left\{2\cos\alpha\,(1+\sin^2\alpha)+\left(\frac{\pi}{2}-\alpha\right)\sin\alpha\right\}$$

$$=\left\{2\cos\alpha\,(1+\sin^2\alpha)+\left(\frac{\pi}{2}-\alpha\right)\sin\alpha\right\}\,l/(4+\pi\sin\alpha)\;\ldots\ldots(6).$$

Taking $p/\rho=C-\frac{1}{2}q^2$, we get on the free stream lines $p/\rho=C-\frac{1}{2}$; and this gives the pressure of the 'dead water' behind the lamina. Therefore the difference of the pressures on opposite sides of the lamina at any point is

$$p' = \tfrac{1}{2}\rho(1-q^2).$$

The resultant thrust on the lamina is therefore

$$\int_{-\frac{1}{2}l}^{\frac{1}{2}l} p'\,dx = \tfrac{1}{2}\rho\int_{-\frac{1}{2}l}^{\frac{1}{2}l}(1-q^2)\,dx$$

$$= \tfrac{1}{2}\rho\int_{-\frac{1}{2}l}^{\frac{1}{2}l}(q^{-1}-q)\,q\,dx.$$

But $\qquad q = \pm(t\cos\alpha + 1 - \sin\alpha\sqrt{1-t^2})/(t+\cos\alpha),$

therefore $\qquad q^{-1} = \pm(t\cos\alpha + 1 + \sin\alpha\sqrt{1-t^2})/(t+\cos\alpha),$

and $\qquad q\,dx = \mp A\,(t+\cos\alpha)\,dt,$

therefore the thrust

$$= -\rho A\int_1^{-1}\sin\alpha\sqrt{1-t^2}\,dt \qquad\text{.....................(7)}$$

$$= \tfrac{1}{2}\pi\rho A\sin\alpha$$

$$= \frac{\pi\rho l\sin\alpha}{4+\pi\sin\alpha} \qquad\text{...(8).}$$

For the distance of the centre of pressure from the end A, we have that the moment of the pressure about the centre

$$= \int_{-\frac{1}{2}l}^{\frac{1}{2}l} xp'\,dx = \tfrac{1}{2}\rho\int_{-\frac{1}{2}l}^{\frac{1}{2}l} x(1-q^2)\,dx.$$

To reduce this integral we notice that it is the same as (7) if we introduce the expression (6) as a factor, then the substitution $t=\sin\chi$ enables us to evaluate the integral at once giving as the result $\frac{3}{16}\pi\rho A^2\sin\alpha\cos\alpha$. But the whole pressure is $\frac{1}{2}\pi\rho A\sin\alpha$, therefore we have for the coordinate of the centre of pressure

$$\bar{x} = \tfrac{3}{8}A\cos\alpha = \tfrac{3}{4}\frac{l\cos\alpha}{4+\pi\sin\alpha} \qquad\text{.....................(9),}$$

on the up-stream side of the middle point.

This problem was discussed at length by Lord Rayleigh as the case of an elongated blade held vertically in a horizontal stream. He obtained results (8) and (9) by Kirchhoff's method and gave tables for their values*.

There is however an objection to the solution of the problem when viewed as one of a plane moving steadily through fluid at rest and carrying with it a region of 'dead water' extending behind it to infinity, namely that the kinetic energy of the dead water would be infinite. But we have seen in **3·72** a reason why the surfaces of discontinuity should be unstable and we shall see later, in the chapters on vortices and viscosity, the way in which the surfaces of discontinuity roll up and in general an eddying wake forms behind the body.

6·54. A variety of cases have been worked out by Michell†, Love‡, Greenhill § and other writers ||, the method has been extended by Hopkinson¶ to include the case of sources and vortices in the liquid, and important

* *Phil. Mag.* II, 1876, p. 430, or *Sci. Papers*, I, p. 286.

† 'On the Theory of Free Stream-Lines', *Phil. Trans.* A, 1890.

‡ *Loc. cit.* p. 135.

§ *Encyc. Brit.* 11th edition, Art. *Hydromechanics.*

|| For a full bibliography of the subject see Love, *Encyc. des Sc. Math.* IV, 18, pp. 118–122, where an account is given of the recent work of T. Levi-Civita, M. Brillouin, H. Villat, U. Cisotti and other writers.

¶ 'Discontinuous Motion involving sources and vortices', *Proc. L.M.S.* 1898.

applications of conformal transformation to curved boundaries have been developed by Leathem* by the introduction of curve factors into Schwarzian transformations.

6·6. The hydrodynamical applications of conformal representation which have received most attention in recent years are concerned with the theory of aerofoils. They are to be found in the Journals of Aeronautical and other Societies and in a large number of books on the subject of Aeronautics†. We must limit ourselves here to the consideration of a few cases of transformation of a state of steady flow about a circle.

6·61. Steady flow about a Circle. Recapitulation. Such a flow is represented by

$$w=(U+iV)(z-z_0)+\frac{(U-iV)a^2}{z-z_0}+\frac{i\kappa}{2\pi}\log(z-z_0)\quad\ldots(1),$$

where z_0 is the centre of the circle and κ the circulation.

The velocity components u, v are given by

$$-u+iv=\frac{\partial\phi}{\partial x}+i\frac{\partial\psi}{\partial x}=\frac{dw}{dz}=U+iV-\frac{(U-iV)a^2}{(z-z_0)^2}+\frac{i\kappa}{2\pi(z-z_0)}\ldots(2),$$

so that the velocity at infinity has components $-U$, V; and if we put $U=Q\cos\beta$, $V=Q\sin\beta$, then Q is the velocity at infinity

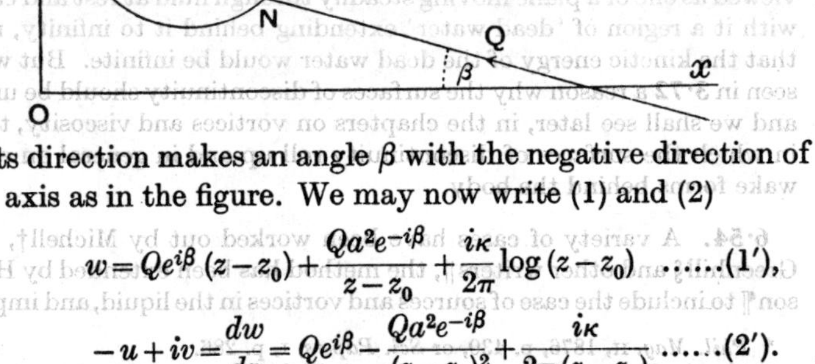

and its direction makes an angle β with the negative direction of the x axis as in the figure. We may now write (1) and (2)

$$w=Qe^{i\beta}(z-z_0)+\frac{Qa^2e^{-i\beta}}{z-z_0}+\frac{i\kappa}{2\pi}\log(z-z_0)\quad\ldots\ldots(1'),$$

and

$$-u+iv=\frac{dw}{dz}=Qe^{i\beta}-\frac{Qa^2e^{-i\beta}}{(z-z_0)^2}+\frac{i\kappa}{2\pi(z-z_0)}\quad\ldots\ldots(2').$$

* *Phil. Trans. R.S.* CCXV, A, 1915, pp. 439–487, and *Phil. Mag.* XXXI, 1916, pp. 190–197.

† E.g. H. Glauert, *Aerofoil and Airscrew Theory*, 1926; and N. Joukowski, *Aérodynamique*, 1916.

If we assume that there is a point of zero velocity on the circle at $z = z_0 + ae^{i(\pi+\alpha)} = z_0 - ae^{i\alpha}$, we have

$$0 = Q\,(e^{i\beta} \mp e^{-i\beta-2i\alpha}) - \frac{i\kappa}{2\pi a} e^{-i\alpha},$$

or

$$\kappa = 4\pi a Q \sin(\alpha+\beta) \quad \text{.................(3).}$$

This (for values of κ not too large) gives two points symmetrically situated about the diameter perpendicular to the stream. These are the points N, N' of **5·25**.

6·62. Joukowski's Condition. When the state of flow is transformed by a relation $t = f(z)$ the velocity in the t plane is given by

$$-u' + iv' = \frac{dw}{dt} = \frac{dw}{dz}\frac{dz}{dt} = (-u+iv)\frac{dz}{dt} \quad \text{........(4).}$$

Now the transformations which prove useful in aerofoil theory are such as give in the t plane the contour of a figure with a 'trailing edge', and this implies a singular point in the transformation, i.e. a point at which dz/dt is infinite. But in an actual state of flow about an aerofoil the velocity is not infinite, and it was pointed out by Joukowski that to get a finite velocity (u', v') at the trailing edge, where dz/dt is infinite, it is necessary that (u, v) should be zero. This implies that the point on the circle which transforms into the edge of the aerofoil must be a point of zero velocity in the motion about the circle, and from (3) this requires that the circulation shall bear a definite ratio to the velocity of the stream. It must be noted that the circulation κ is the same in both planes, but the velocity at infinity in the t plane is $Q\,|dz/dt|$ (t infinite).

6·63. A Joukowski Transformation. Let

$$t = \tfrac{1}{2}\left(z + \frac{a^2}{z}\right) \quad \text{.........................(1)*,}$$

or

$$\xi = \tfrac{1}{2}\left(r + \frac{a^2}{r}\right)\cos\theta, \quad \eta = \tfrac{1}{2}\left(r - \frac{a^2}{r}\right)\sin\theta.$$

This transforms the circle $r = a$ into $\xi = a\cos\theta$, $\eta = 0$, i.e. into the diameter BA of the circle which lies along the real axis taken twice over.

Any contour which surrounds the circle $r = a$ will transform into a contour surrounding the line BA, and may be constructed

* N. Joukowski, *Aérodynamique*, 1916, p. 145.

thus: let P be a point z on a contour to be transformed, P' the inverse of P in the circle $r=a$, then P'' the reflexion of P' in the real axis is the point a^2/z, and if R is the middle point of $P''P$ the vector OR is half the sum of OP and OP'' so that R is the point $\frac{1}{2}\left(z+\frac{a^2}{z}\right)$ or t.

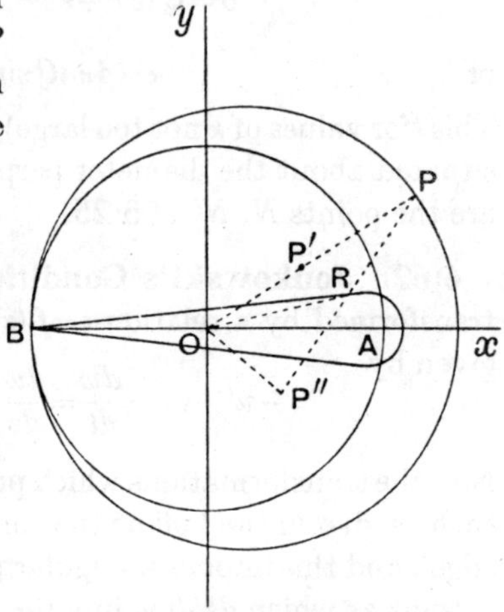

Now take a circle of radius a' in the z plane slightly larger than the circle $r=a$ and touching it at B. The contour obtained by transformation must touch the line BA on both sides at B, and is a 'fish-shaped' figure extending a little beyond the line BA at the end A.

From (1) we have $\frac{dt}{dz}=\frac{1}{2}\left(1-\frac{a^2}{z^2}\right)$, so that the points B and A, i.e. $z=\mp a$, are the infinities of dz/dt. Comparing with **6·61** and **6·62** we see that B should be a point of zero velocity in the motion about the circle. With the centre of the circle as origin, this means that in **6·61** (3) α is zero and

$$\kappa=4\pi a'Q\sin\beta.$$

We might now eliminate z between (1) above and **6·61** (1′), expressing w as a function of t to get the flow about the aerofoil, and then use the theorem of Blasius to determine the resultant thrust upon it due to the motion. But we will reserve these calculations for a more general type of transformation.

6·7. Consider the transformation

$$\frac{t-A'}{t-B'}=\left(\frac{z-A}{z-B}\right)^{2-\frac{\chi}{\pi}} \quad \text{.........................(1)},$$

where A', B' are two points in the t plane which correspond to the points A, B in the z plane. Without loss of generality we may regard the planes as superposed so that the points A', B' coincide with the points A, B. We may call these points the poles, and $2-\frac{\chi}{\pi}$ or k the exponent of the transformation.

If we put $z-A=re^{i\theta}$, $z-B=r'e^{i\theta'}$, $t-A'=\rho e^{i\phi}$ and $t-B'=\rho' e^{i\phi'}$ we see that

$$\phi-\phi'=k(\theta-\theta') \quad \text{..........................(2)}.$$

It follows that if P, P' are corresponding points in the z and t planes, the angle $AP'B = k . APB$.

Now if we apply the transformation to a circle C passing through A, B in the z plane, the corresponding contour in the t plane consists of arcs of circles. And bearing in mind that it is the regions outside the contours which are to be transformed, we may exclude the singular points A and B by drawing small semicircles with centres A and B outside the circle C.

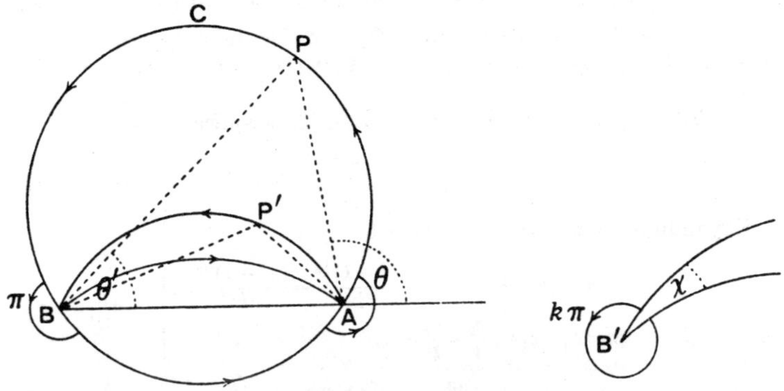

Then when a point P travels round the circle C in the z plane, as P passes B the angle θ' increases by π. so that from (2) the corresponding increase in the angle ϕ' in the t plane is $k\pi$. Consequently the two arcs in the t plane which correspond to the major and minor arcs of the circle C in the z plane intersect at B' and A' at an angle $(2-k)\pi$ or χ. The circle C has thus become a figure C' bounded by two circular arcs.

Further, if we apply the same transformation to a circle C_1 in the z plane which touches C at B and surrounds C, we shall get a contour C_1' in the

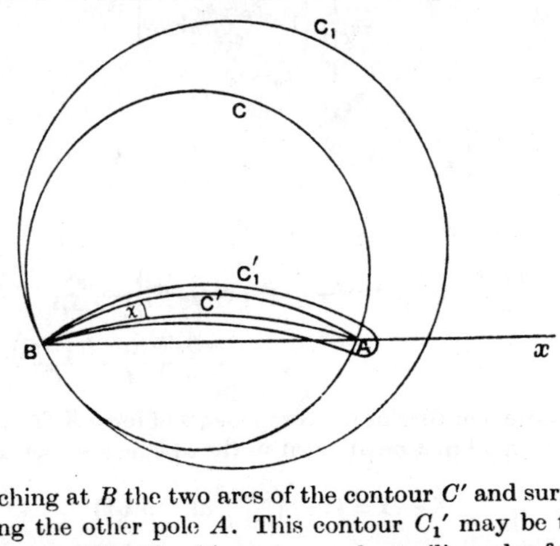

t plane touching at B the two arcs of the contour C' and surrounding C', i.e. enclosing the other pole A. This contour C_1' may be taken as the section of an aerofoil with a blunt nose and a trailing edge formed by the intersection of two surfaces which cut at a finite angle χ at B.

In this general case there is no simple geometrical construction such as we had in **6·63** for points on the contour but an analytical method has been given by Glauert*.

Now suppose that the state of flow represented by **6·61** (1′) exists outside the circle C_1, and that the circulation κ is such that B is a point of zero velocity, i.e. such that, in the notation of **6·61**,

$$\kappa = 4\pi a Q \sin(\alpha + \beta) \quad \text{.......................(3)},$$

when a is now the radius of the circle C_1, and if we take the origin at B, by reference to the figure of **6·61** we see that the centre of the circle C_1 is $z_0 = ae^{i\alpha}$.

Let $BA = 2l$, then, with the origin at B, (1) becomes

$$\frac{t-2l}{t} = \left(\frac{z-2l}{z}\right)^k \quad \text{..........................(4)}.$$

Expanding z in powers of t^{-1} we get

$$\left.\begin{aligned} z &= -(k-1)l + kt - \frac{(k^2-1)l^2}{3kt} - \frac{(k^2-1)l^3}{3kt^2}\dots \\ \text{or}\qquad z &= c_0 + c_1 t + \frac{c_2}{t} + \frac{c_3}{t^2} + \dots \end{aligned}\right\} \quad \text{........(5)},$$

so that

$$\frac{dz}{dt} = c_1 - \frac{c_2}{t^2} - \frac{2c_3}{t^3} - \dots \quad \text{.....................(6)}.$$

Now from **6·61** (2′)

$$\frac{dw}{dz} = Qe^{i\beta} - \frac{Qa^2e^{-i\beta}}{(z-z_0)^2} + \frac{i\kappa}{2\pi(z-z_0)} \quad \text{................(7)},$$

but

$$\begin{aligned} \frac{1}{z-z_0} &= \frac{1}{c_0 - z_0 + c_1 t + \frac{c_2}{t} + \dots} \\ &= \frac{1}{c_1 t}\left\{1 + \frac{c_0 - z_0}{c_1 t} + \dots\right\}^{-1} \\ &= \frac{1}{c_1 t} - \frac{c_0 - z_0}{c_1^2 t^2}\dots, \end{aligned}$$

therefore

$$\frac{dw}{dz} = Q\left(e^{i\beta} - \frac{a^2 e^{-i\beta}}{c_1^2 t^2}\right) + \frac{i\kappa}{2\pi}\left(\frac{1}{c_1 t} - \frac{c_0 - z_0}{c_1^2 t^2}\right) \quad \text{..............(8)}$$

as far as t^{-2}.

Then by multiplying and squaring (6) and (8) we get

$$\left(\frac{dw}{dt}\right)^2 = Q^2\left\{c_1^2 e^{2i\beta} - \frac{2}{t^2}(a^2 + c_1 c_2 e^{2i\beta})\right\} - \frac{\kappa^2}{4\pi^2 t^2} + \frac{i\kappa Q e^{i\beta}}{\pi}\left\{\frac{c_1}{t} - \frac{(c_0 - z_0)}{t^2}\right\} \quad \text{.....(9)}.$$

By the theorem of Blasius the components of force **X, Y** on the contour C_1' due to the fluid motion about it in the t plane are then given by

$$\mathbf{X} - i\mathbf{Y} = \tfrac{1}{2} i\rho \int \left(\frac{dw}{dt}\right)^2 dt \quad (5\cdot61)$$

* H. Glauert, *Reports and Memoranda of the Aeronautical Research Committee*, 911, 1924.

integrated round any surrounding contour. By the theory of residues we get

$$X - iY = \tfrac{1}{2} i\rho \,.\, 2\pi i \,.\, \frac{i\kappa Q e^{i\beta} c_1}{\pi},$$

where from (5) $$c_1 = k.$$

Therefore $$X - iY = -i\rho\kappa Q k (\cos\beta + i\sin\beta)$$

or $$X = \kappa\rho Q k \sin\beta, \quad Y = \kappa\rho Q k \cos\beta \quad \text{..............(10)}.$$

Now Q is the velocity at infinity in the z plane, and since $\frac{dw}{dt} = \frac{dw}{dz}\frac{dz}{dt}$, and from (6) at infinity $\frac{dz}{dt} = c_1 = k$, therefore the velocity at infinity in the t plane is $Q' = kQ$, so that the force components on the aerofoil C_1' in the stream Q' are simply

$$X = \kappa\rho Q' \sin\beta, \quad Y = \kappa\rho Q' \cos\beta \quad \text{..................(11)},$$

i.e. a force $$R = \kappa\rho Q' \quad \text{..............................(12)},$$

at right angles to the stream, as might have been predicted from the theorem of Kutta and Joukowski (5·7).

We remark that since the circulation round a curve is

$$-\int \frac{\partial\phi}{\partial s} ds \quad \text{or} \quad -\int d\phi,$$

therefore as stated in **6·62** the circulations round corresponding contours in the t plane and z plane are the same.

Hence if the circulation be such that the velocity in the t plane is finite at the trailing edge B, (3) is equivalent to

$$\kappa = \frac{4\pi a Q'}{k} \sin(\alpha + \beta)$$

and $$R = \frac{4\pi\rho a Q'^2}{k} \sin(\alpha + \beta) \quad \text{......................(13)}.$$

Again the moment about the origin B of the pressures on the contour C_1' is

$$N = \text{real part of } -\tfrac{1}{2}\rho \int \left(\frac{dw}{dt}\right)^2 t\,dt \quad (5·61)$$

$$= \text{real part of } \pi\rho i \left\{ 2Q^2 (a^2 + c_1 c_2 e^{2i\beta}) + \frac{\kappa^2}{4\pi^2} + \frac{i\kappa Q e^{i\beta}}{\pi} (c_0 - z_0) \right\},$$

where $c_0 = -(k-1) l$, $c_1 = k$, $c_2 = -(k^2 - 1) l^2/3k$ and $z_0 = ae^{i\alpha}$.

Whence we get

$$N = \pi\rho Q \left\{ \tfrac{2}{3} Q (k^2 - 1) l^2 \sin 2\beta + \frac{\kappa}{\pi} (k-1) l \cos\beta + \frac{\kappa}{\pi} a \cos(\alpha + \beta) \right\},$$

or, putting $Q = Q'/k$ and using (3),

$$N = \frac{2\pi\rho Q'^2}{k^2} \left\{ \tfrac{1}{3} (k^2 - 1) l^2 \sin 2\beta + 2(k-1) al \sin(\alpha + \beta) \cos\beta + a^2 \sin 2(\alpha + \beta) \right\} \quad \text{......(14)}.$$

The discussion of this general case is based upon two papers by W. Müller*. It is evident that by varying k, a and α a great variety of contours can be

* *Zeits. f. angew. Math. u. Mech.* 1923, 1924.

obtained; in particular $k=2$ leads to Joukowski aerofoils with cusps as the trailing edges.

It must be observed that the foregoing results have been obtained from Joukowski's hypothesis **6·62** of smooth flow at the trailing edge, and that in the case of real fluids we should need to consider the possible effect of frictional drag on the aerofoil and of the formation of vortices streaming from the edge (**3·72**).

EXAMPLES

1. The irrotational motion in two dimensions of a fluid bounded by the lines $y=0$, $y=b$ is due to a doublet of strength μ at the origin, the axis of the doublet being in the positive direction of the axis of x. Prove that the motion is given by

$$w=\frac{\pi\mu}{2b}\coth\frac{\pi}{2b}z.$$

Sketch the stream lines, and shew that those points where the fluid is moving parallel to the axis of y lie on the curve

$$\cosh(\pi x/b)=\sec(\pi y/b).$$ (Trinity Coll. 1904.)

2. Use the transformation $z'=e^{\frac{\pi z}{a}}$ to find the stream lines of the motion in two dimensions due to a source midway between two infinite parallel boundaries. [Assume the liquid drawn off equally by sinks at the ends of the region.] If the pressure tends to zero at the ends of the streams, prove that the planes are pressed apart with a force which varies inversely as their distance from each other. (M.T. II. 1911.)

3. A source is placed midway between two planes whose distance from one another is $2a$. Find the equation of the stream lines when the motion is in two dimensions; and shew that those particles which at an infinite distance are distant $\frac{1}{2}a$ from one of the boundaries, issued from the source in a direction making an angle $\pi/4$ with it.

4. Fluid motion is taking place in the part of the plane bounded by the real axis and the lines $x=+a$ and $x=-a$, which is due to a source at one corner and a sink at the other corner of the strip, each of strength m; shew that the motion is given by

$$\tanh\frac{w}{4m}=\tan\frac{\pi z}{4a},$$

and that the equation of the stream line which leaves the source at the angle $\pi/4$ to the sides is

$$\cos\frac{\pi x}{2a}=\sinh\frac{\pi y}{2a}.$$ (Trinity Coll. 1907.)

5. Prove that by proper adjustment of the constants $(\alpha, \beta, \gamma, \delta)$ the assumption

$$z=\alpha w+\beta e^{\gamma w}+\delta, \quad (z=x+iy,\ w=\phi+i\psi),$$

may be made to give the solution for the two-dimensional motion of a liquid in a straight pipe of breadth b, and sides $y=\pm\frac{1}{2}b$, extending from $x=-\infty$ to $x=0$, the velocity in the pipe at $x=-\infty$ being V, and the pipe opening into an otherwise unbounded liquid at rest at infinity. Find the values of these constants, assuming that at the point $(0, \frac{1}{2}b)$ the value of ϕ is ϕ_0.

(Trinity Coll. 1903.)

6. Prove in any manner that the velocity potential and stream function of the two-dimensional motion between the walls $y=0$, $y=\pi$, due to a source of strength m at (x_1, y_1) and an equal sink at (x_0, y_0), are given by

$$\phi+i\psi=\frac{m}{2\pi}\log\left[\frac{\{\operatorname{Exp}(x+iy)-\operatorname{Exp}(x_0+iy_0)\}\{\operatorname{Exp}(x+iy)-\operatorname{Exp}(x_0-iy_0)\}}{\{\operatorname{Exp}(x+iy)-\operatorname{Exp}(x_1+iy_1)\}\{\operatorname{Exp}(x+iy)-\operatorname{Exp}(x_1-iy_1)\}}\right].$$

(St John's Coll.)

7. Determine the nature of the fluid motion in the space bounded by

$$y=0, \quad \pi(x^2+y^2)-2y=0,$$

which is given by $\phi+i\psi=\coth(x+iy)^{-1}$. (M.T. 1894.)

8. In the case of uniplanar efflux from a large vessel with two plane sides at right angles and an aperture in the corner equally inclined to the two sides, shew that the coefficient of contraction is

$$\frac{\pi}{\pi+2\sqrt{2}-2\log_e(1+\sqrt{2})},$$

or ·747 (M.T. 1919.)

9. A rectangle open at infinity in the x direction has solid boundaries along $x=0$, $y=0$ and $y=a$. Fluid of amount $2\pi m$ flows into and out of the rectangle at the corners $x=0$, $y=0$ and $x=0$, $y=a$ respectively. Prove that the motion of the fluid is given by

$$w=4m\log\tanh(\pi z/2a).$$

Also shew that half the stream lies between $x=0$ and the stream line which cuts $y=\frac{1}{2}a$ at the point

$$x=\frac{a}{\pi}\log(1+\sqrt{2}).$$ (M.T. 1928.)

10. Prove that for liquid circulating irrotationally under no external forces in the part of the plane between two non-intersecting circles, the pressure on either of the circles is $\pi\rho\kappa^2/c$, where $2c$ is the distance between the limiting points of the circles, and $2\pi\kappa$ the cyclic constant of the motion. (Trinity Coll. 1898.)

11. Shew that the transformations

$$z=\frac{a}{\pi}\{\sqrt{t^2-1}-\sec^{-1}t\}; \quad t=e^{-\frac{\pi w}{aV}},$$

where $z=x+iy$, $w=\phi+i\psi$, give the velocity potential ϕ and the stream function ψ for the flow of a straight river of breadth a running with velocity V at right angles to the straight shore of an otherwise unlimited sheet of water, into which it flows; the motion being treated as two-dimensional. Shew that the real axis in the t-plane corresponds to the whole boundary of the liquid. (Univ. of London, 1910.)

12. What problem is solved by the transformation

$$\frac{d(x+iy)}{dt}=\frac{1}{t-a}\left(\frac{\sqrt{t}+1}{\sqrt{t}-1}\right)^{\frac{1}{2}},$$

$$\phi+i\psi=\log(t-a),$$

where x and y are the Cartesian coordinates of a point and ϕ and ψ the potential and current function respectively? (M.T. 1891.)

13. The sides of a vessel are two planes which extend to infinity in one direction. The straight lines in the section, made by a plane perpendicular to the sides, are inclined at an angle π/n; and they are symmetrically situated with respect to the line joining those extremities that lie in the finite part of the plane of section. Fluid escapes from the orifice, the motion being parallel to the plane of section. Shew that the coefficient of contraction is

$$1\Big/\left(1+\frac{2}{\pi}\int_0^{\frac{\pi}{2}} \sin\frac{\theta}{n}\cot\theta\, d\theta\right).$$

In the case where $n=2$, shew that the coordinates of any point in the free stream line may be expressed as

$$x=2\tanh^{-1}(1+e^{-\frac{1}{2}s})^{\frac{1}{2}}+2\tanh^{-1}(1-e^{-\frac{1}{2}s})^{\frac{1}{2}}-2\{(1+e^{-\frac{1}{2}s})^{\frac{1}{2}}+(1-e^{-\frac{1}{2}s})^{\frac{1}{2}}\},$$

$$y=\pi+2\{(1+e^{-\frac{1}{2}s})^{\frac{1}{2}}-(1-e^{-\frac{1}{2}s})^{\frac{1}{2}}\}-2\tanh^{-1}(1+e^{-\frac{1}{2}s})^{\frac{1}{2}} +2\tanh^{-1}(1-e^{-\frac{1}{2}s})^{\frac{1}{2}},$$

where the middle stream line is the axis of x, the distance along the free stream line from the edge of the nozzle is s, and the scale of measurement is so chosen that the final breadth of the stream is 2π. (M.T. II. 1895.)

14. Liquid moving in the plane (x, y) escapes from an opening between two fixed boundaries given by $y=0$, $x<0$, and $y=h$, $x>b$, the part of the plane for which y is greater than its value on the fixed boundaries being completely filled with liquid which is at rest at infinite distances. Find the equations of the free stream lines, and prove that the ultimate direction of the jet makes with the axis of x an angle α given by the equation

$$\frac{b}{h}=\tfrac{1}{2}\tan\alpha+\frac{1}{\pi}\sec\alpha+\frac{1}{\pi}\log(\tan\tfrac{1}{2}\alpha). \qquad \text{(M.T. II. 1897.)}$$

15. The fixed boundaries of a liquid moving in two dimensions are given by $y=0$ from $x=-\infty$ to $x=0$ and from $x=a$ to $x=\infty$, together with $y=b$ from $x=-\infty$ to $x=\infty$; prove that if c denote the ultimate breadth of the jet escaping through the opening in $y=0$ from $x=0$ to $x=a$, c is given by the relation

$$a=c+\frac{c}{\pi}\left(\frac{2b}{c}+\frac{c}{2b}\right)\log\frac{2b+c}{2b-c};$$

and shew that if $a=b$ the ratio of contraction is approximately 4/7. (M.T. II. 1900.)

16. Discuss the case of a single source on one side of an obstructing line of finite length, when the perpendicular from the source to the line bisects the line, and prove that when the plane of motion bounded by the obstructing line and the free stream lines is conformally represented in the portion of the plane of an auxiliary variable t which is above the real axis, the functions w and Ω are given by equations of the forms

$$\frac{dw}{dt}=\frac{\sin t}{t^2+\beta^2}, \quad \frac{d\Omega}{dt}=\left\{\frac{2t\sqrt{(1+\beta^2)}}{t^2+\beta^2}-\frac{1}{t}\right\}\frac{1}{\sqrt{(1-t^2)}}.$$

Also shew how to obtain equations connecting the length of the obstructing line, the distance of the source from it, the strength of the source, and the velocity along the free stream lines. (M.T. II. 1901.)

17. Prove that the formula

$$\frac{dz}{dw}=A\frac{1-au+\sqrt{(1-a^2)}\sqrt{(1-u^2)}}{u-a}\cdot\frac{1+au+\sqrt{(1-a^2)}\sqrt{(1-u^2)}}{u+a},$$

where $u=e^w$, represents (in two dimensions) the efflux of liquid by a Borda's mouthpiece (inward pointing tube) from the base of a cylindrical vessel, the vessel and the tube being coaxial, and the aperture of the tube at a distance from the base.

Prove that the coefficient of contraction is equal to

$$n-\sqrt{\{n(n-1)\}},$$

where n is the ratio of the breadth of the vessel to that of the tube.

Verify this result from first principles. (M.T. II. 1902.)

18. Shew that, with the usual notation, the substitution

$$w=A\log z_3+B\log(z_3+\lambda),$$

where A, B, λ are appropriate constants and

$$z_3=\{\cosh(\log\zeta)\}^2,$$

gives the flow from a rectangular vessel with two infinite parallel sides and an aperture midway in the third side.

Deduce from this the solution for the two cases (1) flow past a fixed obstacle set perpendicular to an infinite stream, (2) flow through an aperture in an infinite plane wall. (M.T. II. 1906.)

19. Exemplify the treatment of problems in discontinuous two-dimensional liquid motion by investigating the case of a stream whose breadth and velocity at infinity are a and V respectively, whose course is disturbed by a symmetrically placed transverse straight barrier of length b. Shew that the force necessary to keep the barrier in position is

$$\rho a V^2(1-\sin\alpha),$$

where $$b/a=1-\sin\alpha+\frac{1}{\pi}\cos\alpha\log(\cot^2\tfrac{1}{2}\alpha).$$ (M.T. II. 1905.)

20. If a stream of infinite width is obstructed by a lamina with an elevated rim placed transversely, shew that the mean pressure on the lamina is

$$\frac{\pi\rho V^2}{4+\pi}\left\{1+\frac{8+4\pi+2\pi^2}{(4+\pi)^2}\sqrt{(2\epsilon)}\right\},$$

where V is the velocity on the free stream lines, and ϵ is the ratio of the height of the rim to the breadth of the lamina, and higher powers of ϵ are neglected. (Love.)

21. Water escapes, under pressure, from the plane wall of a vessel, by means of a large number of parallel, equal, and equidistant slits. The breadth of each slit is a, and the distance between the centres of consecutive slits is b. Prove that the final breadth c of each issuing jet is given by the equation

$$\frac{a}{c}=1+\frac{2}{\pi}\left(\frac{b}{c}-\frac{c}{b}\right)\tan^{-1}\frac{c}{b}.$$

Calculate the mean pressure on the wall, having given the velocity v of the issuing jets. (M.T. II. 1907.)

22. Shew that the relation $t=z+\frac{1}{z}$ can be used to transform the circumference of a given circle into (i) a circular arc of given angle; (ii) a circular arc whose chord is of any prescribed length not exceeding twice the diameter of the given circle.

23. Explain the derivation of a Joukowski aerofoil by the transformation

$$z'=z+\sum_{r=1}^{n}\frac{a_r}{z^r}$$

applied to a circle of centre z_0 and radius a.

Obtain the lift formula

$$L=4\pi\rho a U^2 \sin(\alpha+\beta)$$

and shew that the moment about the point $z'=z_0$ is

$$M=2\pi\rho b^2 U^2 \sin 2(\alpha+\gamma),$$

where α is the angle of attack and β, b, γ constants of the transformation. (M.T. 1931.)

24. Shew that the relation

$$\frac{t+nc}{t-nc}=\left(\frac{z+c}{z-c}\right)^n$$

may be used for obtaining aerofoil sections with trailing edges of finite angle.

25. Shew that the relation

$$t=z\sin\tfrac{1}{2}\alpha\,\frac{z+a\operatorname{cosec}\frac{1}{2}\alpha}{z+a\sin\frac{1}{2}\alpha},\quad 0<\alpha<\tfrac{1}{2}\pi,$$

where a is real, will transform the circumference of the circle $|z|=a$ into an arc of the same circle subtending an angle 2α at the centre.

Shew that, in a two-dimensional flow about such an arc, the lift produced by a steady current V parallel to the chord of the arc is $2\pi\rho h V^2$, where h is the height of the arc.

26. A flat plate of infinite length and width l is placed in a current of incompressible fluid with its plane at an angle α to the undisturbed stream lines, and its edges perpendicular to them. Determine the resulting flow on the circulation theory, assuming the velocity at the trailing edge is finite. By considering the pressures and velocities over a large cylinder whose axis is the median line of the plate, shew that the forces on the plate are equivalent to a force $\pi\rho U^2 l \sin\alpha$ per unit length perpendicular to the current, acting at a distance $\frac{1}{4}l$ from the leading edge. (M.T. 1926.)

27. A circle $|z|=a$ is transformed into a thin aerofoil section of chord approximately equal to $4a$ by the equations

$$\zeta=z\left(1+\Sigma A_n\frac{a^n}{z^n}\right),$$

$$\zeta'=\zeta+\frac{a^2}{\zeta}.$$

Prove that the lift and moment coefficients are

$$k_L=\pi(\alpha+\beta),$$

$$k_M=\tfrac{1}{4}k_L+\tfrac{1}{4}\pi\beta+\tfrac{1}{8}\pi D_2-\tfrac{1}{4}\pi\alpha C_2,$$

where α is the angle of attack, supposed small, and β, D_2 and C_2 are constants of the aerofoil. (M.T. 1928.)

CHAPTER VII

IRROTATIONAL MOTION IN THREE DIMENSIONS

7·1. It is our purpose now to consider certain special forms of solution of the equation

$$\frac{\partial^2\phi}{\partial x^2}+\frac{\partial^2\phi}{\partial y^2}+\frac{\partial^2\phi}{\partial z^2}=0.$$

We do not propose to enter into a general discussion of spherical and other harmonics such as may be found in many text-books on pure and applied mathematics, and we shall only have occasion to assume an elementary knowledge of these functions.

7·11. Motion of a Sphere through a Liquid at rest at infinity. If the centre of the sphere be moving along a straight line with velocity V, the motion of the liquid will be symmetrical about this line, and Laplace's equation takes the form*

$$\frac{\partial}{\partial r}\left(r^2\frac{\partial\phi}{\partial r}\right)+\frac{\partial}{\sin\theta\,\partial\theta}\left(\sin\theta\frac{\partial\phi}{\partial\theta}\right)=0 \quad \text{...........(1)}.$$

A solution of this equation is known to be

$$\phi=\Sigma\left(A_n r^n+\frac{B_n}{r^{n+1}}\right)P_n \quad \text{.................(2)},$$

where P_n is Legendre's coefficient of order n.

In our special problem if we suppose the centre of the sphere to be passing through the origin we have to satisfy boundary conditions

$$-\frac{\partial\phi}{\partial r}=\text{normal velocity}=V\cos\theta \quad \text{...........(3)},$$

* This may be obtained directly by considering the flow of liquid across the faces of the polar element of volume $r^2\sin\theta\,dr\,d\theta\,d\omega$. The gain of liquid in the element due to the flow in the direction of r is

$$\frac{\partial}{\partial r}\left(\frac{\partial\phi}{\partial r}r^2\sin\theta\,d\theta\,d\omega\right)dr,$$

and the gain due to the flow in the direction perpendicular to r is

$$\frac{\partial}{r\,\partial\theta}\left(\frac{\partial\phi}{r\,\partial\theta}r\sin\theta\,dr\,d\omega\right)r\,d\theta.$$

But the total gain in the element is zero, therefore

$$\frac{\partial}{\partial r}\left(r^2\frac{\partial\phi}{\partial r}\right)+\frac{\partial}{\sin\theta\,\partial\theta}\left(\sin\theta\frac{\partial\phi}{\partial\theta}\right)=0.$$

when r is equal to a, the radius of the sphere, and

$$-\frac{\partial\phi}{\partial r}=0, \text{ at infinity} \quad\text{..................(4).}$$

From (4) it is clear that the solution for ϕ cannot contain positive powers of r, and (3) suggests that we shall take*

$$\phi=\frac{B}{r^2}\cos\theta \quad\text{......................(5)}$$

as the particular form of (2) to suit our conditions, since $P_1=\cos\theta$.

Substituting from (5) in (3) we find that

$$\frac{2B}{a^3}\cos\theta=V\cos\theta,$$

for all values of θ, so that $B=\frac{1}{2}Va^3$.

Hence the velocity potential is given by

$$\phi=\tfrac{1}{2}Va^3r^{-2}\cos\theta \quad\text{....................(6).}$$

To find the lines of flow, at the instant the centre of the sphere is passing through the origin, we have

$$\frac{dr}{\partial\phi/\partial r}=\frac{r\,d\theta}{\partial\phi/r\,\partial\theta},$$

or

$$\frac{dr}{\cos\theta}=\frac{r\,d\theta}{\frac{1}{2}\sin\theta},$$

so that the equation of the lines of flow is

$$r=C\sin^2\theta.$$

7·12. Liquid streaming past a fixed Sphere. If we suppose the sphere to be fixed and the liquid to have a general velocity V, we can obtain the velocity potential from the last case considered by superposing a velocity $-V$ on the sphere and

* The student who is unacquainted with the properties of Legendre's coefficients may proceed thus. The condition (3) suggests that we should try to find a solution of (1) of the form $\phi=f(r)\cos\theta$. We get on substitution

$$\frac{\partial}{\partial r}\left(r^2\frac{\partial f}{\partial r}\right)-2f=0,$$

or

$$r^2\frac{\partial^2 f}{\partial r^2}+2r\frac{\partial f}{\partial r}-2f=0,$$

of which the solution is

$$f=Ar+\frac{B}{r^2}.$$

On account of condition (4) we reject the solution Ar and proceed as above with

$$\phi=Br^{-2}\cos\theta.$$

the liquid. This adds a term Vx, or $Vr\cos\theta$, to the velocity potential, so that now

$$\phi = Vr\cos\theta + \tfrac{1}{2}Va^3r^{-2}\cos\theta.$$

For the stream lines we have

$$\frac{dr}{\left(1-\dfrac{a^3}{r^3}\right)\cos\theta} = \frac{r\,d\theta}{-\left(1+\dfrac{a^3}{2r^3}\right)\sin\theta},$$

or $$-2\cot\theta\, d\theta = \frac{2r^3+a^3}{r^3-a^3}\cdot\frac{dr}{r} = \left(\frac{3r^2}{r^3-a^3}-\frac{1}{r}\right)dr,$$

therefore $$\sin^2\theta = \frac{Cr}{r^3-a^3}.$$

This equation gives, for either this problem or the last, the lines of flow relative to the sphere.

7·13. Equations of Motion of a Sphere. Reverting to the case of a sphere moving in a liquid at rest at infinity, we have to calculate the forces acting on the sphere owing to the presence of the liquid. If the extraneous forces have a potential Ω and act on the sphere and the liquid alike, their resultant effect is, from Hydrostatical considerations, a force equal to the difference between the forces exerted on the sphere and the liquid displaced; i.e. if σ, ρ are the densities of the sphere and the liquid, the resultant extraneous force is $(\sigma-\rho)/\sigma$ times what it would be if the liquid were not present. Omitting the extraneous forces, the pressure is to be found from the equation

$$\frac{p}{\rho} = F(t) + \frac{\partial\phi}{\partial t} - \tfrac{1}{2}q^2 \quad\text{..................(1).}$$

Let the coordinates of the centre of the moving sphere referred to fixed axes be x_0, y_0, z_0 and let $U, V, W \equiv \dot{x}_0, \dot{y}_0, \dot{z}_0$ be the velocities of the centre. Then from **7·11** (6) the velocity potential at a fixed point of space (x, y, z) is

$$\phi = \frac{1}{2}\frac{a^3}{r^2}\left\{U\frac{x-x_0}{r} + V\frac{y-y_0}{r} + W\frac{z-z_0}{r}\right\} \quad\text{.........(2),}$$

where $$r^2 = (x-x_0)^2 + (y-y_0)^2 + (z-z_0)^2$$

and $$r\dot{r} = -(x-x_0)\dot{x}_0 - (y-y_0)\dot{y}_0 - (z-z_0)\dot{z}_0$$
$$= -U(x-x_0) - V(y-y_0) - W(z-z_0).$$

Also $\partial\phi/\partial t$ being the rate of increase of ϕ at the fixed point (x, y, z) we have

$$\frac{\partial\phi}{\partial t}=\frac{a^3}{2r^3}\{\dot{U}(x-x_0)+\ +\}-\frac{a^3}{2r^3}\{U^2+\ +\}+\frac{3a^3}{2r^5}\{U(x-x_0)+\ +\}^2 \quad \text{......(3)}.$$

Again $$\frac{\partial\phi}{\partial x}=\frac{a^3U}{2r^3}-\frac{3a^3}{2r^5}(x-x_0)\{U(x-x_0)+\ +\},$$

so that $$q^2=\left(\frac{\partial\phi}{\partial x}\right)^2+\left(\frac{\partial\phi}{\partial y}\right)^2+\left(\frac{\partial\phi}{\partial z}\right)^2$$

$$=\frac{a^6}{4r^6}(U^2+\ +)+\frac{3a^6}{4r^8}\{U(x-x_0)+\ +\}^2 \quad \text{.........(4)}.$$

Hence on the sphere $r=a$ we have

$$\frac{p}{\rho}=F(t)+\tfrac{1}{2}\{\dot{U}(x-x_0)+\ +\}-\tfrac{5}{8}(U^2+\ +)+\frac{9}{8a^2}\{U(x-x_0)+\ +\}^2 \quad \text{......(5)}.$$

Then the components of the resultant thrust on the sphere due to the motion are given by

$$\mathbf{X}=-\iint lp\,dS=-\iiint\frac{\partial p}{\partial x}\,dx\,dy\,dz \quad \text{(Green's Theorem)}$$

and similar expressions.

Whence we get $\mathbf{X}=-\tfrac{2}{3}\pi\rho a^3\dot{U}$, etc.,

or $$\mathbf{X},\mathbf{Y},\mathbf{Z}=-\tfrac{1}{2}M'(\dot{U},\dot{V},\dot{W}) \quad \text{...............(6)},$$

where M' is the mass of liquid displaced.

It follows that if X', Y', Z' are the components of the extraneous force on the sphere when no liquid is present, and M denotes the mass of the sphere we have equations of motion of the form

$$M\dot{U}=-\tfrac{1}{2}M'\dot{U}+\frac{\sigma-\rho}{\sigma}X'$$

or $$M\dot{U}=\frac{M}{M+\frac{1}{2}M'}\cdot\frac{\sigma-\rho}{\sigma}X',$$

i.e. $$M\dot{U}=\frac{\sigma-\rho}{\sigma+\frac{1}{2}\rho}X' \quad \text{.............................(7)}.$$

Hence the whole effect of the presence of the liquid is to reduce the extraneous force in the ratio $\sigma-\rho:\sigma+\frac{1}{2}\rho$.

Result (6) implies that if the sphere were to move with uniform velocity, the resultant pressure set up by the motion or the resistance to motion would be zero. This is contrary to experience and arises from our hypothesis that the motion is irrotational and the motion of a perfect fluid. In a real fluid the flow does not close up again at the rear of the sphere but separates from the surface leaving an area of suction behind, and in the rear of the sphere there is an eddying wake. In point of fact there is no reason why the result here obtained should resemble observed phenomena because we have assumed that the liquid slips over the surface of the sphere and a real fluid cannot do this.

7·14. We may also obtain result (6) of **7·13** from the principle of energy. From **4·71** the kinetic energy of the liquid is given by

$$T = -\tfrac{1}{2}\rho \iint \phi \frac{\partial \phi}{\partial n}\, dS,$$

integrated over the sphere. So confining ourselves to rectilinear motion with velocity V

$$T = \tfrac{1}{2}\rho \int_0^{\pi} \tfrac{1}{2}aV\cos\theta \,.\, V\cos\theta \,.\, 2\pi a^2 \sin\theta\, d\theta$$

$$= \tfrac{1}{3}\pi\rho a^3 V^2 = \tfrac{1}{4}M'V^2.$$

Therefore the effect of the liquid is to increase the inertia of the sphere by half the mass of liquid displaced. And if X denote the force parallel to the axis of x

$$\frac{d}{dt}(\tfrac{1}{2}MV^2 + \tfrac{1}{4}M'V^2) = \text{rate at which work is being done}$$

$$= XV;$$

so that

$$(M + \tfrac{1}{2}M')\,\dot{V} = X,$$

or

$$M\dot{V} = X - \tfrac{1}{2}M'\frac{dV}{dt},$$

so that the pressure of the liquid, apart from any extraneous force acting on it, is equivalent to a force $\tfrac{1}{2}M'\dot{V}$ opposing the motion.

7·15. Sphere projected in a Liquid under gravity. As an example let us suppose the extraneous force to be gravity. Since there is no horizontal component of extraneous force the horizontal velocity is constant; and as in **7·13** the vertical motion is the same as if the sphere moved in vacuo and gravity were reduced in the ratio $\sigma - \rho : \sigma + \tfrac{1}{2}\rho$. Consequently the centre of the sphere describes a parabola of latus rectum

$$\frac{2\sigma + \rho}{\sigma - \rho}\frac{U^2}{g},$$

where U denotes the horizontal velocity.

7·2. Concentric Spheres. Initial motion. Let there be a sphere of radius a surrounded by a concentric sphere of radius b, the intervening space being filled with liquid. The methods that we have already used will enable us to determine the velocity potential of the *initial* motion when, say, a given velocity is imparted to either of the spheres, or a given impulse is applied to one of the spheres while the other is held fixed, or is free to move.

Suppose the inner sphere receives a velocity V, the outer being fixed. The boundary conditions are

$$-\frac{\partial\phi}{\partial r}=V\cos\theta \quad \text{when } r=a,$$

and

$$-\frac{\partial\phi}{\partial r}=0 \quad \text{when } r=b.$$

Assume that

$$\phi=\left(Ar+\frac{B}{r^2}\right)\cos\theta,$$

then we get

$$-A+\frac{2B}{a^3}=V \quad \text{and} \quad -A+\frac{2B}{b^3}=0.$$

Hence

$$\phi=\frac{Va^3}{b^3-a^3}\left(r+\frac{b^3}{2r^2}\right)\cos\theta.$$

If M be the mass of the sphere and I the impulse necessary to produce the velocity V; we have

$$MV=I-\iint p\cos\theta\, dS,$$

where $p=\rho\phi$ denotes the impulsive pressure of the liquid. Therefore

$$MV=I-\frac{\rho Va^3}{b^3-a^3}\left(a+\frac{b^3}{2a^2}\right)\int_0^\pi \cos^2\theta\,.\,2\pi a^2\sin\theta\, d\theta$$

$$=I-\frac{2\pi\rho a^3(2a^3+b^3)V}{3(b^3-a^3)}.$$

If now the radius b of the outer sphere is increased indefinitely, we get for the limiting value of the impulse necessary to impart a velocity V to the inner sphere

$$I=MV+\tfrac{2}{3}\pi\rho a^3 V,$$

or

$$I=(M+\tfrac{1}{2}M')\,V.$$

Comparing this result with **7·13** we see that the impulse necessary to produce the velocity V is the same whether we regard the liquid as extending to infinity and at rest there, or whether we suppose it to be enclosed by a fixed spherical envelope of infinite radius.

If we calculate the impulsive pressure on the outer sphere, in like manner, we get

$$2\pi\rho a^3b^3V/(b^3-a^3),$$

which tends to the finite limit $2\pi\rho a^3V$, as b tends to infinity.

It can also be shewn by simple calculation that the total momentum of the liquid in the direction of the impulse is $-\frac{4}{3}\pi\rho a^3V$, whatever be the radius of the outer sphere; and thus we have a verification of the dynamical principle that the impulse I is equal, in every case, to the total momentum in the same direction of the solid and the liquid, together with the impulsive pressure on the surrounding sphere.

7·3. Stokes's Stream Function. Motion symmetrical about an axis, the Lines of Motion being in Planes passing through the axis. Let the axis of symmetry be the axis of x and let $\varpi\ (=\sqrt{y^2+z^2})$ denote distance from the axis. Let u, v denote components of velocity in the directions of x and ϖ.

Then the *equation of continuity* may be got by equating to zero the flow out of the annular space obtained by revolving a small rectangle $d\varpi dx$ round the axis. The total flow out parallel to x is $\frac{\partial}{\partial x}(u 2\pi\varpi d\varpi)\,dx$; and parallel to ϖ, the total flow out is

$$\frac{\partial}{\partial \varpi}(v 2\pi\varpi dx)\,d\varpi,$$

so by equating the sum to zero we get for the equation of continuity

$$\frac{\partial}{\partial x}(u\varpi)+\frac{\partial}{\partial \varpi}(v\varpi)=0.$$

This is however the condition that

$$v\varpi dx - u\varpi d\varpi$$

may be an exact differential, and, if we denote this by $d\psi$, we get

$$u=-\frac{1}{\varpi}\frac{\partial\psi}{\partial\varpi}, \quad v=\frac{1}{\varpi}\frac{\partial\psi}{\partial x}.$$

This function ψ is called *Stokes's Stream Function**.

Since the stream lines are given by

$$dx/u = d\varpi/v,$$

or

$$\varpi(v dx - u d\varpi)=0,$$

that is by $d\psi=0$; it follows that the equation

$$\psi = \text{constant}$$

represents the stream lines.

A property of Stokes's stream function is that 2π times the difference of its values at two points in the same meridian plane is equal to the flow across the annular surface obtained by the revolution round the axis of a curve joining the points. For if ds be an element of the curve and θ its inclination to the axis, the flow outwards across the surface of revolution

$$\begin{aligned} &= \int (v\cos\theta - u\sin\theta)\,.\,2\pi\varpi ds \\ &= 2\pi\int\left(\frac{\partial\psi}{\partial x}dx+\frac{\partial\psi}{\partial\varpi}d\varpi\right) \\ &= 2\pi\int d\psi = 2\pi(\psi_2-\psi_1). \end{aligned}$$

* See Stokes's paper 'On the Steady Motion of Incompressible Fluids'. *Trans. Camb. Phil. Soc.* VII, p. 439, or *Math. and Phys. Papers* I ,p. 12.

We might also define the value of Stokes's stream function at any point P as $1/2\pi$ of the amount of flow across a surface got by revolving a curve AP round the axis, A being a fixed point in the meridian plane through P; for this makes

$$\psi = \frac{1}{2\pi}\int_A^P (v\cos\theta - u\sin\theta)\,.\,2\pi\varpi\,ds$$

$$= \int_A^P (v\varpi\,dx - u\varpi\,d\varpi).$$

And by varying the position of P, we get as before

$$u = -\frac{1}{\varpi}\frac{\partial\psi}{\partial\varpi} \quad \text{and} \quad v = \frac{1}{\varpi}\frac{\partial\psi}{\partial x}.$$

Also it is easily seen that the velocity from right to left in the sense indicated in **3·1** across any arc ds is $\partial\psi/\varpi\,\partial s$.

7·31. When the motion is **irrotational**, we have the condition

$$\frac{\partial v}{\partial x} - \frac{\partial u}{\partial \varpi} = 0, \quad (\mathbf{4{\cdot}25}\ (2)),$$

which leads to

$$\frac{\partial^2\psi}{\partial x^2} + \frac{\partial^2\psi}{\partial\varpi^2} - \frac{1}{\varpi}\frac{\partial\psi}{\partial\varpi} = 0 \quad \text{..................(1).}$$

Also, assuming that $u = -\partial\phi/\partial x$ and $v = -\partial\phi/\partial\varpi$, we get from the equation of continuity

$$\frac{\partial^2\phi}{\partial x^2} + \frac{\partial^2\phi}{\partial\varpi^2} + \frac{1}{\varpi}\frac{\partial\phi}{\partial\varpi} = 0 \quad \text{..................(2).}$$

Equations (1) and (2) shew that ϕ and ψ are not interchangeable in the way that applied to the velocity potential and stream function of two-dimensional irrotational motions.

The corresponding equations in polar coordinates (r, θ) are frequently more useful than equations (1) and (2). If we take q_r, q_θ to be the velocities in the directions of dr and $r\,d\theta$, then, since $\varpi = r\sin\theta$ and remembering that the velocity from right to left across ds is $\partial\psi/\omega\,\partial s$, we get

$$q_r = -\frac{1}{r\sin\theta}\frac{\partial\psi}{r\,\partial\theta},$$

and

$$q_\theta = \frac{1}{r\sin\theta}\frac{\partial\psi}{\partial r}.$$

But in irrotational motion

$$q_r = -\frac{\partial\phi}{\partial r} \quad \text{and} \quad q_\theta = -\frac{\partial\phi}{r\,\partial\theta},$$

therefore $$\frac{1}{r^2 \sin\theta}\frac{\partial\psi}{\partial\theta}=\frac{\partial\phi}{\partial r} \quad\text{and}\quad \frac{1}{\sin\theta}\frac{\partial\psi}{\partial r}=-\frac{\partial\phi}{\partial\theta} \quad \text{.........(3).}$$

Hence $$\frac{\partial}{\partial\theta}\left(\frac{1}{r^2\sin\theta}\frac{\partial\psi}{\partial\theta}\right)=\frac{\partial^2\phi}{\partial\theta\,\partial r}=-\frac{\partial}{\partial r}\left(\frac{1}{\sin\theta}\frac{\partial\psi}{\partial r}\right),$$

that is $$r^2\frac{\partial^2\psi}{\partial r^2}+\sin\theta\frac{\partial}{\partial\theta}\left(\frac{1}{\sin\theta}\frac{\partial\psi}{\partial\theta}\right)=0,$$

or, putting $\cos\theta=\mu$,
$$r^2\frac{\partial^2\psi}{\partial r^2}+(1-\mu^2)\frac{\partial^2\psi}{\partial\mu^2}=0 \quad \text{..................(4).}$$

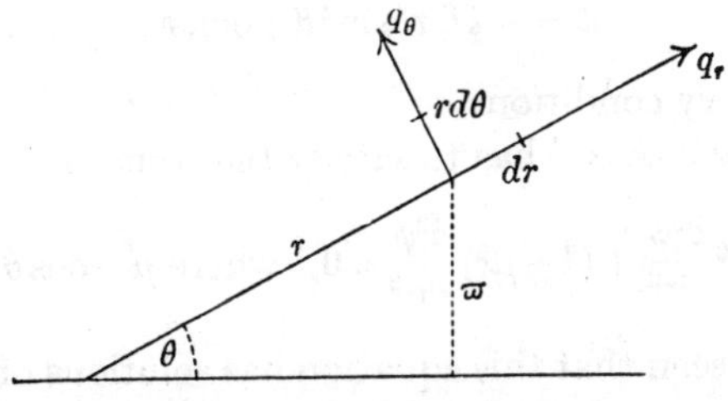

From the equation of continuity in polar coordinates, **1·5** (1), we get the equation for ϕ,

$$\frac{\partial}{\partial r}\left(r^2\frac{\partial\phi}{\partial r}\right)+\frac{\partial}{\partial\mu}\left\{(1-\mu^2)\frac{\partial\phi}{\partial\mu}\right\}=0 \quad \text{...........(5),}$$

remembering that in this case ϕ is a function of r and θ only.

The latter is of course a form of Laplace's equation and has solutions of the forms

$$r^n P_n(\mu) \quad\text{and}\quad r^{-n-1}P_n(\mu).$$

Again from (3) we have

$$\frac{\partial\psi}{\partial\mu}=-r^2\frac{\partial\phi}{\partial r}=-nr^{n+1}P_n \quad\text{or}\quad (n+1)r^{-n}P_n \quad \text{......(6),}$$

and

$$\frac{\partial\psi}{\partial r}=(1-\mu^2)\frac{\partial\phi}{\partial\mu}=(1-\mu^2)r^n\frac{\partial P_n}{\partial\mu} \quad\text{or}\quad (1-\mu^2)r^{-n-1}\frac{\partial P_n}{\partial\mu}\text{...(7).}$$

The last equation gives, on integration, as possible solutions for ψ,

$$\psi=\frac{(1-\mu^2)}{n+1}r^{n+1}\frac{\partial P_n}{\partial\mu} \quad\text{or}\quad -\frac{(1-\mu^2)}{n}\frac{1}{r^n}\frac{\partial P_n}{\partial\mu},$$

it being easy to verify that these forms also satisfy equation (4).

7·32. Applications. Solids of revolution moving along their axes in an infinite Mass of Liquid. If U is the velocity along the axis of x and ds an element of the meridian curve, the normal velocity at any point is

$$U\,\partial\varpi/\partial s \quad \text{or} \quad U\,\partial\,(r\sin\theta)/\partial s;$$

and the normal velocity of the liquid in contact with the surface is $-\partial\psi/\varpi\,\partial s$ or $-\partial\psi/r\sin\theta\,\partial s$. Therefore

$$d\psi = -Ur\sin\theta\,d\,(r\sin\theta),$$

or
$$\psi = -\tfrac{1}{2}Ur^2\sin^2\theta + \text{const.} \qquad \text{..............(1)},$$

is the boundary condition.

We also have that ψ has to satisfy the equation

$$r^2\frac{\partial^2\psi}{\partial r^2} + (1-\mu^2)\frac{\partial^2\psi}{\partial\mu^2} = 0, \quad \text{where } \mu = \cos\theta,$$

and we have seen that this equation has solutions of the types

$$\frac{1-\mu^2}{n+1}r^{n+1}\frac{\partial P_n}{\partial\mu} \quad \text{and} \quad \frac{1-\mu^2}{nr^n}\frac{\partial P_n}{\partial\mu}.$$

The simplest case is that of a sphere of radius a.

Taking $n=1$, we have a solution of the form

$$\psi = A\,(1-\mu^2)/r;$$

then at the boundary we must have

$$A\,(1-\mu^2)/a = -\tfrac{1}{2}Ua^2(1-\mu^2) + C$$

for all values of μ. This requires that

$$C = 0 \quad \text{and} \quad A = -\tfrac{1}{2}Ua^3.$$

Therefore
$$\psi = -\tfrac{1}{2}\frac{Ua^3\sin^2\theta}{r}.$$

But we know that

$$(1-\mu^2)\frac{\partial\phi}{\partial\mu} = \frac{\partial\psi}{\partial r} = \tfrac{1}{2}\frac{Ua^3}{r^2}\sin^2\theta.$$

Therefore
$$\frac{\partial\phi}{\partial\mu} = \tfrac{1}{2}\frac{Ua^3}{r^2},$$

and
$$\phi = \tfrac{1}{2}\frac{Ua^3}{r^2}\cos\theta, \text{ as in } \mathbf{7{\cdot}11}.$$

7·33. Values of Stokes's Stream Function in simple cases.

(1) A *simple source* on the axis of x.

Here, from **3·3** we have $\phi = m/r$; but

$$\frac{\partial\psi}{\partial\mu} = -r^2\frac{\partial\phi}{\partial r} = m.$$

Therefore $$\psi = m\mu = m\cos\theta \text{ or } mx/r.$$

(2) A *doublet* along the axis of x.

Here, from **3·31**, we have $\phi = M\cos\theta/r^2$; but

$$\frac{\partial\psi}{\partial r} = (1-\mu^2)\frac{\partial\phi}{\partial\mu} = \frac{(1-\mu^2)M}{r^2}.$$

Therefore $$\psi = -\frac{M\sin^2\theta}{r}.$$

(3) A *uniform line source* along the axis.

If m is the strength per unit length and the source extends from O to A, we have, at any point $P(\xi, \eta)$,

$$\begin{aligned}\psi &= \int_0^{OA} m\cos\theta\,dx = \int_0^{OA}\frac{m(\xi - x)\,dx}{\sqrt{\{(\xi-x)^2+\eta^2\}}}\\ &= m\{\sqrt{(\xi^2+\eta^2)} - \sqrt{(\xi - OA^2 + \eta^2)}\}\\ &= m(OP - AP).\end{aligned}$$

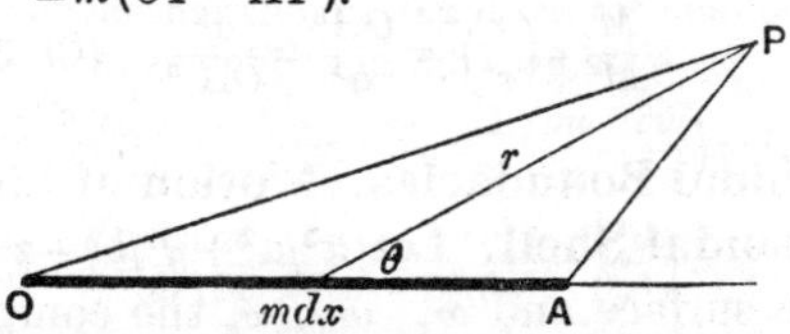

We might also obtain result (2) by differentiating result (1). Thus for a simple source $\psi = mx/r$, therefore for a doublet

$$\begin{aligned}\psi &= -\frac{\partial}{\partial x}\left(\frac{mx}{r}\right)dx\\ &= -m\,dx\left(\frac{1}{r} - \frac{x^2}{r^3}\right) = -\frac{M\sin^2\theta}{r}.\end{aligned}$$

And result (1) might be obtained by considering the flow across a circular area whose centre is on the axis and plane perpendicular to the axis. By definition, taken from right to left, the flow is $2\pi\psi$, and it is also m times the solid angle that the circle subtends at the source, so that having regard to sign

$$2\pi\psi = -2\pi m(1-\cos\theta),$$

or omitting a constant $$\psi = m\cos\theta.$$

7·34. We may now obtain a simple verification of the result of **3·42**, viz. that the image of a simple source m at a distance f $(>a)$ from the centre of a sphere of radius a is a source ma/f at the inverse point and a line sink of strength $-m/a$ per unit length extending from the centre to the inverse point. It is only necessary to shew that the stream function ψ for this arrangement of sources and sink has a constant value on the sphere; and using **7·33** we can easily prove that on the sphere $\psi = -m$.

7·35. A comparison of the stream functions or the velocity potentials due to the motion of a sphere with those produced by a doublet in an infinite mass of liquid, shews that a sphere of radius a moving with velocity U produces the same effect as a doublet of strength $\frac{1}{2}Ua^3$ at its centre. We can now deduce the *stream lines for a sphere in the presence of a doublet.* For if we take two doublets of strengths M and M' at points A, A' on the axis of x with their axes directed towards one another, we have

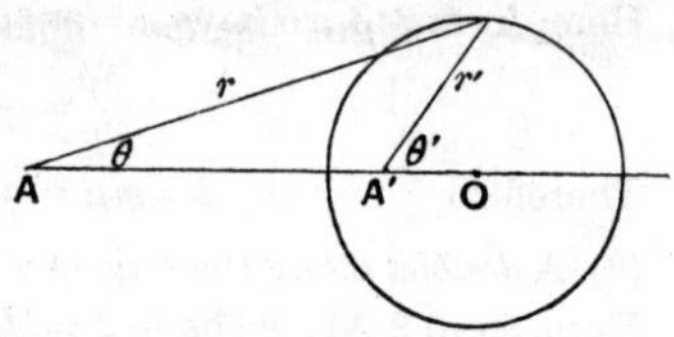

$$\psi = -\frac{M\sin^2\theta}{r} + \frac{M'\sin^2\theta'}{r'}.$$

Hence on the stream line $\psi = 0$

$$\frac{r^3}{r'^3} = \frac{M}{M'} \quad \text{or} \quad \frac{r}{r'} = \left(\frac{M}{M'}\right)^{\frac{1}{3}}.$$

This represents a sphere with regard to which A, A' are inverse points. This sphere may be taken as a solid boundary, and thus we get the stream lines due to a doublet in the presence of a solid sphere. The image is another doublet at the inverse point, such that if O is the centre and a the radius of the sphere

$$\frac{M}{M'} = \left(\frac{r}{r'}\right)^3 = \frac{OA^3}{a^3} = \frac{a^3}{OA'^3}. \quad \text{(Cf. 3·43.)}$$

7·4. Ellipsoidal Boundaries. Motion of Liquid inside a rotating ellipsoidal Shell.

Let $x^2/a^2 + y^2/b^2 + z^2/c^2 = 1$ be the equation of the surface and ω_x, ω_y, ω_z the components of the angular velocity, referred to axes fixed in space and coincident with the axes of the ellipsoid at the instant considered.

The component linear velocities of a point (x, y, z) of the shell are $z\omega_y - y\omega_z$, $x\omega_z - z\omega_x$, $y\omega_x - x\omega_y$; and the direction cosines of the normal are proportional to x/a^2, y/b^2, z/c^2. Hence if ϕ be the velocity potential of the liquid motion the boundary condition is

$$-\frac{x}{a^2}\frac{\partial\phi}{\partial x} - \frac{y}{b^2}\frac{\partial\phi}{\partial y} - \frac{z}{c^2}\frac{\partial\phi}{\partial z}$$

$$= \frac{x}{a^2}(z\omega_y - y\omega_z) + \frac{y}{b^2}(x\omega_z - z\omega_x) + \frac{z}{c^2}(y\omega_x - x\omega_y) \quad \text{...(1)},$$

where

$$x^2/a^2 + y^2/b^2 + z^2/c^2 = 1 \quad \text{.................(2).}$$

To satisfy this assume

$$\phi = Ayz + Bzx + Cxy,$$

this clearly being a solution of Laplace's equation.

The equation (1) then becomes

$$Ayz\left(\frac{1}{b^2}+\frac{1}{c^2}\right)+Bzx\left(\frac{1}{c^2}+\frac{1}{a^2}\right)+Cxy\left(\frac{1}{a^2}+\frac{1}{b^2}\right)$$
$$=yz\omega_x\left(\frac{1}{b^2}-\frac{1}{c^2}\right)+zx\omega_y\left(\frac{1}{c^2}-\frac{1}{a^2}\right)+xy\omega_z\left(\frac{1}{a^2}-\frac{1}{b^2}\right),$$

from which we obtain the values of A, B, C and then*

$$\phi=-\frac{b^2-c^2}{b^2+c^2}\omega_x yz-\frac{c^2-a^2}{c^2+a^2}\omega_y zx-\frac{a^2-b^2}{a^2+b^2}\omega_z xy\quad\text{......(3).}$$

Since this result depends only on the mutual ratios of a, b, c and not on their absolute magnitudes, it follows that the motion is the same in all ellipsoids of the same shape rotating with the same angular velocity.

To find the paths of the particles relative to the ellipsoid. Let (ξ, η, ζ) denote the coordinates of a particle P referred to the axes of the ellipsoid, then the velocities of P referred to axes fixed in space are $\dot{\xi}-\eta\omega_z+\zeta\omega_y$ and similar expressions.

Therefore

$$\dot{\xi}-\eta\omega_z+\zeta\omega_y=-\frac{\partial\phi}{\partial x}=\frac{c^2-a^2}{c^2+a^2}\omega_y\zeta+\frac{a^2-b^2}{a^2+b^2}\omega_z\eta,$$

or

$$\left.\begin{aligned}\dot{\xi}&=a^2(\gamma\eta-\beta\zeta)\\ \dot{\eta}&=b^2(\alpha\zeta-\gamma\xi)\\ \dot{\zeta}&=c^2(\beta\xi-\alpha\eta)\end{aligned}\right\}\quad\text{......................(4),}$$

where $\alpha=\dfrac{2\omega_x}{b^2+c^2}$, $\beta=\dfrac{2\omega_y}{c^2+a^2}$, $\gamma=\dfrac{2\omega_z}{a^2+b^2}$.

Multiply equations (4) by α/a^2, β/b^2, γ/c^2, add and integrate and we get

$$\alpha\xi/a^2+\beta\eta/b^2+\gamma\zeta/c^2=\text{const.}\quad\text{...............(5).}$$

Again multiply the same equations by ξ/a^2, η/b^2, ζ/c^2, add and integrate and we get

$$\xi^2/a^2+\eta^2/b^2+\zeta^2/c^2=\text{const.}\quad\text{...............(6).}$$

The path of the particle therefore lies on the plane (5) and the ellipsoid (6) so that it is an ellipse.

Again, if we assume that equations (4) have solutions of the form

$$\xi=Pe^{ipt},\quad \eta=Qe^{ipt},\quad \zeta=Re^{ipt},$$

* This result was published independently by Beltrami, Bjerknes and Maxwell in 1873. See Hicks, 'Report on Recent Progress in Hydrodynamics', *Brit. Ass. Rep.* 1882, p. 56.

we get by substitution and the elimination of P, Q, R

$$\begin{vmatrix} ip/a^2, & -\gamma & \beta \\ \gamma, & ip/b^2, & -\alpha \\ -\beta, & \alpha, & ip/c^2 \end{vmatrix} = 0,$$

whence
$$p = abc\left(\frac{\alpha^2}{a^2}+\frac{\beta^2}{b^2}+\frac{\gamma^2}{c^2}\right)^{\frac{1}{2}}.$$

Hence every particle of the liquid describes an ellipse relative to the ellipsoid, like a particle moving under a law of force varying as the distance from a fixed point. And the periodic time for each particle is $2\pi/p$, where

$$p = 2abc\left\{\left(\frac{\omega_x/a}{b^2+c^2}\right)^2+\left(\frac{\omega_y/b}{c^2+a^2}\right)^2+\left(\frac{\omega_z/c}{a^2+b^2}\right)^2\right\}^{\frac{1}{2}}.$$

We notice that for a sphere $(a=b=c)$

$$p = (\omega_x{}^2+\omega_y{}^2+\omega_z{}^2)^{\frac{1}{2}},$$

that is, the period of revolution of the liquid relative to the spherical shell is the same as the period of revolution of the shell; which means that the liquid is left at rest in space, the shell revolving alone*.

7·5. Motion of an Ellipsoid in an infinite Mass of Liquid. Before considering the problem it will be convenient to recall from the Theory of Attractions some solutions of Laplace's equation and formulae connected with the ellipsoid.

If $x^2/a^2+y^2/b^2+z^2/c^2=1$ is the equation of the boundary of a solid homogeneous ellipsoid of unit density, its potential at an external point (x, y, z) is

$$V = \pi abc\int_\lambda^\infty\left(1-\frac{x^2}{a^2+u}-\frac{y^2}{b^2+u}-\frac{z^2}{c^2+u}\right)\frac{du}{(a^2+u)^{\frac{1}{2}}(b^2+u)^{\frac{1}{2}}(c^2+u)^{\frac{1}{2}}} \quad \text{......(1)},$$

where λ is the positive root of the equation

$$\frac{x^2}{a^2+\lambda}+\frac{y^2}{b^2+\lambda}+\frac{z^2}{c^2+\lambda}-1=0 \quad \text{..............(2)}.$$

We may write this $V = \pi(\delta-\alpha x^2-\beta y^2-\gamma z^2)$ ....(3),

where
$$\delta = abc\int_\lambda^\infty\frac{du}{\Delta},\quad \alpha = abc\int_\lambda^\infty\frac{du}{(a^2+u)\Delta},\ \text{etc.} \quad \text{......(4)}$$

and
$$\Delta = (a^2+u)^{\frac{1}{2}}(b^2+u)^{\frac{1}{2}}(c^2+u)^{\frac{1}{2}}.$$

* The latter part of this article is based on a paper of Lord Kelvin's, 'On the Motion of a Liquid within an Ellipsoidal Hollow', *Proc. R. Soc. Edin.* XIII, 1885, p. 370, or *Math. and Phys. Papers*, IV, p. 196.

The potential at an internal point is a similar expression with λ put equal to zero, and, with a similar notation, may be denoted by

$$V_0 = \pi(\delta_0 - \alpha_0 x^2 - \beta_0 y^2 - \gamma_0 z^2) \quad \text{............(5)},$$

where $\delta_0, \alpha_0, \beta_0, \gamma_0$ denote what $\delta, \alpha, \beta, \gamma$ become when we put $\lambda = 0$.

The components of attraction at an external point are X, Y, Z, where

$$X = \frac{\partial V}{\partial x} = -2\pi\alpha x + \frac{\partial V}{\partial \lambda}\frac{\partial \lambda}{\partial x}.$$

But $\partial V/\partial\lambda = 0$ in virtue of equation (2), therefore

$$X = -2\pi\alpha x, \quad Y = -2\pi\beta y, \quad Z = -2\pi\gamma z \text{........(6)},$$

where it is to be remembered that α, β, γ are not constants but functions of λ or x, y, z.

We know that V is a solution of Laplace's equation and therefore also so are X, Y, Z.

Now consider an ellipsoid moving with velocity U in the direction of the x axis. The boundary condition is

$$-\frac{x}{a^2}\frac{\partial \phi}{\partial x} - \frac{y}{b^2}\frac{\partial \phi}{\partial y} - \frac{z}{c^2}\frac{\partial \phi}{\partial z} = U\frac{x}{a^2} \quad \text{..............(7)}$$

over the ellipsoid, i.e. where $\lambda = 0$.

Let us try to satisfy this by the assumption

$$\phi = AX.$$

We have
$$\frac{\partial \phi}{\partial x} = -2\pi A\left(\alpha + x\frac{\partial \alpha}{\partial \lambda}\frac{\partial \lambda}{\partial x}\right),$$

but when $\lambda = 0$,
$$\frac{\partial \alpha}{\partial \lambda} = -\frac{1}{a^2},$$

and from (2), by differentiating with regard to x,

$$\frac{2x}{a^2+\lambda} - \frac{\partial \lambda}{\partial x}\left(\frac{x^2}{(a^2+\lambda)^2} + \frac{y^2}{(b^2+\lambda)^2} + \frac{z^2}{(c^2+\lambda)^2}\right) = 0,$$

or
$$\frac{\partial \lambda}{\partial x} = \frac{2p^2 x}{a^2+\lambda},$$

and similarly
$$\frac{\partial \lambda}{\partial y} = \frac{2p^2 y}{b^2+\lambda}, \text{ and } \frac{\partial \lambda}{\partial z} = \frac{2p^2 z}{c^2+\lambda}.$$

Hence when $\lambda = 0$,

$$\frac{\partial \phi}{\partial x} = -2\pi A\left(\alpha_0 - \frac{2p^2 x^2}{a^4}\right).$$

Similarly $$\frac{\partial\phi}{\partial y}=-2\pi A\left(-\frac{2p^2xy}{a^2b^2}\right),$$

and $$\frac{\partial\phi}{\partial z}=-2\pi A\left(-\frac{2p^2xz}{a^2c^2}\right).$$

Therefore, substituting in (7) we get

$$2\pi A\left\{\frac{\alpha_0 x}{a^2}-\frac{2p^2x}{a^2}\left(\frac{x^2}{a^4}+\frac{y^2}{b^4}+\frac{z^2}{c^4}\right)\right\}=\frac{Ux}{a^2},$$

or $$A=\frac{U}{2\pi(\alpha_0-2)}.$$

Hence $$\phi=\frac{U\alpha x}{2-\alpha_0}\quad\text{.......................(8)}$$

gives the velocity potential of the liquid motion*.

If the ellipsoid have a velocity of which U, V, W are the components parallel to the axes, the velocity potential will be

$$\phi=\frac{U\alpha x}{2-\alpha_0}+\frac{V\beta y}{2-\beta_0}+\frac{W\gamma z}{2-\gamma_0}.$$

7·51. Ellipsoid rotating in an infinite Mass of Liquid. Let the ellipsoid turn about the axis of x with angular velocity ω_x.

The component velocities of any point of the ellipsoid are then 0, $-z\omega_x$, $y\omega_x$, so that, with the notation of the last article, the boundary condition is

$$-\frac{x}{a^2}\frac{\partial\phi}{\partial x}-\frac{y}{b^2}\frac{\partial\phi}{\partial y}-\frac{z}{c^2}\frac{\partial\phi}{\partial z}=yz\omega_x\left(\frac{1}{c^2}-\frac{1}{b^2}\right)\quad\text{........(1)},$$

where $\lambda=0$.

To find a solution of Laplace's equation that will satisfy this condition, assume

$$\phi=C(yZ-zY).$$

This makes $$\nabla^2\phi=2C\left(\frac{\partial Z}{\partial y}-\frac{\partial Y}{\partial z}\right)$$

$$=2C\left(\frac{\partial^2 V}{\partial y\,\partial z}-\frac{\partial^2 V}{\partial z\,\partial y}\right)=0,$$

and taking $$\phi=-2\pi Cyz(\gamma-\beta)\quad\text{.................(2)},$$

* This result was first given by Green in his paper 'Researches on the vibration of pendulums in fluid media', *Trans. R. Soc. Edin.* 1833, or *Math. Papers*, p. 315.

and substituting in (1) we get

$$2\pi Cyz\left\{(\gamma-\beta)\left(\frac{1}{b^2}+\frac{1}{c^2}\right)+\frac{x}{a^2}\left(\frac{\partial\gamma}{\partial x}-\frac{\partial\beta}{\partial x}\right)+\frac{y}{b^2}\left(\frac{\partial\gamma}{\partial y}-\frac{\partial\beta}{\partial y}\right)\right.$$
$$\left.+\frac{z}{c^2}\left(\frac{\partial\gamma}{\partial z}-\frac{\partial\beta}{\partial z}\right)\right\}=yz\omega_x\left(\frac{1}{c^2}-\frac{1}{b^2}\right)\text{......(3)},$$

and reducing this as in **7·5**, we get, when $\lambda=0$,

$$2\pi C\left\{(\gamma_0-\beta_0)\left(\frac{1}{b^2}+\frac{1}{c^2}\right)+2\left(\frac{1}{b^2}-\frac{1}{c^2}\right)\right\}=\omega_x\left(\frac{1}{c^2}-\frac{1}{b^2}\right).$$

Therefore
$$\phi=-\frac{\omega_x(\beta-\gamma)yz}{2+\dfrac{b^2+c^2}{b^2-c^2}(\beta_0-\gamma_0)}\text{ ....(4)}$$

is the required velocity potential*, where

$$\beta-\gamma=abc\,(c^2-b^2)\int_\lambda^\infty\frac{du}{(a^2+u)^{\frac{1}{2}}(b^2+u)^{\frac{3}{2}}(c^2+u)^{\frac{3}{2}}},$$

λ being the positive root of

$$\frac{x^2}{a^2+\lambda}+\frac{y^2}{b^2+\lambda}+\frac{z^2}{c^2+\lambda}=1.$$

If the ellipsoid have angular velocities ω_x, ω_y, ω_z about the axes of x, y, z, the velocity potential will be

$$\phi=-\frac{\omega_x(\beta-\gamma)yz}{2+\dfrac{b^2+c^2}{b^2-c^2}(\beta_0-\gamma_0)}-\frac{\omega_y(\gamma-\alpha)zx}{2+\dfrac{c^2+a^2}{c^2-a^2}(\gamma_0-\alpha_0)}-\frac{\omega_z(\alpha-\beta)xy}{2+\dfrac{a^2+b^2}{a^2-b^2}(\alpha_0-\beta_0)}.$$

7·52. Spheroids. For a *prolate* spheroid $b=c<a$, we have

$$\alpha=ab^2\int_\lambda^\infty\frac{du}{(a^2+u)^{\frac{3}{2}}(b^2+u)},$$

and putting $a^2+u=(a^2-b^2)v^2$, we get

$$\alpha=\frac{2ab^2}{(a^2-b^2)^{\frac{3}{2}}}\int_\nu^\infty\frac{dv}{v^2(v^2-1)},$$

where $a^2+\lambda=(a^2-b^2)\nu^2$.

Therefore
$$\alpha=\frac{2(1-e^2)}{e^3}\left(\tfrac{1}{2}\log\frac{\nu+1}{\nu-1}-\frac{1}{\nu}\right)\text{ ....(1)},$$

where e is the eccentricity of the generating ellipse.

* This result is due to Clebsch, see *Crelle*, LIII, p. 287.

Also
$$\beta=\gamma=ab^2\int_\lambda^\infty \frac{du}{(a^2+u)^{\frac{1}{2}}(b^2+u)^2}$$
$$=\frac{2(1-e^2)}{e^3}\int_\nu^\infty \frac{dv}{(v^2-1)^2}$$
$$=\frac{(1-e^2)}{e^3}\left(\frac{\nu}{\nu^2-1}-\tfrac{1}{2}\log\frac{\nu+1}{\nu-1}\right) \quad \text{..............(2).}$$

In this case $\nu=1/e'$, where e' is the eccentricity of the generating ellipse of the confocal spheroid through the external point considered.

For an *oblate* spheroid $a=b>c$, we have
$$\alpha=\beta=a^2c\int_\lambda^\infty \frac{du}{(a^2+u)^2(c^2+u)^{\frac{1}{2}}};$$
and putting $c^2+u=(a^2-c^2)v^2$ we get
$$\alpha=\beta=\frac{2a^2c}{(a^2-c^2)^{\frac{1}{2}}}\int_\nu^\infty \frac{dv}{(1+v^2)^2},$$
where $c^2+\lambda=(a^2-c^2)\nu^2$.

Therefore
$$\alpha=\beta=\frac{(1-e^2)^{\frac{1}{2}}}{e^3}\left(\cot^{-1}\nu-\frac{\nu}{\nu^2+1}\right) \quad \text{..................(3).}$$

Also
$$\gamma=a^2c\int_\lambda^\infty \frac{du}{(a^2+u)(c^2+u)^{\frac{3}{2}}}$$
$$=\frac{2a^2c}{(a^2-c^2)^{\frac{3}{2}}}\int_\nu^\infty \frac{dv}{(1+v^2)v^2}$$
$$=\frac{2(1-e^2)^{\frac{1}{2}}}{e^3}\left(\frac{1}{\nu}-\cot^{-1}\nu\right) \quad \text{.....................(4).}$$

In this case ν is $(1-e'^2)^{\frac{1}{2}}/e'$, where e' has the same meaning as above.

Hence for an *oblate* spheroid moving along the axis with velocity W, we have
$$\phi=\frac{W\gamma z}{2-\gamma_0},$$
where γ has the value given by (4), and γ_0 is the value when $\lambda=0$, or when $\nu=(1-e^2)^{\frac{1}{2}}/e$. Hence
$$\phi=\frac{Wz(1-e^2)^{\frac{1}{2}}}{e^3-(1-e^2)^{\frac{1}{2}}\left(\frac{e}{(1-e^2)^{\frac{1}{2}}}-\sin^{-1}e\right)}\left(\frac{1}{\nu}-\cot^{-1}\nu\right)$$
$$=\frac{Wz}{\sin^{-1}e-e(1-e^2)^{\frac{1}{2}}}\left(\frac{1}{\nu}-\cot^{-1}\nu\right).$$

As a special case we may take $c=0$ or $e=1$, and we get for the case of a **circular disc** moving at right angles to its plane
$$\phi=\frac{2Wz}{\pi}\left(\frac{1}{\nu}-\cot^{-1}\nu\right).$$

In this case $a^2\nu^2=\lambda$ and λ is the positive root of
$$\frac{x^2+y^2}{a^2+\lambda}+\frac{z^2}{\lambda}=1.$$

On the disc itself $z=0$ and $\lambda=0$, so that $\nu=0$, but ϕ has a definite value, for we may write

$$\frac{z}{\nu}=\frac{az}{\lambda^{\frac{1}{2}}}=\pm a\left(1-\frac{x^2+y^2}{a^2}\right)^{\frac{1}{2}}=\pm(a^2-x^2-y^2)^{\frac{1}{2}},$$

so that
$$\phi=\pm\frac{2W}{\pi}(a^2-x^2-y^2)^{\frac{1}{2}},$$

taking the + or − sign on opposite sides of the disc. The normal velocity is $\pm W$, hence for the kinetic energy of the liquid we have

$$\begin{aligned}T&=-\tfrac{1}{2}\rho\iint\phi\frac{\partial\phi}{\partial n}dS\\&=\frac{2W^2}{\pi}\rho\int_0^a(a^2-\varpi^2)^{\frac{1}{2}}.2\pi\varpi\,d\varpi\\&=\tfrac{4}{3}\rho a^3W^2.\end{aligned}$$

We observe that, as is usual in such cases, the theory leads to infinite velocity of the liquid at the edge of the disc.

7·53. Reverting to the case of an ellipsoid moving along one of its axes (**7·5**), we have
$$\phi=\frac{U\alpha x}{2-\alpha_0},$$
and the kinetic energy of the liquid is given by
$$T=-\tfrac{1}{2}\rho\iint\phi\frac{\partial\phi}{\partial n}dS.$$
But on the surface of the ellipsoid the normal velocity $=lU$, where (l, m, n) are the direction cosines of the normal. Therefore
$$T=\tfrac{1}{2}\rho\frac{U^2\alpha_0}{2-\alpha_0}\iint lx\,dS,$$
and this integral is clearly the volume of the ellipsoid, so that
$$T=\tfrac{1}{2}\frac{\alpha_0}{2-\alpha_0}.\tfrac{4}{3}\rho\pi abcU^2,$$
or there is an effective increase in the inertia of the ellipsoid due to the presence of the liquid equal to $\alpha_0/(2-\alpha_0)$ of the mass of liquid displaced.

We shall now shew how the foregoing problems of liquid motion with ellipsoidal boundaries may be treated by a transformation of coordinates.

7·6. Laplace's Equation in Orthogonal Curvilinear Coordinates. Let
$$\lambda=\text{const.},\quad \mu=\text{const.},\quad \nu=\text{const.}$$
be three families of surfaces that cut one another orthogonally at all their points of intersection; λ, μ, ν denoting functions of rectangular coordinates x, y, z.

Let $OABCD$ be a small curvilinear parallelepiped bounded by such surfaces, the opposite faces BC, AD corresponding to λ and $\lambda+\delta\lambda$, and so on; and the edges OA, OB, OC being of lengths δs_1, δs_2, δs_3.

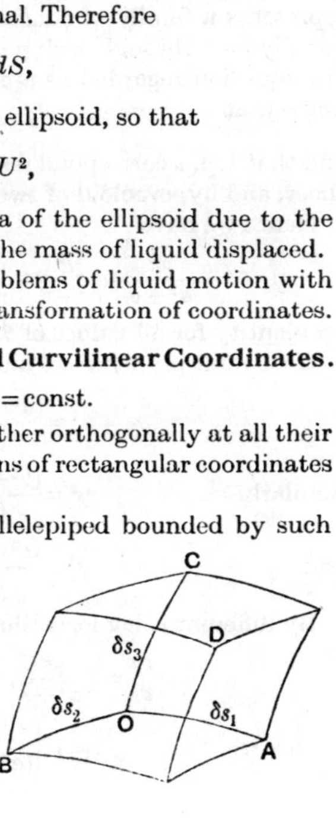

If the coordinates of O are x, y, z those of A are
$$x+\frac{\partial x}{\partial\lambda}\delta\lambda,\quad y+\frac{\partial y}{\partial\lambda}\delta\lambda,\quad z+\frac{\partial z}{\partial\lambda}\delta\lambda.$$

Hence the direction cosines of the normal to the surface $\lambda=$const. are proportional to $\frac{\partial x}{\partial\lambda}, \frac{\partial y}{\partial\lambda}, \frac{\partial z}{\partial\lambda}$, and their values are

$$\left(h_1\frac{\partial x}{\partial\lambda},\quad h_1\frac{\partial y}{\partial\lambda},\quad h_1\frac{\partial z}{\partial\lambda}\right),$$

where

$$\frac{1}{h_1^2}=\left(\frac{\partial x}{\partial\lambda}\right)^2+\left(\frac{\partial y}{\partial\lambda}\right)^2+\left(\frac{\partial z}{\partial\lambda}\right)^2.$$

Also

$$\delta s_1^2=\left\{\left(\frac{\partial x}{\partial\lambda}\right)^2+\left(\frac{\partial y}{\partial\lambda}\right)^2+\left(\frac{\partial z}{\partial\lambda}\right)^2\right\}\delta\lambda^2,$$

so that $\delta s_1=\delta\lambda/h_1$ and similarly $\delta s_2=\delta\mu/h_2$, $\delta s_3=\delta\nu/h_3$.

Now if ϕ is the velocity potential of a liquid motion the total flow of liquid outwards across the surface of the parallelepiped is by **4·52** (i) $-\nabla^2\phi$ times the volume, and we get from the pair of faces BC, AD a contribution

$$-\frac{\partial}{\partial s_1}\left(\frac{\partial\phi}{\partial s_1}\delta s_2\delta s_3\right)\delta s_1 \text{ or } -\frac{\partial}{\partial\lambda}\left(\frac{h_1}{h_2h_3}\frac{\partial\phi}{\partial\lambda}\right)\delta\lambda\,\delta\mu\,\delta\nu,$$

so that by adding similar terms we have

$$\nabla^2\phi=h_1h_2h_3\left\{\frac{\partial}{\partial\lambda}\left(\frac{h_1}{h_2h_3}\frac{\partial\phi}{\partial\lambda}\right)+\frac{\partial}{\partial\mu}\left(\frac{h_2}{h_3h_1}\frac{\partial\phi}{\partial\mu}\right)+\frac{\partial}{\partial\nu}\left(\frac{h_3}{h_1h_2}\frac{\partial\phi}{\partial\nu}\right)\right\}.$$

7·7. Confocal Conicoids. The equation

$$\frac{x^2}{a^2+\theta}+\frac{y^2}{b^2+\theta}+\frac{z^2}{c^2+\theta}-1=0 \quad\text{......................(1)}$$

represents a family of confocal conicoids, of which three cutting orthogonally pass through each point of space. If λ, μ, ν are the three roots of the equation regarded as a cubic in θ, and we assume that $a>b>c$, we know that

$$\infty>\lambda>-c^2>\mu>-b^2>\nu>-a^2,$$

and that λ, μ, ν correspond respectively to an ellipsoid, hyperboloid of one sheet, and hyperboloid of two sheets.

Hence we have

$$\frac{x^2}{a^2+\theta}+\frac{y^2}{b^2+\theta}+\frac{z^2}{c^2+\theta}-1=\frac{(\lambda-\theta)(\mu-\theta)(\nu-\theta)}{(a^2+\theta)(b^2+\theta)(c^2+\theta)} \quad\text{......(2)},$$

an identity for all values of θ.

If we multiply by $a^2+\theta$ and then put $\theta=-a^2$, we get

$$x^2=\frac{(a^2+\lambda)(a^2+\mu)(a^2+\nu)}{(a^2-b^2)(a^2-c^2)}.$$

Similarly

$$y^2=\frac{(b^2+\lambda)(b^2+\mu)(b^2+\nu)}{(b^2-c^2)(b^2-a^2)} \quad\text{......................(3)}$$

and

$$z^2=\frac{(c^2+\lambda)(c^2+\mu)(c^2+\nu)}{(c^2-a^2)(c^2-b^2)}.$$

By differentiating logarithmically we get

$$\frac{\partial x}{\partial\lambda}=\tfrac{1}{2}\frac{x}{a^2+\lambda},\quad \frac{\partial y}{\partial\lambda}=\tfrac{1}{2}\frac{y}{b^2+\lambda},\quad \frac{\partial z}{\partial\lambda}=\tfrac{1}{2}\frac{z}{c^2+\lambda} \quad\text{............(4)},$$

therefore

$$\frac{1}{h_1^2}=\tfrac{1}{4}\left\{\frac{x^2}{(a^2+\lambda)^2}+\frac{y^2}{(b^2+\lambda)^2}+\frac{z^2}{(c^2+\lambda)^2}\right\} \quad\text{..............(5)}.$$

Hence $h_1 = 2p_1$, similarly $h_2 = 2p_2$, and $h_3 = 2p_3$, where p_1, p_2, p_3 are the central perpendiculars on the tangent planes to the ellipsoid and hyperboloids.

Again by differentiating (2) with regard to θ and then putting $\theta = \lambda$ we get

$$\Sigma \frac{x^2}{(a^2+\lambda)^2} = \frac{(\lambda-\mu)(\lambda-\nu)}{(a^2+\lambda)(b^2+\lambda)(c^2+\lambda)},$$

therefore
$$\frac{4}{h_1^2} = \frac{(\lambda-\mu)(\lambda-\nu)}{(a^2+\lambda)(b^2+\lambda)(c^2+\lambda)}.$$

Similarly
$$\frac{4}{h_2^2} = \frac{(\mu-\nu)(\mu-\lambda)}{(a^2+\mu)(b^2+\mu)(c^2+\mu)} \qquad \text{......(6)}$$

and
$$\frac{4}{h_3^2} = \frac{(\nu-\lambda)(\nu-\mu)}{(a^2+\nu)(b^2+\nu)(c^2+\nu)}.$$

In terms of these parameters λ, μ, ν it follows that Laplace's equation takes the form

$$\nabla^2\phi = \frac{-4}{(\mu-\nu)(\nu-\lambda)(\lambda-\mu)} \Sigma(\mu-\nu)\left\{(a^2+\lambda)^{\frac{1}{2}}(b^2+\lambda)^{\frac{1}{2}}(c^2+\lambda)^{\frac{1}{2}}\frac{\partial}{\partial\lambda}\right\}^2 \phi = 0 \qquad \text{......(7).}$$

7·71. We can now find solutions of the preceding equation and give hydrodynamical interpretations to them.

An obvious solution is

$$\phi = \int_\lambda^\infty \frac{du}{(a^2+u)^{\frac{1}{2}}(b^2+u)^{\frac{1}{2}}(c^2+u)^{\frac{1}{2}}}$$

and by assuming the existence of solutions of the form

$$\phi = x\chi(\lambda),$$

and
$$\phi = yz\chi(\lambda),$$

it is easy to shew that there are solutions of the form

$$\phi = x\int_\lambda^\infty \frac{du}{(a^2+u)^{\frac{3}{2}}(b^2+u)^{\frac{1}{2}}(c^2+u)^{\frac{1}{2}}},$$

and
$$\phi = yz\int_\lambda^\infty \frac{du}{(a^2+u)^{\frac{1}{2}}(b^2+u)^{\frac{3}{2}}(c^2+u)^{\frac{3}{2}}}.$$

The last two correspond to the translation and rotation of an ellipsoid and give the same results as were obtained in **7·5, 7·51**; the boundary conditions in this notation being

$$-\frac{\partial\phi}{\partial\lambda} = U\frac{\partial x}{\partial\lambda},$$

and
$$-\frac{\partial\phi}{\partial\lambda} = \omega_x\left(y\frac{\partial z}{\partial\lambda} - z\frac{\partial y}{\partial\lambda}\right),$$

for the two cases. For the details of the work we refer the reader to Lamb's *Hydrodynamics*, pp. 152–155 (6th ed.), from which this investigation is taken.

7·8. Ellipsoid of varying form.

As we saw in **7·71**, or as is clear from the theory of attractions,

$$\phi = C\int_\lambda^\infty \frac{d\lambda}{(a^2+\lambda)^{\frac{1}{2}}(b^2+\lambda)^{\frac{1}{2}}(c^2+\lambda)^{\frac{1}{2}}} \qquad \text{...........(1)}$$

is a solution of Laplace's equation. It clearly vanishes when $\lambda=\infty$ and it is constant over confocal ellipsoids, it may therefore represent the velocity potential of a liquid motion due to an ellipsoid whose surface is changing form. For the velocity at any point being given by

$$-\frac{\partial\phi}{\partial n}=-h_1\frac{\partial\phi}{\partial\lambda}=\frac{Ch_1}{(a^2+\lambda)^{\frac{1}{2}}(b^2+\lambda)^{\frac{1}{2}}(c^2+\lambda)^{\frac{1}{2}}} \quad\text{......(2),}$$

therefore, on any confocal ellipsoid, the velocity varies as the central perpendicular on the tangent plane. Hence the conditions are satisfied by supposing a boundary ellipsoid to vary so as to remain similar to itself keeping its axes fixed in direction. If the axes are changing at the rates $\dot{a}$, $\dot{b}$, $\dot{c}$ the general boundary condition

$$\frac{\partial f}{\partial t}+u\frac{\partial f}{\partial x}+v\frac{\partial f}{\partial y}+w\frac{\partial f}{\partial z}=0$$

becomes in this case

$$\frac{x^2}{a^3}\dot{a}+\frac{y^2}{b^3}\dot{b}+\frac{z^2}{c^3}\dot{c}+\frac{x}{a^2}\frac{\partial\phi}{\partial x}+\frac{y}{b^2}\frac{\partial\phi}{\partial y}+\frac{z}{c^2}\frac{\partial\phi}{\partial z}=0 \quad\text{......(3).}$$

But we have $$\frac{\dot{a}}{a}=\frac{\dot{b}}{b}=\frac{\dot{c}}{c}=K \text{ say,}$$

and on the surface $\lambda=0$, equation (2) becomes $-\dfrac{\partial\phi}{\partial n}=\dfrac{2Cp}{abc}$, therefore, if we take $Kabc=2C$, (3) and (2) are the same.

Another expression for ϕ that will satisfy the general boundary condition (3) is obviously*

$$\phi=-\tfrac{1}{2}\left(\frac{\dot{a}}{a}x^2+\frac{\dot{b}}{b}y^2+\frac{\dot{c}}{c}z^2\right)\text{..................(4),}$$

and it will satisfy Laplace's equation if

$$\frac{\dot{a}}{a}+\frac{\dot{b}}{b}+\frac{\dot{c}}{c}=0 \quad\text{........................(5).}$$

This then is the velocity potential due to an ellipsoid which changes form so that its volume remains constant, for condition (5) is merely the condition that $abc=\text{const.}$

* This result is due to Bjerknes, *Gött. Nachrichten*, 1873, p. 829.

EXAMPLES

1. A solid sphere moves through quiescent frictionless liquid whose boundaries are at a distance from it great compared with its radius. Prove that at each instant the motion in the liquid depends only on the position and velocity of the sphere at that instant. Prove that the liquid streams past the sides of the sphere with half the velocity of the sphere.

(St John's Coll. 1901.)

2. An infinite ocean of an incompressible perfect liquid of density ρ is streaming past a fixed spherical obstacle of radius a. The velocity is uniform and equal to U except in so far as it is disturbed by the sphere, and the pressure in the liquid at a great distance from the obstacle is Π. Shew that the thrust on that half of the sphere on which the liquid impinges is

$$\pi a^2\{\Pi - \rho U^2/16\}.$$

(Trinity Coll. 1900.)

3. A rigid sphere of radius a is moving in a straight line with velocity u and acceleration f through an infinite incompressible liquid, prove that the resultant fluid pressures over the two hemispheres into which the sphere is divided by a diametral plane perpendicular to its direction of motion are $\Pi\pi a^2 \pm \frac{1}{4}Mf - \frac{3}{64}Mu^2/a$, where Π is the pressure at a great distance, and M is the mass of the fluid displaced by the sphere. (M.T. II. 1910.)

4. A solid sphere is moving through frictionless liquid: compare the velocities of slip of the liquid past it at different parts of its surface.

Prove that when the sphere is in motion with uniform velocity U, the pressure at the part of its surface where the radius makes an angle θ with the direction of motion is increased on account of the motion by the amount

$$\tfrac{1}{16}\rho U^2(9\cos 2\theta - 1),$$

where ρ is the density of the liquid. (St John's Coll. 1898.)

5. Find the pressure at any point of a liquid, of infinite extent and at rest at a great distance, through which a sphere is moving under no external forces with constant velocity U, and shew that the mean pressure over the sphere is in defect of the pressure Π at a great distance by $\frac{1}{4}\rho U^2$, it being supposed that Π is sufficiently large for the pressure everywhere to be positive, that is, that $\Pi > \frac{5}{8}\rho U^2$. (M.T. 1908.)

6. An infinite homogeneous liquid is flowing steadily past a rigid boundary consisting partly of the horizontal plane $y=0$, and partly of a hemispherical boss $x^2+y^2+z^2=a^2$, with irrotational motion which tends, at a great distance from the origin, to uniform velocity V parallel to the axis of z. Find the velocity potential and the surfaces of equal pressure.

(St John's Coll. 1905.)

7. A stream of water of great depth is flowing with uniform velocity V over a plane level bottom. A hemisphere of weight w in water and of radius a, rests with its base on the bottom. Prove that the average pressure between the base of the hemisphere and the bottom is less than the fluid pressure at any point of the bottom at a great distance from the hemisphere, if

$$V^2 > 32w/11\pi a^2\rho.$$

(M.T. 1894.)

8. Prove that at a point on a sphere moving through an infinite liquid the pressure is given by the formula

$$(p-p_0)/\rho = \tfrac{1}{2}af\cos\theta_1 + \tfrac{1}{8}v^2(9\cos^2\theta - 5),$$

where v is the velocity, f the acceleration of the sphere, and θ, θ_1 are the angles between the radius and the directions of v, f respectively, and p_0 is the hydrostatic pressure. (St John's Coll. 1909.)

9. When a sphere of radius a moves in an infinite liquid shew that the pressure at any point exceeds what would be the pressure if the sphere were at rest by

$$\frac{a^3}{2r^2}f - \frac{a^3}{8r^6}(4r^3+a^3)q^2 + \frac{3}{8}\frac{a^3}{r^6}(4r^3-a^3)q'^2,$$

where q is the velocity of the sphere and q' and f are the resolved parts of its velocity and acceleration in the direction of r and the density of the liquid is unity. (Coll. Exam. 1894.)

10. A sphere of radius a is in motion in fluid, which is at rest at infinity, the pressure there being Π; determine the pressure at any point of the fluid, and shew that the pressure on the front hemisphere cut off by a plane perpendicular to the direction of motion is the resultant of pressures $\pi a^2(\Pi - \frac{1}{16}\rho V^2)$ and $\frac{1}{3}\pi\rho a^3 f$ in the directions respectively opposite to those of the velocity V, and the acceleration f, of the centre of the sphere.

(Coll. Exam. 1910.)

11. Prove that for liquid contained between two instantaneously concentric spheres, when the outer (radius a) is moving parallel to the axis of x with velocity u and the inner (radius b) is moving parallel to the axis of y with velocity v, the velocity potential is

$$-\frac{1}{a^3-b^3}\left\{a^3ux\left(1+\frac{b^3}{2r^3}\right) - b^3vy\left(1+\frac{a^3}{2r^3}\right)\right\},$$

and find the kinetic energy. (St John's Coll. 1898.)

12. Liquid of density ρ fills the space between a solid sphere of radius a and density ρ' and a fixed concentric spherical envelope of radius b; prove that the work done by an impulse which starts the solid sphere with velocity V is

$$\tfrac{1}{3}\pi a^3 V^2\left(2\rho' + \frac{2a^3+b^3}{b^3-a^3}\rho\right). \qquad \text{(Coll. Exam. 1896.)}$$

13. The space between two concentric spherical shells of radii a and b $(a>b)$ is filled with an incompressible fluid of density ρ and the shells suddenly begin to move with velocities U, V in the same direction: prove that the resultant impulsive pressure on the inner shell is

$$\frac{2\pi\rho b^3}{3(a^3-b^3)}\{3a^3U - (a^3+2b^3)V\}. \qquad \text{(Trinity Coll. 1895.)}$$

14. Incompressible fluid, of density ρ, is contained between two rigid concentric spherical surfaces, the outer one of mass M_1 and radius a, the inner one of mass M_2 and radius b. A normal blow P is given to the outer

surface. Prove that the initial velocities of the two containing surfaces (U for the outer and V for the inner) are given by the equations

$$\left\{M_1+\frac{2\pi\rho a^3(2a^3+b^3)}{3(a^3-b^3)}\right\}U-\frac{2\pi\rho a^3b^3}{a^3-b^3}V=P,$$

$$\left\{M_2+\frac{2\pi\rho b^3(2b^3+a^3)}{3(a^3-b^3)}\right\}V=\frac{2\pi\rho a^3b^3}{a^3-b^3}U.$$

(Trinity Coll. 1896.)

15. A sphere of radius a is placed in an incompressible fluid extending to infinity. Each point of the sphere is moving normally outwards with velocity $\dot{a}$, also the fluid at points very distant from the sphere is moving with velocity V in a given direction. Find the velocity potential at any point of the fluid.

Also prove that the resultant pressure on the sphere is the force $\frac{1}{2}\frac{dM}{dt}V$ in the direction of the stream, where M is the mass of the fluid displaced by the sphere at the instant considered. (Trinity Coll. 1897.)

16. A solid is bounded by the exterior portions of two equal spheres (of radius a) which cut one another orthogonally; and is surrounded by an infinite mass of liquid. If the solid is set in motion with velocity u in the direction of the line of centres, shew that the velocity potential of the resulting motion is

$$\tfrac{1}{2}a^3u\left(\frac{\cos\theta}{r^2}+\frac{\cos\theta'}{r'^2}-\frac{\cos\Theta}{2\sqrt{2}R^2}\right),$$

where r, r', R are the radii vectores of a point, measured respectively from the centres of the two spheres and from the point midway between them, and θ, θ', Θ are the angles which these radii vectores make with the direction of motion of the solid. (Coll. Exam. 1902.)

17. A sphere of radius a is made to move in incompressible perfect fluid with non-uniform velocity u along the x axis. If the pressure at infinity is zero, prove that at a point x in advance of the centre

$$p=\tfrac{1}{2}\rho a^3\left\{\frac{\dot{u}}{x^2}+u^2\left(\frac{2}{x^3}-\frac{a^3}{x^6}\right)\right\}. \quad \text{(M.T. 1928.)}$$

18. Shew that when a sphere of radius a moves with uniform velocity U through a perfect, incompressible, infinite fluid, the acceleration of a particle of fluid at $(r, 0)$ is

$$3U^2\left(\frac{a^3}{r^4}-\frac{a^6}{r^7}\right). \quad \text{(M.T. 1917.)}$$

19. The motion of an incompressible fluid being symmetrical with respect to an axis, and the parts of the velocity resolved along and perpendicularly to a radius vector drawn from a point fixed or moving on the axis in any direction making with the axis an angle θ being U and W, prove that if

$$U=\frac{2C}{r^3}\cos\theta+\frac{C'}{4r^4}(1+3\cos 2\theta),\quad W=\frac{C}{r^3}\sin\theta+\frac{C'}{2r^4}\sin 2\theta,$$

the equation of constancy of mass is satisfied, and $U\,dr+Wr\,d\theta$ is an exact differential, C and C' being either constants or functions of the time.

Shew also that if the fluid be unlimited in extent, and $C'=0$, the assumed motion would be produced by a sphere moving in any manner with its centre on a fixed straight line. (Smith's Prize, 1877.)

20. A doublet of strength M is placed at the point $(0, a, 0)$ with its axis parallel to the axis of z, prove that at points close to the origin the velocity potential of the doublet is approximately

$$\frac{Mz}{a^3}+\frac{3Myz}{a^4},$$

neglecting terms of the order r^3/a^5 and higher powers.

Deduce that if a small sphere of radius c is placed with its centre at the origin, the velocity potential is then increased by the terms

$$\frac{1}{2}\frac{Mc^3}{a^3}\frac{z}{r^3}+2\frac{Mc^5}{a^4}\frac{yz}{r^5}.$$ (Univ. of London, 1911.)

21. Shew that the image of a radial doublet in a sphere is another radial doublet, and compare their magnitudes; shew also that the velocity at any point of the sphere is proportional to ϖr^{-5}, where r is the distance from the doublet, and ϖ the perpendicular on the diameter on which it lies. (Trinity Coll. 1906.)

22. Discuss the motion for which Stokes's stream function is given by

$$\psi=\frac{1}{2}V\{a^4r^{-2}\cos\theta-r^2\}\sin^2\theta,$$

where r is the distance from a fixed point and θ is the angle this distance makes with a fixed direction. (Coll. Exam. 1900.)

23. Find the Stokes stream function ψ where fluid motion is due to a source of strength m (flux $4\pi m$) at a fixed point A, a sink $-m$ at another fixed point B, a translation of the fluid of velocity U in the direction AB being superposed; explain how this solution can be used to deduce the motion of fluid past a certain solid of revolution. If $U=8m/9a^2$, where $AB=2a$, prove that the solid is of axial length $4a$, of equatorial radius approximately $1{\cdot}6a$, and has the same effect on the fluid motion at a great distance as a sphere of radius $a\,(9/2)^{\frac{1}{3}}$. (M.T. 1932.)

24. If AB be a uniform line source, and A and B equal sinks of such strength that there is no total gain or loss of fluid, shew that the stream function at any point is

$$C\{(r_1-r_2)^2-c^2\}\left(\frac{1}{r_1}-\frac{1}{r_2}\right),$$

where c is the length of AB, and r_1, r_2 are the distances of the point considered from A and B. (Univ. of London, 1915.)

25. A solid of revolution is moving along its axis in an infinite liquid; shew that the kinetic energy of the liquid is

$$-\frac{1}{2}\pi\rho\int\frac{\psi}{\varpi}\frac{\partial\psi}{\partial n}\,ds,$$

where ψ is the Stokes stream function of the motion, ϖ the distance of a point from the axis and the integral is taken once round a meridian curve of the solid. Hence obtain the kinetic energy of infinite liquid due to the motion of a sphere through it with velocity V. (Coll. Exam. 1899.)

26. An ellipsoidal cavity (semi-axes a, b, c) in a solid initially at rest is filled with an incompressible frictionless fluid initially at rest. Prove that if the solid be moved with velocities u, v, w parallel to the axes of the

cavity, and be rotated with angular velocities p, q, r round the semi-axes, the angular momentum of the fluid round the semi-axis a at any instant is

$$\tfrac{4}{15}\pi\rho abc\frac{(b^2-c^2)^2}{b^2+c^2}p. \qquad \text{(Trinity Coll. 1902.)}$$

27. A rigid ellipsoidal envelope, without mass, encloses a perfect incompressible fluid of mass M. The equation of the ellipsoid is

$$x^2/a^2+y^2/b^2+z^2/c^2-1=0.$$

An impulsive couple in the plane of xy causes the envelope to rotate initially with angular velocity ω. Find the initial velocity potential of the fluid, and prove that the moment of the couple is

$$\tfrac{1}{5}M\omega(a^2-b^2)^2/(a^2+b^2). \qquad \text{(Trinity Coll. 1910.)}$$

28. Prove that, if two rigid surfaces of revolution, one of which surrounds the other, are moving along their common axis with velocities U_1, U_2 and the space between them is filled with homogeneous liquid, the momentum of the liquid is $M_2U_2-M_1U_1$, where M_1, M_2 are the masses of liquid which either surface would contain.

29. Prove that the same result holds good for surfaces of any form provided that the velocity potential is expressible in the form

$$\{f(x, y, z)+A\}x.$$

30. An ellipsoid surrounded by frictionless homogeneous liquid begins to move in any direction with velocity V. Shew that, if the outer boundary of the liquid is a fixed confocal ellipsoid, the momentum set up in the liquid is $-MV$, where M is the mass of liquid displaced by the ellipsoid.

(M.T. 1924.)

31. If the space between two confocal ellipsoids is filled with liquid, and the inner and outer ellipsoids are suddenly moved with velocities U, U' parallel to the axis of x, prove that the velocity potential of the initial motion is given by

$$\phi=\{(U-U')\alpha-U(\alpha_0'-2\mu)+U'(\alpha_0-2)\}x/\{\alpha_0'-\alpha_0+2-2\mu\},$$

where the notation is that of **7·5**, α_0' is the value of α for the outer ellipsoid, and μ is the ratio of the volume of the inner to the outer ellipsoid.

32. Shew that for a homogeneous solid ellipsoid of mass M rotating about the axis of x, in liquid at rest at infinity, the effective moment of inertia is

$$\tfrac{2}{5}M\left\{b^2+c^2+\frac{\rho}{2\sigma}\cdot\frac{(b^2-c^2)^2(\gamma_0-\beta_0)}{2(b^2-c^2)+(b^2+c^2)(\beta_0-\gamma_0)}\right\}$$

where ρ, σ are the densities of the liquid and solid and β_0, γ_0 have the meanings of **7·5**.

33. Shew that when a circular disc of radius α rotates about a diameter in liquid at rest at infinity the kinetic energy of the liquid is

$$\tfrac{8}{45}\rho a^5\omega^2,$$

ω being the angular velocity of the disc and ρ the density of the liquid.

34. Prove that, when an oblate spheroid of eccentricity $\sin\alpha$ moves parallel to its axis of figure with velocity V in infinite fluid, the kinetic energy of the fluid is

$$\tfrac{1}{2}M'V^2 \frac{\tan\alpha-\alpha}{\alpha-\sin\alpha\cos\alpha},$$

where M' denotes the mass of the displaced fluid. (M.T. II. 1910.)

35. The ellipsoid $x^2/a^2+y^2/b^2+z^2/c^2=1$ is surrounded by an infinite mass of water, and rotates about the axis of x. Prove that the component velocities of any particle of the water, parallel to the axes, will respectively be proportional to

$$\frac{\partial M}{\partial z}-\frac{\partial N}{\partial y},\quad \frac{\partial N}{\partial x}-\frac{\partial L}{\partial z},\quad \frac{\partial L}{\partial y}-\frac{\partial M}{\partial x},$$

where

$$L=\int_\epsilon^\infty \left\{\left(\frac{b^2}{b^2+\psi}-\frac{c^2}{c^2+\psi}\right)\left(1-\frac{x^2}{a^2+\psi}-\frac{y^2}{b^2+\psi}-\frac{z^2}{c^2+\psi}\right)-2\left(\frac{by}{b^2+\psi}\right)^2 + 2\left(\frac{cz}{c^2+\psi}\right)^2\right\}\frac{d\psi}{(a^2+\psi)^{\frac{1}{2}}(b^2+\psi)^{\frac{1}{2}}(c^2+\psi)^{\frac{1}{2}}},$$

$$M=2b^2\int_\epsilon^\infty \frac{xy\,d\psi}{(a^2+\psi)^{\frac{3}{2}}(b^2+\psi)^{\frac{3}{2}}(c^2+\psi)^{\frac{1}{2}}},$$

$$N=-2c^2\int_\epsilon^\infty \frac{zx\,d\psi}{(a^2+\psi)^{\frac{3}{2}}(b^2+\psi)^{\frac{1}{2}}(c^2+\psi)^{\frac{3}{2}}},$$

and ϵ is a positive quantity, given by the equation

$$\frac{x^2}{a^2+\epsilon}+\frac{y^2}{b^2+\epsilon}+\frac{z^2}{c^2+\epsilon}=1.$$

Prove that, if the ellipsoid be a shell filled with water, the values of L, M, N with 0 instead of ϵ for the inferior limit, will similarly determine the velocity of any internal particle of the water. Find the distributions of density, on the surface of the ellipsoid, respectively giving the potentials L, M, N.

(Smith's Prize, 1881.)

CHAPTER VIII

MOTION OF A SOLID THROUGH A LIQUID

8·1. In the foregoing chapters we have considered some simple cases of the motion of a solid through a liquid, chiefly from the kinematical point of view. It is now our purpose to establish dynamical equations for the motion of a solid through an infinite mass of liquid, assuming that the motion of the liquid is due entirely to that of the solid, so that it is irrotational and acyclic. The motion of the liquid is therefore given by a single-valued velocity potential, and by reference to **4·61** we see that the problem is a definite one.

8·11. The dynamical problem possesses features of special interest. It was first solved by Kelvin and Tait* by treating the solid and liquid as one system and using Lagrange's equations and the method of ignoration of coordinates. We shall approach the problem by a different method also due to Lord Kelvin.

8·2. The Impulse. In the general problem which we have to consider, we shall suppose first that the liquid is finite in extent and limited by a *fixed* boundary or envelope, and we shall then proceed to the case of a solid moving in an infinite mass of liquid by supposing the boundary to increase in size until every part of it is at an infinite distance from the moving solid. We saw in **2·71** that any irrotational motion of a liquid may be produced instantaneously from rest by the application of a suitable impulsive pressure at every point of the boundary, and we shall define the *impulse of the motion at any instant* to be the impulsive wrench or system of impulses that, applied to the solid, would generate the motion from rest†. We shall call this briefly 'the impulse'. It is clear that the impulse is equal to the total momentum of the solid and liquid together with the impulsive pressure on the envelope that bounds the liquid.

* *Natural Philosophy*, Art. 320.

† See Lord Kelvin, 'On Vortex Motion', *Trans. R. Soc. Edin.* xxv, 1869, or *Math. and Phys. Papers*, IV, p. 15.

8·21. Example. It will be convenient to recall here the results obtained in a simple case in **7·2**. A solid sphere of radius a moving with velocity V in liquid bounded, at the instant under consideration, by a concentric sphere of radius b. The impulse I necessary to produce the motion instantaneously was calculated and shewn to tend to a definite limit when b is increased to infinity. The impulsive pressure on the envelope was also seen to tend to a definite limit as b is increased to infinity; and the same was shewn to be true of the momentum. We shall see in the next article that the impulse necessary to produce the motion always tends to a definite limit, but except in special cases when the form of the envelope is prescribed the impulsive pressure on the envelope and the momentum are indeterminate.

8·22. The Impulse tends to a definite limit, but the Momentum is generally indeterminate.

We have seen in **4·54** and **4·61** that, whether the surrounding envelope be finite or infinite, if the velocity potential (or impulsive pressure) at each point of the surface of the solid is prescribed, there is only one form of irrotational motion possible. And since any irrotational motion could be produced instantaneously by the application to the solid of a suitable impulsive wrench, and one and only one form of motion can arise from a given impulsive wrench, it follows that, if the envelope be increased indefinitely so that every part of it becomes infinitely distant from the solid, the solid and liquid still having a definite motion, this motion must still be the result of a definite impulse. That is, as the envelope increases without limit the *impulse tends to a definite limit.*

This is not generally true however of the impulsive pressure of the boundary. For the impulsive pressure at a point is measured by $\rho\phi$, and since the envelope is fixed the tubes of flow must all start from and end on the surface of the moving solid, so that at a great distance r from the solid the velocity potential ϕ must be of the same order r^{-2} as the velocity potential due to a doublet. But the element of area of the infinite envelope is of order r^2, so the surface integral of the impulsive pressure on the envelope is in general finite but dependent on the shape of the envelope and therefore indeterminate. Similarly the momentum is in general indeterminate when the mass of liquid is infinite.

8·23. Rate of change of Impulse = external Force.

Considering first the case of a finite mass of liquid and using axes fixed in space, let I_1, I_2 be the x-components of the impulse that would generate the motion from rest and of the impulsive pressure on the envelope at time t; M, X the x-components of the whole momentum and the external force acting on the solid; and (l, m, n) the direction cosines of the outward normal to the element dS of the envelope.

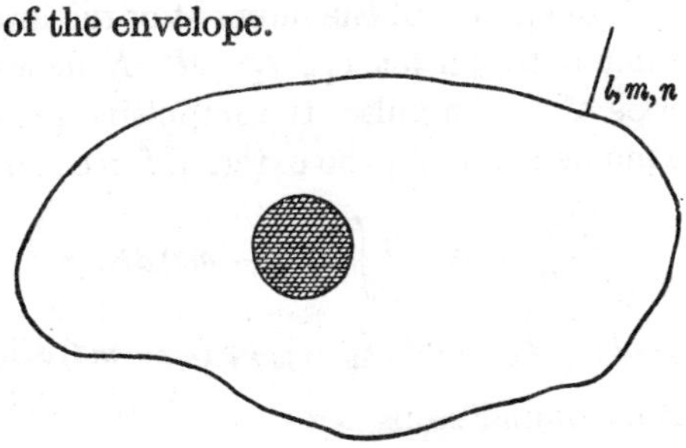

By the ordinary equations of dynamics we have

$$\frac{dM}{dt} = X - \iint p l \, dS,$$

where the integration is over the surface of the envelope.

But $\qquad M = I_1 - I_2$ and $\dfrac{p}{\rho} = \dfrac{\partial \phi}{\partial t} - \tfrac{1}{2}q^2 + F(t),$

where $F(t)$ is an arbitrary function of the time.

Therefore

$$\frac{dI_1}{dt} - \frac{dI_2}{dt} = X - \rho \iint \left\{ \frac{\partial \phi}{\partial t} - \tfrac{1}{2}q^2 + F(t) \right\} l \, dS.$$

But $\qquad I_2 = \iint \rho \phi l \, dS$ and $\dfrac{dI_2}{dt} = \rho \iint \dfrac{\partial \phi}{\partial t} l \, dS;$

also $F(t)$ is constant over the envelope and will give zero result when integrated, so that we get

$$\frac{dI_1}{dt} = X + \tfrac{1}{2}\rho \iint q^2 l \, dS.$$

Now let the envelope increase until every part of it is at an infinite distance from the solid; then, as in the last article, ϕ being of order r^{-2}, q is of order r^{-3} on the surface of the envelope, so that $\iint q^2 l \, dS$ tends to zero, and I_1 tends to a definite limit I, therefore for a solid in an infinite mass of liquid

$$\frac{dI}{dt} = X.$$

As the motion, in general, would require an impulsive wrench to produce it instantaneously, and a linear impulse on the solid might result in an impulsive wrench on the envelope, we must also consider the rate of change of the moment of the impulse.

With a similar notation let I_1', I_2', M', N denote moments about the x axis of the impulse, the impulsive pressure on the envelope, the momentum and the external forces on the solid.

We have $$\frac{dM'}{dt} = N - \iint p\,(ny - mz)\,dS.$$

But $M' = I_1' - I_2'$, and $I_2' = \iint \rho\phi\,(ny - mz)\,dS$,

so that we get by similar steps

$$\frac{dI_1'}{dt} = N + \tfrac{1}{2}\rho \iint q^2\,(ny - mz)\,dS$$

for the case of the finite envelope. When the envelope becomes infinite the surface integral vanishes as before and I_1' tends to a definite limit I', so that

$$\frac{dI'}{dt} = N.$$

8·24. Kinematical Conditions. Before translating the foregoing principles into formal equations of motion, we shall establish some kinematical relations. It will be convenient to take rectangular axes fixed in the body, the origin having velocities u, v, w in the directions of the axes, and the axes having an angular velocity whose components about the axes are p, q, r.

If ϕ be the velocity potential we may write*

$$\phi = u\phi_1 + v\phi_2 + w\phi_3 + p\chi_1 + q\chi_2 + r\chi_3 \qquad \text{.........(1)},$$

where ϕ_1 denotes the velocity potential when the only motion of the body is a translation along the x axis with unit velocity, and χ_1 denotes the velocity potential when the body rotates about the x axis with unit angular velocity, with similar meanings for ϕ_2, ϕ_3, and χ_2, χ_3.

If l, m, n denote the direction cosines of the normal at any point (x, y, z) on the surface of the body, we have

$$-\frac{\partial\phi}{\partial n} = l\,(u - yr + zq) + m\,(v - zp + xr) + n\,(w - xq + yp) \qquad \text{......(2)},$$

* Kirchhoff, *Mechanik*, p. 224.

by equating the normal velocity of the liquid to that of the body. Whence by substituting the value of ϕ from (1) and equating coefficients of u, v, w, p, q, r we get

$$\left.\begin{aligned} -\frac{\partial \phi_1}{\partial n}&=l, & -\frac{\partial \phi_2}{\partial n}&=m, & -\frac{\partial \phi_3}{\partial n}&=n, \\ -\frac{\partial \chi_1}{\partial n}&=ny-mz, & -\frac{\partial \chi_2}{\partial n}&=lz-nx, & -\frac{\partial \chi_3}{\partial n}&=mx-ly \end{aligned}\right\} \quad \ldots(3).$$

We may observe in passing that the values of ϕ_1, χ_1, etc. have been found in the case of an ellipsoid in **7·5** and **7·51**, and that the problem of their determination is a definite one in the general case since they have to satisfy Laplace's equation as well as (3) and their derivatives vanish at infinity, for by hypothesis the liquid is at rest there.

8·3 Equations of Motion. Let $\xi, \eta, \zeta, \lambda, \mu, \nu$ be the components of impulse, and X, Y, Z, L, M, N of the external force system acting on the body at time t referred to axes fixed in the body moving as in **8·24**. At time $t+\delta t$ the coordinates of the origin referred to the axes at time t are $u\,\delta t$, $v\,\delta t$, $w\,\delta t$, and the direction cosines of the axes referred to their former positions are $(1, r\,\delta t, -q\,\delta t)$, $(-r\,\delta t, 1, p\,\delta t)$, $(q\,\delta t, -p\,\delta t, 1)$. Hence by resolving parallel to the new position of the x axis

$$\xi+\delta\xi=\xi+\eta r\,\delta t-\zeta q\,\delta t+X\,\delta t,$$

and by taking moments about the same line

$$\lambda+\delta\lambda=\lambda+\mu r\,\delta t-\nu q\,\delta t+\eta w\,\delta t-\zeta v\,\delta t+L\,\delta t,$$

whence we get the six equations of motion

$$\begin{aligned} \dot\xi-\eta r+\zeta q&=X, & \dot\lambda-\mu r+\nu q-\eta w+\zeta v&=L, \\ \dot\eta-\zeta p+\xi r&=Y, & \dot\mu-\nu p+\lambda r-\zeta u+\xi w&=M, \\ \dot\zeta-\xi q+\eta p&=Z, & \dot\nu-\lambda q+\mu p-\xi v+\eta u&=N. \end{aligned}$$

As suggested by Lord Kelvin, these equations may conveniently be called the Eulerian equations of motion, since they refer to axes fixed in the moving body and correspond precisely to Euler's equations for the rotation of a rigid body*.

* *Math. and Phys. Papers*, IV, p. 70 footnote.

8·31. The Kinetic Energy. The kinetic energy of the liquid, by **4·71**, is given by

$$T = -\tfrac{1}{2}\rho \iint \phi \frac{\partial \phi}{\partial n} dS \quad \text{.....................(1)},$$

where the integration extends to the surface of the moving solid. From **8·24** (1) it follows that T is a homogeneous quadratic function of the velocity components u, v, w, p, q, r, so that we have

$$\begin{aligned} 2T = {} & Au^2 + Bv^2 + Cw^2 + 2A'vw + 2B'wu + 2C'uv \\ & + Pp^2 + Qq^2 + Rr^2 + 2P'qr + 2Q'rp + 2R'pq \\ & + 2p\,(Fu + Gv + Hw) + 2q\,(F'u + G'v + H'w) \\ & + 2r\,(F''u + G''v + H''w) \quad \text{...(2)}, \end{aligned}$$

where the coefficients A, B, etc. by the help of **8·24** (3) can be expressed in the form

$$\left.\begin{aligned} A &= -\rho \iint \phi_1 \frac{\partial \phi_1}{\partial n} dS = \rho \iint l\phi_1 dS, \\ A' &= -\tfrac{1}{2}\rho \iint \left(\phi_2 \frac{\partial \phi_3}{\partial n} + \phi_3 \frac{\partial \phi_2}{\partial n}\right) dS, \\ &= -\rho \iint \phi_2 \frac{\partial \phi_3}{\partial n} dS = -\rho \iint \phi_3 \frac{\partial \phi_2}{\partial n} dS \quad \text{by } \mathbf{4{\cdot}52} \text{ (ii)} \\ &= \rho \iint n\phi_2 dS = \rho \iint m\phi_3 dS, \\ P &= -\rho \iint \chi_1 \frac{\partial \chi_1}{\partial n} dS = \rho \iint \chi_1 (ny - mz)\, dS \\ & \quad \text{etc.} \end{aligned}\right\} \text{...(3)}.$$

The kinetic energy of the solid is also a homogeneous quadratic function of the velocities, so that the whole kinetic energy of the solid and liquid is an expression of the form (2), wherein the coefficients are only represented in part by the expressions (3).

8·32. Impulse in terms of Velocities. It is a well-known dynamical theorem that the work done by an impulse is the product of the impulse and the mean of the velocities of its point of application before and after it acts. Accordingly an extra impulse $\delta\xi$ in the x direction would do work $\delta\xi\,(u + \tfrac{1}{2}\delta u)$, where $u + \delta u$ is the velocity in the same direction after the impulse $\delta\xi$ has taken place; and if $\delta\xi$ be infinitely small we may take $u\,\delta\xi$ to represent the work done or the increase of kinetic energy. Hence when the 'impulse' receives infinitesimal increments $\delta\xi$, $\delta\eta$, $\delta\zeta$, $\delta\lambda$, $\delta\mu$, $\delta\nu$ there is an increase of kinetic energy δT given by

$$\delta T = u\,\delta\xi + v\,\delta\eta + w\,\delta\zeta + p\,\delta\lambda + q\,\delta\mu + r\,\delta\nu \quad \ldots\ldots(1).$$

But

$$\delta T = \frac{\partial T}{\partial u}\delta u + \frac{\partial T}{\partial v}\delta v + \frac{\partial T}{\partial w}\delta w + \frac{\partial T}{\partial p}\delta p + \frac{\partial T}{\partial q}\delta q + \frac{\partial T}{\partial r}\delta r \ldots(2),$$

and if the velocities are all altered in a given ratio it is clear that the impulses will be altered in the same ratio, so that if we write

$$\delta u/u = \delta v/v = \ldots = \delta r/r = \kappa,$$

we must also have

$$\delta\xi/\xi = \delta\eta/\eta = \ldots = \delta\nu/\nu = \kappa.$$

Whence by equating the two expressions for δT in (1) and (2) and substituting from the last equations we get

$$u\xi + v\eta + w\zeta + p\lambda + q\mu + r\nu$$

$$= u\frac{\partial T}{\partial u} + v\frac{\partial T}{\partial v} + w\frac{\partial T}{\partial w} + p\frac{\partial T}{\partial p} + q\frac{\partial T}{\partial q} + r\frac{\partial T}{\partial r} = 2T \ldots(3),$$

since T is a homogeneous function of u, v, etc.

By varying this last equation we get

$$2\delta T = (u\,\delta\xi + \xi\,\delta u) + \ldots + (r\,\delta\nu + \nu\,\delta r);$$

and therefore by subtracting (1)

$$\delta T = \xi\,\delta u + \eta\,\delta v + \zeta\,\delta w + \lambda\,\delta p + \mu\,\delta q + \nu\,\delta r.$$

Comparing the last result with (2), since the small variations are arbitrary, we get

$$\xi, \eta, \zeta, \lambda, \mu, \nu = \frac{\partial T}{\partial u}, \frac{\partial T}{\partial v}, \frac{\partial T}{\partial w}, \frac{\partial T}{\partial p}, \frac{\partial T}{\partial q}, \frac{\partial T}{\partial r} \ldots\ldots(4).$$

These results imply that the components of impulse are linear functions of the components of the velocity, hence the kinetic energy may also be expressed as a homogeneous quadratic function of the components of impulse; and when T is so expressed we get from (1) the reciprocal relations

$$u, v, w, p, q, r = \frac{\partial T}{\partial \xi}, \frac{\partial T}{\partial \eta}, \frac{\partial T}{\partial \zeta}, \frac{\partial T}{\partial \lambda}, \frac{\partial T}{\partial \mu}, \frac{\partial T}{\partial \nu} \ldots(5).$$

8·33. Equations of Motion. The equations of 8·3 now take the form*

$$\frac{d}{dt}\frac{\partial T}{\partial u}=r\frac{\partial T}{\partial v}-q\frac{\partial T}{\partial w}+X,$$

$$\frac{d}{dt}\frac{\partial T}{\partial v}=p\frac{\partial T}{\partial w}-r\frac{\partial T}{\partial u}+Y,$$

$$\frac{d}{dt}\frac{\partial T}{\partial w}=q\frac{\partial T}{\partial u}-p\frac{\partial T}{\partial v}+Z,$$

$$\frac{d}{dt}\frac{\partial T}{\partial p}=r\frac{\partial T}{\partial q}-q\frac{\partial T}{\partial r}+w\frac{\partial T}{\partial v}-v\frac{\partial T}{\partial w}+L,$$

$$\frac{d}{dt}\frac{\partial T}{\partial q}=p\frac{\partial T}{\partial r}-r\frac{\partial T}{\partial p}+u\frac{\partial T}{\partial w}-w\frac{\partial T}{\partial u}+M,$$

$$\frac{d}{dt}\frac{\partial T}{\partial r}=q\frac{\partial T}{\partial p}-p\frac{\partial T}{\partial q}+v\frac{\partial T}{\partial u}-u\frac{\partial T}{\partial v}+N.$$

In the case in which there are no extraneous forces we can obtain three integrals of these equations. Thus if we multiply them by u, v, w, p, q, r and add, we get

$$u\frac{d}{dt}\frac{\partial T}{\partial u}+\ldots+r\frac{d}{dt}\frac{\partial T}{\partial r}=0 \quad\ldots\ldots\ldots\ldots\ldots\ldots(1).$$

But
$$2T=u\frac{\partial T}{\partial u}+\ldots+r\frac{\partial T}{\partial r},$$

therefore
$$2\frac{dT}{dt}=u\frac{d}{dt}\frac{\partial T}{\partial u}+\frac{\partial T}{\partial u}\frac{du}{dt}+\ldots \quad\ldots\ldots\ldots\ldots\ldots(2),$$

but
$$\frac{dT}{dt}=\frac{\partial T}{\partial u}\frac{du}{dt}+\frac{\partial T}{\partial v}\frac{dv}{dt}+\ldots \quad\ldots\ldots\ldots\ldots\ldots(3),$$

and by subtracting (1) and (3) from (2) we get the equation of energy

$$\frac{dT}{dt}=0, \text{ or } T=\text{const.}$$

Again, if we multiply the first three of the equations of motion by $\partial T/\partial u$, $\partial T/\partial v$, $\partial T/\partial w$ and add and integrate, we get

$$\left(\frac{\partial T}{\partial u}\right)^2+\left(\frac{\partial T}{\partial v}\right)^2+\left(\frac{\partial T}{\partial w}\right)^2=\text{const.},$$

or
$$\xi^2+\eta^2+\zeta^2=\text{const.},$$

which represents that the linear component of the impulse or the intensity of the impulsive wrench is constant.

* Kelvin, 'Hydrokinetic solutions and observations', *Phil. Mag.* XLII, p. 362, or *Math. and Phys. Papers*, IV, p. 69. Also Kirchhoff, *Mechanik*, p. 236.

And if we multiply the six equations by

$$\partial T/\partial p, \quad \partial T/\partial q, \quad \partial T/\partial r, \quad \partial T/\partial u, \quad \partial T/\partial v, \quad \partial T/\partial w,$$

we get
$$\frac{\partial T}{\partial u}\frac{\partial T}{\partial p}+\frac{\partial T}{\partial v}\frac{\partial T}{\partial q}+\frac{\partial T}{\partial w}\frac{\partial T}{\partial r}=\text{const.},$$

or
$$\xi\lambda+\eta\mu+\zeta\nu=\text{const.},$$

which represents that the couple component or the pitch of the impulsive wrench is also constant.

8·34. Directions of Permanent Translation. When there are no external forces the equations of motion of **8·33** are satisfied by $p=q=r=0$, provided u, v, w have constant values such that

$$u:v:w=\frac{\partial T}{\partial u}:\frac{\partial T}{\partial v}:\frac{\partial T}{\partial w} \quad \text{.................(1)}.$$

In this case T is a homogeneous function of u, v, w only, of the form

$$2T=Au^2+Bv^2+Cw^2+2A'vw+2B'wu+2C'uv \quad \text{...(2)}.$$

If we regard u, v, w as current coordinates the equation

$$2T=\text{const.}$$

represents an ellipsoid, and the equations (1) determine its principal axes.

Consequently if the body be set moving without rotation in the direction of any one of the axes of this ellipsoid it will continue to move in the same direction without rotation*.

The stability of the motions has been discussed by H. D. Ursell †.

8·4. Hydrokinetic Symmetry. The expression for the kinetic energy in **8·31** contains 21 constants, but the number of terms is reduced in particular cases. Thus the coefficients A', B', C' can always be got rid of by rotating the axes. Also

(1) If the body has three perpendicular planes of symmetry the energy must remain unaltered when the sign of any velocity component is reversed, so that

$$2T=Au^2+Bv^2+Cw^2+Pp^2+Qq^2+Rr^2.$$

(2) If the body is in addition a surface of revolution about Ox, the expression for $2T$ must remain unaltered when we write $v, q, -w, -r$ for w, r, v, q, respectively, for this is equivalent to

* Kirchhoff, *Mechanik*, p. 236.

† 'Motion of a solid through an infinite liquid', *Proc. Camb. Phil. Soc.* XXXVII, p. 150.

turning the axes of yz through a right angle; hence $B = C$ and $Q = R$, so that

$$2T = Au^2 + B(v^2 + w^2) + Pp^2 + Q(q^2 + r^2).$$

The same expression holds when the solid is a right prism whose cross section is a regular polygon*.

(3) When the body is similarly related to the three planes of symmetry as in the case of a sphere or cube we have

$$2T = A(u^2 + v^2 + w^2) + P(p^2 + q^2 + r^2).$$

(4) Another kind of symmetry is that represented by the expression

$$2T = A(u^2 + v^2 + w^2) + P(p^2 + q^2 + r^2) + 2L(up + vq + wr),$$

the form of which is unaltered by any changes in the directions of the axes, and any direction is one of permanent translation. Such a solid is said to be 'helicoidally isotropic†'.

8·5 Applications. Sphere. Taking u, v, w as the components of velocity of the centre of the sphere

$$2T = A(u^2 + v^2 + w^2),$$

where $$\phi = u\phi_1 + v\phi_2 + w\phi_3,$$

and $$\phi_1 = \frac{a^3 x}{2r^3} = \frac{a^3 \cos\theta}{2r^2}, \text{ as in } \mathbf{7{\cdot}11}.$$

Hence
$$\begin{aligned} A &= M + \rho \iint \phi_1 l \, dS \\ &= M + \pi\rho a^3 \int_0^\pi \cos^2\theta \sin\theta \, d\theta \\ &= M + \tfrac{1}{2}M', \end{aligned}$$

where M' is the mass of liquid displaced.

Therefore $$2T = (M + \tfrac{1}{2}M')(u^2 + v^2 + w^2),$$

and $$\xi, \eta, \zeta = (M + \tfrac{1}{2}M')(u, v, w).$$

The equations of motion, in this case, become

$$(M + \tfrac{1}{2}M')(\dot{u}, \dot{v}, \dot{w}) = (X, Y, Z), \text{ as in } \mathbf{7{\cdot}13},$$

where X, Y, Z are the components of external force on the sphere.

If external forces act on the liquid as well, their effect on the sphere is expressed by adding to X, Y, Z the reversed effect that these forces would exert on the liquid displaced by the sphere.

* Larmor, 'On Hydrokinetic Symmetry', *Quart. Journal*, XX, p. 261, or Kirchhoff, *Mechanik*, p. 243.

† See Kelvin, 'Hydrokinetic solutions and observations', *Phil. Mag.* XLII, p. 365, or *Math. and Phys. Papers*, IV, p. 72.

For other special forms see Lamb's *Hydrodynamics*, Art. 126, or Larmor, *loc. cit.*

8·51. Solid of Revolution. Taking the axis of the solid for axis of x, we have

$$2T = Au^2 + B(v^2+w^2) + Pp^2 + Q(q^2+r^2) \quad \text{.........(1).}$$

Assuming that there are no impressed forces, the equations of motion of **8·33** become

$$A\dot{u} = Brv - Bqw \quad \text{..............(2),}$$

$$B\dot{v} = Bpw - Aru \quad \text{..............(3),}$$

$$B\dot{w} = Aqu - Bpv \quad \text{..............(4),}$$

$$P\dot{p} = 0 \quad \text{..............(5),}$$

$$Q\dot{q} = (Q-P)pr + (B-A)uw \quad \text{..............(6),}$$

$$Q\dot{r} = (P-Q)pq + (A-B)uv \quad \text{..............(7).}$$

From (5) we see that p is constant throughout the motion. We can also deduce as in **8·33** three integrals

$$T = \text{const.} \quad \text{..............(8),}$$

$$A^2u^2 + B^2(v^2+w^2) = I^2 \quad \text{..............(9),}$$

and

$$APup + BQ(vq+wr) = IG \quad \text{..............(10),}$$

where I, G are the constant components of the impulsive wrench at any instant.

From (1), (2), (8), (9), (10) we can eliminate v, w, q, r. Thus

$$B^2(v^2+w^2) = I^2 - A^2u^2,$$

$$Q(q^2+r^2) = 2T - Au^2 - B(v^2+w^2) - Pp^2$$

$$= 2T - A^2\left(\frac{1}{A} - \frac{1}{B}\right)u^2 - \frac{I^2}{B} - Pp^2,$$

and

$$BQ(vq+wr) = IG - APup,$$

therefore

$$A^2\dot{u}^2 = B^2(rv-qw)^2$$

$$= B^2\{(v^2+w^2)(q^2+r^2) - (vq+wr)^2\}$$

$$= \frac{I^2-A^2u^2}{Q}\left\{2T - A^2\left(\frac{1}{A}-\frac{1}{B}\right)u^2 - \frac{I^2}{B} - Pp^2\right\} - \left(\frac{IG-APup}{Q}\right)^2,$$

a polynomial of the fourth degree in Au so that Au is an elliptic function of the time.

Again, if we put $v/w = \tan\psi$, we have

$$(v^2+w^2)\dot{\psi} = w\dot{v} - v\dot{w}$$

$$= p(v^2+w^2) - Au(qv+rw)/B, \text{ from (3) and (4).}$$

Therefore

$$\dot{\psi} = p - \frac{Au}{Q}\cdot\frac{IG-APup}{I^2-A^2u^2}.$$

Thus, having expressed u in terms of the time, the last relation gives v/w and (9) gives v^2+w^2, then p being constant (8) and (10) determine q and r, so that all the velocity components are determined.

The evaluation in terms of elliptic functions was first performed by Kirchhoff, and the problem has been discussed at length by Greenhill* and others.

* *American Journal of Mathematics*, 1898, 1906.

8·52. Solid of Revolution—Quadrantal Pendulum. The case considered in **8·51** is much simplified if the axis of the solid moves in a fixed plane. Taking this as the plane xy we have $w=p=q=0$, and the equations of **8·51** become

$$A\dot{u}=Brv, \quad B\dot{v}=-Aru, \quad Q\dot{r}=(A-B)\,uv,$$

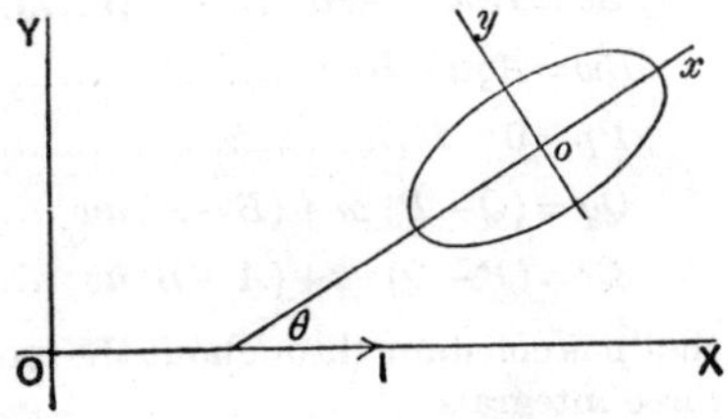

the three integrals reducing to two

$$Au^2+Bv^2+Qr^2=\text{const.},$$

and
$$A^2u^2+B^2v^2=I^2,$$

the third being an identity, as the 'impulse' at any instant consists of a single impulsive force I.

Let $\mathbf{x}, \mathbf{y}$ be the coordinates of the centre of gravity o of the solid referred to axes fixed in the given plane whereof the $\mathbf{x}$ axis coincides with the line of the impulse I and makes an angle θ with ox.

Then
$$r=\dot{\theta}, \quad Au=I\cos\theta, \quad Bv=-I\sin\theta,$$

so that the first two equations of motion are satisfied identically, expressing the fact that the impulse is fixed in magnitude and direction. The third equation gives

$$Q\ddot{\theta}+\frac{A-B}{AB}I^2\cos\theta\sin\theta=0 \quad \text{.....................(1)},$$

or, if we write $2\theta=\phi$,
$$\ddot{\phi}+\frac{A-B}{ABQ}I^2\sin\phi=0 \quad \text{..........................(2)},$$

shewing that the motion corresponds to that of a simple pendulum, the body moving according to the same law through a quadrant on each side of its mean position, as the common pendulum with reference to a half circle on each side. A body moving in such a manner is called a *Quadrantal Pendulum**. This motion is acquired by a solid of revolution in an infinite mass of liquid when it is given a rotation about an axis perpendicular to its axis of figure, or simply projected without rotation.

The body, as it moves, may make complete revolutions or it may oscillate about a mean position.

(i) In the case of complete revolutions we may write the first integral of (1)
$$\dot{\theta}^2=\omega^2(1-\kappa^2\sin^2\theta),$$

where ω is the value of $\dot{\theta}$ in the position $\theta=0$ and

$$\omega^2\kappa^2=(A-B)\,I^2/ABQ \quad \text{.....................(3)}.$$

* Kelvin and Tait, *Natural Philosophy*, § 322.

Hence $$\omega t=\int_0^\theta \frac{d\theta}{(1-\kappa^2\sin^2\theta)^{\frac{1}{2}}}$$

$$=\int_0^\zeta \frac{d\zeta}{(1-\zeta^2)^{\frac{1}{2}}(1-\kappa^2\zeta^2)^{\frac{1}{2}}},\text{ where } \zeta=\sin\theta.$$

Therefore $$\sin\theta=\zeta=\operatorname{sn}\omega t \qquad\qquad (4),$$

where κ, as given by (3), is the modulus of the elliptic function.

(ii) In the case of oscillations through an angle 2α about the position $\theta=0$, we may write the first integral of (1)

$$\dot\theta^2=\omega^2\left(1-\frac{\sin^2\theta}{\sin^2\alpha}\right),$$

where $$\sin^2\alpha=\frac{ABQ}{A-B}\cdot\frac{\omega^2}{I^2} \qquad\qquad (5).$$

Therefore $$\omega t=\int_0^\theta \frac{\sin\alpha\,.\,d\theta}{(\sin^2\alpha-\sin^2\theta)^{\frac{1}{2}}},\text{ or if } \sin\theta=\zeta\sin\alpha,$$

$$=\sin\alpha\int_0^\zeta \frac{d\zeta}{(1-\zeta^2)^{\frac{1}{2}}(1-\sin^2\alpha\,.\,\zeta^2)^{\frac{1}{2}}},$$

so that $$\sin\theta=\zeta\sin\alpha=\sin\alpha\operatorname{sn}(\omega t\operatorname{cosec}\alpha) \qquad\qquad (6),$$

where $\sin\alpha$, as given by (5), is the modulus of the elliptic function.

To find the path of the centre of gravity we have

$$\dot{\mathbf{x}}=u\cos\theta-v\sin\theta=I\left(\frac{\cos^2\theta}{A}+\frac{\sin^2\theta}{B}\right),$$

and $$\dot{\mathbf{y}}=u\sin\theta+v\cos\theta=I\left(\frac{1}{A}-\frac{1}{B}\right)\sin\theta\cos\theta.$$

Hence in case (i)

$$\dot{\mathbf{x}}=I\left\{\frac{1}{A}+\left(\frac{1}{B}-\frac{1}{A}\right)\sin^2\theta\right\}$$

$$=I\left\{\frac{1}{A}+\left(\frac{1}{B}-\frac{1}{A}\right)\operatorname{sn}^2\omega t\right\},\text{ from (4)},$$

$$=I\left\{\frac{1}{A}+\left(\frac{1}{B}-\frac{1}{A}\right)\frac{1-\operatorname{dn}^2\omega t}{\kappa^2}\right\}$$

$$=I\left(\frac{1}{A}+\frac{A-B}{AB\kappa^2}\right)-\frac{I(A-B)}{AB\kappa^2}\operatorname{dn}^2\omega t$$

$$=\left(\frac{I}{A}+\frac{Q\omega^2}{I}\right)-\frac{Q\omega}{I}\operatorname{dn}^2\omega t,\text{ from (3)}.$$

Therefore $$\mathbf{x}=\left(\frac{I}{A}+\frac{Q\omega^2}{I}\right)t-\frac{Q\omega}{I}E(\omega t,\kappa),$$

where E is the elliptic integral of the second kind.

Similarly $$\dot{\mathbf{y}}=I\left(\frac{1}{A}-\frac{1}{B}\right)\operatorname{sn}\omega t\operatorname{cn}\omega t,\text{ from (4)},$$

therefore $$\mathbf{y}=I\frac{(A-B)}{AB}\frac{\operatorname{dn}\omega t}{\omega\kappa^2}$$

$$=\frac{Q\omega}{I}\operatorname{dn}\omega t,\text{ from (3)}.$$

In case (ii), in like manner, putting v for ωt cosec α,

$$\dot{\mathbf{x}} = I\left(\frac{1}{A} + \frac{A-B}{AB}\sin^2\alpha\,\mathrm{sn}^2 v\right)$$

$$= \frac{I}{A} + \frac{Q\omega^2}{I}\frac{1-\mathrm{dn}^2 v}{\sin^2\alpha}, \text{ from (5).}$$

Therefore $\mathbf{x} = \left(\dfrac{I}{A} + \dfrac{Q\omega^2}{I\sin^2\alpha}\right)t - \dfrac{Q\omega}{I\sin\alpha}E(\omega t\,\mathrm{cosec}\,\alpha, \sin\alpha).$

Similarly
$$\dot{\mathbf{y}} = -I\frac{(A-B)}{AB}\sin\alpha\,\mathrm{sn}\,v\,\mathrm{dn}\,v,$$

therefore
$$\mathbf{y} = I\frac{(A-B)}{AB\omega}\sin^2\alpha\,\mathrm{cn}\,v$$

$$= \frac{Q\omega}{I}\mathrm{cn}\,(\omega t\,\mathrm{cosec}\,\alpha, \sin\alpha).$$

In either case we see that the velocity of the centre of gravity consists of a constant part in a fixed direction together with periodic parts along and perpendicular to this direction.

There is an intermediate case in which

$$ABQ\omega^2 = (A-B)I^2,$$

corresponding to $\kappa = 1$, or $\alpha = \pi/2$; then we have

$$\dot{\theta} = \omega\cos\theta,$$

so that
$$\omega t = \log\tan(\tfrac{1}{4}\pi + \tfrac{1}{2}\theta).$$

Also
$$\dot{\mathbf{x}} = \frac{I}{A} + I\frac{(A-B)}{AB}\sin^2\theta$$

$$= \frac{I}{A} + \frac{Q\omega^2}{I}\tanh^2\omega t,$$

and therefore
$$\mathbf{x} = \left(\frac{I}{A} + \frac{Q\omega^2}{I}\right)t - \frac{Q\omega}{I}\tanh\omega t.$$

Also
$$\dot{\mathbf{v}} = -I\frac{(A-B)}{AB}\sin\theta\cos\theta$$

$$= -\frac{Q\omega^2}{I}\tanh\omega t\,\mathrm{sech}\,\omega t,$$

so that
$$\mathbf{y} = \frac{Q\omega}{I}\mathrm{sech}\,\omega t.$$

In case (i) the curve described by the centre of gravity does not cross the line of the impulse, but in case (ii) the curve is a sinuous one crossing the line of the impulse at regular intervals, the points of crossing marking the extreme positions of the axis of the solid in its swing about its mean position.

8·53. Cylinder. In the two-dimensional motion of an infinitely long cylinder in an infinite mass of liquid, the expression for the kinetic energy included between two planes perpendicular to the length of the cylinder at unit distance apart is

$$2T = Au^2 + Bv^2 + Qr^2,$$

with the same notation as in the last article. The motion of the cylinder is therefore given by the results of the preceding article. The curves described by the centre of the cylinder are to be found in Lamb's *Hydrodynamics*, 1932, p. 176.

8·54. Stability. Let us consider the stability of a solid of revolution moving uniformly along its axis of figure. In the equations of **8·51** we may put $u=u_0+u'$ and regard u', v, w, p, q, r as small, then we get

$$A\dot{u}'=0, \quad B\dot{v}=-Aru_0, \quad B\dot{w}=Aqu_0,$$
$$P\dot{p}=0, \quad Q\dot{q}=(B-A)\,u_0 w, \quad Q\dot{r}=(A-B)\,u_0 v.$$

Hence
$$B\ddot{v}+\frac{A\,(A-B)}{Q}\,u_0^2 v=0,$$

with similar equations for w, q and r.

Therefore the motion is not stable unless $A>B$.

For an ellipsoid we have

$$A=M+\rho\iint \phi_1 l\,dS$$
$$=M+\rho\iint \frac{\alpha_0 x}{2-\alpha_0}\,l\,dS, \; (7\cdot 5)$$

and
$$\iint xl\,dS=\tfrac{4}{3}\pi abc,$$

so that
$$A=M+\tfrac{4}{3}\pi\rho abc\,\frac{\alpha_0}{2-\alpha_0},$$

similarly
$$B=M+\tfrac{4}{3}\pi\rho abc\,\frac{\beta_0}{2-\beta_0}.$$

Hence we have $A>B$, provided $\alpha_0>\beta_0$, where α_0, β_0 are as defined in **7·5**. And $\alpha_0>\beta_0$ requires that $a<b$; thus it follows that when an oblate spheroid moves uniformly along its axis the motion is stable, but for a prolate spheroid the motion is unstable. This accords with the observed tendency of a body to turn its flat side or its length across the direction of its motion.

8·55. Stability increased by Rotation. Now let us suppose that the solid of revolution is moving with velocity u_0 along its axis and angular velocity p_0 about its axis. When a slight disturbance takes place we may put $u=u_0+u'$, $p=p_0+p'$ and regard u', v, w, p', q, r as small. The equations of motion of **8·51** become

$$A\dot{u}'=0, \quad B\dot{v}=Bp_0 w-Au_0 r, \quad B\dot{w}=Au_0 q-Bp_0 v,$$
$$P\dot{p}'=0, \quad Q\dot{q}=(Q-P)\,p_0 r+(B-A)\,u_0 w, \quad Q\dot{r}=(P-Q)\,p_0 q+(A-B)\,u_0 v.$$

These give $u'=\text{const.}$, $p'=\text{const.}$, and if we assume that

$$v=\lambda_1 e^{i\sigma t}, \quad w=\lambda_2 e^{i\sigma t}, \quad q=\lambda_3 e^{i\sigma t}, \quad r=\lambda_4 e^{i\sigma t},$$

we get
$$Bi\sigma\lambda_1-Bp_0\lambda_2+Au_0\lambda_4=0,$$
$$Bi\sigma\lambda_2-Au_0\lambda_3+Bp_0\lambda_1=0,$$
$$Qi\sigma\lambda_3+(P-Q)\,p_0\lambda_4+(A-B)\,u_0\lambda_2=0,$$
$$Qi\sigma\lambda_4-(P-Q)\,p_0\lambda_3-(A-B)\,u_0\lambda_1=0.$$

The elimination of $\lambda_1, \lambda_2, \lambda_3, \lambda_4$ gives a biquadratic for σ, which resolves into two quadratics

$$BQ\sigma^2\pm B\,(P-2Q)\,p_0\sigma-\{B\,(P-Q)\,p_0^2+A\,(A-B)\,u_0^2\}=0,$$

and the condition for real roots, which must be satisfied for small oscillations, is that

$$(P-2Q)^2 p_0^2+4Q\,(P-Q)\,p_0^2+4\frac{A}{B}(A-B)\,Qu_0^2$$

should be positive, or that $P^2p_0^2+4\dfrac{A}{B}(A-B)Qu_0^2$ should be positive.

This condition is always satisfied if $A>B$; and when $B>A$ the condition can be satisfied by making p_0 large enough. That is, an elongated projectile can be made to move in the direction of its axis by giving it a sufficiently great angular velocity. This explains the necessity for the rifling of guns. But Ursell has shewn that for some bodies a steady translation can be made unstable by rotation, *loc. cit.* p. 195.

8·56. Steady motion of Solid of Revolution in a Helical Path. As in **8·51** when there are no external forces we have

$$A\dot{u}=B(rv-qw),\quad B\dot{v}=Bpw-Aru,\quad B\dot{w}=Aqu-Bpv,$$

$$P\dot{p}=0,\quad Q\dot{q}=(Q-P)pr+(B-A)uw,\quad Q\dot{r}=(P-Q)pq+(A-B)uv.$$

If we make the hypothesis that $rv-qw=0$ the equations are satisfied by

$$u=\text{const., and } v^2+w^2=\text{const.},$$

and we have also $\quad p=\text{const., and } q^2+r^2=\text{const.}$

Let F, G be the impulsive force and couple that constitute the impulsive wrench at any instant; since there are no forces the axis OZ of this wrench is fixed in space. Let O' be the centre of gravity of the body, $O'O$ perpendicular to OZ and F, G' the force and couple components of the impulse referred to O' as origin. Then ξ, η, ζ are the components of F and λ, μ, ν those of G', where

$$\xi,\ \eta,\ \zeta=Au,\ Bv,\ Bw,$$

and

$$\lambda,\ \mu,\ \nu=Pp,\ Qq,\ Qr.$$

Since $rv=qw$, the direction of the motion of O' given by (u, v, w) is coplanar with F and G', i.e. in a plane perpendicular to OO'. Therefore OO' is of constant length.

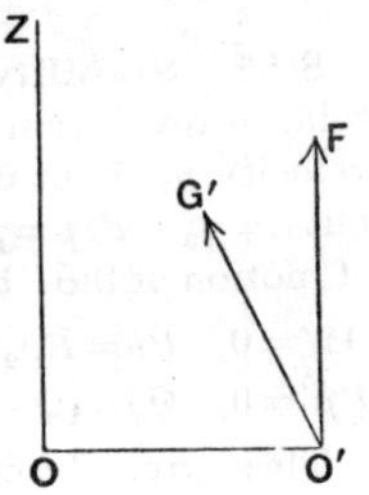

Again, if U denote the velocity of O', so that

$$U^2=u^2+v^2+w^2,$$

the angle ϕ between U and F is given by

$$\cos\phi=\frac{Au^2+B(v^2+w^2)}{UF}=\text{const.}$$

Therefore O' describes a helix round the axis OZ of the impulse, the velocity parallel to OZ being

$$U\cos\phi,$$

and the plane ZOO' turning round OZ with angular velocity

$$U\sin\phi/OO'.$$

The axis of the solid of revolution, its direction cosines being (1, 0, 0), and the instantaneous axis of rotation (p, q, r) are also clearly coplanar with F, G' and make constant angles with OZ. Hence the motion is a steady motion.

8·57. Steady motion of Isotropic Helicoid under no forces. In this case

$$2T=A(u^2+v^2+w^2)+P(p^2+q^2+r^2)+2L(up+vq+wr)$$
$$=AU^2+P\Omega^2+2L\Omega U\cos\theta,$$

where U, Ω are the resultant linear and angular velocities and θ the angle between the direction of U and the axis of Ω.

Representing the impulsive wrench as in the last article, we have for the components of F and G',

$$\xi, \eta, \zeta = A(u, v, w) + L(p, q, r),$$
$$\lambda, \mu, \nu = P(p, q, r) + L(u, v, w).$$

Therefore F is the resultant of vectors AU and $L\Omega$, and G' is the resultant of vectors $P\Omega$ and LU.

Hence as in the last article the directions of the vectors U and Ω must lie in the plane of F, G', i.e. in the plane through O' perpendicular to OO'. As before OO' is of constant length, and therefore G' and the angle $G'O'F$ are constant and therefore U and Ω are constant and make constant angles with F.

As in **8·56** O' describes a helix.

Also U is the resultant of

$$\frac{PF}{AP-L^2} \quad \text{and} \quad -\frac{LG'}{AP-L^2},$$

and if the angle $FO'G' = \alpha$, $G' \cos\alpha = G$, and $G' \sin\alpha = F.OO'$.

Hence the velocity of O' parallel to OZ is

$$U\cos\beta = \frac{PF}{AP-L^2} - \frac{LG'\cos\alpha}{AP-L^2} = \frac{PF-LG}{AP-L^2},$$

where β is the angle between U and OZ; and the angular velocity about OZ is

$$\frac{U\sin\beta}{OO'} = -\frac{LG'\sin\alpha}{OO'(AP-L^2)} = -\frac{LF}{AP-L^2}.$$

Hence the pitch of the helix is $(LG-PF)/LF$.

Since Ω is the resultant of $\dfrac{AG'}{AP-L^2}$ and $-\dfrac{LF}{AP-L^2}$ it is also completely determined when the impulse and the distance of the centre of gravity from the impulse are known, and thus the motion is completely determined in terms of these data*.

8·6. Two Spheres.

Though the general discussion of the motion of two or more solids through a liquid may be regarded as beyond the scope of this book, there are some special cases which are capable of treatment by fairly simple methods so far as approximate results are concerned. The first of these is the motion of two spheres, moving (1) in their line of centres, (2) in parallel directions at right angles to their line of centres.

8·61. *Two spheres moving in their line of centres.*

Let A, B be the centres, a, b the radii, c the distance AB and U, U' the velocities of A along AB and of B along BA. Let (r, θ), (r', θ') be polar coordinates of a point P measured as in the figure.

* For a method of constructing an isotropic helicoid see Kelvin, 'Hydrokinetic solutions and observations', *Phil. Mag.* XLII, or *Math. and Phys. Papers*, IV, p. 73. For other cases of motion of an isotropic helicoid see Miss Fawcett, 'Note on the motion of solids in a liquid', *Quart. Journal*, XXVI, p. 231.

The velocity potential will be of the form

$$U\phi + U'\phi',$$

and the kinetic energy of the liquid will be given by

$$2T = LU^2 + 2MUU' + NU'^2 \quad \text{.....................(1)},$$

where as in 8·31

$$L = -\rho \iint \phi \frac{\partial \phi}{\partial n} dS_A, \quad M = -\rho \iint \phi \frac{\partial \phi'}{\partial n} dS_B, \quad N = -\rho \iint \phi' \frac{\partial \phi'}{\partial n} dS_B \quad \text{......(2)}.$$

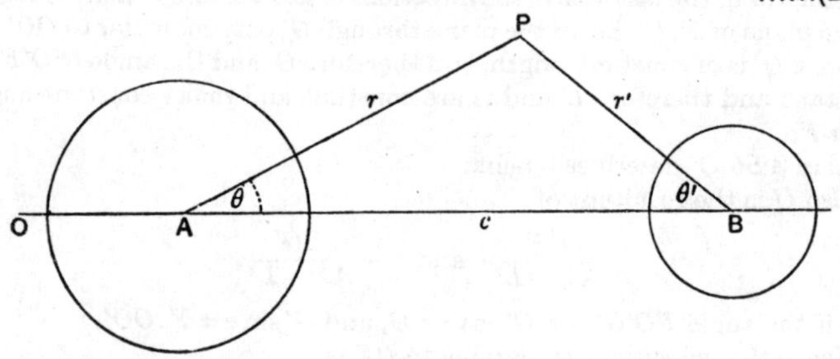

To find the values of ϕ, ϕ' we might use the method of successive images, each sphere when alone in the liquid producing the same effect as a doublet; but it is simpler to proceed as follows.

The boundary conditions to be satisfied are

$$\frac{\partial \phi}{\partial r} = -\cos\theta \text{ over } A, \quad \text{and} \quad \frac{\partial \phi}{\partial r'} = 0 \text{ over } B;$$

$$\frac{\partial \phi'}{\partial r} = 0 \text{ over } A, \quad \text{and} \quad \frac{\partial \phi'}{\partial r'} = -\cos\theta' \text{ over } B.$$

If the sphere A were alone in the liquid, moving with unit velocity, we should have a velocity potential

$$\phi_1 = \tfrac{1}{2}\frac{a^3}{r^2}\cos\theta,$$

which would make $\partial\phi_1/\partial r = -\cos\theta$ over A.

Now
$$\frac{\cos\theta}{r^2} = \frac{r\cos\theta}{r^3} = \frac{c - r'\cos\theta'}{\{c^2 - 2r'c\cos\theta' + r'^2\}^{\frac{3}{2}}}$$
$$= \frac{1}{c^2}\left(1 + \frac{2r'}{c}\cos\theta' + \dots\right).$$

Hence, near B, we have

$$\phi_1 = \tfrac{1}{2}\frac{a^3}{c^2}\left(1 + \frac{2r'\cos\theta'}{c}\right),$$

giving a normal velocity over $B = -\dfrac{a^3}{c^3}\cos\theta'$.

This normal velocity might be cancelled by the addition of a velocity potential

$$\phi_2 = \tfrac{1}{2}\frac{a^3b^3}{c^3}\frac{\cos\theta'}{r'^2};$$

and, as above, the value of this near A is

$$\phi_2 = \tfrac{1}{2}\frac{a^3b^3}{c^5}\left(1 + \frac{2r\cos\theta}{c}\right),$$

giving a normal velocity over $A = -\dfrac{a^3b^3}{c^6}\cos\theta$.

This normal velocity might be cancelled by the addition of a velocity potential

$$\phi_3 = \tfrac{1}{2}\frac{a^6 b^3}{c^6}\frac{\cos\theta}{r^2}\text{, and so on.}$$

To this order of approximation, i.e. neglecting a^6b^3/c^9, we have

$$\phi = \phi_1 + \phi_2 + \phi_3;$$

and, on A, $$\phi = \text{const.} + \tfrac{1}{2}a\left(1 + 3\frac{a^3b^3}{c^6}\right)\cos\theta \quad\text{.................(3)},$$

while, on B, $$\phi = \text{const.} + \tfrac{3}{2}\frac{a^3}{c^3}b\cos\theta' \quad\text{.................(4)}.$$

Hence $$L = -\rho\int_0^\pi \phi\frac{\partial\phi}{\partial r}2\pi a^2\sin\theta d\theta$$

$$= \pi\rho a^3\left(1 + 3\frac{a^3b^3}{c^6}\right)\int_0^\pi \cos^2\theta\sin\theta d\theta$$

$$= \tfrac{2}{3}\pi\rho a^3\left(1 + 3\frac{a^3b^3}{c^6}\right).$$

Similarly $$M = 2\pi\rho\frac{a^3b^3}{c^3} \quad\text{and}\quad N = \tfrac{2}{3}\pi\rho b^3\left(1 + 3\frac{a^3b^3}{c^6}\right) \quad\text{........(5)}.$$

If we put $U = U'$ and $a = b$, the motion is symmetrical about the plane bisecting AB at right angles, which may be taken as a fixed boundary. Hence for the motion of a sphere at right angles to a fixed plane boundary at distance $h = \tfrac{1}{2}c$, the kinetic energy of the liquid being half that just obtained is given by

$$2T = \tfrac{2}{3}\pi\rho a^3U^2\left(1 + \tfrac{3}{8}\frac{a^3}{h^3} + \dots\right) \quad\text{.....................(6)}.$$

If m, m' are the masses of the spheres, for the whole kinetic energy in the general case we have

$$2T = (L + m)U^2 + 2MUU' + (N + m')U'^2 \quad\text{...........(7)}.$$

If we now assume that Lagrange's equations* may be applied to the whole system and let x, x' denote the distances OA, OB, where O is an origin on the line of centres, we have

$$2T = (L + m)\dot{x}^2 - 2M\dot{x}\dot{x}' + (N + m')\dot{x}'^2 \quad\text{...........(8)},$$

and $x' - x = c$, so that

$$\left.\begin{aligned} &\frac{d}{dt}\{(L+m)\dot{x} - M\dot{x}'\} + \tfrac{1}{2}\left(\frac{\partial L}{\partial c}\dot{x}^2 - 2\frac{\partial M}{\partial c}\dot{x}\dot{x}' + \frac{\partial N}{\partial c}\dot{x}'^2\right) = X \\ \text{and}\quad &\frac{d}{dt}\{-M\dot{x} + (N+m')\dot{x}'\} - \tfrac{1}{2}\left(\frac{\partial L}{\partial c}\dot{x}^2 - 2\frac{\partial M}{\partial c}\dot{x}\dot{x}' + \frac{\partial N}{\partial c}\dot{x}'^2\right) = X' \end{aligned}\right\}\text{...(9)},$$

where X, X' are the forces acting on the spheres in the x direction.

To a first approximation, assuming that a and b are small compared to c, and retaining only the most important terms, we have

$$\frac{\partial L}{\partial c} = 0, \quad \frac{\partial M}{\partial c} = -6\pi\rho\frac{a^3b^3}{c^4}, \quad \frac{\partial N}{\partial c} = 0 \quad\text{..............(10)}.$$

* For the justification of this assumption reference may be made to Lamb's *Hydrodynamics*, chap. VI and Kelvin and Tait's *Natural Philosophy*, §§ 319, 320.

(α) If the spheres both move with constant velocity the force necessary to maintain the motion of A is

$$X=-\frac{dM}{dt}\dot{x}'-\frac{\partial M}{\partial c}\dot{x}\dot{x}'=-\frac{\partial M}{\partial c}\dot{c}\dot{x}'-\frac{\partial M}{\partial c}\dot{x}\dot{x}'=-\frac{\partial M}{\partial c}\dot{x}'^2=6\pi\rho\frac{a^3b^3}{c^4}\dot{x}'^2 \quad \text{......(11).}$$

This force is directed towards B and depends only on the velocity of B, so that two spheres projected towards one another would appear to repel one another.

(β) If the spheres perform small oscillations about fixed positions, we may put

$$x=\lambda\cos pt,$$

$$x'=c+\lambda'\cos(pt+\epsilon).$$

The mean value of X is then the mean value of

$$-\frac{\partial M}{\partial c}\lambda\lambda' p^2\sin pt\sin(pt+\epsilon),$$

which

$$=3\pi\rho\frac{a^3b^3}{c^4}\lambda\lambda' p^2\cos\epsilon \quad \text{........................(12).}$$

The force is therefore repulsive if the difference of phase ϵ is less than a quarter period, and attractive if more than a quarter period.

(γ) Let $U=U'$ and $a=b$ so that the motion is symmetrical about the plane bisecting AB at right angles, then this plane may be taken as a fixed boundary, and we conclude from (α) that a sphere moving at right angles to a fixed plane boundary is repelled from the boundary.

8·62. *Two spheres moving in parallel directions at right angles to the line joining them.*

Let V, V' denote the velocities, and with the same notation, but measuring θ, θ' as in the figure, the velocity potential is

$$V\phi+V'\phi',$$

where

$$\frac{\partial\phi}{\partial r}=-\cos\theta \text{ over } A, \quad \text{and} \quad \frac{\partial\phi}{\partial r'}=0 \text{ over } B;$$

$$\frac{\partial\phi'}{\partial r}=0 \text{ over } A, \quad \text{and} \quad \frac{\partial\phi'}{\partial r'}=-\cos\theta' \text{ over } B.$$

As before, a velocity potential

$$\phi_1=\tfrac{1}{2}\frac{a^3}{r^2}\cos\theta$$

would make $\partial\phi_1/\partial r=-\cos\theta$ over A.

And, near B, we have

$$\phi_1=\tfrac{1}{2}\frac{a^3}{r^3}r\cos\theta=\tfrac{1}{2}\frac{a^3}{c^3}r'\cos\theta',$$

giving a normal velocity over $B=-\tfrac{1}{2}\dfrac{a^3}{c^3}\cos\theta'$.

This normal velocity might be cancelled by the addition of a velocity potential

$$\phi_2=\tfrac{1}{4}\frac{a^3}{c^3}\frac{b^3}{r'^2}\cos\theta';$$

and the value of this near A is

$$\phi_2 = \tfrac{1}{4}\frac{a^3b^3}{c^6}r\cos\theta,$$

giving a normal velocity over $A = -\tfrac{1}{4}\dfrac{a^3b^3}{c^6}\cos\theta$.

This normal velocity might be cancelled by the addition of a velocity potential

$$\phi_3 = \tfrac{1}{8}\frac{a^3b^3}{c^6}\frac{a^3}{r^2}\cos\theta, \text{ and so on.}$$

To this order of approximation, i.e. neglecting a^6b^3/c^9, we have

$$\phi = \phi_1 + \phi_2 + \phi_3;$$

and, on A,

$$\phi = \tfrac{1}{2}a\left(1 + \tfrac{3}{4}\frac{a^3b^3}{c^6}\right)\cos\theta \quad \text{.....................(13)},$$

while, on B,

$$\phi = \tfrac{3}{4}\frac{a^3}{c^3}b\cos\theta' \quad \text{.................................(14)}.$$

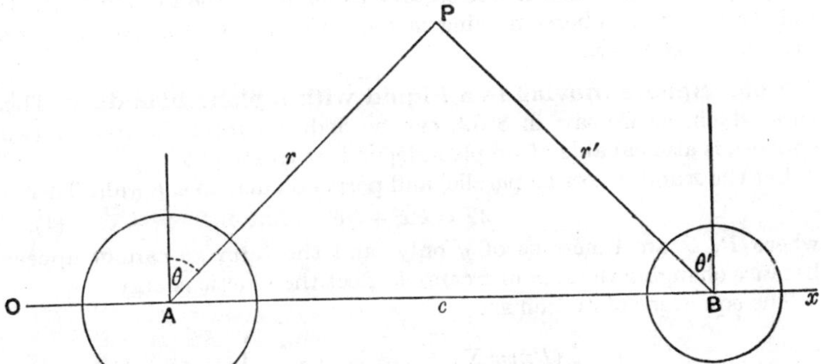

Hence if the kinetic energy of the liquid be given by

$$2T = L'V^2 + 2M'VV' + N'V'^2 \quad \text{..................(15)},$$

we have

$$\left.\begin{aligned} L' &= -\rho\iint \phi\frac{\partial\phi}{\partial n}dS_A \\ &= \pi\rho a^3\left(1 + \tfrac{3}{4}\frac{a^3b^3}{c^6}\right)\int_0^\pi \cos^2\theta\sin\theta\,d\theta \\ &= \tfrac{2}{3}\pi\rho a^3\left(1 + \tfrac{3}{4}\frac{a^3b^3}{c^6}\right); \\ M' &= -\rho\iint \phi\frac{\partial\phi'}{\partial n}dS_B = \pi\rho\frac{a^3b^3}{c^3}, \\ N' &= -\rho\iint \phi'\frac{\partial\phi'}{\partial n}dS_B = \pi\rho b^3\left(1 + \tfrac{3}{4}\frac{a^3b^3}{c^6}\right) \end{aligned}\right\} \text{...........(16)}.$$

(similarly, for M'.)

If we put $V = V'$ and $a = b$ the motion will be symmetrical about the plane bisecting AB at right angles, so that the kinetic energy of the liquid due to the motion of a sphere parallel to a fixed plane boundary at distance $h = c/2$, being half the kinetic energy in the last case, is given by

$$2T = \tfrac{2}{3}\pi\rho a^3V^2\left(1 + \tfrac{3}{16}\frac{a^3}{h^3} + \dots\right) \quad \text{.................(17)}.$$

Reverting to the case of the two spheres, for the whole kinetic energy we may write

$$2T = (L'+m)V^2 + 2M'VV' + (N'+m')V'^2 \quad \text{.........(18)},$$

and taking an origin O on the line of centres so that if $OA = x$ and $OB = x'$, $x' - x = c$, L', M', N' are functions of c or $x' - x$, and retaining only the most important terms,

$$\frac{\partial L'}{\partial c} = 0, \quad \frac{\partial M'}{\partial c} = -3\pi\rho\frac{a^3b^3}{c^4}, \quad \frac{\partial N'}{\partial c} = 0 \quad \text{...........(19)}.$$

Hence the equation of motion

$$\frac{d}{dt}\left(\frac{\partial T}{\partial \dot{x}}\right) - \frac{\partial T}{\partial x} = X$$

gives

$$X = \frac{\partial M'}{\partial c} VV'$$

$$= -3\pi\rho\frac{a^3b^3}{c^4} VV' \quad \text{........................(20)}$$

as the force in direction AB necessary to maintain the motion of A. It follows that two spheres moving in the same direction in parallel lines attract one another.

8·63. Sphere moving in a Liquid with a plane boundary. This case which, as we saw in **8·62**, can be deduced from the case of two spheres, is also capable of simple independent treatment.

Let the x and y axes be parallel and perpendicular to the wall. Then

$$2T = P\dot{x}^2 + Q\dot{y}^2 \quad \text{..........................(1)},$$

where P, Q are functions of y only, and the term $\dot{x}\dot{y}$ cannot appear because changing the sign of $\dot{x}$ cannot affect the kinetic energy.

The equations of motion are

$$\left.\begin{aligned} &\frac{d}{dt}(P\dot{x}) = X \\ &\frac{d}{dt}(Q\dot{y}) - \tfrac{1}{2}\left(\frac{\partial P}{\partial y}\dot{x}^2 + \frac{\partial Q}{\partial y}\dot{y}^2\right) = Y \end{aligned}\right\} \quad \text{.................(2)},$$

where X, Y are the forces in the directions of x and y.

If there are no external forces and the sphere is moving at right angles to the wall, $\dot{x} = 0$ and, since the kinetic energy is constant, therefore

$$Q\dot{y}^2 = \text{const.} \quad \text{............................(3)}.$$

But from (17) of **8·62** and (6) of **8·61**

$$\left.\begin{aligned} P &= m + \tfrac{2}{3}\pi\rho a^3\left(1 + \tfrac{3}{16}\frac{a^3}{y^3}\right) \\ Q &= m + \tfrac{2}{3}\pi\rho a^3\left(1 + \tfrac{3}{8}\frac{a^3}{y^3}\right) \end{aligned}\right\} \quad \text{.....................(4)},$$

so that P and Q both decrease as y increases, therefore $\dot{y}$ increases as y increases or the sphere has an acceleration *from* the wall.

Again, if the sphere moves parallel to the wall, so that $\dot{y} = 0$, there must be a constraining force

$$Y = -\tfrac{1}{2}\frac{\partial P}{\partial y}\dot{x}^2$$

$$= \tfrac{3}{16}\frac{\pi\rho a^6}{y^4}\dot{x}^2 \quad \text{.............................(5)}$$

acting away from the wall, so that the sphere is attracted towards the wall.

This problem was discussed by Stokes*, who obtained results (6) of **8·61** and (17) of **8·62** by a somewhat similar method. Some results were given by Kelvin and Tait†, and for further information on the subject of the motion of two spheres reference may be made to papers by W. M. Hicks‡, R. A. Herman§, and A. B. Basset||.

EXAMPLES

1. A homogeneous liquid is contained between two concentric spherical rigid envelopes of given masses; these bounding surfaces are set in motion, the one with velocity U, and the other with velocity V, in perpendicular directions; find the impulses which must be applied to the envelopes to produce the motion, and determine the motion of the fluid at any point. (Coll. Exam. 1893.)

2. The space between two coaxial cylindrical shells of radii a, b is filled with incompressible liquid of density ρ. The outer shell, of radius a, is suddenly made to move with velocity U: shew that the impulsive force per unit length necessary to be applied to the inner cylinder to hold it at rest is $2\pi\rho a^2b^2U/(a^2-b^2)$. (Trinity Coll. 1901.)

3. A uniform sphere is surrounded by a uniform incompressible fluid of the same density, initially at rest and extending through all space. The sphere is set in motion by a blow P along a diameter. Prove that its resulting velocity is $\frac{2}{3}P/M$, where M is its mass. (Trinity Coll. 1909.)

4. An incompressible perfect fluid of mass m is contained between two rigid concentric spherical envelopes, the outer of radius b and mass M, the inner of radius a and of no mass. The system is started from rest by an impulse normal to the outer envelope. Prove that the initial momentum is shared between the envelope and the fluid in the ratio of $M(2a^3+b^3)$ to mb^3. (Trinity Coll. 1904.)

5. A sphere of radius a is made to describe a circle uniformly in an infinite fluid at rest at infinity; find the pressure at any point of the sphere, and shew that the resultant pressure on it is a force $(2\pi/3)\,\rho a^3c\omega^2$ towards the centre of the circle, where a is the radius of the sphere, c the radius of the circle described by its centre, ω the angular velocity. (Trinity Coll. 1907.)

6. A solid body is moved in any manner in an unlimited liquid, find the motion set up and shew that if the body be moved with unit velocity along Ox, the momentum set up parallel to Oy is equal to that set up parallel to Ox by moving the body with unit velocity along Oy. Also if the body be turned round Ox with unit angular velocity the momentum generated parallel to Oy is equal to the angular momentum generated around Ox by moving the body with unit velocity parallel to Oy.

* 'On some cases of fluid motion', *Trans. Camb. Phil. Soc.* VIII, or *Math. and Phys. Papers*, I, pp. 47–49.

† *Natural Philosophy*, §§ 320, 321.

‡ 'Motion of Two Spheres in a Fluid', *Phil. Trans.* 1880, p. 455.

§ 'On the motion of Two Spheres in Fluid and Allied Problems', *Quart. Journal*, XXII, p. 204.

|| 'On the Motion of Two Spheres in a Liquid', *Proc. L.M.S.* XVIII, p. 369.

7. A pendulum with an elliptical cylindrical cavity filled with liquid, the generating lines of the cylinder being parallel to the axis of suspension, performs finite oscillations under gravity. If l be the length of the equivalent pendulum, and l' the length of the equivalent pendulum when the liquid is solidified, find l and l', and prove that

$$l'-l=\frac{m}{M+m}\frac{a^2b^2}{a^2+b^2}\cdot\frac{1}{h},$$

where M is the mass of the pendulum, m of the liquid, h the distance of the centre of gravity of the whole mass from the axis of suspension, and a, b the semi-axes of the elliptic cylinder. (M.T. 1878.)

8. A pendulum, of mass M, with an ellipsoidal cavity (semi-axes a, b, c) filled with liquid of mass m, oscillates about a horizontal axis parallel to the c-axis of the ellipsoid; prove that the length of the equivalent simple pendulum is

$$[MK^2+m\{d^2+(a^2-b^2)^2/5(a^2+b^2)\}]/(M+m)\,l,$$

where K is the radius of gyration of M about the axis of suspension, d the distance of the centre of the ellipsoid and l the distance of the centre of gravity of the whole mass from the same axis. (Coll. Exam. 1898.)

9. In the midst of an infinite mass of homogeneous incompressible liquid at rest is a spherical surface of radius a, which is suddenly strained into an equal spheroid of small ellipticity. Find the kinetic energy contained between the given surface and an imaginary concentric spherical surface of radius c; and shew that if the imaginary surface were a real boundary surface which could not be deformed, the kinetic energy in this case would be to that in the former case in the ratio

$$c^5(3a^5+2c^5):2(c^5-a^5)^2. \qquad \text{(M.T. 1878.)}$$

10. Find the ratio of the kinetic energy of the infinite liquid surrounding an oblate spheroid, moving with a given velocity in its equatorial plane, to the kinetic energy of the spheroid; and denoting this by P, prove that if the spheroid swing as the bob of a pendulum under gravity, the distance between the axis of suspension and the axis of the spheroid being c, the length of the simple equivalent pendulum is

$$\frac{(1+P)\,c+\frac{2}{5}a^2/c}{1-\sigma/\rho},$$

where a is the equatorial radius, ρ the density of the spheroid and σ that of the liquid. (M.T. 1879.)

11. A sphere of radius a immersed in an infinite mass of liquid with a plane boundary is set in motion with velocity V towards the boundary. Shew that, if the boundary is at a distance c such that $(a/c)^3$ is negligible, the impulsive thrust on the boundary is $2\pi\rho a^3V$. Also find the momentum set up in the liquid. (M.T. 1925.)

12. A small sphere of radius a is moving with uniform velocity U in liquid of density ρ at rest at an infinite distance, in a direction at right angles to an infinite plane boundary. Shew that, when it is at a distance c

from the boundary, the pressure at a point on the boundary at distance ξ from the centre of the sphere is

$$\Pi - \frac{\rho U^2 a^3}{\xi^3}\left(1 - \frac{3c^2}{\xi^2}\right),$$

where Π is the pressure at an infinite distance, and higher powers of a than a^3 are neglected. (M.T. 1920.)

13. Shew that for a rigid body moving under no external forces in infinite fluid at rest at infinity there are:

(i) three directions of permanent translation;

(ii) three permanent screw motions such that the corresponding impulsive wrench reduces to a couple.

Shew further that in general the impulsive wrenches needed to start the motions in (i) do not reduce to single forces, but that if the body has a plane of symmetry the motions (i) can be started by single impulsive forces and the screw motions (ii) consist of pure rotations. (M.T. 1925.)

14. An elliptic cylindrical shell, the mass of which may be neglected, is filled with water, and placed on a horizontal plane very nearly in the position of unstable equilibrium with its axis horizontal, and is then let go. When it passes through the position of stable equilibrium, find the angular velocity of the cylinder (i) when the horizontal plane is perfectly smooth, (ii) when it is perfectly rough; and prove that in these two cases the squares of the angular velocities are in the ratio

$$(a^2-b^2)^2+4b^2(a^2+b^2):(a^2-b^2)^2,$$

$2a$ and $2b$ being the axes of the cross section of the cylinder. (M.T. 1886.)

15. A solid ellipsoid of uniform density is set rotating in an infinite liquid about one of its axes by a given impulsive couple; find its angular velocity. (M.T. 1882.)

16. A cylinder is moving in an infinite fluid, and the motion is defined by u, v, ω; shew how to reduce the kinetic energy to its simplest form.

If $2T = Au^2 + 2Huv + Bv^2 + K\omega^2$ and there are no forces, prove the equation

$$K\ddot{\theta} + J^2\{(A-B)\sin\theta\cos\theta + H(\cos^2\theta - \sin^2\theta)\}/(AB - H^2) = 0,$$

where J is the resultant momentum (linear). (St John's Coll. 1895.)

17. An infinite elliptic cylinder of density σ is moving through incompressible fluid of density ρ that extends to infinity and is at rest there. Shew that if a, b be the semi-axes and $c^2 = a^2 - b^2$,

$$2T = \pi(\rho b^2 + \sigma ab)\,U^2 + \pi(\rho a^2 + \sigma ab)\,V^2 + \pi[\rho c^4/8 + \sigma ab\,(a^2+b^2)/4]\,\omega^2,$$

and that at any time $\left\{\dfrac{c^2}{8} + \dfrac{\sigma}{\rho}\dfrac{ab}{4}\dfrac{a^2+b^2}{a^2-b^2}\right\}\ddot{\theta} + UV = 0,$

where U, V are the velocities of the centre along the axes and θ the angle turned through by the transverse axis. (Trinity Coll. 1894.)

18. A prolate spheroid is moving through fluid with velocity u in the direction of its axis; shew that the motion is unstable, but that it will be stable if the spheroid is at the same time spinning about its axis with an angular velocity greater than $\frac{2u}{A}\left\{\frac{BP}{Q}(Q-P)\right\}^{\frac{1}{2}}$, where P and Q are the effective inertias of the spheroid along the axis of revolution and a perpendicular axis respectively, and A, B are the effective moments of inertia about those axes. (M.T. 1892.)

19. A solid ellipsoid of density σ is placed inside a fixed concentric, confocal and similarly situated ellipsoidal shell and the space between them is filled with fluid of density ρ. Supposing that the whole matter attracts according to the Newtonian Law, and that σ is greater than ρ, shew that when the solid ellipsoid is slightly displaced parallel to its greatest axis, the time of a small oscillation is given by

$$\frac{4\pi^2}{T^2}=\frac{\pi\rho(\sigma-\rho)A}{\dfrac{\sigma+\rho}{2}-\dfrac{\rho abc}{abc(2-A')-a'b'c'(2-A)}},$$

where a, b, c and a', b', c' are the semi-axes of the outer and inner ellipsoids and

$$A=\int_0^\infty \frac{abc\,d\lambda}{(a^2+\lambda)^{\frac{3}{2}}(b^2+\lambda)^{\frac{1}{2}}(c^2+\lambda)^{\frac{1}{2}}},$$

with a similar expression for A'. (M.T. 1881.)

20. If a thin ellipsoidal shell without mass be filled with water, and set in motion about its centre as a fixed point, prove that its subsequent motion will be determined by three equations of the form

$$\frac{(b^2-c^2)^2}{b^2+c^2}\frac{d\omega_1}{dt}+\frac{(b^2-c^2)(b^2c^2+c^2a^2+a^2b^2-3a^4)}{(c^2+a^2)(a^2+b^2)}\omega_2\omega_3=L.$$

21. If A and B be the forces required to act for unit of time in order to generate unit velocity perpendicular and parallel respectively to the axis of an ellipsoid of revolution in an infinite mass of homogeneous frictionless liquid, and if G be the couple required to act for unit of time in order to generate unit angular velocity about an equatorial axis, prove that the kinetic energy T of the ellipsoid and liquid is

$$\tfrac{1}{2}(Au^2+Av^2+Bw^2+G\omega_1{}^2+G\omega_2{}^2+C\omega_3{}^2),$$

with Euler's notation, C being the polar moment of inertia of the ellipsoid.

Express T in terms of Lagrange's coordinates $x, y, z, \theta, \phi, \psi$; and prove that if the axis of z be parallel to the impressed impulse F, then

$$\dot{x}=-F\left(\frac{1}{A}-\frac{1}{B}\right)\sin\theta\cos\theta\cos\psi,\quad \dot{y}=-F\left(\frac{1}{A}-\frac{1}{B}\right)\sin\theta\cos\theta\sin\psi,$$

$$\dot{z}=F\left(\frac{\sin^2\theta}{A}+\frac{\cos^2\theta}{B}\right),\quad \dot{\phi}+\cos\theta\dot{\psi}=\omega_3,\quad G\sin^2\theta\dot{\psi}+C\omega_3\cos\theta=E,$$

$$G\dot{\theta}^2+G\sin^2\theta\dot{\psi}^2+C\omega_3{}^2+F^2\left(\frac{\sin^2\theta}{A}+\frac{\cos^2\theta}{B}\right)=2T,$$

where ω_3, E, T are constants; the last three equations being the same for a solid of revolution with a bar of soft iron in its axis, moving about its centre in a uniform magnetic field. (M.T. 1877.)

22. A rigid body immersed in a homogeneous incompressible liquid at rest extending to infinity is set in motion by an impulsive couple: prove that its subsequent motion relative to a certain point O fixed in it is the same as if a certain ellipsoid, fixed in it with its centre at O, rolled on a fixed plane; and express geometrically the variable velocity of translation necessary to complete the representation of the actual motion. (Lamb.)

23. The presence of an infinite liquid increases the apparent inertia of a moving sphere by half the mass of the liquid displaced. Shew that this increase is raised in the ratio $1+3a^3/8\xi^3:1$ nearly, if the liquid is bounded by an infinite plane perpendicular to the direction of motion, and at a great distance ξ from the centre of the sphere, whose radius is a.

(Trinity Coll. 1895.)

24. Two infinite parallel circular cylinders in an infinite fluid are projected (i) in opposite directions along a line at right angles to their axes, (ii) in the same direction perpendicular to this line. Prove that they experience in the two cases respectively a mutual repulsion and a mutual attraction. (Trinity Coll. 1894.)

25. A sphere of mass M, displacing a mass M' of fluid, is projected with velocity V normally to an infinite rigid plane with which it is in contact; shew that its limiting velocity is

$$V\left[1+\frac{3M'}{2M+M'}\sum_2^\infty \frac{1}{n^3}\right]^{\frac{1}{2}}. \qquad \text{(Trinity Coll. 1898.)}$$

26. Find the complete system of images which will represent the motion of a sphere perpendicular to an infinite bounding plane; and shew that, if the density of the sphere be the same as that of the fluid, the ratio of the velocity of the sphere at impact to its velocity at an infinite distance from the plane is

$$1:\left(\sum_1^\infty \frac{1}{n^3}\right)^{\frac{1}{2}}. \qquad \text{(M.T. 1889.)}$$

27. Find the nature of the interaction between two spheres moving in a liquid of infinite extent (i) when the spheres each make small vibrations along the line of centres, (ii) when one vibrates and the other is at rest. [Take the kinetic energy of the system to be

$$\tfrac{1}{2}(Lu^2-2Muv+Nv^2),$$

where

$$L=m+\tfrac{2}{3}\pi\rho a^3\left(1+\frac{3a^3b^3}{c^6}\right), \quad M=2\pi\rho\frac{a^3b^3}{c^3},$$

$$N=m'+\tfrac{2}{3}\pi\rho b^3\left(1+\frac{3a^3b^3}{c^6}\right);$$

m, m' are the masses, a, b the radii, and u, v the velocities of the spheres, c the distance between their centres, and only the lowest powers of a/c and b/c are retained.]

Mention some experimental evidence of the results obtained.

(M.T. 1911.)

28. (a) Investigate the condition of stability of the motion of an elongated solid of revolution with a plane of symmetry at right angles to its axis of figure moving parallel to its axis of figure and rotating about that axis.

(b) Prove that, when this condition is satisfied, there are possible two states of steady motion in which the velocities of translation and rotation are constant and the directions of translation and rotation are in a plane through the axis of figure and make constant angles with that axis while the plane in question rotates uniformly around the axis.

(c) Prove that the two modes of simple harmonic oscillation about the state of steady motion described in (a) are really steady motions of the types described in (b), the angles made with the axis of figure by the directions of translation and rotation being small.

(M.T. 1904.)

CHAPTER IX

VORTEX MOTION

9·1. So far we have confined our attention almost entirely to cases involving irrotational motion only. But we saw (**4·1**) that the most general displacement of a fluid involves rotation of which the component angular velocities at a point (x, y, z) are

$$\xi = \frac{1}{2}\left(\frac{\partial w}{\partial y} - \frac{\partial v}{\partial z}\right), \quad \eta = \frac{1}{2}\left(\frac{\partial u}{\partial z} - \frac{\partial w}{\partial x}\right), \quad \zeta = \frac{1}{2}\left(\frac{\partial v}{\partial x} - \frac{\partial u}{\partial y}\right),$$

where u, v, w are the components of linear velocity at the point. We also saw (**2·51**, **2·6** and **4·24**) that if at any instant the motion of a fluid mass is irrotational under the action of conservative forces it remains irrotational for all time. In this chapter we shall consider the theory of rotational or vortex motion. The theory is due to Helmholtz whose epoch-making paper was published in 1858*. It was afterwards developed by Kelvin†, Kirchhoff and other writers.

9·11. It is important to realize at the outset that some portions of a fluid mass may possess rotation while others are moving irrotationally.

Lines drawn in the fluid so as at every point to coincide with the instantaneous axis of rotation of the corresponding fluid element are called **vortex lines** (*Wirbellinien*).

Portions of the fluid bounded by vortex lines drawn through every point of an infinitely small closed curve are called **vortex filaments** (*Wirbelfäden*), or simply **vortices**, and the boundary of a vortex filament is called a **vortex tube**.

9·12. The theory will shew that elements of fluid which at any time belong to one vortex line, however they may be translated, remain on the same vortex line, or that the vortex lines move with the fluid. Also that the product of the section and angular velocity of a vortex filament is constant throughout its

* *Crelle's Journal*, LV, 'Ueber Integrale der hydrodynamischen Gleichungen welche den Wirbelbewegungen entsprechen'. A translation by Tait was published in *Phil. Mag.* XXXIII, Fourth Series, p. 485.

† 'Vortex Motion', *Trans. R. Soc. Edin.* XXV, 1869, p. 217, or *Math. and Phys. Papers*, IV, p. 13.

whole length and constant for all time. Hence vortex filaments must either form closed curves or have their ends on the bounding surface of the fluid. A vortex in perfect fluid is therefore permanent and indestructible; and the enunciation of these properties by Helmholtz suggested to Lord Kelvin the idea that vortex rings are the only true atoms, inasmuch as the generation or destruction of vortex motion in a perfect fluid can only be an act of creative power*, a theory long since abandoned.

9·2. Kelvin's Proofs. To prove the properties just enunciated:

(1) *The product of the cross section and angular velocity at any point on a vortex filament is constant all along the vortex filament and for all time.*

By Stokes's Theorem (**4·2**) the circulation round any closed curve is equal to

$$2\iint (l\xi + m\eta + n\zeta)\, dS,$$

where ξ, η, ζ are the components of spin, and l, m, n are direction cosines of the normal to an element dS of a surface bounded by the curve. If the curve be a reducible circuit drawn on the surface of a vortex tube the circulation will be zero, because at every point of such a surface

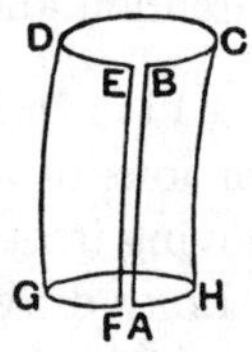

$$l\xi + m\eta + n\zeta = 0.$$

Let the circuit be $ABCDEFGHA$ as in the figure, where $FGHA$ and $EDCB$ are two cross sections of the vortex tube. Then since the circulation round $ABCDEFGHA$ is zero and the contributions of AB, EF are equal and opposite, it follows that

$$\text{flow round } FGHA = \text{flow round } EDCB,$$

or, ultimately,

$$\text{circulation round } AGHA = \text{circulation round } BDCB.$$

But, as in **4·2**, if ω denote the angular velocity and σ the cross section of the vortex tube supposed small, the circulation round this section is $2\omega\sigma$. Hence this product is constant for all sections, and we shall take it as a measure of the **strength of the vortex.**

Again, from **4·23**, when the forces have a single-valued potential and the density is a function of the pressure the circulation in any closed circuit moving with the fluid is constant for

* 'On Vortex Atoms', *Phil. Mag.* XXXIV, 1867, p. 15, or *Math. and Phys. Papers*, IV, p. 1.

all time. And if we apply this to any circuit embracing the vortex it follows that the strength of the vortex is constant for all time.

It is clear also that the circulation in any circuit is the sum of the strengths of the vortices that it embraces.

(2) *The vortex lines move with the fluid.*

It is clear from the formula $2\iint(l\xi+m\eta+n\zeta)\,dS$ for circulation in a closed circuit, that if the circulation is zero in every circuit that can be drawn on a certain surface no vortex lines can cut the surface, and any that meet the surface must lie wholly upon it, for we must have $l\xi+m\eta+n\zeta=0$ at every point of the surface. Consider a surface S composed of vortex lines at time t. The circulation in any circuit C on this surface is zero. At time $t+\delta t$ the particles that formed the surface S now lie on another surface S', and the circuit C moving with the particles now lies on S' and the circulation in it is still zero and this being true for all such circuits on S', the surface S' must be composed of vortex lines. Hence any surface composed of vortex lines, as it moves with the fluid, continues to be composed of vortex lines. The intersection of two such surfaces must always be a vortex line and so we arrive at the theorem that vortex lines move with the fluid.

The foregoing proofs are due to Lord Kelvin. The proof given by Helmholtz is less satisfactory but we reproduce it here on account of its historical interest.

9·21. Helmholtz's Proof. Let ω denote the resultant spin at any point on a vortex line and $\epsilon\omega$ a small element of length of the vortex line. The projections of this element on the axes are

$$\delta x,\ \delta y,\ \delta z=\epsilon\xi,\ \epsilon\eta,\ \epsilon\zeta \quad \text{..................(1)}.$$

The rate at which δx increases as the fluid moves is the difference in the values of u at the ends of that element. Therefore

$$\frac{D\delta x}{Dt}=\frac{\partial u}{\partial x}\delta x+\frac{\partial u}{\partial y}\delta y+\frac{\partial u}{\partial z}\delta z=\epsilon\left(\xi\frac{\partial u}{\partial x}+\eta\frac{\partial u}{\partial y}+\zeta\frac{\partial u}{\partial z}\right)$$

$$=\epsilon\frac{D\xi}{Dt},\ \text{from } \mathbf{2{\cdot}6\ (1)};$$

or

$$\frac{D}{Dt}(\delta x-\epsilon\xi)=0 \quad \text{....................(2)}.$$

Helmholtz infers from (2)* that relations (1) continue to be

* Dr Goldstein points out that this inference is equivalent to the assumption that if $f'(x)=0$ when $f(x)=0$, then if $f(x)$ vanishes for some value x_0 of x it is identically zero; which is false.

true as time advances; or, as the particles composing a vortex line move, their join is still the instantaneous axis of rotation, which means that 'each vortex line remains composed of the same elements of fluid, and swims forward with them in the fluid'.

Now, regarding the element of length of a vortex line as the join of two definite particles or elements of fluid, we have seen that ξ, η, ζ vary as the projections of this element of length on the coordinate axes, hence the resultant angular velocity in a defined element varies as the distance between this and its neighbour along the axis of rotation.

Now, regarding the fluid as incompressible, consider a short length of a vortex filament. Its volume is constant as it moves in the fluid because it is always composed of the same elements of fluid, but the angular velocity varies directly as its length, therefore the product of the angular velocity and the cross section in a portion of vortex filament containing the same element of fluid, remains constant during the motion of that element.

Again from the expression for ξ, η, ζ in terms of u, v, w we get

$$\frac{\partial \xi}{\partial x}+\frac{\partial \eta}{\partial y}+\frac{\partial \zeta}{\partial z}=0.$$

But $$\iint (l\xi+m\eta+n\zeta)\,dS=\iiint\left(\frac{\partial \xi}{\partial x}+\frac{\partial \eta}{\partial y}+\frac{\partial \zeta}{\partial z}\right)dx\,dy\,dz$$
$$=0;$$

where the surface integral extends to any portion of the fluid bounded by a surface S. Applying this to the surface of a portion of a vortex filament cut off by cross sections of area σ, σ', the integral over the curved surface is zero and the result reduces to

$$\omega\sigma=\omega'\sigma',$$

where ω, ω' are the angular velocities.

That is, the product $\omega\sigma$ is constant throughout the whole length of any one vortex filament.

9·22. Third Proof from Cauchy's Equations. A third proof follows very simply from Cauchy's equations of **2·51**, viz.

$$\frac{\xi}{\rho}=\frac{\xi_0}{\rho_0}\frac{\partial x}{\partial a}+\frac{\eta_0}{\rho_0}\frac{\partial x}{\partial b}+\frac{\zeta_0}{\rho_0}\frac{\partial x}{\partial c},\ \text{etc.}$$

For, the initial equations of a vortex line are

$$\frac{da}{\xi_0}=\frac{db}{\eta_0}=\frac{dc}{\zeta_0}=\frac{\lambda}{\rho_0};$$

and x, y, z being the coordinates at any time of the particle originally at a, b, c,

$$dx = \frac{\partial x}{\partial a}da + \frac{\partial x}{\partial b}db + \frac{\partial x}{\partial c}dc$$

$$= \frac{\lambda}{\rho_0}\left(\xi_0\frac{\partial x}{\partial a} + \eta_0\frac{\partial x}{\partial b} + \zeta_0\frac{\partial x}{\partial c}\right)$$

$$= \frac{\lambda\xi}{\rho}, \text{ from above;}$$

therefore $$\frac{dx}{\xi} = \frac{dy}{\eta} = \frac{dz}{\zeta} = \frac{\lambda}{\rho};$$

that is, the moving element whose projections on the axes have become dx, dy, dz is still part of a vortex line; or the vortex lines move with the fluid.

Again, if ds be the length of the element and ω the angular velocity and ds_0, ω_0 their initial values

$$\frac{ds}{\omega} = \frac{dx}{\xi} = \ldots = \frac{\lambda}{\rho}, \text{ and } \frac{ds_0}{\omega_0} = \frac{da}{\xi_0} = \ldots = \frac{\lambda}{\rho_0}.$$

But if σ, σ_0 denote the cross sections of the filament, the mass of the element being constant, $$\rho\sigma ds = \rho_0\sigma_0 ds_0,$$

therefore $\omega\sigma = \omega_0\sigma_0$, or the strength of the vortex filament is constant with regard to the time. That it is constant along the filament can then be proved as before.

9·3. Rectilinear Vortices. Before going further into the general theory of vortex motion we shall consider the case of rectilinear vortices in homogeneous liquid, which is capable of simple independent treatment.

Suppose a number of straight parallel vortex filaments either in an indefinitely extended mass of liquid, or in a mass bounded by two planes perpendicular to the filaments.

Taking the axis of z parallel to the filaments, we have

$$w = 0, \quad \frac{\partial u}{\partial z} = 0, \quad \text{and} \quad \frac{\partial v}{\partial z} = 0,$$

so that $$\xi = 0, \quad \eta = 0, \quad \text{and} \quad 2\zeta = \frac{\partial v}{\partial x} - \frac{\partial u}{\partial y}.$$

The equation of the lines of motion is

$$v\,dx - u\,dy = 0,$$

and it follows from the equation of continuity that $v\,dx - u\,dy$ is a perfect differential $d\psi$; hence, as before,

$$u = -\frac{\partial\psi}{\partial y}, \quad v = \frac{\partial\psi}{\partial x},$$

and $$\frac{\partial^2\psi}{\partial x^2} + \frac{\partial^2\psi}{\partial y^2} = 2\zeta \quad \ldots\ldots\ldots\ldots\ldots\ldots\ldots(1),$$

and the lines of motion are given by $\psi = \text{const.}$

Except along the vortex filaments the motion is irrotational and ζ is zero; and the form of the equation for ψ shews that ψ may be regarded as the potential at any point of an infinite medium, the density of which is zero, except along the vortex filaments, which may be regarded as straight gravitating rods of density $-\zeta/2\pi$.

Hence the x, y differential coefficients of ψ are the components of the attractions of these rods parallel to the axes*.

Supposing that only a single vortex filament is in existence at the point (a, b) and that $da\,db$ is its areal section, we get for the velocity components at a point (x, y) at distance r from (a, b)

$$u=-\frac{\partial\psi}{\partial y}=\frac{2da\,db}{r}\left(\frac{-\zeta}{2\pi}\right)\frac{y-b}{r}=-\frac{\zeta\,da\,db}{\pi}\cdot\frac{y-b}{r^2},$$

and
$$v=\frac{\partial\psi}{\partial x}=-\frac{2da\,db}{r}\left(\frac{-\zeta}{2\pi}\right)\frac{x-a}{r}=\frac{\zeta\,da\,db}{\pi}\cdot\frac{x-a}{r^2}.$$

From this it follows that the resultant velocity q is perpendicular to r, and that

$$q=\frac{\zeta\,da\,db}{\pi r};$$

or, if κ is the strength of the vortex,

$$q=\frac{\kappa}{2\pi r},$$

the direction of q being in the sense of the rotation ζ. And for a single vortex

$$\psi=\frac{\kappa}{2\pi}\log r \qquad \text{.......................(2).}$$

We might also obtain (2) from the simpler consideration that *outside* a single vortex, ψ being a function of r only, we have from (1)

$$\frac{\partial^2\psi}{\partial r^2}+\frac{1}{r}\frac{\partial\psi}{\partial r}=0,$$

so that
$$\psi=C\log r;$$

and the motion outside the vortex being irrotational there is a velocity potential

$$\phi=-C\theta.$$

But the strength κ of the vortex is the circulation or decrease in ϕ in making one turn round the vortex, so that

$$2\pi C=\kappa$$

and
$$\psi=\frac{\kappa}{2\pi}\log r.$$

* The attraction of an infinitely long thin rod at distance r from itself is $2m/r$ perpendicular to the rod, m being the mass of unit length.

The irrotational motion outside the vortex is therefore given by

$$w = \frac{i\kappa}{2\pi} \log z;$$

and if there be any number of vortex filaments of strength κ_s at $c_s\,(= a_s + ib_s)$, $s = 1, 2, 3, \ldots$, then the motion outside the filaments is given by

$$w = \sum_s \frac{i\kappa_s}{2\pi} \log (z - c_s),$$

and the velocity components may be written down as the sums of the components due to the separate vortices in the forms

$$u = -\sum_s \frac{\kappa_s}{2\pi} \frac{y - b_s}{r_s^{\,2}}, \qquad v = \sum_s \frac{\kappa_s}{2\pi} \frac{x - a_s}{r_s^{\,2}},$$

or deduced from $u - iv = -\dfrac{dw}{dz} = -\sum\limits_s \dfrac{i\kappa_s}{2\pi} \dfrac{1}{z - c_s}$.

9·31. In the case of any number of filaments, if u_s, v_s denote the velocity components of the filament of strength κ_s, the expressions

$$\Sigma\,(\kappa_s u_s) \quad \text{and} \quad \Sigma\,(\kappa_s v_s)$$

will both vanish, for they consist of pairs of terms of the forms

$$\kappa_1 \frac{\kappa_2}{2\pi} \frac{x_1 - x_2}{r^2} \quad \text{and} \quad \kappa_2 \frac{\kappa_1}{2\pi} \frac{x_2 - x_1}{r^2}.$$

Hence regarding κ as a mass, the centre of gravity of the vortex filaments remains stationary during their motions about one another.

A single rectilinear vortex in an unlimited mass of liquid therefore remains stationary; and when such a vortex is in the presence of other vortices it has no tendency to move of itself but its motion through the liquid is entirely due to the velocities caused by the other vortices.

9·32. Consider the case of two vortex filaments of strengths κ_1, κ_2 and of small section at distance a apart. Each will produce a motion of the other perpendicular to the line joining them. If they meet the plane xy in A, B, the point O that divides AB in the ratio $\kappa_2 : \kappa_1$ will remain at rest and, the velocities of A and B being $\kappa_2/2\pi a$ and $\kappa_1/2\pi a$ respectively, the line AB will revolve with angular velocity $(\kappa_1 + \kappa_2)/2\pi a^2$, the vortices describing circles round O.

If the strengths of the vortices are equal but of opposite sign, say κ and $-\kappa$, O is at infinity and the vortices move in parallel directions with the same velocity $\kappa/2\pi a$.

If r_1, r_2 are the distances of a point P from A, B and θ_1, θ_2 their inclinations to BA, the velocities are $\kappa/2\pi r_1$, $\kappa/2\pi r_2$ at right angles to AP, BP. So the velocity along the tangent to the circle APB is

$$\frac{\kappa}{2\pi r_1}\sin\theta_2-\frac{\kappa}{2\pi r_2}\sin\theta_1=0.$$

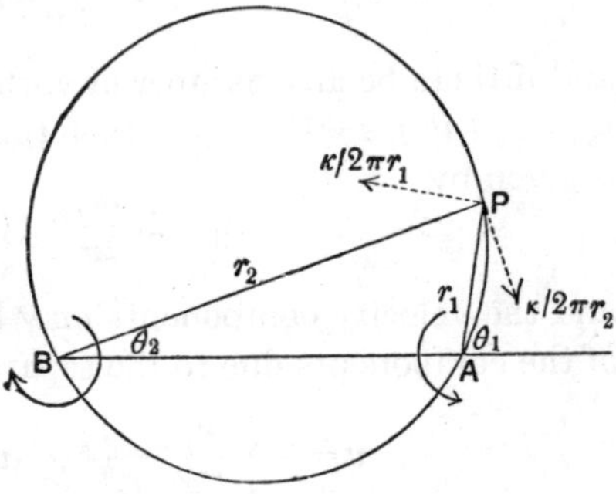

Hence the stream line through P cuts the circle APB orthogonally; that is the stream lines are the coaxial circles having A, B as limiting points.

This is also evident from the fact that

$$\psi=\frac{\kappa}{2\pi}\log\frac{r_1}{r_2}.$$

Such a pair of vortices may be called a **vortex pair.**

The reader will notice an analogy between a vortex filament and an electric current. The straight current of strength i produces a magnetic field in which the force at distance r is $2i/r$ at right angles to r and to the current. And two equal and opposite parallel currents produce a magnetic field in which the lines of force are coaxial circles corresponding to the stream lines in the case just considered.

To return to the case of the vortices, it is clear that there is no flow across a plane bisecting AB at right angles so that this might be made a rigid boundary; and consequently a single rectilinear vortex parallel to a plane boundary and at distance c from it will move parallel to the boundary with uniform velocity $\kappa/4\pi c$.

The image of such a vortex with regard to a parallel plane is therefore an equal vortex symmetrically placed, the rotation of the two being in opposite senses.

The velocity half way between the vortices being due to both of them is $\kappa/\pi c$, so the vortex moves parallel to the plane with one quarter of the velocity of the liquid at the boundary.

9·33. As a further example we may obtain the motion of a vortex pair moving directly towards or from a parallel plane boundary or of a single vortex in a corner between planes meeting at right angles. The figure shews the necessary arrangement of images, and for the velocity of the vortex at A (x, y) due to the other three, we have components

$$u=\frac{\kappa}{2\pi AB}-\frac{\kappa}{2\pi AB'}\cdot\frac{AB}{AB'}=\frac{\kappa}{4\pi}\cdot\frac{x^2}{y(x^2+y^2)},$$

and
$$v=\frac{-\kappa}{2\pi AA'}+\frac{\kappa}{2\pi AB'}\cdot\frac{AA'}{AB'}=-\frac{\kappa}{4\pi}\cdot\frac{y^2}{x(x^2+y^2)}.$$

For the path of the vortex A, we have

$$\dot{x}=u \quad\text{and}\quad \dot{y}=v,$$

so that
$$\frac{dx}{x^3}=-\frac{dy}{y^3},$$

whence by integration $\frac{1}{x^2}+\frac{1}{y^2}=\frac{1}{a^2}$,

or in polar coordinates $r \sin 2\theta = 2a$;

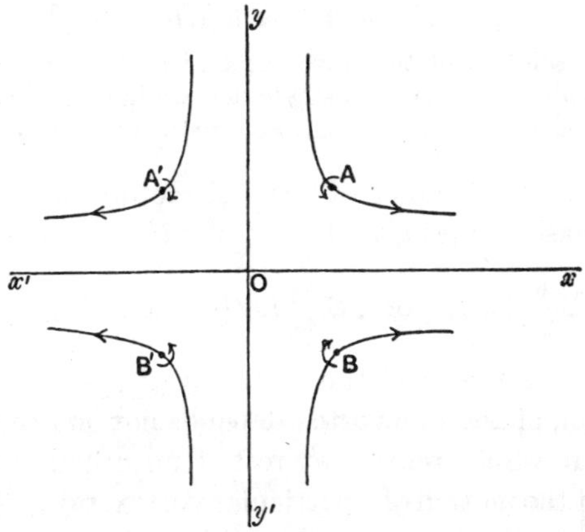

which represents a Cotes spiral with asymptotes parallel to the axes.

Also since $$x\dot{y}-y\dot{x}=-\kappa/4\pi,$$

the vortices describe the Cotes spiral in the same way as a particle under a central force, which can easily be seen to be a repulsion directed from the origin and varying as the inverse cube of the distance.

9·34. A rectilinear vortex within a circular cylinder of liquid will remain at rest if it lies along the axis, but not in any parallel position. It follows from **9·32** that the image is an equal and opposite vortex so situated that the vortices cut a cross section of the cylinder in inverse points.

Thus if C be the centre and A, B a pair of inverse points, we have seen that the stream lines due to equal and opposite vortices through A and B are coaxial circles having A, B as limiting points, so the cylinder in question will satisfy the condition for stream lines.

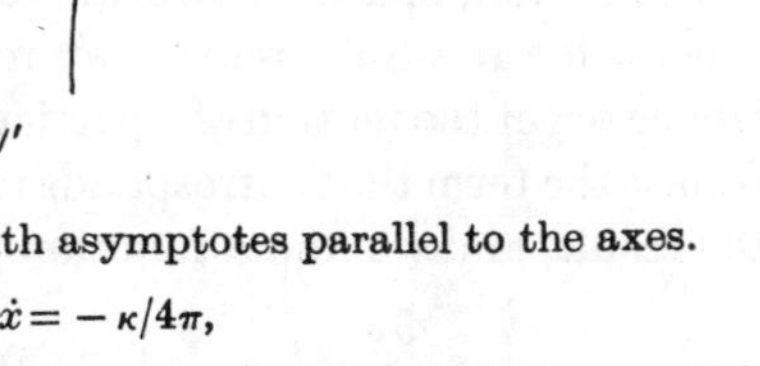

The velocities of the vortices are both equal to $\kappa/2\pi AB$ so they will not remain on the same radial plane through C, and the motions of the liquid inside and outside the cylinder only correspond at the instant under consideration. But so far as the motion inside the cylinder goes the vortex A describes a circle round C with uniform velocity $\kappa/2\pi AB$ or $\kappa . CA/2\pi(c^2-CA^2)$, c being the radius of the cylinder.

In the problem of the vortex B in liquid outside the cylinder, we notice that the foregoing solution with the image vortex at A implies a circulation κ round the cylinder due to the vortex A; but we want a solution in which the only circulation is due to the vortex B, and we can get this by superposing the motion due to another vortex $-\kappa$ at C. This will make the vortex B describe a circle round the cylinder with velocity (counter-clockwise)

$$\frac{\kappa}{2\pi AB}-\frac{\kappa}{2\pi CB}=\frac{\kappa c^2}{2\pi CB\,(CB^2-c^2)}.$$

To get the solution of the corresponding problem when there is an arbitrary circulation κ' round the cylinder, we have only to superpose a vortex of strength κ' at C, adding $\kappa'/2\pi CB$ to the velocity of the vortex B*.

9·35. For any number of parallel rectilinear vortices in an unlimited mass of liquid, we have a stream function

$$\psi=\Sigma\frac{\kappa}{2\pi}\log r, \quad \text{or} \quad \Sigma\frac{\kappa_1}{4\pi}\log\{(x-x_1)^2+(y-y_1)^2\},$$

where κ_1 is the strength of the vortex at (x_1, y_1).

The motion of any one vortex depends not on itself but on the others, for it would remain at rest if no others were present. Hence to get the motion of a particular vortex, say κ_1, we subtract from ψ the term that corresponds to this vortex, then if $\psi'(x,y)$ be the result, and we find a function $\chi(x_1, y_1)$ such that

$$-\frac{\partial\chi}{\partial y_1}=\left(-\frac{\partial\psi'}{\partial y}\right)_1, \quad \text{and} \quad \frac{\partial\chi}{\partial x_1}=\left(\frac{\partial\psi'}{\partial x}\right)_1,$$

these are the components of the velocity of the vortex, and $\chi(x_1, y_1)$ may be regarded as a stream function giving the motion of the vortex.

For example, if there be a vortex of strength κ at (x_1, y_1) and the axis of x be a boundary of the liquid, there is an image $-\kappa$ at $(x_1, -y_1)$, and

$$\psi=\frac{\kappa}{4\pi}\log\{(x-x_1)^2+(y-y_1)^2\}-\frac{\kappa}{4\pi}\log\{(x-x_1)^2+(y+y_1)^2\}.$$

Hence, in this case,

$$\psi'(x, y)=-\frac{\kappa}{4\pi}\log\{(x-x_1)^2+(y+y_1)^2\}.$$

Therefore
$$-\frac{\partial\chi}{\partial y_1}=\frac{\kappa}{4\pi y_1} \quad \text{and} \quad \frac{\partial\chi}{\partial x_1}=0,$$

so that the stream function for the motion of the vortex is

$$\chi(x_1, y_1)\equiv-\frac{\kappa}{4\pi}\log y_1,$$

or the path of the vortex is given by

$$y_1=\text{constant},$$

as we know from the discussion of **9·32**.

* See F. A. Tarleton, 'On a problem in vortex motion', *Proc. R.I.A.* Third Series, II, p. 617.

9·36. Use of Conformal Transformation. The method of **6·1–6·21** is also applicable when parallel rectilinear vortices exist in the liquid; and regarding the problem as one of two-dimensional motion, as in **6·12**, if a vortex Π of strength κ exists in one liquid at a point whose coordinates are (ξ_1, η_1), there will be a vortex P of equal strength at the corresponding point (x_1, y_1) of the other liquid; for the strength is $-\int d\phi$ taken round a small curve surrounding the vortex; and ϕ having the same value at corresponding points in the two liquids, the integral must have the same value when taken round corresponding curves. These vortices however do not necessarily continue to move so as to occupy corresponding points; but we may deduce the motion of one when we know that of the other. Thus, if $\psi(\xi, \eta)$ denote the stream function of the first motion, the path of the vortex Π will be given by a stream function $\chi(\xi_1, \eta_1)$ deduced, as in **9·35**, by omitting from ψ the term

$$\frac{\kappa}{4\pi}\log\{(\xi-\xi_1)^2+(\eta-\eta_1)^2\},$$

or the real part of $$\frac{\kappa}{2\pi}\log(t-t_1),$$

where $t=\xi+i\eta$.

Similarly in the transformed motion there will be a stream function $\chi'(x_1, y_1)$ for the motion of the vortex P obtained from ψ in the same way by the omission of the term

$$\frac{\kappa}{4\pi}\log\{(x-x_1)^2+(y-y_1)^2\},$$

or the real part of $$\frac{\kappa}{2\pi}\log(z-z_1).$$

Hence it follows that $\chi'=\chi+\chi''$, where χ'' is such that

$$\frac{\partial\chi''}{\partial y_1}=\text{the real part of}\left[\frac{\partial}{\partial y}\frac{\kappa}{2\pi}\log\frac{t-t_1}{z-z_1}\right]_{z=z_1}$$

Now $\dfrac{\partial}{\partial y}=i\dfrac{d}{dz}$, and we assume that $t-t_1$ is expansible in powers of $z-z_1$, so that

$$t-t_1=(z-z_1)\left(\frac{dt}{dz}\right)_1+\frac{(z-z_1)^2}{2!}\left(\frac{d^2t}{dz^2}\right)_1+\ldots\ldots;$$

therefore we require

$$\text{the real part of}\ \frac{i\kappa}{2\pi}\left[\frac{d}{dz}\log\left\{\left(\frac{dt}{dz}\right)_1+\tfrac{1}{2}(z-z_1)\left(\frac{d^2t}{dz^2}\right)_1+\ldots\ldots\right\}\right]_1,$$

or of $$\frac{i\kappa}{2\pi}\left[\tfrac{1}{2}\left(\frac{d^2t}{dz^2}\right)_1\Big/\left(\frac{dt}{dz}\right)_1+\text{positive powers of }(z-z_1)\right]_1;$$

that is, the real part of $$\frac{i\kappa}{4\pi}\left(\frac{d^2t}{dz^2}\right)_1\Big/\left(\frac{dt}{dz}\right)_1,$$

which is $$\frac{\kappa}{4\pi}\frac{\partial}{\partial y_1}\log\left|\frac{dt}{dz}\right|_1.$$

Hence $$\chi''=\frac{\kappa}{4\pi}\log\left|\frac{dt}{dz}\right|_1,$$

and $$\chi'(x_1, y_1)=\chi(\xi_1, \eta_1)+\frac{\kappa}{4\pi}\log\left|\frac{dt}{dz}\right|_1 \quad\text{.........(1)*.}$$

9·37. Examples. (i) *To find the path of a rectilinear vortex in the angle between two planes to which it is parallel.*

Let π/n be tne angle between the planes.

The transformation suitable to this case is

$$\xi+i\eta=c\left(\frac{x+iy}{c}\right)^n \quad\text{or}\quad t=c\left(\frac{z}{c}\right)^n \quad\text{..................(1);}$$

or, in polar coordinates, $\rho=c\,(r/c)^n$, $\omega=n\theta$.

This transforms the ξ axis ($\omega=0$, $\omega=\pi$) into the straight lines $\theta=0$, $\theta=\pi/n$.

The stream function due to a vortex Π at (ξ_1, η_1) in liquid bounded by the ξ axis is, as in **9·35**,

$$\psi=\frac{\kappa}{4\pi}\log\frac{(\xi-\xi_1)^2+(\eta-\eta_1)^2}{(\xi-\xi_1)^2+(\eta+\eta_1)^2} \quad\text{......................(2).}$$

Therefore the stream function due to a vortex P at (x_1, y_1) or (r_1, θ_1) in liquid bounded by $\theta=0$, $\theta=\pi/n$ is

$$\psi=\frac{\kappa}{4\pi}\log\frac{r^{2n}+r_1^{2n}-2r^nr_1^n\cos n(\theta-\theta_1)}{r^{2n}+r_1^{2n}-2r^nr_1^n\cos n(\theta+\theta_1)}.$$

Again $$|dt/dz|=d\rho/dr=n\,(r/c)^{n-1};$$

so that for the path of P

$$\chi'(x_1, y_1)=\chi(\xi_1, \eta_1)+\frac{\kappa}{4\pi}\log r_1^{n-1},$$

where, as in **9·35**, $$\chi(\xi_1, \eta_1)=-\frac{\kappa}{4\pi}\log\eta_1.$$

Therefore $$\chi'(x_1, y_1)=-\frac{\kappa}{4\pi}\log r_1^n\sin n\theta_1+\frac{\kappa}{4\pi}\log r_1^{n-1}$$

$$=-\frac{\kappa}{4\pi}\log r_1\sin n\theta_1,$$

neglecting constant terms.

Hence the path of P is $r_1\sin n\theta_1=\text{const.}$,

which is a Cotes spiral.

* This theorem was enunciated by Routh—'Some Applications of Conjugate Functions', *Proc. L.M.S.* XII, 1881, p. 83.

This agrees with **9·33** for the case $n=2$. The same problem might be solved directly by a series of images provided n is an integer, but this restriction is not necessary in the method used above*.

(ii) *There is a rectilinear vortex in liquid filling the space between two parallel planes. To find the paths of the particles.*

The relation
$$\xi+i\eta=e^{p(x+iy)},$$
or
$$\xi=e^{px}\cos py,\quad \eta=e^{px}\sin py,$$
transforms the ξ axis $\eta=0$ into the lines $y=0$, $y=\pi/p$.

Taking a vortex of strength κ at a suitable point (ξ_1, η_1) with the ξ axis as boundary, we get a corresponding vortex at (x_1, y_1) between the parallel planes $y=0$, $y=\pi/p$.

As before the stream function of the original motion is
$$\psi=\frac{\kappa}{4\pi}\log\frac{(\xi-\xi_1)^2+(\eta-\eta_1)^2}{(\xi-\xi_1)^2+(\eta+\eta_1)^2};$$
and we get an expression for the stream function between the parallel planes by substituting for ξ, η in terms of x, y. Thus if the distance between the planes be c and the vortex be midway between them we have $p=\pi/c$, and $y_1=c/2$, and if we take the y axis through the vortex we also have $x_1=0$, and therefore $\xi_1=0$ and $\eta_1=1$.

Hence we get
$$\frac{e^{2\pi x/c}\cos^2\pi y/c+(e^{\pi x/c}\sin\pi y/c-1)^2}{e^{2\pi x/c}\cos^2\pi y/c+(e^{\pi x/c}\sin\pi y/c+1)^2}=\text{const.}$$
which reduces to $\cosh\pi x/c=A\sin\pi y/c$ and this represents the paths of the particles†.

9·4. An infinite Row of parallel rectilinear Vortices of the same Strength κ at a distance a apart. Considering $2n+1$ vortices, taking the origin at the middle one and the axis of x through the centres of their sections

we have from **9·3** $w=\dfrac{i\kappa}{2\pi}\log z\,(z^2-a^2)\;\ldots\ldots(z^2-n^2a^2)$,

or
$$w=\frac{i\kappa}{2\pi}\log\frac{\pi z}{a}\left(1-\frac{z^2}{a^2}\right)\cdots\left(1-\frac{z^2}{n^2a^2}\right)+\text{const.}\quad\ldots\ldots(1).$$
When $n\to\infty$ for an infinite row, this becomes
$$w=\frac{i\kappa}{2\pi}\log\sin\frac{\pi z}{a}\quad\ldots\ldots\ldots\ldots\ldots\ldots(2).$$

Then, for the velocity components
$$u-iv=-\frac{dw}{dz}=-\frac{i\kappa}{2a}\cot\frac{\pi z}{a}=-\frac{i\kappa}{2a}\,\frac{\cos\frac{\pi}{a}(x+iy)\sin\frac{\pi}{a}(x-iy)}{\sin\frac{\pi}{a}(x+iy)\sin\frac{\pi}{a}(x-iy)},$$

* Greenhill, *Quart. Journal*, xv, p. 15, 'Plane Vortex motion'.

† For other examples of this method and the extension of the method by inversion, see Routh, *Proc. L.M.S.* xii, p. 81.

so that

$$u = -\frac{\kappa}{2a}\frac{\sinh 2\pi y/a}{\cosh 2\pi y/a - \cos 2\pi x/a}, \quad v = \frac{\kappa}{2a}\frac{\sin 2\pi x/a}{\cosh 2\pi y/a - \cos 2\pi x/a} \qquad \text{......(3).}$$

By considering the effect produced by pairs of vortices at equal distances from a given vortex, it follows that the vortices remain at rest.

9·41. A double Line of Vortices. Consider two such lines of vortices at a distance b apart, symmetrically placed with regard to the plane mid-way between them and such that the rotation in the two rows is in opposite senses.

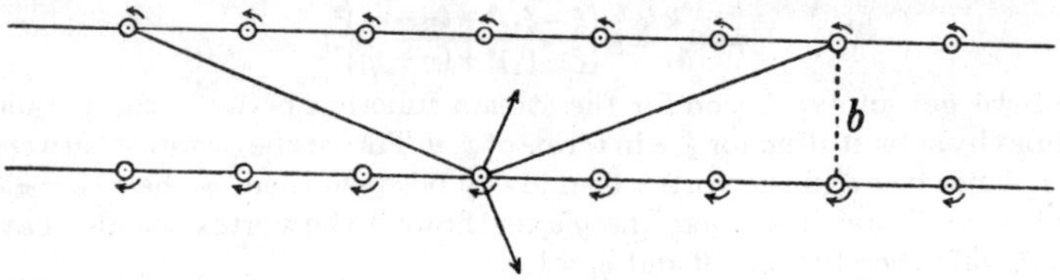

It follows from **9·4** that neither row has any effect in producing velocity in itself; and by considering the effect on a chosen vortex of equidistant vortices in the other row we see that the resultant velocity is along the rows. Its magnitude is obtained from **9·4** (3) by putting $x = na$, $y = -b$, or by summing the effects of all the vortices thus

$$\begin{aligned} U &= \sum_{n=-\infty}^{\infty} \frac{\kappa}{2\pi}\frac{b}{b^2+n^2a^2} \\ &= \frac{\kappa}{2\pi}\left\{\frac{1}{b} + \frac{2b}{b^2+a^2} + \frac{2b}{b^2+2^2a^2} + \dots\right\} \\ &= \frac{\kappa}{2a}\coth\frac{\pi b}{a}. \end{aligned}$$

9·42. A Kármán Street. This is a double line of vortices similar to the last save that each vortex is opposite to the point midway between two vortices of the opposite row.

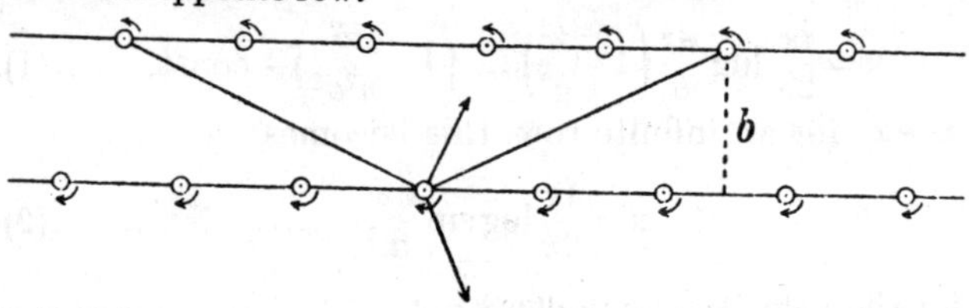

As in **9·41** it is evident that the system moves along the rows and that the velocity, from **9·4** (3) by putting $x = (n+\frac{1}{2})a$, $y = -b$, or by summing thus, is

$$\begin{aligned} U &= \sum_{n=-\infty}^{\infty} \frac{\kappa}{2\pi}\frac{b}{b^2+(n+\frac{1}{2})^2a^2} \\ &= \frac{\kappa}{\pi}\left\{\frac{4b}{4b^2+a^2} + \frac{4b}{4b^2+3^2a^2} + \frac{4b^2}{4b^2+5^2a^2} + \dots\right\} \\ &= \frac{\kappa}{2a}\tanh\frac{\pi b}{a}. \end{aligned}$$

This arrangement is called after Th. von Kármán who first discussed the stability of such arrangements and pointed out that a double trail of vortices of this kind is often formed when a body like a flat plate moves broadside through a fluid. This arrangement is under certain conditions stable, whereas the single row of **9·4** and the double row of **9·41** are unstable. A discussion of the stability with references to papers by von Kármán, Kelvin and Rosenhead may be found in Lamb's *Hydrodynamics**.

9·5. Rectilinear Vortex with circular section. We shall consider now some cases of vortices with finite cross section. Let the section be a circle of radius a, and suppose the spin to be uniform and equal to ζ throughout the whole section, the vortex being rectilinear.

The equations for the stream function are

$$\frac{\partial^2\psi}{\partial x^2}+\frac{\partial^2\psi}{\partial y^2}=2\zeta, \text{ inside the vortex,}$$

and
$$\frac{\partial^2\psi}{\partial x^2}+\frac{\partial^2\psi}{\partial y^2}=0, \text{ outside the vortex.}$$

These are equivalent to

$$\frac{\partial^2\psi}{\partial r^2}+\frac{1}{r}\frac{\partial\psi}{\partial r}=2\zeta, \text{ when } r<a \quad\quad(1),$$

and
$$\frac{\partial^2\psi}{\partial r^2}+\frac{1}{r}\frac{\partial\psi}{\partial r}=0, \text{ when } r>a \quad\quad(2).$$

The complete integral of (2) is

$$\psi=C\log r+D,$$

and a particular integral of (1) is

$$\psi=\tfrac{1}{2}\zeta r^2,$$

therefore, when $r<a$, $\quad \psi=A\log r+B+\tfrac{1}{2}\zeta r^2 \quad\quad(3),$

and, when $r>a$, $\quad \psi=C\log r+D \quad\quad(4).$

Since ψ is not to be infinite when $r=0$ we must have $A=0$. And if the motion is continuous at the surface we have ψ and the tangential velocity $\partial\psi/\partial r$ continuous so that

$$B+\tfrac{1}{2}\zeta a^2=C\log a+D,$$

and
$$\zeta a=C/a.$$

Hence neglecting an additive constant we have, when $r<a$,

$$\psi=-\tfrac{1}{2}\zeta(a^2-r^2) \quad\quad(5),$$

and, when $r>a$, $\quad \psi=\zeta a^2\log r/a \quad\quad(6).$

The velocity is wholly transversal both inside and outside the vortex, its values being ζr and $\zeta a^2/r$.

Outside the vortex the motion is irrotational and the velocity potential can be found by taking

$$w=i\zeta a^2\log z/a,$$

for this gives the correct value for ψ. Hence we have

$$\phi=-\zeta a^2\theta,$$

a many-valued function as we should expect, the motion being cyclic. If κ denote the circulation or the strength of the vortex, $\kappa=2\pi a^2\zeta$, so that

$$\phi=-\frac{\kappa\theta}{2\pi} \quad \text{and} \quad \psi=\frac{\kappa}{2\pi}\log r,$$

as for a thin filament.

* § 156, 1932.

To find the pressure. Outside the vortex we have

$$\frac{p}{\rho}=\frac{\partial\phi}{\partial t}-\tfrac{1}{2}q^2+F(t),$$

or, since the motion is steady, and $q=\zeta a^2/r$ or $\kappa/2\pi r$,

$$\frac{p}{\rho}=\frac{\Pi}{\rho}-\frac{\kappa^2}{8\pi^2r^2},\quad \text{when } r>a,$$

where Π is the value of p when r is infinite.

Inside the vortex we have the case of a liquid rotating uniformly with angular velocity ζ, so that

$$\frac{dp}{\rho}=\zeta^2 r\,dr,$$

or

$$\frac{p}{\rho}=\tfrac{1}{2}\zeta^2r^2+\frac{P}{\rho},$$

where P is the pressure at the centre of the vortex. Since the values of p are equal when $r=a$, therefore

$$P=\Pi-\kappa^2\rho/4\pi^2a^2.$$

Hence when $r<a$

$$\frac{p}{\rho}=\frac{\Pi}{\rho}-\frac{\kappa^2}{4\pi^2a^2}+\frac{\kappa^2r^2}{8\pi^2a^4},$$

shewing that if $\Pi<\kappa^2\rho/4\pi^2a^2$, there will be a value of $r<a$ for which p becomes negative, implying that a cylindrical hollow must exist inside the vortex.

It is possible to have cyclic irrotational motion surrounding a hollow cylindrical space. The necessary condition is $p=0$ when $r=a$; that is

$$\Pi=\kappa^2\rho/8\pi^2a^2.$$

The oscillations of vortices of the forms just considered were discussed by Lord Kelvin*.

9·51. Rankine's Combined Vortex consists of a circular vortex with axis vertical in a mass of liquid moving irrotationally under the action of gravity. The kinematical equations are as in the case just considered, and if a is the radius the pressure equations are

$$\frac{p}{\rho}=\text{const.}-\frac{\kappa^2}{8\pi^2r^2}-gz,\quad \text{when } r>a,$$

and

$$\frac{p}{\rho}=\text{const.}+\frac{\kappa^2r^2}{8\pi^2a^4}-gz,\quad \text{when } r<a.$$

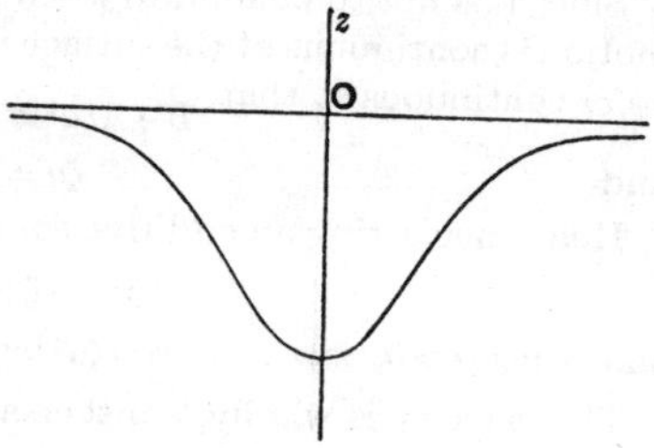

The free surface has a depression or dimple over the top of the vortex as shewn in the figure. The equations of the free surface, obtained by making p constant, are

$$z=\frac{\kappa^2}{8\pi^2a^4g}\left(a^2-\frac{a^4}{r^2}\right)+C,\quad \text{when } r>a \quad \text{..............(1)},$$

and

$$z=\frac{\kappa^2}{8\pi^2a^4g}(r^2-a^2)+C,\quad \text{when } r<a \quad \text{..............(2)},$$

the constants being arranged to preserve continuity when $r=a$.

* 'Vibrations of a columnar Vortex', *Phil. Mag.* x, 1880, p. 155, or *Math. and Phys. Papers*, IV, p. 152.

Taking the origin in the general level of the free surface, in (1) we can put $z=0$ when $r=\infty$, so that

$$C=-\kappa^2/8\pi^2a^2g.$$

Then in (2) by putting $r=0$ we get the depth of the central depression given by

$$-z=\kappa^2/4\pi^2a^2g.$$

9·52. Elliptic Section. To shew that a rectilinear vortex whose cross section is an ellipse and whose spin is constant can maintain its form rotating as if it were a solid cylinder in an infinite liquid*.

We have seen in **5·35**, that if a rigid elliptic cylinder of semi-axes a, b rotates with uniform angular velocity ω in an infinite mass of liquid the stream function for cyclic irrotational motion with circulation κ is

$$\psi=\tfrac{1}{4}\omega(a+b)^2e^{-2\xi}\cos 2\eta+\kappa\xi/2\pi \quad \text{.................(1).}$$

In this case $\kappa=2\pi\zeta ab$, where ζ is the constant spin.

Inside the vortex we have $\dfrac{\partial^2\psi}{\partial x^2}+\dfrac{\partial^2\psi}{\partial y^2}=2\zeta$ ....(2),

with a boundary condition that the velocity of the liquid normal to the boundary is equal to that of the boundary, that is

$$\frac{ux}{a^2}+\frac{vy}{b^2}=-\omega y\frac{x}{a^2}+\omega x\frac{y}{b^2} \quad \text{.......................(3).}$$

Assume that $\psi=\zeta(Ax^2+By^2)$ ....(4),

then from (2) and (3) we have

$$A+B=1,\ \text{and}\ Aa^2-Bb^2=\omega(a^2-b^2)/2\zeta \quad \text{...........(5).}$$

The further condition of continuity of the tangential velocity at the boundary makes the values of $\partial\psi/\partial\xi$ obtained from (1) and (4) the same.

Putting $x=c\cosh\xi\cos\eta$, $y=c\sinh\xi\sin\eta$ in (4), this gives at the boundary

$$\begin{aligned}-\tfrac{1}{2}\omega(a+b)^2e^{-2\xi}\cos 2\eta+\zeta ab&\\ =\zeta c^2\cosh\xi\sinh\xi\{A+B+(A-B)\cos 2\eta\}&\end{aligned}$$

for all values of η from 0 to 2π.

Equating coefficients of $\cos 2\eta$ we get

$$-\tfrac{1}{2}\omega(a+b)^2e^{-2\xi}=\zeta c^2(A-B)\cosh\xi\sinh\xi,$$

but on the boundary $a=c\cosh\xi$, $b=c\sinh\xi$, and $a-b=ce^{-\xi}$, therefore

$$A-B=-\frac{\omega}{2\zeta}\cdot\frac{a^2-b^2}{ab} \quad \text{.......................(6).}$$

From (5) and (6) we find

$$Aa=Bb=ab/(a+b),$$

and

$$\omega=\frac{2ab}{(a+b)^2}\zeta.$$

This gives the velocity of rotation of the cylinder as a whole in terms of the spin and eccentricity of the section.

* Kirchhoff, *Mechanik*, p. 261; see also Love, 'On the Stability of certain Vortex Motions', *Proc. L.M.S.* xxv 1893.

To find the paths of the particles. If x, y are coordinates of a particle of the vortex referred to the axes of the cross section

$$\dot{x}-\omega y=u=-\frac{\partial\psi}{\partial y}=-2\zeta By=-y\omega(a+b)/b,$$

and
$$\dot{y}+\omega x=v=\frac{\partial\psi}{\partial x}=2\zeta Ax=x\omega(a+b)/a.$$

Therefore $\dot{x}=-\omega ya/b$ and $\dot{y}=\omega xb/a$,
which lead on integration to

$$x=La\cos(\omega t+\epsilon),\quad y=Lb\sin(\omega t+\epsilon),$$

so that the paths of the particles of the vortex relative to the boundary are similar ellipses, and the period of the relative motion is the same as that of the rotation of the cylinder.

9·6. Uniqueness Theorem. If an infinite mass of liquid filling all space be at rest at infinity we conclude from **4·6** that the liquid must either be at rest everywhere, or that, if in motion, its motion cannot be irrotational at every point.

We shall now prove that in such a liquid at rest at infinity the motion is determinate when we know the values of the components of spin ξ, η, ζ at all points. For if possible let there be two sets of values u_1, v_1, w_1 and u_2, v_2, w_2 of the velocity components each satisfying the equation of continuity and the equations

$$\frac{\partial w}{\partial y}-\frac{\partial v}{\partial z}=2\xi,\quad \frac{\partial u}{\partial z}-\frac{\partial w}{\partial x}=2\eta,\quad \frac{\partial v}{\partial x}-\frac{\partial u}{\partial y}=2\zeta,$$

at all points of space and vanishing at infinity.

Then the differences $u'=u_1-u_2$, $v'=v_1-v_2$, $w'=w_1-w_2$ also satisfy the equation of continuity and

$$\frac{\partial w'}{\partial y}-\frac{\partial v'}{\partial z}=0, \text{ etc.}$$

at all points of space and vanish at infinity. That is, u', v', w' are velocity components of irrotational motion of a liquid filling all space and vanish at infinity. Hence we must have $u'=v'=w'=0$ everywhere, and therefore there is only one motion satisfying the prescribed conditions.

A similar argument would prove that the motion of a liquid contained in a limited simply-connected region is determinate when the motion of the boundary and the components of spin are known. For a multiply-connected region a knowledge of the circulations in the several independent circuits must be included in the given conditions.

9·61. In general there may be several contributory causes that go to produce motion at a point in a fluid; for example, the presence of sources and sinks or the motions of boundaries or immersed solids or the presence of one or more vortices in a fluid result in a general motion of the fluid. The velocities due to the several causes may be superposed and it is our purpose now to find expressions for the components of velocity u, v, w at any point in a liquid due to given vortices, i.e. in terms of given components of spin ξ, η, ζ.

9·62. To find u, v, w from ξ, η, ζ. The liquid being incompressible the flow across any two surfaces having the same curve for boundary will be the same, and therefore depends only on the form of the boundary. If we assume that this flow can be represented by a line integral round the boundary, we get an equation

$$\iint (lu + mv + nw)\,dS = \int (F\,dx + G\,dy + H\,dz),$$

where F, G, H are components of a certain vector.

But from **4·2**

$$\int (F\,dx + G\,dy + H\,dz)$$
$$= \iint \left\{ l\left(\frac{\partial H}{\partial y} - \frac{\partial G}{\partial z}\right) + m\left(\frac{\partial F}{\partial z} - \frac{\partial H}{\partial x}\right) + n\left(\frac{\partial G}{\partial x} - \frac{\partial F}{\partial y}\right)\right\} dS,$$

hence we must have

$$u = \frac{\partial H}{\partial y} - \frac{\partial G}{\partial z}, \quad v = \frac{\partial F}{\partial z} - \frac{\partial H}{\partial x}, \quad w = \frac{\partial G}{\partial x} - \frac{\partial F}{\partial y} \quad \ldots\ldots(1),$$

or as it may be expressed more briefly

$$u, v, w = \text{curl}\,(F, G, H).$$

It is clear that the values of u, v, w given by (1) satisfy the equation of continuity; and substituting in the values for ξ, η, ζ we get

$$2\xi = \frac{\partial w}{\partial y} - \frac{\partial v}{\partial z} = \frac{\partial}{\partial x}\left(\frac{\partial F}{\partial x} + \frac{\partial G}{\partial y} + \frac{\partial H}{\partial z}\right) - \nabla^2 F \quad \ldots\ldots\ldots(2),$$

and similar expressions for 2η, 2ζ.

Hence the assumptions of equations (1) will be justified if we can find F, G, H so as to satisfy the four equations

$$\frac{\partial F}{\partial x} + \frac{\partial G}{\partial y} + \frac{\partial H}{\partial z} = 0 \quad \ldots\ldots\ldots\ldots\ldots\ldots\ldots(3),$$

$$\nabla^2 F = -2\xi, \quad \nabla^2 G = -2\eta, \quad \nabla^2 H = -2\zeta \quad \ldots\ldots(4).$$

The last equations can be satisfied by assuming F, G, H to be potential functions due to distributions of gravitating matter of volume densities $\xi/2\pi$, $\eta/2\pi$, $\zeta/2\pi$ respectively. We then have

$$\left.\begin{aligned} F &= \frac{1}{2\pi}\iiint \frac{\xi'}{r}\,dx'\,dy'\,dz' \\ G &= \frac{1}{2\pi}\iiint \frac{\eta'}{r}\,dx'\,dy'\,dz' \\ H &= \frac{1}{2\pi}\iiint \frac{\zeta'}{r}\,dx'\,dy'\,dz' \end{aligned}\right\} \quad \ldots\ldots\ldots\ldots\ldots\ldots(5)$$

for the values of F, G, H at the point (x, y, z), where

$$r^2 = (x-x')^2 + (y-y')^2 + (z-z')^2,$$

and ξ', η', ζ' are components of spin of the element $dx'\,dy'\,dz'$ at (x', y', z'), and the range of integration may be taken as extending throughout the whole liquid, though the integrand is zero at all points at which there is no spin.

To complete the solution we must shew that the expressions (5) satisfy (3).

We have
$$\begin{aligned} \frac{\partial F}{\partial x} &= \frac{1}{2\pi}\iiint \xi' \frac{\partial}{\partial x}\left(\frac{1}{r}\right) dx'\,dy'\,dz' \\ &= -\frac{1}{2\pi}\iiint \xi' \frac{\partial}{\partial x'}\left(\frac{1}{r}\right) dx'\,dy'\,dz'; \end{aligned}$$

and integrating by parts

$$\frac{\partial F}{\partial x} = -\frac{1}{2\pi}\iint \frac{l\xi'}{r}\,dS + \frac{1}{2\pi}\iiint \frac{1}{r}\frac{\partial \xi'}{\partial x'}\,dx'\,dy'\,dz'.$$

Therefore

$$\begin{aligned} \frac{\partial F}{\partial x} + \frac{\partial G}{\partial y} + \frac{\partial H}{\partial z} = &-\frac{1}{2\pi}\iint \frac{1}{r}(l\xi' + m\eta' + n\zeta')\,dS \\ &+\frac{1}{2\pi}\iiint \frac{1}{r}\left(\frac{\partial \xi'}{\partial x'} + \frac{\partial \eta'}{\partial y'} + \frac{\partial \zeta'}{\partial z'}\right) dx'\,dy'\,dz', \end{aligned}$$

where (l, m, n) are direction cosines of the normal to the element dS of the boundary of the liquid.

Now the vortex filaments are all either closed or end on the surface S of the liquid, and in the latter case we can always continue these filaments either on the surface S or outside it until they return into themselves so that a greater space exists bounded by a surface S', in which exist only re-entrant vortex filaments.

Without loss of generality we may suppose the boundary to be of this kind, and then at every point on it either $\xi=\eta=\zeta=0$, or else

$$l\xi+m\eta+n\zeta=0,$$

so that the surface integral in the last equation vanishes. And since

$$\frac{\partial\xi}{\partial x}+\frac{\partial\eta}{\partial y}+\frac{\partial\zeta}{\partial z}$$

vanishes identically at all points of the liquid, as can be seen by substituting for ξ, η, ζ in terms of u, v, w, therefore the volume integral vanishes also. Hence

$$\frac{\partial F}{\partial x}+\frac{\partial G}{\partial y}+\frac{\partial H}{\partial z}=0.$$

We have therefore shewn that (3) is the necessary and sufficient condition that the expressions (5) for F, G, H in terms of the spin shall give a consistent set of values for the velocity components when substituted in (1). But it must be observed that these expressions only constitute a particular solution of the equations, and that without invalidating the solution we might add to F, G, H respectively three functions of the form $\partial\chi/\partial x$, $\partial\chi/\partial y$, $\partial\chi/\partial z$ provided $\nabla^2\chi=0$.

It must not be assumed however that there is a possible motion corresponding to any arbitrary distribution of spin components, for unless the components of velocity u, v, w and the pressure p are continuous they do not in general represent a possible state of the liquid. We shall refer later to one possible state of discontinuity under the head of vortex sheets.

9·63. Each element of rotating Liquid produces a Velocity in every other element of the Liquid Mass. In (1) of **9·62** let us substitute from (5) so much of the values of F, G, H as are contributed by the element $dx'\,dy'\,dz'$ and call the resulting components of velocity at (x, y, z) $\delta u, \delta v, \delta w$. We have

$$\left.\begin{aligned}\delta u&=-\frac{1}{2\pi}\{(y-y')\,\zeta'-(z-z')\,\eta'\}\frac{dx'\,dy'\,dz'}{r^3}\\ \delta v&=-\frac{1}{2\pi}\{(z-z')\,\xi'-(x-x')\,\zeta'\}\frac{dx'\,dy'\,dz'}{r^3}\\ \delta w&=-\frac{1}{2\pi}\{(x-x')\,\eta'-(y-y')\,\xi'\}\frac{dx'\,dy'\,dz'}{r^3}\end{aligned}\right\}\quad\ldots\ldots(1).$$

Hence $\qquad (x-x')\,\delta u+(y-y')\,\delta v+(z-z')\,\delta w=0,$

so that the resultant of δu, δv, δw is at right angles to r. Also

$$\xi'\,\delta u+\eta'\,\delta v+\zeta'\,\delta w=0;$$

and this resultant is therefore also at right angles to the axis of spin at (x', y', z').

Lastly $\delta q=\{(\delta u)^2+(\delta v)^2+(\delta w)^2\}^{\frac{1}{2}}=\dfrac{dx'\,dy'\,dz'}{2\pi r^2}\,\omega'\sin\nu$ ...(2),

where ω' is the resultant of ξ', η', ζ' and ν is the angle between r and the axis of spin at (x', y', z').

Hence each rotating element A of liquid implies in each other element B of the same liquid mass a velocity whose direction is perpendicular to the plane through B and the axis of rotation of A, its magnitude being given by the result (2). If the element at A be a length $\delta s'$ of a vortex filament of strength κ we have

$$\omega'\,dx'\,dy'\,dz'=\tfrac{1}{2}\kappa\,\delta s',$$

so that we may write the result

$$\delta q=\frac{\kappa}{4\pi}\cdot\frac{\sin\nu\,\delta s'}{r^2}.$$

9·64. The reader familiar with the theory of electromagnetism will again recognise the analogy to which reference was made in **9·32**. The vortices correspond to electric currents and the liquid velocities to magnetic force due to the currents. The relations between ξ, η, ζ and u, v, w are analogous to

(electric current) = curl (magnetic force);

the result of **9·63** corresponds to the force on a magnetic pole due to an element of an electric current, and in **9·62** the vector (F, G, H) corresponds to the vector potential of magnetic induction.

9·65. If the fluid be not incompressible we may write the equation of continuity

$$\frac{\partial u}{\partial x}+\frac{\partial v}{\partial y}+\frac{\partial w}{\partial z}=-\frac{1}{\rho}\frac{D\rho}{Dt}.$$

But if v be the volume of a small element of fluid its mass ρv is invariable, so that

$$0=\frac{D(\rho v)}{Dt}=v\frac{D\rho}{Dt}+\rho\frac{Dv}{Dt};$$

therefore
$$\frac{\partial u}{\partial x}+\frac{\partial v}{\partial y}+\frac{\partial w}{\partial z}=\frac{1}{v}\frac{Dv}{Dt}=\theta,\text{ say},$$

where θ denotes the 'expansion' or rate of increase of volume at (x, y, z).

The expansion will cause extra terms in the expressions for the velocities; the expansion of an element $dx'\,dy'\,dz'$ being equivalent to a simple source of strength $\frac{\theta'}{4\pi}\,dx'\,dy'\,dz'$ at (x', y', z').

This gives rise to a velocity potential whose value at (x, y, z) is

$$\phi = \frac{1}{4\pi}\iiint \frac{\theta'}{r}\,dx'\,dy'\,dz', \text{ by } \mathbf{3{\cdot}3},$$

and the complete expressions for the velocity are

$$u = -\frac{\partial\phi}{\partial x} + \frac{\partial H}{\partial y} - \frac{\partial G}{\partial z},$$

$$v = -\frac{\partial\phi}{\partial y} + \frac{\partial F}{\partial z} - \frac{\partial H}{\partial x},$$

$$w = -\frac{\partial\phi}{\partial z} + \frac{\partial G}{\partial x} - \frac{\partial F}{\partial y}.$$

9·66. Velocity Potential due to a Vortex in incompressible fluid. Considering a single re-entrant vortex filament of strength κ, we may write the expressions (1) of **9·63**

$$\delta u = -\frac{\kappa}{4\pi r^3}\{(y-y')\,dz' - (z-z')\,dy'\}, \text{ etc.}$$

by putting $\xi', \eta', \zeta' = \omega'\,(dx'/ds', dy'/ds', dz'/ds')$,

and $\omega'\,dx'\,dy'\,dz' = \frac{1}{2}\kappa\,ds'$.

Hence
$$u = -\frac{\kappa}{4\pi}\int\left\{\frac{\partial}{\partial y'}\left(\frac{1}{r}\right)dz' - \frac{\partial}{\partial z'}\left(\frac{1}{r}\right)dy'\right\},$$

where the integration is taken round the filament.

By Stokes's Theorem this line integral is equal to a surface integral over any surface bounded by the filament. Thus if we write

$$u = \frac{\kappa}{4\pi}\int (X\,dx' + Y\,dy' + Z\,dz'),$$

we also have

$$u = \frac{\kappa}{4\pi}\iint\left\{l\left(\frac{\partial Z}{\partial y'} - \frac{\partial Y}{\partial z'}\right) + m\left(\frac{\partial X}{\partial z'} - \frac{\partial Z}{\partial x'}\right) + n\left(\frac{\partial Y}{\partial x'} - \frac{\partial X}{\partial y'}\right)\right\}dS'.$$

But
$$X = 0, \quad Y = \frac{\partial}{\partial z'}\left(\frac{1}{r}\right), \quad Z = -\frac{\partial}{\partial y'}\left(\frac{1}{r}\right);$$

therefore
$$\frac{\partial Z}{\partial y'} - \frac{\partial Y}{\partial z'} = -\left(\frac{\partial^2}{\partial y'^2} + \frac{\partial^2}{\partial z'^2}\right)\frac{1}{r} = \frac{\partial^2}{\partial x'^2}\frac{1}{r};$$

$$\frac{\partial X}{\partial z'} - \frac{\partial Z}{\partial x'} = \frac{\partial^2}{\partial x'\,\partial y'}\left(\frac{1}{r}\right); \text{ and } \frac{\partial Y}{\partial x'} - \frac{\partial X}{\partial y'} = \frac{\partial^2}{\partial x'\,\partial z'}\left(\frac{1}{r}\right).$$

Hence $u=\frac{\kappa}{4\pi}\iint\left(l\frac{\partial}{\partial x'}+m\frac{\partial}{\partial y'}+n\frac{\partial}{\partial z'}\right)\frac{\partial}{\partial x'}\left(\frac{1}{r}\right)dS'\cdot$

or since
$$\frac{\partial}{\partial x'}\left(\frac{1}{r}\right)=-\frac{\partial}{\partial x}\left(\frac{1}{r}\right),$$
$$u=-\frac{\kappa}{4\pi}\frac{\partial}{\partial x}\iint\left(l\frac{\partial}{\partial x'}+m\frac{\partial}{\partial y'}+n\frac{\partial}{\partial z'}\right)\frac{1}{r}dS',$$
and similar expressions for v and w.

The velocity potential from which u, v, w are derived is therefore
$$\phi=\frac{\kappa}{4\pi}\iint\left(l\frac{\partial}{\partial x'}+m\frac{\partial}{\partial y'}+n\frac{\partial}{\partial z'}\right)\frac{1}{r}\,dS'$$
$$=\frac{\kappa}{4\pi}\iint\frac{\cos\theta\,dS'}{r^2}\quad\text{.............................(1)},$$
where θ is the angle between the normal (l, m, n) to the element dS' and the line r joining (x, y, z) and (x', y', z').

This result may clearly be written
$$\phi=\kappa\Omega/4\pi\quad\text{......................(2)},$$
where Ω is the solid angle subtended at the point (x, y, z) by a surface having the vortex filament for edge.

This potential function is clearly a cyclic quantity increasing by the cyclic constant κ every time the path of a moving point completes a circuit linked with the vortex, for in these circumstances the solid angle increases by 4π. It resembles the magnetic potential due to an electric current in a closed circuit or to a magnetic shell.

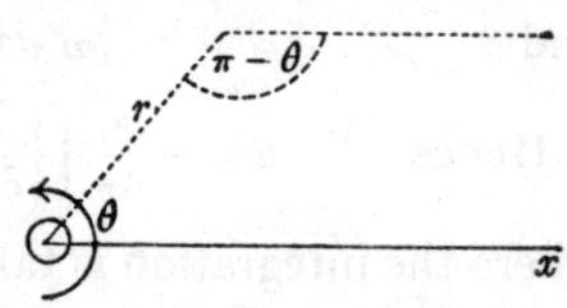

For a single rectilinear vortex we may take
$$\Omega=2(\pi-\theta)$$
and
$$\phi=\kappa(\pi-\theta)/2\pi,$$
making the velocity $-\partial\phi/r\,\partial\theta=\kappa/2\pi r$, as before.

9·67. From **3·31** and **9·66** (1) we see that the velocity potential is what would be produced by a distribution of doublets over the surface S' of strength $\kappa/4\pi$ per unit area with their axes all normal to the surface and directed to the same side of the surface. This can easily be understood from the fact that the stream lines all thread the vortex cutting across any surface bounded by it, and the motion might conceivably be produced by a giving out of liquid normally on one side of such a surface and the absorption of it at the same rate on the other side, combined with a suitable flow parallel to the surface in order to give the stream lines their actual directions at each point of the surface.

9·7. Vortex Sheets. Suppose that a surface exists in a fluid over which the normal component of velocity is continuous but the tangential component has different values on opposite sides of the surface.

Consider a small circuit consisting of two lines of length ds drawn on opposite sides of the surface and having their extremities joined by two infinitely shorter lines dn normal to the surface. Let the lines ds be in the direction of the relative velocity, which is clearly tangential to the surface and of magnitude

$$\{(u-u')^2+(v-v')^2+(w-w')^2\}^{\frac{1}{2}},$$

if u, v, w and u', v', w' denote the components on opposite sides of the surface.

If q, q' denote the components of velocity in direction ds on opposite sides of the surface the circulation in the small circuit is $(q-q')\,ds$. But $q-q'$ is clearly the relative velocity, so that the circulation is also

$$\{(u-u')^2+(v-v')^2+(w-w')^2\}^{\frac{1}{2}}\,ds.$$

This may be regarded as due to a stratum of vortices whose axes are at right angles to the direction of the relative velocity. If ω be the spin at the point considered, the circulation is $2\omega\,ds\,dn$, so that

$$2\omega\,dn=\{(u-u')^2+(v-v')^2+(w-w')^2\}^{\frac{1}{2}},$$

and the components of spin ξ, η, ζ are given by

$$\xi\,(u-u')+\eta\,(v-v')+\zeta\,(w-w')=0$$

and

$$l\xi+m\eta+n\zeta=0,$$

where l, m, n are direction cosines of the normal to the surface.

Here dn is infinitely small and ξ, η, ζ are infinitely great but such that the products $\xi\,dn, \eta\,dn, \zeta\,dn$ are finite.

Thus the surface of discontinuity may be regarded as a surface covered with vortex filaments, the spin at any point being given by the foregoing expressions and the discontinuity in the tangential velocity may be regarded as due to this vortex sheet.

9·71. Uniform plane Sheet. Consider the case of uniform flow parallel to the axis of y with velocity v where $z>0$ and v' where $z<0$. The axes of the vortices are then parallel to Ox, and if κ is the strength of the vortex sheet per unit breadth parallel to Oy, positive when the sense is that of circulation from Oy towards Oz, then $\kappa=v'-v$.

The strip of the vortex sheet of breadth dy at a distance y from Ox will produce at the point $(0, 0, z)$ a velocity $\kappa\,dy/2\pi r$, where $r=(y^2+z^2)^{\frac{1}{2}}$ is the distance of the point from the strip; and by taking strips equidistant from the point it is easy to see that the resultant velocity at the point due to the vortex sheet is parallel to yO provided $z>0$, and of magnitude

$$\frac{\kappa}{2\pi}\int_{-\infty}^{\infty}\frac{z\,dy}{z^2+y^2}=\tfrac{1}{2}\kappa=\tfrac{1}{2}(v'-v).$$

While for $z<0$ there is an equal velocity parallel to Oy.

If now we regard this vortex sheet as superposed upon a uniform flow with velocity $\frac{1}{2}(v+v')$ parallel to Oy through all space, we see that the result is two uniform streams with velocities v, v' respectively on either side of the plane xy.

Looking back now to the case of the infinite row of parallel vortices of **9·4**, we see that if in (3) we make $y=\pm\infty$ we get $u=\mp\frac{1}{2}\kappa/a$, $v=0$ and a comparison shews that at a great distance the infinite row of parallel vortices is equivalent to a plane vortex sheet of strength κ/a per unit breadth.

9·72. Production of Vorticity. We saw in Chapters V and VI that when a solid moves through a fluid the 'lifting force', in the two-dimensional case, depends on the existence of a circulation about the solid. Experience shews that such forces and circulations actually exist, and the question arises how does such circulation come into being and to what extent is the Kelvin Helmholtz theory of the permanence of vortices in accordance with observed facts. In the first place it must be observed that *permanence of irrotational motion* established in **2·51** and **4·24** refers not to regions of space but to portions of matter, and that the correct inference to be drawn in relation to motion started from rest in perfect fluid is not that vorticity cannot arise but that it can only occur in sheets, i.e. in the surface of separation of definite masses of fluid*. Then it must be remembered that actual observations are made with *real* fluids in which there is viscosity, and, as we shall see later, viscosity plays an important part in the production of circulation or vorticity. We have had occasion to consider several cases of fluid motion involving surfaces of discontinuity of tangential velocity, beginning with **3·72** where it was explained how such a surface comes to be unstable. We have now seen that such a surface is a vortex sheet, and that the production of such a sheet in perfect fluid is not inconsistent with the theory. When a stream is obstructed by a body like a flat plate across the stream or a bluff body like a circular cylinder the surfaces of discontinuity or vortex sheets behind the body commonly roll up on themselves and produce

* For this observation I am indebted to Dr Goldstein.

a Kármán street of more or less concentrated vortices. When an aerofoil meets a stream and divides it (in the two-dimensional case) the two portions of the stream which pass above and below the body meet again behind it; vorticity is produced by viscosity in a thin layer of the fluid surrounding the body and is shed off behind the body; this vorticity collects into a single vortex and moves away from the aerofoil leaving behind it a state of steady flow. The region of space which includes the body is cyclic and, when the vortex is cast off behind the body, a circulation is set up round the body equal and opposite to that of the vortex, so that the total circulation in a circuit which embraces the body and the vortex remains zero*.

9·73. Extension of the Theorem of Kutta and Joukowski. It should be remarked that the proof of the theorem of Kutta and Joukowski (5·7) involves the hypothesis that there are no singularities such as vortices in the fluid surrounding the body. We shall shew how the formula for thrust must be modified when sources and vortices are present.

Referring to **5·61** and **5·7**, suppose that in the finite part of the plane round the cylinder C' there is also an arbitrary distribution of sources and vortices, giving rise to an additional motion represented by

$$w = -\sum^{r} m_r \log(z-a_r) + \sum^{s} \frac{i\kappa_s}{2\pi}\log(z-c_s), \quad \begin{array}{l} r=1,2,\ldots \\ s=1,2,\ldots \end{array} \quad \ldots(1);$$

let the velocity of the steady stream at infinity have components $-U$, $-V$, and as before let κ denote the circulation about the cylinder. These additional terms in w give rise to singularities in the integrand in $\int\left(\frac{dw}{dz}\right)^2 dz$ lying between the contours C' and C. Hence we cannot proceed from **5·61** (4) to **5·61** (5); but we can use the second theorem in **5·6** in the form

$$\int_C \left(\frac{dw}{dz}\right)^2 dz - \int_{C'}\left(\frac{dw}{dz}\right)^2 dz$$

$$= 2\pi i\left\{\text{sum of residues of } \left(\frac{dw}{dz}\right)^2 \text{ at poles between } C' \text{ and } C\right\} \quad \ldots\ldots(2).$$

* For further information see Ewald, Pöschl and Prandtl, *The Physics of Solids and Fluids*, 1930, p. 326, and von Kármán and Levi-Civita, *Vorträge...Hydro- u. Aerodynamik*, 1924, papers by Prandtl, Trefftz and Wieselsberger.

The whole motion about the fixed cylinder C' may be represented by

$$w=(U-iV)z+\frac{i\kappa}{2\pi}\log z+w_0-\sum^{r} m_r\log(z-a_r)+\sum^{s}\frac{i\kappa_s}{2\pi}\log(z-c_s) \quad \text{......(3),}$$

so that

$$-u+iv=\frac{dw}{dz}=U-iV+\frac{i\kappa}{2\pi z}+\frac{dw_0}{dz}-\Sigma\frac{m_r}{z-a_r}+\Sigma\frac{i\kappa_s}{2\pi(z-c_s)} \quad \text{......(4),}$$

where w_0 must be such that $\left|\frac{dw_0}{dz}\right|\to 0$ as $|z|\to\infty$. It is clear that the poles of $\left(\frac{dw}{dz}\right)^2$ which lie between C' and an infinite circle C are at the sources and sinks, and that the residue at a_r is

$$-2m_r\begin{Bmatrix}\text{sum of all terms on the right of (4)}\\ \text{except that which contains } a_r\end{Bmatrix}_{z\to a_r}$$

$=2m_r(u_r-iv_r)$, where u_r, v_r are the components of velocity at a_r omitting the source m_r in the calculation.

Similarly at the vortex at c_s the residue is $\frac{-i\kappa_s}{\pi}(u_s-iv_s)$, where u_s, v_s are defined in the same way. Then from **5·61** (4) and from (2) above

$$\begin{aligned}\mathbf{X}-i\mathbf{Y}&=\tfrac{1}{2}i\rho\int_{C'}\left(\frac{dw}{dz}\right)^2dz\\ &=\tfrac{1}{2}i\rho\int_{C}\left(\frac{dw}{dz}\right)^2dz+\pi\rho\,\{\text{sum of residues}\}\\ &=\tfrac{1}{2}i\rho\int_{C}\left(\frac{dw}{dz}\right)^2dz+\pi\rho\left\{2\Sigma m_r(u_r-iv_r)-\Sigma\frac{i\kappa_s}{\pi}(u_s-iv_s)\right\}\end{aligned} \quad \text{......(5).}$$

Substituting from (4) in the last integral and integrating round an infinite circle on which $|dw_0/dz|$ is zero, we get

$$\tfrac{1}{2}i\rho\int_{C}\left(\frac{dw}{dz}\right)^2dz=-\pi\rho(U-iV)\left\{\frac{i\kappa}{\pi}-2\Sigma m_r+\Sigma\frac{i\kappa_s}{\pi}\right\},$$

so that

$$\begin{aligned}\mathbf{X}-i\mathbf{Y}=-i\rho\kappa(U-iV)&+2\pi\rho\Sigma m_r(u_r-iv_r+U-iV)\\ &-i\rho\Sigma\kappa_s(u_s-iv_s+U-iV)\end{aligned} \quad \text{......(6)*.}$$

It is obvious that **5·7** is a particular case of this theorem.

* This generalization is due to M. Lagally, *Münchener Ber.* 1921, quoted in *Handbuch der Physik*, VII, p. 88, J. Springer, 1927.

9·8. Kinetic Energy of a system of Vortices. The kinetic energy of a fluid is T, where

$$2T=\rho\iiint(u^2+v^2+w^2)\,dx\,dy\,dz,$$

which by **9·65** becomes

$$2T=\rho\iiint\left\{u\left(-\frac{\partial\phi}{\partial x}+\frac{\partial H}{\partial y}-\frac{\partial G}{\partial z}\right)+v\left(-\frac{\partial\phi}{\partial y}+\frac{\partial F}{\partial z}-\frac{\partial H}{\partial x}\right)\right.$$
$$\left.+w\left(-\frac{\partial\phi}{\partial z}+\frac{\partial G}{\partial x}-\frac{\partial F}{\partial y}\right)\right\}dx\,dy\,dz.$$

Integrating this by parts we get

$$2T=-\rho\iint\phi\frac{\partial\phi}{\partial n}dS+\rho\iiint\phi\nabla^2\phi\,dx\,dy\,dz$$
$$+\rho\iint\{l(Hv-Gw)+m(Fw-Hu)+n(Gu-Fv)\}\,dS$$
$$+2\rho\iiint(F\xi+G\eta+H\zeta)\,dx\,dy\,dz,$$

where the surface integrals extend to the whole boundary of the liquid and the triple integrals are taken throughout the volume.

If we suppose that the liquid extends to infinity and is at rest there and that the vortices are all within a finite distance of the origin, then as in **4·6** the first integral vanishes, the second is zero because $\nabla^2\phi=0$, and the third is zero because at points on the infinitely distant boundary F, G, H are ultimately of order $1/R^2$, and u, v, w of order $1/R^3$. Therefore

$$T=\rho\iiint(F\xi+G\eta+H\zeta)\,dx\,dy\,dz.$$

Substituting the values of F, G, H from **9·62** we get

$$T=\frac{\rho}{2\pi}\iiiint\!\!\iint\frac{\xi\xi'+\eta\eta'+\zeta\zeta'}{r}\,dx\,dy\,dz\,dx'\,dy'\,dz',$$

where each volume integral extends through the whole space occupied by the vortices.

Another form, in which we integrate by filaments, may be obtained thus. If ds, ds' are elements of length of two filaments, σ, σ' their cross sections, ω, ω' the corresponding angular velocities and ϵ the angle between ds and ds', the elements of volume are

σds and $\sigma' ds'$, and the integrand is $\omega\omega' \cos\epsilon/r$, so if we write $2\omega\sigma = \kappa$ and $2\omega'\sigma' = \kappa'$, we get

$$T = \frac{\rho}{4\pi} \Sigma \kappa\kappa' \iint \frac{\cos\epsilon}{r} \, ds\, ds',$$

where the integration is along the filaments and the summation includes each pair of filaments once. This formula corresponds to that obtained by F. Neumann for the energy of electric currents.

9·81. Kinetic Energy Constant. We can also shew that the kinetic energy is constant when no extraneous forces act.

The equations of motion are

$$\frac{D}{Dt}(u, v, w) = -\frac{1}{\rho}\left(\frac{\partial p}{\partial x}, \frac{\partial p}{\partial y}, \frac{\partial p}{\partial z}\right).$$

Multiplying these by u, v, w and adding we get

$$\tfrac{1}{2}\rho \frac{D}{Dt}(u^2 + v^2 + w^2) = -\left(u\frac{\partial p}{\partial x} + v\frac{\partial p}{\partial y} + w\frac{\partial p}{\partial z}\right).$$

If now we multiply by $dx\,dy\,dz$ and integrate over any region we get

$$\frac{DT}{Dt} = -\iiint \left(u\frac{\partial p}{\partial x} + v\frac{\partial p}{\partial y} + w\frac{\partial p}{\partial z}\right) dx\,dy\,dz$$

$$= \iint (lu + mv + nw)\, p\, dS,$$

integrated over the boundary of the region.

Let the boundary extend to infinity, and enclose all the vortices, then since

$$\frac{p}{\rho} = \frac{\partial \phi}{\partial t} - \tfrac{1}{2}q^2 + F(t),$$

therefore at a great distance R from the vortices p will be finite
9·66 and $lu + mv + nw$ of order $1/R^3$ while dS is of order R^2. Hence the expression for DT/Dt vanishes and we have

$$T = \text{constant}.$$

9·82. Circular Vortex Rings. We have already seen (**9·67**) that a vortex ring produces the same effect as a sheet of doublets bounded by the ring, so that at points whose distance from a circular vortex is great compared with the radius, we might as a first approximation replace the vortex ring by a doublet perpendicular to its plane. For more detailed treatment we proceed as follows.

When the vortex lines are circles in planes parallel to the yz plane with centres on the axis of x, we may use Stokes's stream function and write, in the notation of **7·3**,

$$u = -\frac{1}{\varpi}\frac{\partial \psi}{\partial \varpi}, \quad v = \frac{1}{\varpi}\frac{\partial \psi}{\partial x};$$

then by **4·25** the spin ω at the point (x, ϖ) is given by

$$2\omega = \frac{\partial v}{\partial x} - \frac{\partial u}{\partial \varpi} = \frac{1}{\varpi}\left\{\frac{\partial^2 \psi}{\partial x^2} + \frac{\partial^2 \psi}{\partial \varpi^2} - \frac{1}{\varpi}\frac{\partial \psi}{\partial \varpi}\right\} \quad \text{..............(1).}$$

Let us consider the case of a single circular vortex filament. We may transform the expressions for F, G, H, namely

$$\frac{1}{2\pi}\iiint \frac{\xi'}{r}\,dx'\,dy'\,dz', \text{ etc.,}$$

by taking σ as the cross section and ds the length of an element of the filament, and putting $\xi', \eta', \zeta' = l\omega, m\omega, n\omega$, where ω is the spin and (l, m, n) are the direction cosines of the vortex line, so that

$$\xi'\,dx'\,dy'\,dz' = l\omega\sigma\,ds = \tfrac{1}{2}\kappa l\,ds = \tfrac{1}{2}\kappa\,dx',$$

κ denoting the strength of the vortex.

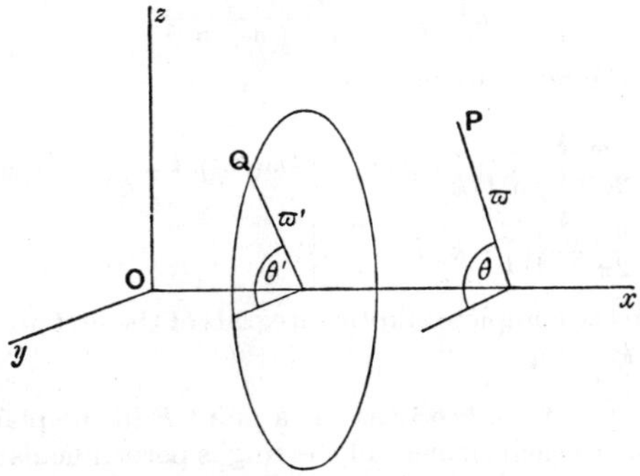

Hence $\quad F, G, H = \dfrac{\kappa}{4\pi}\displaystyle\int \frac{dx'}{r},\ \frac{\kappa}{4\pi}\int \frac{dy'}{r},\ \frac{\kappa}{4\pi}\int \frac{dz'}{r}.$

Now let the filament have its centre on the x axis and be parallel to the plane yz.

Let (x', y', z') be any point Q on the filament, where

$$y' = \varpi' \cos\theta', \quad z' = \varpi' \sin\theta'.$$

Let P be the point (x, y, z), where

$$y = \varpi\cos\theta, \quad z = \varpi\sin\theta,$$

then $\quad r^2 = (x - x')^2 + \varpi^2 + \varpi'^2 - 2\varpi\varpi'\cos(\theta - \theta').$

We get

$$F = 0, \quad G = -\frac{\kappa\varpi'}{4\pi}\int_0^{2\pi} \frac{\sin\theta'}{r}\,d\theta', \quad H = \frac{\kappa\varpi'}{4\pi}\int_0^{2\pi} \frac{\cos\theta'}{r}\,d\theta',$$

as the values at P.

Hence the vector whose components are F, G, H lies in a plane parallel to yz and its component in the direction ϖ is

$$G\cos\theta + H\sin\theta = \frac{\kappa\varpi'}{4\pi}\int_0^{2\pi} \frac{\sin(\theta - \theta')}{r}\,d\theta' = 0,$$

so that the vector is perpendicular to ϖ as well as to x. If we denote its value by A, we have

$$A = H\cos\theta - G\sin\theta = \frac{\kappa\varpi'}{4\pi}\int_0^{2\pi} \frac{\cos(\theta - \theta')}{r}\,d\theta'.$$

Remembering that the line integral of this vector round any curve represents the flow across a surface bounded by the curve, by taking a circle of radius ϖ with centre on Ox, we get

$$2\pi\varpi A = \text{flow through the circle} = -2\pi\psi \;(\mathbf{7\cdot 3}),$$

the flow being from left to right in the figure.

Therefore

$$\psi = -\varpi A = -\frac{\kappa\varpi\varpi'}{4\pi}\int_0^{2\pi}\frac{\cos(\theta-\theta')\,d\theta'}{\{(x-x')^2+\varpi^2+\varpi'^2-2\varpi\varpi'\cos(\theta-\theta')\}^{\frac{1}{2}}},$$

and since the range of integration is round a circle we may clearly write ϵ for $\theta'-\theta$, so that

$$\psi = -\frac{\kappa\varpi\varpi'}{4\pi}\int_0^{2\pi}\frac{\cos\epsilon\,d\epsilon}{\{(x-x')^2+\varpi^2+\varpi'^2-2\varpi\varpi'\cos\epsilon\}^{\frac{1}{2}}}.$$

Putting
$$k^2 = \frac{4\varpi\varpi'}{(x-x')^2+(\varpi+\varpi')^2}$$

and $\epsilon = \pi - 2\phi$ the result reduces to

$$\psi = -\frac{\kappa(\varpi\varpi')^{\frac{1}{2}}}{2\pi}\int_0^{\frac{\pi}{2}}\left\{\left(\frac{2}{k}-k\right)(1-k^2\sin^2\phi)^{-\frac{1}{2}}-\frac{2}{k}(1-k^2\sin^2\phi)^{\frac{1}{2}}\right\}d\phi$$

$$= -\frac{\kappa(\varpi\varpi')^{\frac{1}{2}}}{2\pi}\left\{\left(\frac{2}{k}-k\right)K-\frac{2}{k}E\right\},$$

where K, E are the complete elliptic integrals of the first and second order with modulus k.

9·83. It is clear from **9·63** that at a point P in the plane of the ring the velocity due to each element of the ring is perpendicular to the plane, hence there can be no radial velocity at any point in the plane of the ring. The radius of the ring is therefore constant, for it could not vary without causing radial velocity in the particles close to it.

To find the motion of the ring, we observe that near the ring $x = x'$, and $\varpi = \varpi'$ nearly, so that $k = 1$ nearly, and ψ becomes infinitely great. The determination of the velocity depends on the form of the section of the ring; an exact expression for the case of a circular section was given by Lord Kelvin*, but we can obtain approximate results for the velocity in the neighbourhood of the ring as follows.

If k' denote the complementary modulus

$$k'^2 = \frac{(x-x')^2+(\varpi-\varpi')^2}{(x-x')^2+(\varpi+\varpi')^2},$$

then k' tends to zero as the point (x, ϖ) approaches the ring.

For small values of k', i.e. when k is nearly unity,

$$K = \log 4/k' \quad\text{and}\quad E = 1, \text{ approximately}\dagger.$$

Hence
$$\psi = -\frac{\kappa(\varpi\varpi')^{\frac{1}{2}}}{2\pi}\log\frac{4}{k'}$$

is the principal part of ψ when k' is small.

* *Phil. Mag.* XXXIII, Fourth Series, 1867, p. 511; see also Lamb's *Hydrodynamics*, § 163.

† Cayley's *Elliptic Functions*, p. 54.

And taking this value for ψ we have

$$u=-\frac{1}{\varpi}\frac{\partial\psi}{\partial\varpi}=\frac{\kappa}{4\pi\varpi}\left(\frac{\varpi'}{\varpi}\right)^{\frac{1}{2}}\log\frac{4}{k'}-\frac{\kappa}{2\pi}\left(\frac{\varpi'}{\varpi}\right)^{\frac{1}{2}}\frac{d}{d\varpi}\log k'.$$

But
$$\frac{d}{d\varpi}\log k'=\frac{\varpi-\varpi'}{(x-x')^2+(\varpi-\varpi')^2}-\frac{\varpi+\varpi'}{(x-x')^2+(\varpi+\varpi')^2};$$

and, if we take the value for a point on the ring for which $\varpi=\varpi'$ and $x=x'+\epsilon$ say, where ϵ is small, being commensurable with the linear dimensions of the section of the ring, we get

$$k'=\frac{\epsilon}{2\varpi'},$$

and
$$\frac{d}{d\varpi}\log k'=-\frac{2\varpi'}{\epsilon^2+4\varpi'^2}.$$

Hence the principal part of the velocity parallel to the axis is

$$u=\frac{\kappa}{4\pi\varpi'}\log\frac{8\varpi'}{\epsilon}.$$

For a ring of small section this implies a large velocity and we conclude that a thin circular ring will move along its axis with a large approximately constant velocity.

The direction of the velocity is to the side to which the fluid flows through the ring.

For a complete investigation reference may be made to Lamb's *Hydrodynamics* (*loc. cit.*).

9·84. We shall conclude with some observations on the motion of two circular vortex rings moving on the same axis, taken from Helmholtz's paper on vortex motion. "We can now see generally how two ring-formed vortex-filaments having the same axis would mutually affect each other, since each, in addition to its proper motion, has that of its elements of fluid as produced by the other. If they have the same direction of rotation they travel in the same direction; the foremost widens and travels more slowly, the pursuer shrinks and travels faster, till finally if their velocities are not too different, it overtakes the first and penetrates it. Then the same game goes on in the opposite order, so that the rings pass through each other alternately.

"If they have equal radii and equal and opposite angular velocities, they will approach each other and widen one another; so that finally, when they are very near each other, their velocity of approach becomes smaller and smaller, and their rate of widening faster and faster. If they are perfectly symmetrical, the velocity of fluid elements midway between them parallel to the axis is zero. Here then we might imagine a rigid plane to be inserted, which would not disturb the motion, and so obtain the case of a vortex ring which encounters a fixed plane.

"In addition it may be noticed that it is easy in nature to study these motions of circular vortex rings, by drawing rapidly for a short space along the surface of a fluid a half-immersed circular disk, or the nearly semicircular point of a spoon, and quickly withdrawing it. There remain in the fluid half vortex rings whose axis is in the free surface. The free surface forms a bounding plane of the fluid through the axis, and thus there is no essential change in the motion. These vortex rings travel on, widen when they come to a wall, and are widened or contracted by other vortex rings, exactly as we have deduced from theory*."

9·9. Steady Motion. When the external forces have a potential Ω the general equations of motion are of the form

$$\frac{\partial u}{\partial t}+u\frac{\partial u}{\partial x}+v\frac{\partial u}{\partial y}+w\frac{\partial u}{\partial z}=-\frac{\partial \Omega}{\partial x}-\frac{1}{\rho}\frac{\partial p}{\partial x},$$

and similar equations.

And if we put $$\chi=\int\frac{dp}{\rho}+\tfrac{1}{2}q^2+\Omega,$$

the foregoing equations may be written

$$\frac{\partial u}{\partial t}-v\left(\frac{\partial v}{\partial x}-\frac{\partial u}{\partial y}\right)+w\left(\frac{\partial u}{\partial z}-\frac{\partial w}{\partial x}\right)$$
$$=-\frac{\partial \Omega}{\partial x}-\frac{1}{\rho}\frac{\partial p}{\partial x}-\left(u\frac{\partial u}{\partial x}+v\frac{\partial v}{\partial x}+w\frac{\partial w}{\partial x}\right),$$

or $$\frac{\partial u}{\partial t}-2v\zeta+2w\eta=-\frac{\partial \chi}{\partial x}\qquad(1),$$

and similar equations.

When the motion is steady we have

$$\frac{\partial u}{\partial t}=0,\quad \frac{\partial v}{\partial t}=0,\quad \frac{\partial w}{\partial t}=0;$$

therefore

$$\frac{\partial \chi}{\partial x}=2(v\zeta-w\eta),\quad \frac{\partial \chi}{\partial y}=2(w\xi-u\zeta),\quad \frac{\partial \chi}{\partial z}=2(u\eta-v\xi).$$

Hence $$\xi\frac{\partial \chi}{\partial x}+\eta\frac{\partial \chi}{\partial y}+\zeta\frac{\partial \chi}{\partial z}=0,$$

and $$u\frac{\partial \chi}{\partial x}+v\frac{\partial \chi}{\partial y}+w\frac{\partial \chi}{\partial z}=0.$$

Therefore $\chi=$const. represents a surface the normal to which at any point is at right angles to both the vortex line and the stream line through the point. That is, there exists in the liquid a family of surfaces $\chi=$const. each covered by a network of vortex lines and stream lines.

In the special case in which the motion is irrotational, however, χ is constant throughout the whole liquid.

* See also a paper by Love, 'On the motion of paired vortices', *Proc. L.M.S.* 1894, p. 185.

If for an instant we take the axis of x normal to the surface $\chi=$ const., we must have $u=0$, $\xi=0$, and if $\partial\nu$ is an element of the normal to the surface

$$\frac{\partial\chi}{\partial\nu}=\frac{\partial\chi}{\partial x}=2(v\zeta-w\eta)=2q\omega\sin\theta \quad \text{............(2)},$$

where θ is the angle between the direction of the velocity q and the axis of spin ω, i.e. the angle between the stream line and the vortex line.

Hence we have as the conditions for steady motion that it must be possible to draw a family of surfaces in the liquid each covered by a network of stream lines and vortex lines and such that at every point of a surface $q\omega\sin\theta\,\partial\nu$ is constant, where $\partial\nu$ is the normal distance between the surface and the next consecutive surface of the family*.

In two-dimensional liquid motion it is obvious that $q\,\partial\nu$ is constant along a stream line, therefore the condition for steady motion is that the spin ζ shall be constant along a stream line. This will be the case if we put $2\zeta=f(\psi)$, where ψ is the stream function and f an arbitrary constant.

But $$2\zeta=\frac{\partial v}{\partial x}-\frac{\partial u}{\partial y}=\frac{\partial^2\psi}{\partial x^2}+\frac{\partial^2\psi}{\partial y^2};$$

therefore for two-dimensional steady motion we have to satisfy

$$\frac{\partial^2\psi}{\partial x^2}+\frac{\partial^2\psi}{\partial y^2}=f(\psi)\dagger \quad \text{.....................(3)}.$$

This is clearly satisfied whenever the stream lines are concentric circles with the origin as centre. Another case is where the stream lines are a system of similar and similarly situated ellipses or hyperbolas; thus

$$\psi=\tfrac{1}{2}(ax^2+2bxy+cy^2)$$

makes $\nabla^2\psi=a+c$, so that equation (3) is satisfied, and the spin $\zeta=\frac{1}{2}(a+c)$ is uniform.

In like manner a system of equal parabolas having the same axis may be seen to satisfy the conditions for stream lines in steady motion.

9·91. Steady Motion symmetrical in Planes through an Axis.

If the motion is symmetrical about the x axis and ϖ denotes distance from the axis, we clearly have $q\,.\,2\pi\varpi\,\partial n$ constant along a stream line, for this represents the flow between two

* Lamb, 'On the conditions for Steady Motion of a Fluid', *Proc. L.M.S.* IX, p. 91, or *Hydrodynamics*, § 165.

† Stokes, 'On the Steady Motion of Incompressible Fluids', *Trans. Camb. Phil. Soc.* VII, p. 439, or *Math. and Phys. Papers*, I, p. 1.

consecutive stream surfaces of revolution. But we must also have $q\omega\,\partial n$ constant over such a surface from **9·9** (2), because from symmetry the vortex rings must have their centres on the x axis and their planes perpendicular to it, so that they cut the stream lines at right angles. Therefore ω/ϖ must be constant along a stream line. This is satisfied by making $2\omega = \varpi f(\psi)$, where f is an arbitrary function of Stokes's stream function ψ. Hence from **9·82** (1) we have

$$\frac{\partial^2\psi}{\partial x^2}+\frac{\partial^2\psi}{\partial \varpi^2}-\frac{1}{\varpi}\frac{\partial\psi}{\partial\varpi}=\varpi^2 f(\psi) \quad\text{.................(1)}$$

as the necessary condition.

An example in which this condition is satisfied is Hill's 'Spherical Vortex*'.

EXAMPLES

1. Assuming that, in an infinite unbounded mass of incompressible fluid, the circulation in any closed circuit is independent of the time, shew that the angular velocity of any element of the fluid moving rotationally varies as the length of the element measured in the direction of the axis of rotation. (M.T. 1880.)

2. If $u=\dfrac{ax-by}{x^2+y^2}$, $v=\dfrac{ay+bx}{x^2+y^2}$, and $w=0$, investigate the nature of the motion of the liquid.

3. When an infinite liquid contains two parallel equal and opposite rectilinear vortices at a distance $2b$, prove that the stream lines relative to the vortices are given by the equation

$$\log\frac{x^2+(y-b)^2}{x^2+(y+b)^2}+\frac{y}{b}=C,$$

the origin being the middle point of the join, which is taken for axis of y.

4. In the last example, if the vortices are of the same strength, and the spin is in the same sense in both, shew that the relative stream lines are given by

$$\log(r^4+b^4-2b^2r^2\cos 2\theta)-r^2/2b^2=\text{constant},$$

θ being measured from the join of the vortices, the origin being its middle point.

Shew also that the surfaces of equipressure at any instant are given by

$$r^4+b^4-2b^2r^2\cos 2\theta=\lambda(r^2\cos 2\theta+a^2).$$

(Coll. Exam. 1913.)

5. An infinitely long line vortex of strength m, parallel to the axis of z, is situated in infinite liquid bounded by a rigid wall in the plane $y=0$. Prove

* 'On a Spherical Vortex', *Phil. Trans.* A, 1894, or see Lamb's *Hydrodynamics*, § 165.

that, if there be no field of force, the surfaces of equal pressure are given by

$$\{(x-a)^2+(y-b)^2\}\{(x-a)^2+(y+b)^2\}=C\{y^2+b^2-(x-a)^2\},$$

where (a, b) are the coordinates of the vortex, and C is a parametric constant. (Univ. of London, 1909.)

6. If n rectilinear vortices of the same strength κ are symmetrically arranged as generators of a circular cylinder of radius a in an infinite liquid, prove that the vortices will move round the cylinder uniformly in time $8\pi^2a^2/(n-1)\kappa$, and find the velocity at any point of the liquid.

7. When a pair of equal and opposite rectilinear vortices are situated in a long circular cylinder at equal distances from its axis, shew that the path of each vortex is given by the equation

$$(r^2\sin^2\theta-b^2)(r^2-a^2)^2=4a^2b^2r^2\sin^2\theta,$$

θ being measured from the line through the centre perpendicular to the join of the vortices. (Greenhill.)

8. Obtain the distribution of the velocity round a straight vertical vortex core in liquid: and find the form of the dimple where the core meets the free surface. (St John's Coll. 1897.)

9. Find the motion of a straight vortex filament in an infinite region bounded by an infinite plane wall to which the filament is parallel, and prove that the pressure defect at any point of the wall due to the filament is proportional to $\cos^2\theta\cos 2\theta$, where θ is the inclination of the plane through the filament and the point to the plane through the filament perpendicular to the wall. (M.T. 1912.)

10. If a rectilinear vortex moves parallel to two rigid planes which intersect at right angles, prove that on the line of intersection of the planes the excess of pressure due to the vortex varies inversely as the square of the distance of the vortex from the line of intersection.

(Univ. of London, 1915.)

11. If (r_1, θ_1), (r_2, θ_2) ... be polar coordinates at time t of a system of rectilinear vortices of strength $\kappa_1, \kappa_2, \ldots$, prove that

$$\Sigma\kappa r^2=\text{const.} \quad \text{and} \quad \Sigma\kappa r^2\dot{\theta}=\frac{1}{2\pi}\Sigma\kappa_1\kappa_2. \qquad \text{(Kirchhoff.)}$$

12. The space enclosed between the planes $x=0$, $x=a$, $y=0$ on the positive side of $y=0$ is filled with uniform incompressible liquid. A rectilinear vortex parallel to the axis of z has coordinates (x', y'). Determine the velocity at any point of the liquid and shew that the path of the vortex is given by

$$\cot^2\frac{\pi x}{a}+\coth^2\frac{\pi y}{a}=\text{constant}. \qquad \text{(M.T. 1899.)}$$

13. An elliptic cylinder is filled with liquid which has molecular rotation ω at every point, and whose particles move in planes perpendicular to the axis; prove that the stream lines are similar ellipses described in periodic time $\dfrac{\pi}{\omega}\cdot\dfrac{a^2+b^2}{ab}$. (M.T. 1876.)

14. An infinite row of equidistant rectilinear vortices are at a distance a apart. The vortices are of the same numerical strength κ but they are alternately of opposite signs. Find the complex function that determines the velocity potential and stream function. Shew that the vortices remain at rest and draw the stream lines. Shew also that, if α be the radius of a vortex, the amount of flow between any vortex and the next is

$$(\kappa/\pi)\log\cot(\pi\alpha/2a). \qquad \text{(M.T. 1925.)}$$

15. In an incompressible fluid the vorticity at every point is constant in magnitude and direction; shew that the components of velocity u, v, w are solutions of Laplace's equation. (Trinity Coll. 1906.)

16. Prove that, in the *steady* motion of an incompressible liquid, under the action of conservative forces, we have

$$\xi\frac{\partial u}{\partial x}+\eta\frac{\partial u}{\partial y}+\zeta\frac{\partial u}{\partial z}=0,$$

and two more similar equations in v, w.

Hence shew that if u, v, w are *linear* functions of x, y, z, then

$$\xi u+\eta v+\zeta w=0,$$

and that there are two and only two possible cases:

(i) an irrotational motion with a velocity potential which is *any* solid harmonic of degree two in x, y, z,

(ii) a rotational motion which may, by choice of axes, be reduced to the form

$$u=ax+(h-\zeta)y, \quad v=(h+\zeta)x-ay, \quad w=0.$$

Find the lines of flow in case (ii); and shew that the motion is periodic if

$$\zeta^2>(a^2+h^2). \qquad \text{(St John's Coll. 1902.)}$$

17. Prove that the kinetic energy of a vortex system of finite dimensions in an infinite liquid at rest at infinity can be expressed in the form

$$2\rho\iiint\{u(y\zeta-z\eta)+v(z\xi-x\zeta)+w(x\eta-y\xi)\}\,dx\,dy\,dz.$$

18. Prove that a thin cylindrical vortex of strength σ, running parallel to a plane boundary at distance a will travel with velocity $\sigma/4\pi a$: and shew that a stream of fluid will flow past between the travelling vortex and the boundary of total amount $\dfrac{\sigma}{2\pi}\left\{\log\left(\dfrac{2a}{c}\right)-\dfrac{1}{2}\right\}$ per unit length along the vortex, when c is the (small) radius of the cross section of the vortex. (M.T. 1916.)

19. If, with the usual notation, $u\,dx+v\,dy+w\,dz=d\theta+\lambda\,d\chi$ where θ, λ, χ are functions of x, y, z and t, prove that the vortex lines at any time are the lines of intersection of the surfaces $\lambda=$ const. and $\chi=$ const. (Coll. Exam. 1912.)

20. Prove that the necessary and sufficient condition that the vortex lines may be at right angles to the stream lines is

$$u, v, w=\mu\left(\frac{\partial\phi}{\partial x}, \frac{\partial\phi}{\partial y}, \frac{\partial\phi}{\partial z}\right),$$

where μ and ϕ are functions of x, y, z, t.

21. Prove that in regions remote from a single thin vortex ring the stream lines approximate to the curves $r \operatorname{cosec}^2 \theta = \text{const.}$, where r denotes the distance of a point P from the centre O of the ring, and θ the angle which the line OP makes with the axis of the ring. (M.T. II. 1910.)

22. Find the motion of the liquid around a closed vortex filament, shewing its equivalence to a double sheet of sources and sinks: deduce that the image of a circular filament moving in infinite liquid surrounding a rigid sphere is another filament; compare the circulations. Describe the behaviour of the filament as it approaches the sphere. (M.T. 1911.)

23. Shew that if the velocity is stationary at a point on a stream line in the steady motion of a liquid, the stream line is a geodesic on a member of the family of surfaces that contains the stream lines and vortex lines. (Greenhill.)

24. A straight cylindrical vortex column of uniform vorticity ζ is surrounded by an infinite quantity of fluid moving irrotationally which is at rest at infinity, prove that the difference between the kinetic energy included between two planes at right angles to the axis of the cylinder and separated by unit distance when the cross section of the cylinder is an ellipse and when it is a circle of equal area A is

$$\frac{\rho}{\pi}\zeta^2 A^2 \log \frac{a+b}{2\sqrt{ab}},$$

where ρ is the density of the fluid and a and b the semi-axes of the ellipse. (M.T. 1887.)

25. If the velocities at a point in a liquid in motion under a system of external forces having a potential be expressed by

$$u = -\frac{\partial\phi}{\partial x} + \lambda\frac{\partial\chi}{\partial x}, \quad v = -\frac{\partial\phi}{\partial y} + \lambda\frac{\partial\chi}{\partial y}, \quad w = -\frac{\partial\phi}{\partial z} + \lambda\frac{\partial\chi}{\partial z},$$

prove that the result of operating with $\frac{D}{Dt}$ on the identity

$$\frac{\xi}{\rho}\frac{\partial\lambda}{\partial x} + \frac{\eta}{\rho}\frac{\partial\lambda}{\partial y} + \frac{\zeta}{\rho}\frac{\partial\lambda}{\partial z} = 0,$$

where ξ, η, ζ are the rotations, gives, after a reduction,

$$\left(\xi\frac{\partial}{\partial x} + \eta\frac{\partial}{\partial y} + \zeta\frac{\partial}{\partial z}\right)\frac{D\lambda}{Dt} = 0. \quad \text{(Dublin Univ. 1911.)}$$

26. If

$$u = -\phi_x + H_y - G_z, \quad v = -\phi_y + F_z - H_x, \quad w = -\phi_z + G_x - F_y,$$

where $\phi_x = \partial\phi/\partial x$, etc.,

prove that $\iiint (u^2+v^2+w^2)\,dx\,dy\,dz$ taken through any portion of space within which ϕ, F, G, H and all their differential coefficients are finite and continuous, equals

$$\iiint (\phi_1{}^2 + F_1{}^2 + G_1{}^2 + H_1{}^2 - J^2)\,dx\,dy\,dz,$$

taken through the same space, together with $\iint \chi\, dS$ taken over the boundary, where $\phi_1{}^2 = \phi_x{}^2 + \phi_y{}^2 + \phi_z{}^2$, with similar values for F_1, G_1, H_1, $J = F_x + G_y + H_z$, and χ is to be found. (Dublin Univ. 1911.)

27. A liquid extending to infinity moves under the influence of a finite system of vortices: find the force and couple resultants of the system of impulses which would produce the motion. (Dublin Univ. 1907.)

28. Shew that every irrotational motion, whether cyclic or acyclic, of a liquid occupying a given region, can be produced by a proper distribution of vortex sheets on the boundaries, and shew how to determine this distribution. (Dublin Univ. 1907.)

29. A liquid, extending to infinity, moves under the influence of a sphere composed of circular vortex rings whose planes are perpendicular to the axis of x, whose centres lie on this axis, and in which the molecular angular velocity in each ring is proportional to its radius.

If the components u, v, w of the velocity of the liquid are expressed by the equations

$$u = -\frac{\partial \phi}{\partial x} + \frac{\partial H}{\partial y} - \frac{\partial G}{\partial z}, \text{ etc.},$$

find F, G, H at a point outside the sphere. (Dublin Univ. 1907.)

30. Shew that the motion of the liquid outside a certain surface surrounding a circular vortex ring the radius of whose core is small compared with the radius of its aperture, is the same as that due to the motion of this surface through the liquid with the velocity of translation of the ring.

Find the equation to this surface and the length of the axis of the ring ntercepted by it. (M.T. 1892.)

31. A cylindrical vortex sheet in infinite liquid is such that the vortex lines are generators of the cylinder and the vorticity at any point is $2U \sin \theta$, where θ is the angle measured from a fixed plane through the axis of the cylinder. Prove that the vortex sheet moves through the liquid with velocity U parallel to the fixed plane. (M.T. 1924.)

32. When the motion of an infinite liquid is due to a single circular vortex ring, in which the spin at any point is proportional to the distance from the straight axis, and the section is taken to be a circle of radius small compared with the radius of the aperture, obtain an expression for the velocity at any point of the fluid parallel to the straight axis.

Prove that the fluid carried forward with the ring is or is not ring-shaped according as the ratio of the radius of the section to the radius of the aperture is less or greater than a certain fraction, and find an approximation to this fraction. (M.T. 1897.)

33. A uniform incompressible perfect liquid extends to infinity and is at rest there. Within it is a spherical vortex sheet of radius a with its vortex lines arranged in parallel circles, on the axis of which is a fixed point C at a distance $c\ (<a)$ from the centre; the strength of the sheet at any point P is $m \sin \phi$, where ϕ is the angle between CP and the axis of the circles. Shew that the velocity at a point on the axis at a distance $r\ (>a)$ from the centre is

$$2m \sum_{n=2}^{n=\infty} \frac{n(n-1)}{2n-1}\left(\frac{a^2}{2n-3} - \frac{c^2}{2n+1}\right)\frac{ac^{n-2}}{r^{n+1}}. \qquad \text{(M.T. 1900.)}$$

CHAPTER X

WAVES

10·1. THE dynamics of wave motion is of great importance in physical investigations, as wave motion constitutes one of the principal modes of transmission of energy. The energy received from the sun is transmitted by waves in the ether, the energy of sound by air waves, the practical applications of electric waves are now spread the world over and the theory of waves occupies an important place in the field of research into the constitution of matter. In the present chapter we shall only consider water waves, which, though most familiar, are not the easiest to discuss mathematically.

10·11. The oscillatory nature of Wave Motion. By a wave we mean the continuous transference of a particular state or form from one part of a medium to another. This does not imply the transference of the medium itself from one place to another but merely the propagation through it of a particular form, state or condition. Thus in water waves, the fact that small bodies floating on the water are not borne onwards by the waves is an indication that the elevated masses of water are not moving forward bodily, and that it is only the unevenness of the surface that is moving from place to place. As the waves pass a floating body it appears to be carried forwards a small distance on the crest of a wave and backwards when in the trough of the wave so that on the whole each wave leaves the position of the body very little altered.

The following explanation of how water waves can be maintained by small oscillatory movements of each particle of water is due to Airy*.

Let $ABCDEFG$ represent the outline at one instant and $abcdefg$ an instant later; we want to shew that the displacement of the contour of the surface can be produced by a small oscillatory movement of each particle of water.

Draw vertical lines to the bottom of the water and suppose the particles in each vertical line to be moving in the direction of the arrows in the

* Article 'Waves and Tides', *Ency. Metrop.* 1845.

figure; that is, all particles below the crest of the wave are moving forwards, all below the hollows are moving backwards, and all below the midway points A, C, E, G are for the moment stationary. And suppose the velocities of the horizontal motion of the particles in the vertical lines intermediate to those drawn in the figure are intermediate to the velocities of the particles in the lines drawn in the figure. This supposition will account for the motion of the wave or shape. For, take points B_0, B_1 near to B; C_0, C_1 near to C, etc.: draw lines from them to the bottom and consider the horizontal motion of the particles in those lines. B_0 and B_1 are both between the principal point of backward motion B and points of rest A, C, therefore the particles below B_0 and those below B_1 will be moving backwards and with nearly the same speed, so that the intermediate surface at B will not be sensibly elevated or depressed inasmuch as the vertical boundaries B_0B_0' and B_1B_1' of the included column of water will after a short time be at the same distance apart as at present. But the particles in the line C_0C_0' are between a point of rest C and a point

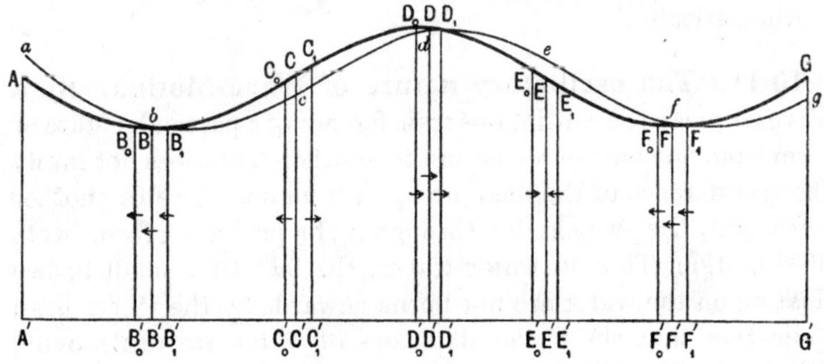

of backward motion B and therefore are moving backwards, those in the line C_1C_1' are between a point of rest and a point of forward motion D and therefore are moving forwards; consequently the vertical boundaries C_0C_0', C_1C_1' of the included column are separating and therefore the surface at C will drop and after a short time will be found depressed to c. In like manner it will be found that the particles in D_0D_0' and D_1D_1' are moving forwards with nearly the same velocity so that in the intermediate part at D there is no sensible alteration of level. But in E_0E_0' the particles are moving forwards and in E_1E_1' backwards resulting in a raising of the level from E to e. Pursuing this reasoning it will be evident that the continuous horizontal motion of the wave or shape forwards is entirely accounted for by the rising of some portions of the surface and the falling of others and that these risings and fallings may be considered as the effect of small horizontal motions of the water, some forwards and others backwards. And as in the progress of the waves, the same particles are alternately in the crest and in the hollow of the wave, every particle will be alternately moving forwards and backwards and alternately upwards and downwards, that is the particles are oscillating while the waves advance continually in the same direction.

10·12. Mathematical representation of Wave Motion. Graphically the equation

$$y=f(x-ct) \quad \text{.......................(1)}$$

represents a wave motion, in which a curve of the form $y=f(x)$ moves in the positive direction of the x axis with velocity c. For if in (1) we increase t by t' and x by ct' we leave the ordinate y unaltered.

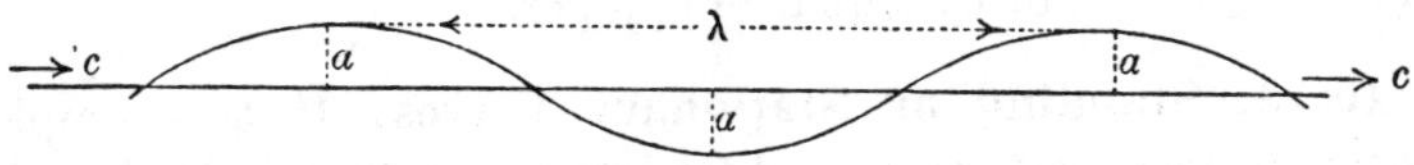

A *simple harmonic progressive wave* is represented by a curve of sines moving with definite velocity in the direction of its length. Thus the equation

$$y=a\sin(mx-nt+\epsilon) \quad \text{..................(2)}$$

represents a wave moving in the positive direction of the x axis with velocity n/m, called the **velocity of propagation,** c say. The distance between two consecutive crests of the curve is $2\pi/m$; this is called the **wave length** and denoted by λ. The **period** of the wave is $2\pi/n$ or λ/c, for the wave at time $t=2\pi/n$ presents the same appearance relative to the origin as at time $t=0$, each crest in this interval moving forward a distance λ, i.e. to the position occupied at the beginning of the interval by the next consecutive crest.

The maximum value of y, viz. a, is called the **amplitude.**

Equation (2) may also be written

$$y=a\sin\frac{2\pi}{\lambda}(x-ct+\epsilon') \quad \text{.................(3),}$$

or

$$y=a\sin 2\pi\left(\frac{x}{\lambda}-\frac{t}{\tau}+\epsilon'\right) \quad \text{.................(4),}$$

where in the latter case τ denotes the period λ/c.

The reciprocal of the period is called the **frequency;** it denotes the number of oscillations per second.

Phase. In equation (2) ϵ represents the phase of the wave at the instant from which t is measured. If we compare the equations

$$y=a\sin(mx-nt),$$

and

$$y=a\sin(mx-nt+\epsilon),$$

we see that both represent wave motions having the same amplitude, wave length and period, but that they differ in phase. As regards position the one is a distance ϵ/m in advance of the other, or as regards time the one has a start of ϵ/n from the other. Strictly speaking the difference of phase is a number ϵ, representing radians, but in such a case as we are considering it is not unusual to speak of the phase in terms of either distance or time; thus, if $\epsilon = \pi/2$, one wave is one-quarter of a wave length in front of the other; or, in terms of time, one is one-quarter of a period ahead of the other, and we may say that the phases differ by a quarter of a wave length or by a quarter of a period.

10·13. Standing or Stationary Waves. If two simple harmonic progressive waves of the same amplitude, wave length and period travel in opposite directions the resulting disturbance of the medium is represented by the equation

$$\begin{aligned} y &= a \sin(mx - nt) + a \sin(mx + nt) \\ &= 2a \sin mx \cos nt. \end{aligned}$$

Such a wave is called a standing or stationary wave. At any instant the equation represents a sine curve but the amplitude $2a \cos nt$ varies continuously. The points of intersection of the curve with the x axis are fixed points called *nodes*.

In the same way a progressive wave system can be regarded as the combination of two systems of standing waves of the same amplitude, wave length and period, the crests and troughs of one system coinciding with the nodes of the other and their phases differing by a quarter period.

For if $y_1 = a \sin mx \cos nt$ be one of the standing waves the other must be $y_2 = a \cos mx \sin nt$, and by combining the two we get $y = y_1 \pm y_2 = a \sin(mx \pm nt)$ representing a progressive wave.

10·14. We propose to consider waves in incompressible liquid under the action of gravity. Such waves in water are generally produced by disturbing forces such as wind pressure, by the relative motion of a body such as a ship on the water, or by such causes as irregularities in the bed of a stream, so that, neglecting viscosity, the motion is irrotational. Roughly speaking the cases that we shall consider fall into two classes: (1) *Long waves in shallow water*, where the depth of the water is small compared to the wave length and the disturbance affects the motion of the whole of the liquid; (2) *Surface waves*, where the wave length may be small compared to the depth so that the effects of the disturbance cease to be appreciable below a certain depth.

10·2. Long Waves. Let us consider the case of waves travelling along a straight canal of uniform section. Take the axis of x in the direction of the length of the canal and y vertically upwards and let η be the elevation of the free surface above the equilibrium level at the point whose abscissa is x at time t. If the wave length be large in comparison with the mean depth, the vertical acceleration can be neglected in comparison with the horizontal, so that as far as vertical forces are concerned we may regard the liquid as in equilibrium and take for the pressure at any point the statical pressure due to the depth below the free surface.

Therefore the pressure p at a point (x, y) is given by

$$p - p_0 = g\rho\,(y_0 + \eta - y) \quad \text{.................(1)},$$

where y_0 is the ordinate of the undisturbed free surface and p_0 is the pressure above the liquid supposed constant. Hence we get

$$\frac{\partial p}{\partial x} = g\rho\,\frac{\partial \eta}{\partial x} \quad \text{...........................(2)},$$

and as this is independent of y, and the horizontal acceleration of an element depends on the difference of pressure at its ends, i.e. $\frac{\partial p}{\partial x}dx$, it follows that the horizontal acceleration of all points in the same vertical cross section of the canal is the same, and consequently that points that are once in a vertical plane are always in a vertical plane.

Considering a small horizontal cylinder PP' of liquid of length dx' the difference of pressure at its ends is $g\rho\,\frac{\partial \eta}{\partial x'}dx'$. And if x be the abscissa of the vertical plane of particles through P in its equilibrium position and ξ the horizontal displacement of this plane of particles,

$$x' = x + \xi \quad \text{..........................(3)}$$

and the horizontal acceleration is $\partial^2\xi/\partial t^2$.

If κ be the cross section of the cylinder PP', the mass is $\kappa\rho\,dx'$ and the equation of motion is

$$\rho\kappa\,dx'\,\frac{\partial^2\xi}{\partial t^2} = -g\rho\kappa\,\frac{\partial \eta}{\partial x'}\,dx',$$

or

$$\frac{\partial^2\xi}{\partial t^2} = -g\,\frac{\partial \eta}{\partial x'} \quad \text{.......................(4)}.$$

If now we suppose the motion to be small and neglect the squares of small quantities, we get from (3) and (4)

$$\frac{\partial^2 \xi}{\partial t^2} = -g\frac{\partial \eta}{\partial x} \quad \text{.....................(5).}$$

We have now to form the equation of continuity. Let A be the area of the cross section of the canal, and b the breadth at the surface. In the position of equilibrium the volume of liquid between the planes x and $x+dx$ is $A\,dx$. At time t the distance between the bounding planes of this liquid is $dx+\frac{\partial \xi}{\partial x}dx$, and the area of the cross section of the liquid is $A+b\eta$; therefore

$$(A+b\eta)\left(dx+\frac{\partial \xi}{\partial x}dx\right)=A\,dx.$$

Neglecting the product of the small quantities this becomes

$$A\frac{\partial \xi}{\partial x}+b\eta=0 \quad \text{.....................(6),}$$

and we therefore obtain from (5)

$$\frac{\partial^2 \xi}{\partial t^2}=\frac{gA}{b}\frac{\partial^2 \xi}{\partial x^2} \quad \text{.....................(7).}$$

To integrate this equation we write

$$gA/b=c^2,$$

and

$$x-ct=x_1, \quad x+ct=x_2,$$

so that

$$\frac{\partial}{\partial t}=-c\frac{\partial}{\partial x_1}+c\frac{\partial}{\partial x_2}, \text{ and } \frac{\partial}{\partial x}=\frac{\partial}{\partial x_1}+\frac{\partial}{\partial x_2},$$

reducing equation (7) to the form

$$\frac{\partial^2 \xi}{\partial x_1 \partial x_2}=0,$$

the solution of which is

$$\xi=f(x_1)+F(x_2),$$

where f, F are arbitrary functions.

Hence the solution of (7) is

$$\xi=f(x-ct)+F(x+ct) \quad \text{.................(8),}$$

representing two waves travelling in opposite directions with velocity $c=(gA/b)^{\frac{1}{2}}$.

If the canal be of rectangular section and depth h the wave velocity is $(gh)^{\frac{1}{2}}$; i.e. a velocity due to half the depth of the liquid.

The displacement being given by (8), the elevation η is given by

$$\eta=-\frac{A}{b}\frac{\partial\xi}{\partial x},\quad\text{from (6)},$$

that is,
$$\eta=-\frac{A}{b}f'(x-ct)-\frac{A}{b}F'(x+ct).$$

We should expect the expression for η to contain two arbitrary functions because the elimination of ξ between (5) and (6) shews that η satisfies the same equation (7) as ξ.

The particle velocity $\dot{\xi}$ is given by

$$\dot{\xi}=-cf'(x-ct)+cF'(x+ct).$$

The meaning of the solution that we have obtained is not that the hypothesis of the existence of a 'long wave' involves a complicated motion represented by arbitrary functions, but that all possible motions subject to the limitations we have imposed are included in the general solution (8); and the forms of the functions f, F to suit any special case must be determined from given initial conditions. A discussion of the adaptation of the solution to special cases will be given in a later chapter. At present we will confine our attention to the determination of the motion of the individual particles.

10·21. Assuming the canal to be of rectangular section it is clear that the particles move in planes parallel to the length of the canal. A vertical column bounded by two such planes and two others at right angles to them remains a vertical column on a rectangular base, but the area of this base changes during the motion and the height of any particle in the column changes in such a way that the volume of the part of the column below the particle is unaltered; hence the vertical displacement of any particle is proportional to its height above the base. Therefore when the motion of a particle at the surface is known the motion of any particles in the same vertical line is found by diminishing the vertical displacement in a given ratio without altering the horizontal displacement.

To trace the motion of a surface particle when a progressive wave passes over it in either direction, we may take

$$\xi=f(x-ct).$$

Then from **10·2** (6), putting $A=bh$, we have

$$\eta=-h\frac{\partial\xi}{\partial x}=-hf'(x-ct)=\frac{h}{c}\dot{\xi},$$

or

$$\dot{\xi}=c\frac{\eta}{h}\quad\text{.........................(1).}$$

The particle is at rest until the wave reaches it, then it moves forward as well as upward with a velocity proportional to the elevation of the wave above the equilibrium level; when the crest of the wave reaches the particle the upward motion ceases but the horizontal velocity is a maximum, η then decreases and ξ increases less rapidly and as the wave leaves the particle $\eta=0$ so that the particle is at the same height from the bottom as before; but $\xi=\frac{1}{h}\int\eta c\,dt=\frac{1}{bh}\int b\eta c\,dt$ and when the wave has passed the particle this expression represents the total volume of the elevated water divided by the sectional area of the canal. Hence the particle is finally deposited in front of its initial position by this distance.

If the wave consists of a single depression instead of an elevation, everything is the same as before except that the particle moves backwards instead of forwards.

10·22. To recapitulate—the results of the foregoing Articles have been obtained on the hypothesis that the motions are so small that squares and products of ξ and η can be neglected, and that the vertical acceleration can be neglected in comparison with the horizontal. We may observe that if we consider the passage of a wave consisting of a single elevation of length λ and maximum elevation η the time taken to pass a particular particle is λ/c, where c is the velocity, so that the vertical velocity will be of order $\eta c/\lambda$, and the vertical acceleration of order $\eta c^2/\lambda^2$. But from **10·21** (1) the maximum horizontal velocity is $c\eta/h$, and taking $c^2=gh$, we get that the ratio of the maximum vertical and horizontal velocities is of order h/λ, and the vertical acceleration being of order $g\eta h/\lambda^2$ can be neglected if h/λ is a small quantity. This shews that waves of the type described are propagated only when h/λ is small, and justifies the application to them of the term 'long waves'.

The foregoing discussion is based on an article by Stokes*.

* 'On Waves', *Camb. and Dub. Math. Journal*, IV, p. 219, or *Math. and Phys. Papers*, II, p. 222.

10·23. Long Waves—general equation. Reverting to **10·2**, if we form the equation of motion for the liquid which in equilibrium occupies the space between two cross sections at a distance dx, x and $x+dx$ being the abscissae in the undisturbed state, and $x+\xi$ and $x+\xi+dx+\frac{\partial \xi}{\partial x}dx$ the abscissae at time t, of the bounding planes, the mass is $\rho A\,dx$ and the equation of motion

$$\rho A dx \frac{\partial^2 \xi}{\partial t^2} = -\frac{\partial p}{\partial x} dx\,(A+b\eta);$$

where as before
$$\frac{\partial p}{\partial x} = g\rho \frac{\partial \eta}{\partial x},$$

so that the equation of motion is

$$\frac{\partial^2 \xi}{\partial t^2} = -g\frac{\partial \eta}{\partial x}\left(1+\frac{b\eta}{A}\right) \qquad \text{.....................(1).}$$

The equation of continuity is

$$(A+b\eta)\left(dx+\frac{\partial \xi}{\partial x}dx\right) = A\,dx,$$

or
$$\frac{b\eta}{A} = -\frac{\partial \xi}{\partial x}\left(1+\frac{\partial \xi}{\partial x}\right)^{-1} \qquad \text{.....................(2);}$$

and the elimination of η between (1) and (2) gives

$$\frac{\partial^2 \xi}{\partial t^2} = g\frac{A}{b}\frac{\partial^2 \xi}{\partial x^2}\left(1+\frac{\partial \xi}{\partial x}\right)^{-3} \qquad \text{..................(3)*.}$$

Our former equation is an approximation to this in which the squares of small quantities are neglected. Airy's discussion of this equation shews that waves cannot be propagated to infinity without change of form.

10·24. Long Waves—another method. In any case in which waves are propagated in one direction only without change of form, the problem of finding the velocity of propagation can be simplified by imposing on the whole mass of liquid a velocity equal and opposite to the velocity of propagation of the waves, the wave form having the same relative velocity as before becomes fixed in space, and the problem becomes one of steady motion.

In the case of long waves, neglecting the vertical velocity, let c denote the velocity of propagation, and u the small additional velocity due to the wave motion at points where the elevation is η.

* Airy, 'Tides and Waves', *Encyc. Metrop.* 1845.

The equation of continuity is

$$(A+b\eta)(c+u)=Ac \quad \text{.................(1)},$$

where A is the area of the cross section and b the breadth at the surface.

If δp denote the excess of pressure due to the wave motion we have

$$\frac{\delta p}{\rho}+g\eta+\tfrac{1}{2}(c+u)^2=\tfrac{1}{2}c^2 \quad \text{.................(2)},$$

therefore

$$\delta p=\tfrac{1}{2}\rho c^2\left\{1-\frac{A^2}{(A+b\eta)^2}\right\}-g\rho\eta$$

$$=\left\{\tfrac{1}{2}c^2\frac{(2Ab+b^2\eta)}{(A+b\eta)^2}-g\right\}\rho\eta \quad \text{..............(3)}.$$

If η be small compared to A/b, this reduces to

$$\delta p=\{c^2 b/A-g\}\rho\eta,$$

and if $c^2=gA/b$ the surface pressure is constant to a first approximation, so that a free surface is possible. This value of c gives the velocity of propagation of a long wave in still water, or the velocity of the stream for a stationary long wave.

Assuming that $c^2=gA/b$ and substituting in (3) we get

$$\delta p=-\frac{3g\rho b\eta^2}{2A}$$

as the second approximation, shewing that the pressure is defective at all parts of the wave at which η is not zero. Hence, *unless η^2 can be neglected, it is impossible to satisfy the condition of a free surface for a stationary long wave; that is, it is impossible for a long wave whose height is not small compared to the depth of the water to be propagated in still water without change of type.*

From (3) we see that δp will vanish if

$$c^2=\frac{2g(A+b\eta)^2}{2Ab+b^2\eta};$$

and since

$$\frac{2g(A+b\eta)^2}{2Ab+b^2\eta}-\frac{gA}{b}=g\eta\frac{(3A+2b\eta)}{2A+b\eta},$$

it follows that if η is positive everywhere the conditions for the propagation of the wave are more nearly satisfied by taking a value of c greater than $(gA/b)^{\frac{1}{2}}$, and if η is negative everywhere a value less than $(gA/b)^{\frac{1}{2}}$. Hence an elevation in the surface travels rather faster than a depression*.

* Lord Rayleigh, 'On Waves', *Phil. Mag.* I, 1876, p. 257 or *Sci. Papers*, I, p. 251.

10·25. Energy of a Long Wave. For a wave in a canal of rectangular section the potential energy is due to the elevation or depression of the water above the mean level, and for a unit breadth of the wave the potential energy is therefore

$$\tfrac{1}{2}g\rho\int\eta^2dx,$$

where η is the elevation at x and the integration is over the whole length of the wave.

The kinetic energy is $\quad \tfrac{1}{2}\rho h\int\xi^2dx$

for the same range of integration.

For a wave travelling in one direction, we have, as in **10·21** (1),

$$\xi=\pm\frac{c}{h}\eta,$$

and since $c^2=gh$, therefore the above expressions are equal and at any instant the energy of the wave is half potential and half kinetic.

10·26. Examples of the artifice of Steady Motion.

(i) *Long waves at the common surface of two liquids of different densities in a horizontal pipe.*

The artifice consists in reversing the velocity of propagation c of the waves on the whole mass of liquid. The wave form then becomes fixed in space and the liquids move below and above it with general velocity c. Let ρ, ρ' denote the densities, A, A' the cross sections of the two liquid streams and b the breadth of the common surface. The problem is to express c in terms of these data. If for example the liquids were separated by a thin rigid sheet of the prescribed wave form, then they might be forced through the pipe at any speeds, but we want to find the particular speed c at which, if they move, their common surface would keep the prescribed form without the aid of the material surface of separation. Then when this velocity is again reversed on the whole mass the liquids are stationary save for the wave motion.

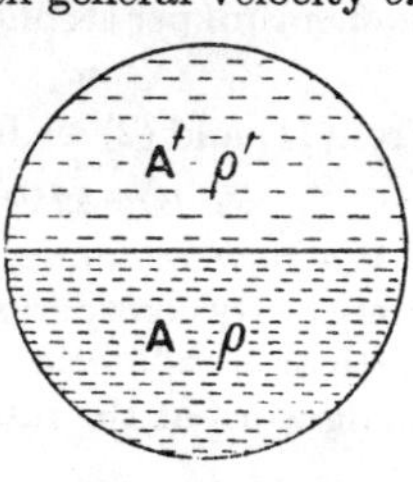

Let η denote the elevation of the common surface due to the wave motion, and u, u' the small additional velocities due to the wave motions in the two liquids.

The equations of continuity in the two liquids are

$$(A+b\eta)(c+u)=Ac$$

and

$$(A'-b\eta)(c+u')=A'c,$$

or

$$\left.\begin{aligned}Au+bc\eta&=0\\ A'u'-bc\eta&=0\end{aligned}\right\}\quad\text{..............................(1).}$$

(the second line of (1) preceded by "and")

Also, if δp, $\delta p'$ denote increments of pressure close to the common surface in the two liquids due to the waves,

$$\frac{\delta p}{\rho}+g\eta+\tfrac{1}{2}(c+u)^2=\tfrac{1}{2}c^2$$

and
$$\frac{\delta p'}{\rho'}+g\eta+\tfrac{1}{2}(c+u')^2=\tfrac{1}{2}c^2,$$

hence, if we ignore surface tension so that $\delta p=\delta p'$, we have

$$g(\rho-\rho')\eta=(\rho' u'-\rho u)c \quad\text{.......................(2).}$$

Then by eliminating η, u and u' from (1) and (2) we get

$$c^2=g(\rho-\rho')/b\left(\frac{\rho}{A}+\frac{\rho'}{A'}\right) \quad\text{....................(3)*.}$$

(ii) *A 'bore' or a 'wave' invading a region of still water.*

Consider the case of a steady stream in which there is a transition from a uniform velocity u and depth h to a uniform velocity u' and depth h'. There is an equation of continuity

$$hu=h'u' \quad\text{...............................(1).}$$

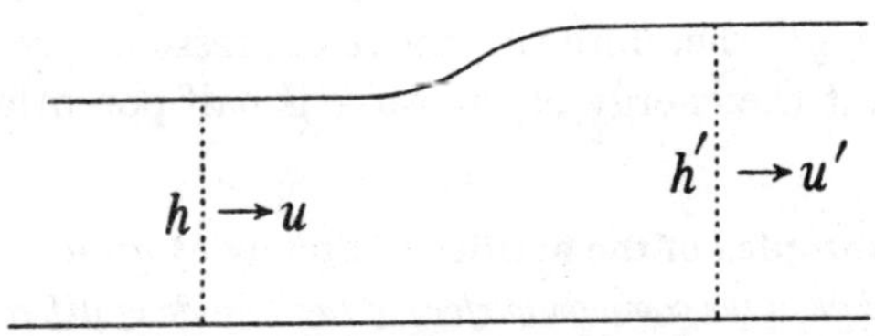

Taking the density as unity and unit breadth of the stream, the mean pressures over the two cross sections are $\tfrac{1}{2}gh$ are $\tfrac{1}{2}gh'$ and the total pressures are $\tfrac{1}{2}gh^2$ and $\tfrac{1}{2}gh'^2$, so that, by considering the change of momentum per second, we have

$$hu(u-u')=\tfrac{1}{2}g(h'^2-h^2) \quad\text{......................(2).}$$

From (1) and (2) we find that

$$u^2=\tfrac{1}{2}g(h+h')h'/h \quad\text{and}\quad u'^2=\tfrac{1}{2}g(h+h')h/h' \quad\text{...........(3).}$$

We now examine how far this is compatible with conservation of energy. By considering the energy changes per unit time between the two vertical cross sections we have that the pressures on the boundary are doing work at the rate

$$\tfrac{1}{2}gh^2u-\tfrac{1}{2}gh'^2u'=\tfrac{1}{2}ghu(h-h') \quad\text{.................(4).}$$

But the centre of gravity of the liquid entering the region is raised by an amount $\tfrac{1}{2}(h'-h)$, so that the potential energy is increasing at a rate

$$\tfrac{1}{2}ghu(h'-h) \quad\text{............................(5);}$$

and subtracting (5) from (4) leaves a rate of working

$$ghu(h-h') \quad\text{...............................(6)}$$

available for increasing the kinetic energy. But the kinetic energy leaving the region per unit time exceeds that entering it by

$$\tfrac{1}{2}h'u'^3-\tfrac{1}{2}hu^3=\tfrac{1}{2}hu(u'^2-u^2)$$

which from (3)
$$=\tfrac{1}{4}gu(h+h')^2(h-h')/h' \quad\text{.......................(7).}$$

* Greenhill, 'Wave Motion in Hydrodynamics', *Amer. Journ. of Math.* IX, 1887.

We now find that expression (6) exceeds (7) by an amount

$$\tfrac{1}{4}gu\,(h'-h)^3/h' \qquad \text{......(8)}$$

which implies a dissipation of surplus energy when h' is greater than h.

Now suppose that a velocity u from right to left is impressed on the whole mass of liquid. We then have the case of a wave invading a region of still water. The velocity with which the wave travels is u, and the velocity of the liquid particles is $u-u'$ in the direction of propagation. This is positive or negative according as h' is greater or less than h, i.e. according as the wave is an elevation or depression. In the case of an elevation or 'bore' expression (8) gives the rate at which energy is dissipated at the transition; and since this expression is negative when h' is less than h, it follows that a depression cannot be propagated without a supply of energy*.

10·3. Surface Waves. We shall next consider waves due to small oscillatory motions which take place at and near the surface of an unlimited sheet of water where the depth may be considerable compared to the wave length. We no longer neglect the vertical acceleration, but we suppose that the squares of the velocities of the particles are negligible. The motion is supposed to be two-dimensional, the ridges and hollows of the waves being all parallel to one another. The axis of x is taken in the undisturbed surface in the direction of propagation of the waves and the axis of y vertically upwards. The motion being such as could be produced from rest by natural forces is irrotational and the velocity potential ϕ has to satisfy the equations

$$\frac{\partial^2\phi}{\partial x^2}+\frac{\partial^2\phi}{\partial y^2}=0 \qquad \text{......(1),}$$

throughout the liquid, and $\dfrac{\partial\phi}{\partial n}=0$ ......(2),

at a fixed boundary.

The pressure is given by

$$\frac{p}{\rho}=\frac{\partial\phi}{\partial t}-gy-\tfrac{1}{2}q^2+F(t) \qquad \text{......(3).}$$

The free surface is a surface of equipressure $p=$ const., therefore as in **1·6** (3)

$$\frac{\partial p}{\partial t}+u\frac{\partial p}{\partial x}+v\frac{\partial p}{\partial y}=0,$$

or writing $-\partial\phi/\partial x$ for u and $-\partial\phi/\partial y$ for v we have

$$\frac{\partial p}{\partial t}-\frac{\partial\phi}{\partial x}\frac{\partial p}{\partial x}-\frac{\partial\phi}{\partial y}\frac{\partial p}{\partial y}=0 \qquad \text{......(4)}$$

at the free surface.

* Rayleigh 'On the Theory of Long Waves and Bores', *Proc. R.S.* A, XC, 1914 or *Sci. Papers*, VI, p. 250.

If now we suppose the motion so small that the squares of small quantities (e.g. velocities) can be neglected we may neglect q^2 in (3), and if we also regard the arbitrary function $F(t)$ as absorbed in $\partial\phi/\partial t$ and then substitute the value of p from (3) in (4) we get

$$\frac{\partial^2\phi}{\partial t^2} - \frac{\partial\phi}{\partial x}\frac{\partial^2\phi}{\partial x\,\partial t} - \frac{\partial\phi}{\partial y}\left(\frac{\partial^2\phi}{\partial y\,\partial t} - g\right) = 0,$$

or, neglecting the second and third terms which are of the same order as q^2,

$$\frac{\partial^2\phi}{\partial t^2} + g\frac{\partial\phi}{\partial y} = 0 \quad \text{......................(5).}$$

This condition holds at the free surface.

If η denote the elevation of the free surface at time t above the point whose abscissa is x, the equation of the free surface will be of the form

$$\eta - f(x, t) = 0,$$

and this being a boundary must satisfy the condition **1·6** (3). Hence

$$\frac{\partial f}{\partial t} + u\frac{\partial f}{\partial x} - v = 0.$$

But $\partial f/\partial t$ is $\dot{\eta}$, and $\partial f/\partial x$ or $\partial\eta/\partial x$ is the tangent of the slope of the free surface which by hypothesis is small so that the second term can be neglected and the equation becomes

$$\dot{\eta} = v = -\frac{\partial\phi}{\partial y} \quad \text{......................(6),}$$

at the free surface.

Hence in a wave motion in which the squares of the velocities can be neglected the velocity potential must be a solution of Laplace's equation which makes $\partial\phi/\partial n = 0$ as a fixed boundary and satisfies (5) and (6) at the free surface of the liquid.

10·31. Let us apply these conditions to investigate the propagation of simple harmonic waves of type

$$\eta = a\sin(mx - nt) \quad \text{....................(7)}$$

at the surface of water of uniform depth h, either of unlimited extent or contained in a canal with parallel vertical sides at right angle to the ridges and hollows.

If we assume that there is a solution of the form

$$\phi = f(y)\cos(mx - nt)$$

and substitute in (1) we obtain

$$\frac{\partial^2 f}{\partial y^2} - m^2 f = 0,$$

so that $$f(y) = Ae^{my} + Be^{-my},$$

and $$\phi = (Ae^{my} + Be^{-my}) \cos(mx - nt).$$

This value of ϕ must satisfy (2), i.e. $\partial\phi/\partial y = 0$ when $y = -h$.

Hence $$Ae^{-mh} = Be^{mh} = \tfrac{1}{2}C, \text{ say,}$$

so that $$\phi = C \cosh m(y+h) \cos(mx - nt) \quad \text{...........(8).}$$

Again if we substitute this value in the surface condition (5) putting $y = 0$, we get $$n^2 = gm \tanh mh \quad \text{....................(9).}$$

Now if $c\,(=n/m)$ denote the velocity of propagation and $\lambda\,(=2\pi/m)$ denote the wave length, it follows that

$$c^2 = \frac{g}{m} \tanh mh = \frac{g\lambda}{2\pi} \tanh \frac{2\pi h}{\lambda} \quad \text{...........(10).}$$

The constant C of (8) can be expressed in terms of the amplitude a of the wave by substituting from (7) and (8) in (6). Thus putting $y = 0$ we have
$$-na = -mC \sinh mh,$$

so that $$\phi = \frac{na}{m} \frac{\cosh m(y+h)}{\sinh mh} \cos(mx - nt),$$

or using (9) $$\phi = \frac{ga}{n} \frac{\cosh m(y+h)}{\cosh mh} \cos(mx - nt) \quad \text{........(11).}$$

10·32. The Paths of the Particles. If $\mathbf{x}, \mathbf{y}$ be the coordinates of a particle relative to its mean position (x, y), neglecting the squares of small quantities we may write

$$\dot{\mathbf{x}} = -\frac{\partial\phi}{\partial x} = na \frac{\cosh m(y+h)}{\sinh mh} \sin(mx - nt),$$

$$\dot{\mathbf{y}} = -\frac{\partial\phi}{\partial y} = -na \frac{\sinh m(y+h)}{\sinh mh} \cos(mx - nt).$$

Whence, by integrating, we get

$$\mathbf{x} = a \frac{\cosh m(y+h)}{\sinh mh} \cos(mx - nt),$$

$$\mathbf{y} = a \frac{\sinh m(y+h)}{\sinh mh} \sin(mx - nt);$$

so that the particle describes the ellipse

$$\mathbf{x}^2/\cosh^2 m(y+h) + \mathbf{y}^2/\sinh^2 m(y+h) = a^2/\sinh^2 mh$$

about its mean position. For a given particle $mx - nt$ plays the part of the eccentric angle in the ellipse; so that the eccentric angle increases at a uniform rate, as in an orbit described under a central force varying as the distance.

The distance between the foci $2a$ cosech mh is the same for all such ellipses, their major axes are horizontal, and both axes decrease as the depth of the particle increases, the minor axis vanishing when $y=-h$.

10·33. Deep Water. If the depth h of the water be sufficiently great in comparison with λ for e^{-mh} to be neglected, then in **10·31** we must have $B=0$, so that we have instead of (8)

$$\phi = Ae^{my}\cos(mx-nt) \quad\text{.................(8')},$$

and instead of (9)

$$n^2 = gm \quad\text{...........................(9')},$$

or

$$c^2 = \frac{g\lambda}{2\pi} \quad\text{...........................(10')}.$$

Also if $\eta = a\sin(mx-nt)$ is the free surface we get from (6), $na=mA$, so that

$$\phi = \frac{na}{m}e^{my}\cos(mx-nt),$$

or

$$\phi = \frac{ga}{n}e^{my}\cos(mx-nt) \quad\text{.................(11')}.$$

Following the method of **10·32** we get in this case for the displacement of a particle from its mean position

$$\mathbf{x} = ae^{my}\cos(mx-nt), \quad \mathbf{y} = ae^{my}\sin(mx-nt),$$

and the path of the particle is a circle

$$\mathbf{x}^2 + \mathbf{y}^2 = a^2e^{2my},$$

described with uniform angular velocity n, which in this case is equal to $(gm)^{\frac{1}{2}}$ or $(2\pi g/\lambda)^{\frac{1}{2}}$.

10·34. Wave Length and Wave Velocity. A comparison of **10·31** (10) and **10·33** (10′) shews that in what we have described as 'deep water' we have taken the factor $\tanh(2\pi h/\lambda)$ to be unity. Now when $x=\pi$ then $\tanh x = 1$, with an error of less than 1 per cent., and, for larger values of x, $\tanh x$ is more nearly equal to unity. It follows that it is only necessary for the depth h to exceed half the wave length for the circumstances to be such as we have described as 'deep water', and in all such cases the wave velocity is given by (10′) and is independent of the depth.

Now reverting to the formula **10·31** (10), let us consider the function

$$y = \frac{\tanh x}{x} \quad \text{.........................(1)},$$

for which

$$\frac{1}{y}\frac{dy}{dx} = -\frac{1}{x} + \frac{2}{\sinh 2x}.$$

Since

$$\sinh 2x = 2x + (2x)^3/3! + \ldots,$$

therefore, for $x > 0$, the derivative of y is negative, or y increases as x decreases.

Hence for water of a given depth h, the velocity c of wave propagation increases as the wave length λ increases; but by expanding $\tanh(2\pi h/\lambda)$ we see that the value of c will not exceed $(gh)^{\frac{1}{2}}$, which is the value previously found for long waves.

Also in (1) there is only one value of x corresponding to each value of y, therefore there is only one wave length corresponding to a given velocity, and every velocity up to $(gh)^{\frac{1}{2}}$ is the velocity of some wave.

The supposition of simple harmonic waves is the simplest that can be made and since by Fourier's Theorem functions can be expanded in series of sines and cosines it follows that waves of a general type can be regarded as the result of the superposition of a number of simple harmonic waves.

10·35. Standing or Stationary Waves. The velocity potential for a system of stationary waves can be deduced from **10·31** by regarding the system as the result of the superposition of two such trains of waves as we have just been considering moving in opposite directions as explained in **10·13**. Thus corresponding to a wave profile

$$\eta = a \sin mx \cos nt \quad \text{.....................(1)}$$

we shall have

$$\phi = \frac{na}{m}\frac{\cosh m(y+h)}{\sinh mh}\sin mx \sin nt \quad \text{...........(2)},$$

or

$$\phi = \frac{ga}{n}\frac{\cosh m(y+h)}{\cosh mh}\sin mx \sin nt \quad \text{...........(3)},$$

for ϕ clearly satisfies **10·3** (1) and (2), and η and ϕ together satisfy (6) of the same article.

It is not necessary to regard standing waves as a case of superposition of progressive waves, we might investigate this form for ϕ independently, starting with an assumption

$$\phi = f(y) \sin mx \sin nt,$$

and proceeding as in **10·31** we get the same equation for f as before, and hence the result follows as in that article.

For *Standing waves in deep water*, as in **10·33**, equations (2) and (3) above take the forms

$$\phi = \frac{na}{m} e^{my} \sin mx \sin nt,$$

and
$$\phi = \frac{ga}{n} e^{my} \sin mx \sin nt.$$

10·36. Paths of the Particles in Stationary Waves. With the same notation as in **10·32** we have

$$\dot{\mathbf{x}} = -\frac{\partial \phi}{\partial x} = -na\frac{\cosh m(y+h)}{\sinh mh} \cos mx \sin nt,$$

and
$$\dot{\mathbf{y}} = -\frac{\partial \phi}{\partial y} = -na\frac{\sinh m(y+h)}{\sinh mh} \sin mx \sin nt,$$

so that, by integration

$$\mathbf{x} = a\frac{\cosh m(y+h)}{\sinh mh} \cos mx \cos nt,$$

and
$$\mathbf{y} = a\frac{\sinh m(y+h)}{\sinh mh} \sin mx \cos nt.$$

Hence
$$\mathbf{y}/\mathbf{x} = \tanh m(y+h) \tan mx,$$

and since this is independent of t the motion of each particle is rectilinear, the direction varying from vertical beneath the crests and troughs ($mx = (\kappa + \frac{1}{2})\pi$), to horizontal beneath the nodes ($mx = \kappa\pi$).

10·37. Stationary Waves in a limited area. We have supposed that the liquid is unlimited in the direction of the axis of x, so that there is no restriction on the value of m. But if the liquid be confined in a canal with closed vertical ends, say at $x = -\frac{1}{2}l$ and $x = \frac{1}{2}l$, then there is a restriction on the value of m, for as we shall see only waves of a certain length can exist in such a canal. The extra condition is that $\partial\phi/\partial x = 0$ when $x = -\frac{1}{2}l$ and $x = \frac{1}{2}l$. Taking the form for ϕ in **10·35** (2) we have to satisfy $\cos mx = 0$ at $x = -\frac{1}{2}l$ and $x = \frac{1}{2}l$. This requires that $ml = (2s+1)\pi$, where s is any integer, and then possible wave lengths are included in the formula $\lambda = 2l/(2s+1)$.

Standing waves are really the principal or normal modes of free oscillation of (usually) a restricted system, and from this point of view the periods are fundamental and they determine the possible wave lengths.

10·38. The Energy of Progressive Waves. Considering a train of progressive waves at the surface of water of depth h, given, as in **10·31**, by

$$\eta = a \sin(mx - nt) \quad \text{.....................(1)}$$

and
$$\phi = \frac{ga}{n} \frac{\cosh m(y+h)}{\cosh mh} \cos(mx - nt) \quad \text{...........(2),}$$

if we calculate the energy of the water between two vertical planes parallel to the direction of propagation at unit distance apart, we have, for a single wave length, the potential energy

$$V = \tfrac{1}{2} g\rho \int_0^\lambda \eta^2 \, dx$$

$$= \tfrac{1}{4} g\rho a^2 \lambda; \text{ since } \lambda = 2\pi/m.$$

The kinetic energy is given by

$$T = \tfrac{1}{2}\rho \int_0^\lambda \int_{-h}^\eta \left\{ \left(\frac{\partial \phi}{\partial x}\right)^2 + \left(\frac{\partial \phi}{\partial y}\right)^2 \right\} dx\, dy$$

and, as in **4·52**, this may be transformed to

$$T = -\tfrac{1}{2}\rho \int \phi \frac{\partial \phi}{\partial n} ds,$$

integrated along the profile of a wave length, where ∂n is measured along the normal into the water. To the order of small quantities we are using this may be written

$$T = \tfrac{1}{2}\rho \int_0^\lambda \left(\phi \frac{\partial \phi}{\partial y}\right)_{y=0} dx$$

$$= \tfrac{1}{2} g\rho a^2 \int_0^\lambda \cos^2(mx - nt) \, dx$$

$$= \tfrac{1}{4} g\rho a^2 \lambda.$$

Hence it follows that the total energy per wave length is $\tfrac{1}{2} g\rho a^2 \lambda$, and that *it is half kinetic and half potential.*

Also considering any length in the water, in direction of the wave propagation, which is either an exact number of wave lengths or is so long that the energy of a fractional part of a wave length may be neglected in comparison with the energy of the whole, it follows that it is correct to say that *the energy of a progressive train of waves is half kinetic and half potential.*

10·39. The Energy of Stationary Waves may be calculated in the same way. Thus if we take

$$\eta = a \sin mx \cos nt,$$

and
$$\phi = \frac{ga}{n} \frac{\cosh m(y+h)}{\cosh mh} \sin mx \sin nt,$$

as in **10·35**, we find for the potential energy of a wave length

$$V = \tfrac{1}{4} g\rho a^2 \lambda \cos^2 nt,$$

and for the kinetic energy

$$T = \tfrac{1}{4} g\rho a^2 \lambda \sin^2 nt.$$

Hence the total energy per wave length at any time is $\frac{1}{4} g\rho a^2 \lambda$ and the amounts of kinetic and potential energy change continuously with the time.

10·4. Progressive Waves reduced to a case of Steady Motion. The method of **10·24** and **10·26**, of finding the velocity of propagation, namely, imposing on the whole mass a velocity equal and opposite to the velocity of propagation of the waves, may also be applied to the case of progressive waves considered in **10·31**. The wave form having the same relative velocity as before becomes fixed in space and the problem becomes one of steady motion. As the problem is a two-dimensional one it only remains to determine suitable expressions for the velocity potential and stream function so that the free surface and the bottom of the liquid may satisfy the conditions for stream lines.

Consider the relation

$$w = cz + P \cos mz - iQ \sin mz,$$

or
$$\phi + i\psi = c(x+iy) + P \cos m(x+iy) - iQ \sin m(x+iy).$$

It gives

$$\left.\begin{aligned} \phi &= cx + (P \cosh my + Q \sinh my) \cos mx \\ \text{and} \quad \psi &= cy - (P \sinh my + Q \cosh my) \sin mx \end{aligned}\right\} \quad \ldots\ldots\ldots(1).$$

These expressions satisfy Laplace's equation and give the general superposed velocity $-c$.

For the bottom to be a stream line we must have ψ constant when $y = -h$, so that $-P \sinh mh + Q \cosh mh = 0$.

Hence the expressions (1) may be written

$$\left.\begin{aligned} \phi &= cx + A \cosh m(y+h) \cos mx \\ \psi &= cy - A \sinh m(y+h) \sin mx \end{aligned}\right\} \quad \ldots\ldots\ldots\ldots\ldots(2).$$

If the free surface be a simple sine curve $\eta = a \sin mx$, equations (2) will make this the stream line $\psi = 0$ provided

$$ca - A \sinh mh = 0 \quad \text{.....................(3)},$$

neglecting squares of small quantities.

Again, the formula for pressure is

$$\frac{p}{\rho} + gy + \tfrac{1}{2}\left\{\left(\frac{\partial\phi}{\partial x}\right)^2 + \left(\frac{\partial\phi}{\partial y}\right)^2\right\} = \text{const.}$$

At the free surface this becomes

$$\frac{p}{\rho} + ga \sin mx + \tfrac{1}{2}c^2\{1 - 2ma \coth mh \sin mx\} = \text{const.},$$

neglecting a^2.

But p is constant at the free surface, therefore the coefficient of $\sin mx$ must vanish, that is

$$g = mc^2 \coth mh,$$

or

$$c^2 = \frac{g\lambda}{2\pi} \tanh \frac{2\pi h}{\lambda} \quad \text{.......................(4)}.$$

Another way of regarding this problem is as follows.

Imagine a straight horizontal pipe of rectangular section, the upper surface of which has small corrugations of the form $\eta = a \sin 2\pi x/\lambda$. Water filling this pipe can be made to flow along it at any speed, but we have found in (4) the particular speed that the water must have if the removal of the corrugated upper surface of the pipe would leave the water flowing with the corrugations in its surface unaltered.

We observe that the expression for ϕ in (2) is the steady motion value, and the expression (11) of **10·31** corresponding to the progressive waves can be obtained from (2) and (3) by reimposing the velocity c, which amounts to omitting the term cx and writing $mx - nt$ for mx.

10·41. The same on Deep Water. In this case we may take

$$\left.\begin{aligned} \phi &= cx + Ae^{my} \cos mx \\ \psi &= cy - Ae^{my} \sin mx \end{aligned}\right\} \quad \text{.................(1)},$$

and

with a free surface

$$\eta = a \sin mx \quad \text{.......................(2)}.$$

The free surface is the stream line $\psi = 0$, if

$$ca = A \quad \text{.........................(3)},$$

so that

$$\left.\begin{aligned} \phi &= cx + cae^{my} \cos mx \\ \psi &= cy - cae^{my} \sin mx \end{aligned}\right\} \quad \text{.................(4)}.$$

and

The formula for the pressure

$$\frac{p}{\rho}+gy+\tfrac{1}{2}\left\{\left(\frac{\partial\phi}{\partial x}\right)^2+\left(\frac{\partial\phi}{\partial y}\right)^2\right\}=\text{const.}$$

becomes

$$\frac{p}{\rho}+gy+\tfrac{1}{2}c^2\{1-2mae^{my}\sin mx+m^2a^2e^{2my}\}=\text{const.} \quad \text{...(5).}$$

If we neglect the last term on the left, this equation may be written

$$\frac{p}{\rho}+y\,(g-mc^2)+mc\psi=\text{const.} \quad \text{..............(6).}$$

This equation not only gives

$$c^2=g/m \quad \text{..........................(7)}$$

at the free surface ($p=$const. and $\psi=0$), but also shews that, if $c^2=g/m$, the pressure is constant along each stream line. It follows that the solution contained in (4) and (7) can be applied to the case of any number of liquids of different densities arranged one above the other in horizontal strata including the case of liquid of continuously varying density since there is no limit to the thinness of a stratum, the only limitations being that the upper surface is free and the total depth infinite*.

10·42. Waves at the Common Surface of two Liquids. Suppose a liquid of density ρ' and depth h' to be moving with velocity V' over another liquid of density ρ and depth h moving in the same direction with velocity V; the liquids being bounded above and below by two fixed horizontal planes.

Let c be the velocity of propagation of oscillatory waves at the common surface in the direction in which the liquids are moving. Taking the axis of x in this direction in the undisturbed common surface and y vertically upwards, as in the last article, let us make the motion steady by superposing on the whole mass the velocity $-c$ thereby bringing the wave form to rest in space.

Let
$$\left.\begin{aligned}\phi&=-(V-c)\,x+A\cosh m\,(y+h)\cos mx\\ \psi&=-(V-c)\,y-A\sinh m\,(y+h)\sin mx\end{aligned}\right\} \quad \text{.........(1)}$$
and

relate to the lower liquid, and

$$\left.\begin{aligned}\phi'&=-(V'-c)\,x+A'\cosh m\,(y-h')\cos mx\\ \psi'&=-(V'-c)\,y-A'\sinh m\,(y-h')\sin mx\end{aligned}\right\} \quad \text{......(2)}$$
and

* See Lamb's *Hydrodynamics*, § 233, from which this article is taken.

relate to the upper. These expressions for ψ and ψ' clearly make the boundaries $y=-h$, $y=h'$ stream lines; and if $\eta=a\sin mx$ gives the displacement of the common surface and the liquids do not separate this must be a stream line for both surfaces. We can satisfy this condition by taking the stream line to be $\psi=\psi'=0$, which gives

$$\left.\begin{aligned}-(V-c)\,a-A\ \sinh mh&=0\\ \text{and}\qquad -(V'-c)\,a+A'\sinh mh'&=0\end{aligned}\right\}\quad\text{...........(3)},$$

neglecting the squares of small quantities.

The expressions for the pressure are

$$\frac{p}{\rho}+gy+\tfrac{1}{2}\left\{\left(\frac{\partial\phi}{\partial x}\right)^2+\left(\frac{\partial\phi}{\partial y}\right)^2\right\}=\text{const.},$$

and

$$\frac{p'}{\rho'}+gy+\tfrac{1}{2}\left\{\left(\frac{\partial\phi'}{\partial x}\right)^2+\left(\frac{\partial\phi'}{\partial y}\right)^2\right\}=\text{const.}$$

At the common surface, neglecting a^2, these become

$$\frac{p}{\rho}+ga\sin mx+\tfrac{1}{2}(V-c)^2(1-2am\coth mh\sin mx)=\text{const.},$$

$$\frac{p'}{\rho'}+ga\sin mx+\tfrac{1}{2}(V'-c)^2(1+2am\coth mh'\sin mx)=\text{const.},$$

and $p=p'$.

Hence we must have

$$g(\rho-\rho')=(V-c)^2 m\rho\coth mh+(V'-c)^2 m\rho'\coth mh'\ \text{...(4)}.$$

This equation determines the velocity of propagation c of waves of length $2\pi/m$ at the common surface of two streams whose velocities are V, V'; but it may also be regarded as the condition for stationary waves at the common surface of two streams whose velocities are $V-c$ and $V'-c$.

It should be noticed that in any such case as the above, even when V and V' are both zero, the tangential velocities on opposite sides of the surface of separation are different so that this surface constitutes a vortex sheet.

10·43. Special Cases. (i) If the liquids are at rest save for the wave motion, the wave velocity is given by

$$c^2=\frac{g}{m}\,\frac{\rho-\rho'}{\rho\coth mh+\rho'\coth mh'}\ \text{.....................(1)}.$$

Since there is no real value for c when $\rho'>\rho$, this indicates that when $\rho'>\rho$ the equilibrium position is unstable.

(ii) If in addition the depths of the liquid are so large compared to the wave length that we may put $\coth mh = \coth mh' = 1$, then

$$c^2 = \frac{g}{m}\frac{\rho - \rho'}{\rho + \rho'} \quad \text{.............................(2).}$$

(iii) The foregoing results obtained for incompressible liquids will be applicable to the case of waves propagated along the surface of water exposed to the air, provided that in considering the effect of the air we neglect terms which, in comparison with those retained, are of the order of the ratio of the lengths of the waves considered to the length of a wave of sound of the same period in air. Thus, in (1), making $h' = \infty$ we have

$$c^2 = \frac{g}{m} \tanh mh \left\{1 - (1 + \tanh mh)\frac{\rho'}{\rho}\right\}, \text{ approx.} \quad \text{........(3).}$$

These results were obtained by Stokes*.

10·44. It has been shewn by Greenhill† that if the velocities V, V' of the currents make angles α, α' with the direction of wave propagation, equation (4) of **10·42** only needs modifying by the insertion of $V\cos\alpha$, $V'\cos\alpha'$ instead of V, V', the components $V\sin\alpha$, $V'\sin\alpha'$ of the currents perpendicular to the direction of propagation of the waves having no effect upon the determination of c.

10·45. Upper Surface free. Another case of interest is that in which the surface of the upper liquid is free; e.g. a layer of oil upon water or of fresh water upon salt water.

With the notation of **10·42** but assuming the liquids to be at rest save for the wave motion, we assume a common velocity of wave propagation c at the free surface of the upper liquid and at the common surface and reverse this velocity on the whole mass so that the motion becomes steady. We may then take

$$\psi = cy - A \sinh m\,(y + h) \sin mx \quad \text{...........(1)}$$

in the lower liquid, and

$$\psi' = cy - (B \cosh my + C \sinh my) \sin mx \quad \text{........(2)}$$

in the upper.

The bottom $y = -h$ is then a stream surface $\psi = -ch$, and if the common surface is

$$\eta = a \sin mx \quad \text{.......................(3)}$$

it is also the stream surface $\psi = \psi' = 0$,

if
$$\left.\begin{aligned} ca - A \sinh mh &= 0 \\ ca - B &= 0 \end{aligned}\right\} \quad \text{.....................(4).}$$
and

Also the free surface $\quad y = h' + b \sin mx$ ....(5)

* 'On the Theory of Oscillatory Waves', *Trans. Camb. Phil. Soc.* VIII, p. 441, or *Math. and Phys. Papers*, I, p. 197.

† 'Hydromechanics', *Encyc. Brit.* 11th edition.

is a stream surface $\psi' = \text{const.}$ if

$$cb - (B \cosh mh' + C \sin mh') = 0 \quad \text{.........(6).}$$

The equations for the pressure in the lower and upper liquids are

$$\frac{p}{\rho} + gy + \tfrac{1}{2}\left\{\left(\frac{\partial \psi}{\partial x}\right)^2 + \left(\frac{\partial \psi}{\partial y}\right)^2\right\} = \text{const.} \quad \text{.........(7)}$$

and

$$\frac{p'}{\rho'} + gy + \tfrac{1}{2}\left\{\left(\frac{\partial \psi'}{\partial x}\right)^2 + \left(\frac{\partial \psi'}{\partial y}\right)^2\right\} = \text{const.} \quad \text{.........(8).}$$

Substituting from (1) and (2), noting that A, B and C are of order a, neglecting squares of small quantities and equating the values of p and p' at the common surface, we get

$$ga(\rho - \rho') - cm(\rho A \cosh mh - \rho' C) = 0 \quad \text{.........(9);}$$

and, using (4) and (6), this gives

$$g(\rho - \rho') = c^2 m\left\{\rho \coth mh + \rho' \coth mh' - \rho' \frac{b}{a} \operatorname{cosech} mh'\right\} \text{...(10).}$$

Then using the fact that p' is constant at the free surface we get

$$gb = cm(B \sinh mh' + C \cosh mh'),$$

and, from (4) and (6),

$$g = c^2 m\left(\coth mh' - \frac{a}{b} \operatorname{cosech} mh'\right) \quad \text{.........(11).}$$

The elimination of the ratio $a:b$ from (10) and (11) gives the equation for c, viz.

$$c^4 m^2 (\rho \coth mh \coth mh' + \rho')$$
$$- c^2 m g \rho (\coth mh + \coth mh') + g^2(\rho - \rho') = 0 \text{ ...(12);}$$

and the ratio of the amplitudes of the waves is given from (11) by

$$\frac{b}{a} = \frac{c^2 m}{c^2 m \cosh mh' - g \sinh mh'} \quad \text{.........(13).}$$

From (12) we see that there are two possible velocities of propagation for a given wave length, provided $\rho > \rho'$.

In the particular case in which the lower liquid is 'deep' we put $\coth mh = 1$. The roots of (12) are then

$$c^2 = \frac{g}{m} \text{ and } c^2 = \frac{g}{m} \frac{\rho - \rho'}{\rho \coth mh' + \rho'},$$

the forms being in agreement with the case dealt with in **10·41**. The ratios of the amplitudes of the upper and lower waves in the two cases are

$$e^{mh'} \text{ and } -\left(\frac{\rho}{\rho'} - 1\right) e^{-mh'}.$$

10·46. Stability. The motion considered in **10·42** is really a case of small oscillations about a state of steady motion. To examine the stability of the motion, we have a quadratic equation (4) for the velocity of wave propagation c and we require that the roots of this quadratic should be real.

The condition for real or imaginary roots in c is

$$m^2 (V\rho \coth mh + V'\rho' \coth mh')^2 \gtrless m(\rho \coth mh + \rho' \coth mh')$$
$$\times \{m\rho V^2 \coth mh + m\rho' V'^2 \coth mh' - g(\rho - \rho')\},$$

or $$g(\rho-\rho')(\rho \coth mh + \rho' \coth mh')$$
$$\gtrless m\rho\rho' \coth mh \coth mh' (V - V')^2.$$

This means that the stream motion is stable or unstable according as

$$(V - V')^2 \lessgtr \frac{\rho \coth mh + \rho' \coth mh'}{\rho\rho' \coth mh \coth mh'} \cdot \frac{g(\rho - \rho')}{m}.$$

We remark that if $\rho < \rho'$, that is, if the upper liquid is denser than the lower, there is instability for all wave lengths. The same is true when $\rho = \rho'$, that is when two streams of the same liquid are flowing with different velocities and a horizontal common surface.

In fact when $\rho = \rho'$ and the depths are so great that

$$\coth mh = \coth mh' = 1,$$

we get $$c = \tfrac{1}{2}\{(V + V') \pm i(V - V')\}.$$

We may consider the case $V = V'$ by first putting $V' = V(1 + \alpha)$ and then making α tend to zero.

The common surface in the steady motion being given by $\eta = a \sin mx$, for progressive waves the corresponding form is $\eta = a \sin(mx - nt)$, when

$$n = mc = \tfrac{1}{2} mV \{2 + \alpha \mp i\alpha\}.$$

Hence $$\eta = a \sin m \{x - Vt - \tfrac{1}{2}\alpha(1 \mp i) Vt\},$$

and as α tends to zero we may write this

$$\eta = a \sin m(x - Vt) - \tfrac{1}{2} am\alpha(1 \pm i) Vt \cos m(x - Vt),$$

or $$\eta = a \sin m(x - Vt) - bmVt \cos m(x - Vt).$$

This shews that the corrugations of the surface increase in height indefinitely with t.

This case is of special interest as it explains the flapping of sails and flags. The uniform medium can be regarded as divided by a thin membrane on both sides of which the medium moves with the same velocity, the motion is unstable and a slight disturbance will result in a larger departure from the steady motion. This and other cases were considered by Lord Rayleigh in a paper 'On the Instability of Jets*'.

10·5. Group Velocity. In general when waves are started by a local disturbance such as, for example, the dropping of a stone into a pond or the motion of a boat through water, the successive waves have different lengths and are propagated with different velocities. Let us examine the phenomena that arise from the simultaneous motion in the same direction over the same water of two simple harmonic trains of waves of the same amplitude and slightly different wave lengths.

We may write for the elevation at any point

$$\eta = a\sin(mx-nt)+a\sin(m'x-n't)$$
$$= 2a\cos\tfrac{1}{2}\{(m-m')x-(n-n')t\}\sin\tfrac{1}{2}\{(m+m')x-(n+n')t\}.$$

If $m=m'$ nearly, $(m-m')x$ varies with x much more slowly than does $(m+m')x$, so it is convenient at any instant to regard the equation as representing a sinuous curve obtained by drawing the curve $\eta = 2a\sin\frac{1}{2}\{(m+m')x-(n+n')t\}$ and multiplying the ordinates by $\cos\frac{1}{2}\{(m-m')x-(n-n')t\}$. Hence the result represents a train of waves whose amplitude

$$2a\cos\tfrac{1}{2}\{(m-m')x-(n-n')t\}$$

is periodic, varying between 0 and $2a$. The profile of this train will be a group of sinuosities of amplitude gradually increasing from zero to $2a$ and then decreasing to zero followed by a succession of equal groups. The appearance on the water will be that of alternate groups of waves separated by intervals of nearly still water.

The distance between the centres of two successive groups is $2\pi/(m-m')$ and the time occupied in moving this distance is $2\pi/(n-n')$ so that the velocity of propagation of the groups is given by

$$U = \frac{n-n'}{m-m'},$$

or

$$U = \frac{dn}{dm} \quad\text{.............................(1)†},$$

* *Proc. L.M.S.* x, 1879, p. 4, or *Sci. Papers*, I, p. 361. On the general question of stability and instability of a perfect fluid see a paper by W. McF. Orr, *Proc. R.I.A.* XXVII, p. 9.

† The theory of group velocity is generally attributed to Stokes, who set a question on it in the Smith's Prize Examination in 1876, *Math. and Phys. Papers*, V, p. 362, but the result (1) appears to have been obtained first by Hamilton in a paper on 'Researches respecting vibration connected with the Theory of Light', *Proc. R.I.A.* I, p. 341. For this reference the author is indebted to Professor Sir Joseph Larmor.

when the difference of the wave lengths of the original trains is small.

But the velocity of propagation of a single wave is

$$c=\frac{n}{m};$$

therefore $$U=\frac{d}{dm}(mc)=c+m\frac{dc}{dm} \quad\text{.................(2);}$$

or if λ be the wave length $(2\pi/m)$,

$$U=c-\lambda\frac{dc}{d\lambda} \quad\text{.......................(3).}$$

Thus it appears that the group velocity, in general, differs from the velocity of propagation of the separate waves. This is in accordance with the results of observation, for when the eye views a group of waves advancing over deep sea water, single waves are seen to advance through the group, their amplitudes increasing and then dying away as they give place to others.

In the case of waves on the surface of water of depth h, we have

$$c^2=(g/m)\tanh mh,$$

so that $$U=\tfrac{1}{2}c\,(1+2mh\operatorname{cosech} 2mh).$$

Hence the ratio of the group velocity to the wave velocity is $\frac{1}{2}+\frac{mh}{\sinh 2mh}$. When h is small compared with the wave length this ratio is unity, and as h increases to infinity the ratio decreases to $\frac{1}{2}$; or the group velocity for deep sea waves is half the wave velocity.

We shall see later that the group phenomenon is not peculiar to water waves, but occurs in sound waves where the phenomenon is known as 'beats'. It can exist in all forms of waves.

10·51. The theory of group velocity has been treated in a more general manner by Lord Rayleigh*. We assume that a disturbance travelling in one dimension can be resolved by Fourier's theorem into infinite trains of waves of harmonic type and of various amplitudes and wave lengths. Thus the only case in which we can expect a simple result is that in which a considerable number of consecutive waves are sensibly of a given harmonic type, though the wave length and amplitude may vary within moderate limits at points whose distance amounts to a large multiple of λ.

* 'On the Velocity of Light', *Nature*, xxv, p. 52, or *Sci. Papers*, I, p. 540.

Assuming that the complete expression by Fourier's series involves only wave lengths which differ but little from one another, we may write

$$\eta = a_1 \sin\{(m+\delta m_1)x-(n+\delta n_1)t+\epsilon_1\}$$
$$+a_2\sin\{(m+\delta m_2)x-(n+\delta n_2)t+\epsilon_2\}+\ldots$$
$$=\sin(mx-nt)\,\Sigma a_1\cos(x\,\delta m_1-t\,\delta n_1+\epsilon_1)$$
$$+\cos(mx-nt)\,\Sigma a_1\sin(x\,\delta m_1-t\,\delta n_1+\epsilon_1).$$

Also by hypothesis $\dfrac{\delta n_1}{\delta m_1}=\dfrac{\delta n_2}{\delta m_2}=\ldots=\dfrac{dn}{dm}$,

and the first term in the expression for η represents a simple train of type $\sin(mx-nt)$ with varying amplitude $\Sigma a_1\cos(x\,\delta m_1-t\,\delta n_1+\epsilon_1)$, and the amplitude itself is propagated as a wave with velocity dn/dm; and similarly the second term. Hence we arrive at the idea of groups of waves of a more general kind, but the velocity of propagation is given by the same formula as in the special case considered in **10·5**.

10·52. Transmission of Energy. We have seen in **10·38** how to calculate the energy of a progressive wave. In a progressive wave the wave form advances with a definite velocity but it does not follow that this is the rate of transmission of energy, for it is the particles of water that possess the energy and there is no reason to suppose that they hand on the energy at the same rate as the wave form advances. This question was discussed by Prof. Osborne Reynolds, in a paper* from which we borrow some illustrations: If a number of small balls are suspended by threads so that the balls all hang in a row, the threads being of the same length; and if the balls be then set swinging in succession in planes perpendicular to the row, as by running the finger along them, the motion will present the appearance of a series of waves propagated from one end of the row to the other, but in reality each pendulum swings independently of its neighbour and there is *no communication of energy*. If however the balls are connected by an elastic string and any one be given a transverse motion, it will communicate its motion to the others, so that now there is a transmission of energy and the rate at which the first ball gives up energy to the others will clearly depend on the tension of the string.

As another illustration: If a rope be laid out on the ground in a straight line with one end fixed and an upward jerk be given to the other end, a wriggle will travel along the rope to the other end leaving the rope straight and at rest on the ground behind it.

* 'On the Rate of Progression of Groups of Waves and the Rate at which Energy is Transmitted by Waves', *Nature*, XVI, 1877, p. 343.

This is a case in which the energy is transmitted at the same rate as the wave.

The particular case with which we are concerned, that of surface waves on water, is a case intermediate between the two just considered; energy is transmitted but at a rate less than the wave velocity.

10·53. Rate of Transmission of Energy in simple harmonic Surface Waves. The rate of transmission of energy is measured by taking a vertical section of the liquid at right angles to the direction of propagation and determining the rate at which the pressure on one side of this section is doing work on the liquid on the other side.

Considering liquid of depth h, we have, as in **10·31**,

$$\phi = \frac{ga}{n}\frac{\cosh m(y+h)}{\cosh mh}\cos(mx-nt).$$

And neglecting squares of small quantities the variable part of the pressure is given by $\delta p = \rho\,\partial\phi/\partial t$,

and the horizontal velocity is $-\partial\phi/\partial x$.

Hence the work done in unit time or the energy carried across unit width of the section is

$$W = -\int_{-h}^{0}\delta p\frac{\partial\phi}{\partial x}dy \quad \text{.......................................(1)}$$

$$= \frac{g^2\rho a^2 m}{n}\frac{\sin^2(mx-nt)}{\cosh^2 mh}\int_{-h}^{0}\cosh^2 m(y+h)\,dy$$

$$= \frac{g^2\rho a^2 m}{n}\frac{\sin^2(mx-nt)}{\cosh^2 mh}\left(\frac{\sinh 2mh}{4m}+\frac{h}{2}\right),$$

and since $n^2 = gm\tanh mh$, this may be written

$$W = \tfrac{1}{2}g\rho a^2\frac{n}{m}(1+2mh\,\mathrm{cosech}\,2mh)\sin^2(mx-nt) \quad \text{...(2)}.$$

We note that in (1) the integral should be taken between the limits $-h$ and η, but the range 0 to η will only add a term in a^3 to the result.

The mean value of the expression (2) over a complete period or any number of complete periods, or any interval that is so long compared to a period that the part corresponding to the fractional part of a period can be neglected in comparison with the whole, is

$$\tfrac{1}{4}g\rho a^2 \frac{n}{m}(1+2mh\operatorname{cosech}2mh)^*.$$

Referring to **10·5**, since $n/m=c$, this expression for the energy transmitted in unit time is equal to

$$\tfrac{1}{2}g\rho a^2 \times \text{group velocity}.$$

And from **10·38**, $\frac{1}{2}g\rho a^2$ is the whole energy per unit length at any instant. *Hence the energy is transmitted at a rate equal to the group velocity.*

10·6. Capillary Waves. When surface tension is taken into account, the surface conditions $p=\text{const.}$ (**10·3**) and $p=p'$ (**10·42**) no longer hold good. They must be replaced by the condition that, if T denotes the surface tension or energy per unit area due to capillary forces, the difference of the pressures on opposite sides of the surface is given by†

$$T\left(\frac{1}{\rho}+\frac{1}{\rho'}\right),$$

where ρ and ρ' are the principal radii of curvature of the surface.

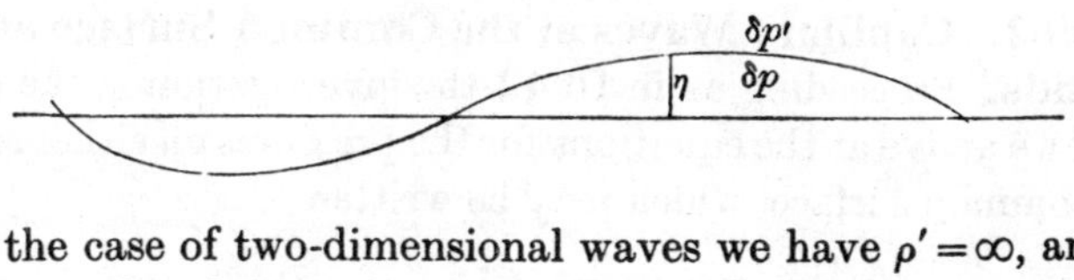

In the case of two-dimensional waves we have $\rho'=\infty$, and, if η denote the elevation, $1/\rho=-d^2\eta/dx^2$, neglecting squares of small quantities. So if δp, $\delta p'$ denote the variable parts of the pressure below and above the surface, as in the figure, we have

$$T\frac{d^2\eta}{dx^2}+\delta p-\delta p'=0 \quad \text{.....................(1)}$$

as the surface condition.

10·61. Capillary Waves on a canal of Uniform Depth. Taking the case considered in **10·31** and **10·4**, let us use the method of **10·4**, reducing the problem to one of steady motion

* Lord Rayleigh, 'On Progressive Waves', *Proc. L.M.S.* IX, 1877, p. 21, or *Sci. Papers*, I, p. 322, or *Theory of Sound*, I, Appendix.

† *Vide Hydrostatics*, Art. 101.

by superposing a velocity $-c$ on the whole mass, where c is the velocity of propagation. As in **10·4**, we have

$$\psi = cy - A \sinh m(y+h) \sin mx,$$

and for the free surface $\quad \eta = a \sin mx,$

provided $\quad ca - A \sinh mh = 0.$

And the variable part of the pressure is given by

$$\frac{\delta p}{\rho} + ga \sin mx + \tfrac{1}{2}c^2 (1 - 2ma \coth mh \sin mx) = \text{const.}$$

But from **10·6** (1), since in this case we regard the air pressure as constant, we have

$$\delta p = -T \frac{d^2\eta}{dx^2} = Tam^2 \sin mx.$$

Substituting this value in the last equation and equating to zero the coefficient of $\sin mx$, we get

$$c^2 = \left(\frac{g}{m} + \frac{Tm}{\rho}\right) \tanh mh \quad \text{.................(1).}$$

When h is large compared to the wave length this becomes

$$c^2 = \frac{g}{m} + \frac{Tm}{\rho} \quad \text{.......................(2).}$$

10·62. Capillary Waves at the Common Surface of two Liquids. Proceeding as in **10·42** the investigation is the same until we arrive at the equations for the pressures on either side of the common surface, which may be written

$$\frac{\delta p}{\rho} + ga \sin mx + \tfrac{1}{2}(V-c)^2 (1 - 2am \coth mh \sin mx) = \text{const.},$$

and

$$\frac{\delta p'}{\rho'} + ga \sin mx + \tfrac{1}{2}(V'-c)^2 (1 + 2am \coth mh' \sin mx) = \text{const.},$$

where $\quad T\dfrac{d^2\eta}{dx^2} + \delta p - \delta p' = 0, \text{ and } \eta = a \sin mx.$

Hence $\quad \delta p - \delta p' = Tam^2 \sin mx,$

and by eliminating δp, $\delta p'$, we get

$$Tm^2 + g(\rho - \rho') = (V-c)^2 m\rho \coth mh + (V'-c)^2 m\rho' \coth mh' \quad \text{......(1).}$$

As a special case, if the liquids are so deep compared to the wave length that we may put $\coth mh = \coth mh' = 1$, and the liquids are undisturbed save for the wave motion, then the velocity of propagation c_0 is given by

$$c_0^2 = \frac{g}{m}\frac{\rho-\rho'}{\rho+\rho'} + \frac{Tm}{\rho+\rho'} \quad \text{.....................(2).}$$

Again we get the case of the effect of wind on deep water, regarding air as incompressible, if we retain V' but put $V=0$, (1) reducing to

$$Tm + (\rho-\rho')g/m = c^2\rho + (V'-c)^2\rho',$$

or

$$c^2 - \frac{2\rho'}{\rho+\rho'}V'c + \frac{\rho'}{\rho+\rho'}V'^2 - c_0^2 = 0,$$

where c_0 denotes the velocity of propagation when there is no wind.

This gives

$$c = \frac{\rho' V'}{\rho+\rho'} \pm \left\{c_0^2 - \frac{\rho\rho' V'^2}{(\rho+\rho')^2}\right\}^{\frac{1}{2}} \quad \text{..............(3).}$$

This result was obtained by Lord Kelvin*, who considered some special cases as follows: For a given wave length $2\pi/m$, the wave velocity c is greatest when the wind velocity $V' = c_0(1+\rho'/\rho)^{\frac{1}{2}}$, c having then the same value as V'. Hence it follows that 'with wind of any other speed than that of the waves, their speed is less. For instance, the wave speed with no wind, which is c_0, is less by approximately $\rho'/2\rho$ of c_0 (i.e. about $\frac{1}{1650}$ of c_0) than the speed when the wind is with the waves and of their speed. The explanation clearly being that when the air is motionless relatively to the wave crests and hollows its inertia is not called into play'.

From (3) we draw the following conclusions:

'(1) When $\quad V'/c_0 = \left(1+\frac{\rho}{\rho'}\right)^{\frac{1}{2}} = 28\cdot7\,(1+\frac{1}{1650})$,

one of the values of c is zero, that is to say, static corrugations of wave length $2\pi/m$, would be equilibrated by wind of velocity $c_0(1+\rho/\rho')^{\frac{1}{2}}$.

But the equilibrium would be unstable.

(2) When $\quad V'/c_0 = (\rho+\rho')/(\rho\rho')^{\frac{1}{2}} = 28\cdot7\,(1+\frac{1}{825})$,

the two values of c are equal.

(3) When $\quad V'/c_0 > (\rho+\rho')/(\rho\rho')^{\frac{1}{2}}$,

both values of c are imaginary, and therefore the wind would blow into spin-drift waves of length $2\pi/m$ or shorter.

Looking back to (2), we see that it gives a minimum value for c_0 equal to

$$\sqrt{\frac{2\sqrt{gT(1-\rho'/\rho)}}{1+\rho'/\rho}}.$$

Hence the water with a plane level surface would be unstable, even if air were frictionless, when the velocity of the wind exceeds

$$\sqrt{\frac{2\sqrt{gT(1-\rho'^2/\rho^2)}}{\rho'/\rho}},$$

* Letter to Prof. Tait, August 16, 1871. Printed in *Math. and Phys. Papers*, IV, p. 76, also in *Baltimore Lectures*, p. 590.

10·63. Ripples. Referring to the case of **10·61** the wave velocity c is given by

$$c^2 = \frac{g}{m} + \frac{Tm}{\rho} \quad \text{.........................(1)},$$

from which we see that there are in general two values of m corresponding to a given value of c; i.e. two different wave lengths corresponding to a given wave velocity.

Also, since (1) can be written

$$c^2 = \left\{\sqrt{\frac{g}{m}} - \sqrt{\left(\frac{Tm}{\rho}\right)}\right\}^2 + 2\sqrt{\left(\frac{gT}{\rho}\right)} \quad \text{............(2)},$$

therefore c^2 has a unique minimum value $2\sqrt{(gT/\rho)}$ corresponding to the value $m = \sqrt{(g\rho/T)}$. We shall denote these special values by c_m and m_m.

Again the frequency $n/2\pi$ of the oscillations is given by

$$n^2 = c^2m^2 = gm + Tm^3/\rho \quad \text{..................(3)}.$$

It follows that as the wave length $2\pi/m$ decreases from ∞ to 0 the frequency $n/2\pi$ continually increases, but the wave velocity c decreases to a minimum value c_m and then increases again; i.e. waves cannot be propagated at less than a certain minimum velocity in terms of g, T and ρ.

Again writing (1) in the form

$$c^2 = \frac{g\lambda}{2\pi} + \frac{2\pi T}{\lambda\rho} \quad \text{......................(4)},$$

and putting $\lambda_m = 2\pi/m_m$, it is clear that the product of the roots of this quadratic in λ is $\lambda_m{}^2$, so that corresponding to any value of c greater than c_m there are two values of λ one greater and the other less than λ_m. Also for large values of λ the gravitational term on the right of (4) preponderates, and for small values of λ the second term depending on surface tension is the more effective. Lord Kelvin defined a *ripple* as any wave on water whose length is less than the critical value λ_m or $2\pi\sqrt{(T/g\rho)}$. Thus the ripple length corresponding to a given velocity is the smaller root of (4).

At the common surface of two fluids exactly similar considerations follow from the consideration of the roots of equation **10·62** (2).

Ripples may be seen in front of any solid in motion cutting the surface of water. If ρ' denotes the density of air and ρ that of water, the ripple length is the smaller root of the quadratic

$$\frac{g\lambda}{2\pi}\frac{\rho-\rho'}{\rho+\rho'} + \frac{2\pi T}{\lambda(\rho+\rho')} = c_0{}^2 \quad \text{......................(5)},$$

where c_0 is the velocity of the solid. 'The latter may be a sailing-vessel or a row-boat, a pole held vertically and carried horizontally, an ivory pencil-case, a penknife-blade, either edge or flat side foremost, or (best) a fishing-line kept approximately vertical by a lead weight hanging down below water, while carried along at about half a mile per hour by a becalmed vessel*.'

Again the group velocity U is given (**10·5** (3)) by

$$U = c - \lambda \frac{dc}{d\lambda},$$

and from (4) this gives

$$U = c\left\{1 - \frac{1}{2}\frac{\frac{g\lambda}{2\pi} - \frac{2\pi T}{\lambda\rho}}{\frac{g\lambda}{2\pi} + \frac{2\pi T}{\lambda\rho}}\right\},$$

so that U is greater or less than c according as λ is less or greater than λ_m. We have also the limiting values

$$U = \tfrac{1}{2}c \text{ for waves of great length,}$$

$$U = \tfrac{3}{2}c \text{ for the shortest ripples.}$$

10·7. Waves due to a given Local Disturbance on the Surface of Water. We shall consider first a simple case where the liquid is limited by vertical planes, distant l apart, parallel to the crests of the waves, and suppose that the motion starts from rest with a given initial elevation

$$\eta = f(x).$$

The motion is therefore irrotational and if the liquid were unlimited in extent there would be no limitation on the lengths of the waves but the motion would be the result of the superposition of waves of infinite variety of lengths. In this case, as we shall see, there is a limitation on the possible wave lengths. If h be the depth of the liquid, a suitable solution of Laplace's equation for the velocity potential is

$$\phi = A\cosh m(y+h)\cos mx \sin nt, \quad \text{where } n^2 = mg\tanh mh,$$

making ϕ zero when $t=0$, also when $y=-h$. But we also require that $\partial\phi/\partial x = 0$ when $x=0$ and when $x=l$; and this makes $\sin ml = 0$, or $ml = s\pi$, where s is an integer.

Again the pressure equation

$$\frac{p}{\rho} = \frac{\partial\phi}{\partial t} - gy - \tfrac{1}{2}q^2 + F(t)$$

* Letter from Lord Kelvin to Prof. Tait, of date August 23, 1871, *loc. cit.* p. 287.

gives initially at the free surface

$$\frac{\partial\phi}{\partial t}=g\eta=gf(x).$$

And the most general expression for ϕ is

$$\phi=\sum_{s=1}^{\infty}A_s\cosh\frac{s\pi}{l}(y+h)\cos\frac{s\pi x}{l}\sin nt,$$

and, substituting this value in the last equation, we get

$$\sum_{s=1}^{\infty}nA_s\cosh\frac{s\pi h}{l}\cos\frac{s\pi x}{l}=gf(x).$$

But by Fourier's Theorem we have

$$f(x)=\frac{1}{l}\int_0^l f(v)\,dv+\frac{2}{l}\sum_{s=1}^{\infty}\cos\frac{s\pi x}{l}\int_0^l f(v)\cos\frac{s\pi v}{l}\,dv,$$

and by equating the coefficients of $\cos s\pi x/l$ in the two series we get

$$nA_s\cosh\frac{s\pi h}{l}=\frac{2g}{l}\int_0^l f(v)\cos\frac{s\pi v}{l}\,dv,$$

so that

$$\phi=\frac{2g}{l}\sum_{s=1}^{\infty}\frac{\cosh\dfrac{s\pi(y+h)}{l}}{n\cosh\dfrac{s\pi h}{l}}\cos\frac{s\pi x}{l}\int_0^l f(v)\cos\frac{s\pi v}{l}\,dv\sin nt,$$

where $$n^2=\frac{s\pi g}{l}\tanh\frac{s\pi h}{l}.$$

If we require the form of the surface at any subsequent time, the relation

$$\dot\eta=-\left(\frac{\partial\phi}{\partial y}\right)_{y=0}$$

gives $$\eta=\frac{2}{l}\sum_{s=1}^{\infty}\cos\frac{s\pi x}{l}\int_0^l f(v)\cos\frac{s\pi v}{l}\,dv\cos nt.$$

10·71. We may now consider the case in which the liquid is unlimited in extent, the initial disturbance being of the same type as before, that is, given by

$$\eta=f(x),$$

so that we are still dealing with two-dimensional motion. To simplify the expressions we shall suppose the depth of the liquid to be infinite, then from **10·35** we can write down as a typical solution for a wave of length $2\pi/m$ the equations

$$\eta = \frac{\sin}{\cos} mx \cos nt,$$

and $$\phi = \frac{g}{n} e^{my} \frac{\sin}{\cos} mx \sin nt,$$

where $$n^2 = gm.$$

To obtain general expressions which embrace the superposition of all such solutions and give the initial values

$$\eta = f(x), \quad \phi = 0,$$

we must make use of Fourier's double integral theorem

$$f(x) = \frac{1}{\pi}\int_0^\infty dm \int_{-\infty}^{\infty} f(\alpha) \cos m(x-\alpha)\, d\alpha,$$

and the required expressions are

$$\eta = \frac{1}{\pi}\int_0^\infty dm \int_{-\infty}^{\infty} f(\alpha) \cos nt \cos m(x-\alpha)\, d\alpha,$$

$$\phi = \frac{g}{\pi}\int_0^\infty dm \int_{-\infty}^{\infty} f(\alpha) \frac{\sin nt}{n} e^{my} \cos m(x-\alpha)\, d\alpha;$$

for these expressions clearly satisfy all the conditions specified, and as an additional verification they make $\dot{\eta} = -(\partial\phi/\partial y)_{y=0}$, in virtue of the relation $n^2 = gm$.

10·72. A similar method may be adopted when the surface is initially horizontal but subject to initial impulsive pressure. Thus we may suppose that initially

$$\phi = F(x) \quad \text{and} \quad \eta = 0.$$

Then, taking as the typical solution

$$\phi = e^{my} \frac{\sin}{\cos} mx \cos nt,$$

$$\eta = -\frac{n}{g} \frac{\sin}{\cos} mx \sin nt,$$

where $$n^2 = mg,$$

we have for the general solution

$$\phi = \frac{1}{\pi}\int_0^\infty dm \int_{-\infty}^{\infty} F(\alpha) \cos nt\, e^{my} \cos m(x-\alpha)\, d\alpha,$$

$$\eta = -\frac{1}{g\pi}\int_0^\infty dm \int_{-\infty}^{\infty} F(\alpha)\, n \sin nt \cos m(x-\alpha)\, d\alpha.$$

For a full discussion of these results see Lamb's *Hydrodynamics*, §§ 238–240 and Lord Kelvin's papers on 'Deep-Water Waves*'.

10·8. Gerstner's Trochoidal Waves. An exact solution of the equations representing wave motion on the surface of deep water was discovered by Gerstner in 1802 and re-discovered by Rankine in 1863†, but the motion represented is *rotational* and cannot therefore be brought about by natural causes in frictionless liquid.

Consider the equations

$$\left.\begin{aligned} x &= a + \frac{1}{\kappa} e^{\kappa b} \sin \kappa (a+ct) \\ y &= b - \frac{1}{\kappa} e^{\kappa b} \cos \kappa (a+ct) \end{aligned}\right\} \quad \ldots\ldots\ldots\ldots\ldots\ldots(1),$$

where the Lagrangian notation is employed, a and b being parameters which specify a particular particle whose coordinates are x, y at time t.

Since $$\frac{\partial (x, y)}{\partial (a, b)} = 1 - e^{2\kappa b} \quad \ldots\ldots\ldots\ldots\ldots\ldots\ldots\ldots(2),$$

therefore the equation of continuity of **1·4** is satisfied. The equations of motion of **2·5**, in this case, become

$$\frac{1}{\rho}\frac{\partial p}{\partial a} + g\frac{\partial y}{\partial a} = -\frac{\partial^2 x}{\partial t^2}\frac{\partial x}{\partial a} - \frac{\partial^2 y}{\partial t^2}\frac{\partial y}{\partial a},$$

and $$\frac{1}{\rho}\frac{\partial p}{\partial b} + g\frac{\partial y}{\partial b} = -\frac{\partial^2 x}{\partial t^2}\frac{\partial x}{\partial b} - \frac{\partial^2 y}{\partial t^2}\frac{\partial y}{\partial b};$$

or $$\frac{\partial}{\partial a}\left(\frac{p}{\rho} + gy\right) = \kappa c^2 e^{\kappa b} \sin \kappa (a+ct),$$

and $$\frac{\partial}{\partial b}\left(\frac{p}{\rho} + gy\right) = \kappa c^2 e^{2\kappa b} - \kappa c^2 e^{\kappa b} \cos \kappa (a+ct).$$

If we multiply these equations by da, db, add and integrate we get

$$\frac{p}{\rho} = \text{const.} - g\left\{b - \frac{1}{\kappa} e^{\kappa b} \cos \kappa (a+ct)\right\}$$
$$- c^2 e^{\kappa b} \cos \kappa (a+ct) + \tfrac{1}{2} c^2 e^{2\kappa b} \quad \ldots(3).$$

* *Phil. Mag.* June, October 1904, June 1905, January 1907, or *Math. and Phys. Papers*, IV, pp. 338–456.

† 'On the Exact Form of Waves near the Surface of Deep Water', *Phil. Trans.* 1863, p. 127.

At the free surface the pressure must be constant, which requires that

$$c^2 = g/\kappa \quad \text{.........................(4).}$$

Now the periodic form of equations (1) shews that they represent a wave motion, the waves of length $2\pi/\kappa$ being propagated with velocity c in the negative direction of the x axis; and the relation (4) shews that the velocity is what we have previously found for deep-water waves.

If we substitute from (4) in (3) we get

$$\frac{p}{\rho} = \text{const.} - gb + \tfrac{1}{2}c^2 e^{2\kappa b} \quad \text{..................(5),}$$

shewing that p is constant when b is constant.

To shew that the motion is rotational, we have

$$\left.\begin{aligned} u &= \dot{x} = ce^{\kappa b}\cos\kappa(a+ct) \\ \text{and} \qquad v &= \dot{y} = ce^{\kappa b}\sin\kappa(a+ct) \end{aligned}\right\} \quad \text{..................(6),}$$

and the spin is given by $2\omega = \dfrac{\partial v}{\partial x} - \dfrac{\partial u}{\partial y}$.

But $$\frac{\partial v}{\partial x} = \frac{\partial(v, y)}{\partial(a, b)} \bigg/ \frac{\partial(x, y)}{\partial(a, b)} \quad \text{and} \quad \frac{\partial u}{\partial y} = \frac{\partial(x, u)}{\partial(a, b)} \bigg/ \frac{\partial(x, y)}{\partial(a, b)};$$

therefore $$2\omega \frac{\partial(x, y)}{\partial(a, b)} = \frac{\partial(v, y)}{\partial(a, b)} - \frac{\partial(x, u)}{\partial(a, b)},$$

and on substituting from (1), (2) and (6) we get

$$\omega = -c\kappa e^{2\kappa b}/(1 - e^{2\kappa b}) \quad \text{................(7).}$$

From (1) it is clear that the path of the particle (a, b) is a circle of radius $\kappa^{-1}e^{\kappa b}$.

The curves of equipressure are the paths of the particles when the motion is made steady by superposing the velocity $-c$, that is they are given by

$$x = a + \frac{1}{\kappa}e^{\kappa b}\sin\kappa a, \quad y = b - \frac{1}{\kappa}e^{\kappa b}\cos\kappa a,$$

or, putting $\kappa a = \theta$,

$$x = \kappa^{-1}\theta + \kappa^{-1}e^{\kappa b}\sin\theta, \quad y = b - \kappa^{-1}e^{\kappa b}\cos\theta.$$

These equations, for any constant value of b, represent a trochoid traced by a point at distance $\kappa^{-1}e^{\kappa b}$ from the centre of a circle of radius κ^{-1} which rolls on the under side of the line $y = b + \kappa^{-1}$. Any one such trochoid may be taken to represent a possible form of the free surface, the extreme case corresponding to $b = 0$ being a cycloid with cusps upwards*.

* For a diagram see Lamb's *Hydrodynamics*, § 251.

10·9. Miscellaneous Problems. *Stationary waves in running water. A stream flowing with uniform velocity over a corrugated bed whose section is a sine curve.*

The waves produced in a stream by obstacles or by inequalities in its bed have been discussed at length by Lord Rayleigh* and Lord Kelvin†.

Taking axes as usual, let the bed of the stream be given by

$$y = -h + k\sin mx,$$

and let V be the mean velocity.

The conditions of the problem will be satisfied by the equations

$$\phi = -Vx + (A\cosh my + B\sinh my)\cos mx \quad \text{........(1)},$$

and

$$\psi = -Vy - (A\sinh my + B\cosh my)\sin mx \quad \text{........(2)},$$

provided they make the bed a stream line and the free surface a surface of constant pressure as well as a stream line.

The condition that the bed

$$y = -h + k\sin mx$$

may be a stream line is that

$$-V(-h + k\sin mx) - (-A\sinh mh + B\cosh mh)\sin mx$$

may be constant for all values of x.

Therefore

$$kV = A\sinh mh - B\cosh mh \quad \text{....................(3)}.$$

If we assume for the free surface

$$\eta = a\sin mx \quad \text{..............................(4)},$$

this will be the stream line $\psi = 0$, provided

$$-Va - B = 0 \quad \text{..............................(5)}.$$

Again the pressure equation in the steady motion is

$$\frac{p}{\rho} + gy + \tfrac{1}{2}q^2 = \text{const.} \quad \text{........................(6)},$$

and at the free surface p is constant, so that by substitution from (1) and (4) in (6), neglecting squares of small quantities, we must have

$$ga\sin mx + VAm\sin mx = \text{const.}$$

for all values of x.

Therefore

$$ga + VAm = 0 \quad \text{..........................(7)},$$

and from (3), (5) and (7) we get A, B and a, and the free surface is given by

$$\eta = \frac{k}{\cosh mh - g/mV^2 . \sinh mh}\sin mx \quad \text{..............(8)}.$$

Taking k to be positive, the multiplier of $\sin mx$ in the last expression is positive or negative according as V^2 is greater or less than $(g/m)\tanh mh$. That is, according as V is greater or less than the velocity in still water of depth h of waves of the same length $2\pi/m$ as the corrugations. In the former case the ridges and hollows of the free surface are vertically over the ridges and hollows of the bed of the stream, and in the latter case the ridges of the free surface are over the hollows of the bed.

If $V^2 = (g/m)\tanh mh$ the amplitude a cannot be small and the hypothesis on which we have obtained the result (8) no longer holds good.

* 'The Form of Standing Waves on the surface of Running Water', *Proc. L.M.S.* xv, p. 69, or *Sci. Papers*, II, p. 258.

† 'On Stationary Waves in Flowing Water', *Phil. Mag.* Oct. 1886, or *Math. and Phys. Papers*, IV, p. 270.

10·91. *If water flows along a rectangular canal which consists of two uniform portions of slightly different breadths, with a gradual transition, the free surface will be lower where the canal is narrower, or contrariwise, according as $u^2 \lessgtr gh$, where u is the mean velocity, and h the mean depth. [The motion is supposed to be steady.]* (M.T. 1912.)

Let A, B denote points on the free surface of the two portions, $h+\alpha$, $h+\alpha'$ the depths, $b+\beta$, $b+\beta'$ the breadths, and $u+v$, $u+v'$ the velocities in the two portions, b denoting the mean breadth.

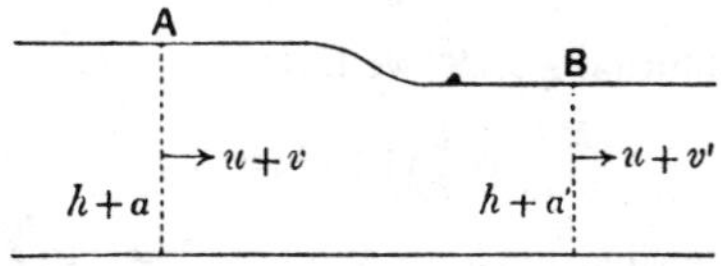

From continuity we have

$$(h+\alpha)(b+\beta)(u+v) = hbu = (h+\alpha')(b+\beta')(u+v').$$

Therefore $\quad v = -u\left(\frac{\beta}{b}+\frac{\alpha}{h}\right)$, and $v' = -u\left(\frac{\beta'}{b}+\frac{\alpha'}{h}\right)$.

If p, p' denote the pressures at A and B

$$\frac{p-p'}{\rho} = -(h+\alpha-h-\alpha')g - \tfrac{1}{2}\{(u+v)^2-(u+v')^2\}.$$

But the pressures at A and B on the free surface are equal, therefore

$$0 = -(\alpha-\alpha')g - u(v-v')$$

and

$$(\alpha-\alpha')\left(g-\frac{u^2}{h}\right) = \frac{u^2}{b}(\beta-\beta').$$

Hence $\alpha-\alpha'$ and $\beta-\beta'$ have the same or opposite signs according as $u^2 \lessgtr gh$, i.e. the free surface is lower where the canal is narrower or contrariwise according as $u^2 \lessgtr gh$.

10·92. Canal of Variable Section. With the notation of **10·2**, the same considerations give an equation of motion

$$\frac{\partial^2 \xi}{\partial t^2} = -g\frac{\partial \eta}{\partial x} \quad \text{..............................(1)}$$

and an equation of continuity

$$\left(A+\xi\frac{\partial A}{\partial x}+b\eta\right)\left(dx+\frac{\partial \xi}{\partial x}dx\right) = A\,dx,$$

since A is now a function of x, or

$$\frac{\partial}{\partial x}(A\xi)+b\eta = 0 \quad \text{..........................(2).}$$

Also, since A and b are functions of x only, (1) may be written

$$\frac{\partial^2 (A\xi)}{\partial t^2} = -gA\frac{\partial \eta}{\partial x} \quad \text{..........................(3),}$$

so that, by eliminating $A\xi$ from (2) and (3), we get

$$\frac{\partial^2 \eta}{\partial t^2} = \frac{g}{b}\frac{\partial}{\partial x}\left(A\frac{\partial \eta}{\partial x}\right) \quad \text{..........................(4).}$$

Example. *A canal of uniform depth h and length 2l is widest at the midpoint and tapers uniformly to a point at each end. Shew that two types of free oscillation can exist and that their periods $2\pi/\sigma$ are derived from the roots of the following equations:*

$$J_0\{l\sigma(gh)^{-\frac{1}{2}}\}=0 \quad \text{and} \quad J_1\{l\sigma(gh)^{-\frac{1}{2}}\}=0,$$

where J_0 and J_1 represent Bessel Functions of order 0 and 1. (M.T. 1923.)

In this case (4) becomes $\dfrac{\partial^2\eta}{\partial t^2}=\dfrac{g}{b}\dfrac{\partial}{\partial x}\left(hb\dfrac{\partial\eta}{\partial x}\right)$

and for a harmonic solution $\eta \propto e^{i\sigma t}$ we have

$$\frac{gh}{b}\frac{\partial}{\partial x}\left(b\frac{\partial\eta}{\partial x}\right)+\sigma^2\eta=0.$$

Now with the origin at one end $b \propto x$ and from $x=0$ to $x=l$ the equation becomes

$$\frac{\partial^2\eta}{\partial x^2}+\frac{1}{x}\frac{\partial\eta}{\partial x}+k^2\eta=0, \quad \text{where} \quad k^2=\sigma^2/gh.$$

This is Bessel's equation of order 0 and has a solution

$$\eta=CJ_0(kx).$$

It is clear that there may be oscillations in which η has equal and opposite values at corresponding points in the two halves of the canal and vanishes at the centre $x=l$, and for these $J_0(kl)=0$ or $J_0\{l\sigma(gh)^{-\frac{1}{2}}\}=0$. And there may also be oscillations which are symmetrical about the centre for which $\partial\eta/\partial x=0$ at $x=l$. For these $J_0'(kl)=0$, but $J_0'=-J_1$, so that

$$J_1\{l\sigma(gh)^{-\frac{1}{2}}\}=0.$$

EXAMPLES

1. Find the velocity of ocean rollers, 20 yards long from crest to crest, in miles per hour. (St John's Coll. 1901.)

2. The crests of rollers which are directly following a ship 220 ft. long are observed to overtake it at intervals of $16\frac{1}{2}$ seconds and it takes a crest 6 seconds to run along the ship. Find the length of the waves and the speed of the ship. (M.T. 1921.)

3. Find the type of waves that would travel on deep water at 30 knots. How much is the velocity of the waves affected by the presence of the atmosphere above the water, its density being ·0013? (St John's Coll. 1897.)

4. A fixed buoy in deep water is observed to rise and fall twenty times in a minute, prove that the velocity of the waves is about ten and a half miles per hour. (Coll. Exam. 1907.)

5. Shew that when irrotational waves of length λ are propagated in water of infinite depth, the pressure at any particle of the water is the same as it was in the equilibrium position of the particle when the water was at rest. (Coll. Exam. 1908.)

6. From considerations of dimensions alone shew that the period of oscillatory waves in a deep cylindrical tank varies as the square root of the diameter and inversely as the square root of the intensity of gravity. (M.T. 1879.)

7. If a horizontal rectangular canal of great depth has two vertical barriers at a distance l apart, prove that the periods of oscillation of the water are $2\sqrt{\pi l}/\sqrt{sg}$, where s is a positive integer; and that corresponding to any mode, all the particles of fluid oscillate in straight lines of length inversely proportional to $\exp(s\pi z/l)$, where z is the depth. (Coll. Exam. 1906.)

8. If in the irrotational motion of homogeneous liquid in two dimensions under gravity there be a free surface exposed to an atmosphere of constant pressure, shew that there must be a surface of equal pressure at which

$$g\frac{\partial\phi}{\partial y}+\frac{\partial^2\phi}{\partial t^2}-2\left\{\frac{\partial\phi}{\partial x}\frac{\partial^2\phi}{\partial x\,\partial t}+\frac{\partial\phi}{\partial y}\frac{\partial^2\phi}{\partial y\,\partial t}\right\}$$
$$-\left\{\left(\frac{\partial\phi}{\partial x}\right)^2\frac{\partial^2\phi}{\partial x^2}+2\frac{\partial\phi}{\partial x}\frac{\partial\phi}{\partial y}\frac{\partial^2\phi}{\partial x\,\partial y}+\left(\frac{\partial\phi}{\partial y}\right)^2\frac{\partial^2\phi}{\partial y^2}\right\}=0.$$

Work out the case $\phi=b\cos\pi x/a\cosh\pi(y+h)/a\sin pt$ and give it a possible physical realisation; b being so small that its square is negligible. (St John's Coll. 1906.)

9. When simple harmonic waves of length λ are propagated over the surface of deep water, prove that, at a point whose depth below the undisturbed surface is h, the pressure at the instants when the disturbed depth of the point is $h+\eta$ bears to the undisturbed pressure at the same point the ratio

$$1+\frac{\eta}{h}e^{-2\pi h/\lambda}:1,$$

atmospheric pressure and surface tension being neglected. (M.T. 1913.)

10. Let a shallow trough be filled with oil and water, and let the depth of the water be k and its density σ, and the depth of the oil h and its density ρ. Then shew that if g be gravity, and v the velocity of propagation of long waves,

$$v^2/g=\tfrac{1}{2}(h+k)+\tfrac{1}{2}\{(h-k)^2+4hk\rho/\sigma\}^{\frac{1}{2}}.$$

Note that there may be slipping between the two fluids. (M.T. 1882.)

11. Two fluids of densities ρ_1, ρ_2 have a horizontal surface of separation but are otherwise unbounded. Shew that when waves of small amplitude are propagated at their common surface, the particles of the two fluids describe circles about their mean positions; and that at any point of the surface of separation where the elevation is η, the particles on either side have a relative velocity $4\pi c\eta/\lambda$. (Trinity Coll. 1907.)

12. If a canal of rectangular section contain a depth h of liquid of density ρ on which is superposed a depth h' of liquid of density ρ', the free surface of the latter being exposed to constant atmospheric pressure, prove that the velocities of propagation of waves of length $2\pi/m$ are given by $c^2=gu/m$, where

$$\rho(u\coth mh-1)(u\coth mh'-1)=\rho'(1-u^2).$$

(Coll. Exam. 1907.)

13. Two-dimensional waves of length $2\pi/m$ are produced at the surface of separation of two liquids which are of densities ρ, ρ' ($\rho > \rho'$) and depths h, h' confined between two fixed horizontal planes. Prove that, if the potential energy is reckoned zero in the position of equilibrium, the total energy of the lower liquid is to that of the upper in the ratio

$$\rho\{(2\rho-\rho')\coth mh+\rho'\coth mh'\}:\rho'\{(\rho-2\rho')\coth mh'-\rho\coth mh\}.$$

(M.T. 1899.)

14. If there be two liquids in a straight canal of uniform section, of densities σ_1, σ_2 and depths l_1, l_2, shew that the velocity c of propagation of long waves is given by the equation

$$\left(\frac{c^2}{l_1 g}-1\right)\left(\frac{c^2}{l_2 g}-1\right)=\frac{\sigma_1}{\sigma_2},$$

where $\sigma_2 > \sigma_1$, and it is assumed that the liquids do not mix.

(St John's Coll. 1900.)

15. An open rectangular box of length a contains two liquids of densities ρ, ρ' and depths h, h' respectively, that of density ρ being at the bottom. Prove that the periods of oscillation when the liquids are slightly disturbed so that there is no motion perpendicular to the sides of the box are determined by equations of the type

$$\left(p^2\coth\frac{n\pi h}{a}-\frac{gn\pi}{a}\right)\left(p^2\coth\frac{n\pi h'}{a}-\frac{gn\pi}{a}\right)+\frac{\rho'}{\rho}\left(p^4-\frac{g^2n^2\pi^2}{a^2}\right)=0,$$

where n is an integer. (M.T. 1906.)

16. A layer of fluid of density ρ_2 and thickness h separates two fluids of densities ρ_1 and ρ_3, extending to infinity in opposite directions. If waves of length λ, large compared with h, be set up in the fluid, shew that their velocity of propagation is either

$$\left\{\frac{g\lambda}{2\pi}\frac{\rho_3-\rho_1}{\rho_3+\rho_1}\right\}^{\frac{1}{2}}\quad\text{or}\quad\left\{gh\frac{(\rho_2-\rho_1)(\rho_3-\rho_2)}{\rho_2(\rho_3-\rho_1)}\right\}^{\frac{1}{2}}.$$

(Trinity Coll. 1906.)

17. A canal, of infinite length and rectangular section, is of uniform depth h and breadth b in one part but changes gradually to uniform depth h' and breadth b' in another part. An infinite train of simple harmonic waves travelling in one direction only is propagated along the canal. Prove that, if a, a' are the heights and $2\pi/m$, $2\pi/m'$ the lengths of the waves in the two uniform portions,

$$m\tanh mh=m'\tanh m'h',$$

and

$$a^2b\,\text{sech}^2\,mh\,(\sinh 2mh+2mh)=a'^2b'\,\text{sech}^2\,m'h'\,(\sinh 2m'h'+2m'h').$$

(Coll. Exam. 1903.)

18. Shew that, if the velocity of the wind is just great enough to prevent the propagation of waves of length λ against it, the velocity of propagation of waves with the wind is $2c\{\sigma/(1+\sigma)\}^{\frac{1}{2}}$, where σ is the specific gravity of air and c is the wave velocity when no air is present.

(Coll. Exam. 1897.)

19. Find the velocity of straight ripples of length λ, on water of density ρ, surmounted by air of density ρ', as maintained by gravity and the surface tension τ, and if $\tau = 81$ c.g.s. for water, find for what wave length the velocity of propagation is least, and also the value of this minimum velocity. (St John's Coll. 1899.)

20. The velocity of propagation of capillary waves of length $2\pi/m$ along a uniform canal of depth h is c, and ρ is the density of the liquid. Shew that, if the waves are produced by a distribution of external surface pressure of the type $P \sin m(x - Vt)$ travelling with a velocity V greater than c, then the form of the surface is given by

$$\eta = a \sin m(x - Vt),$$

where

$$a = P \tanh mh/\rho m(V^2 - c^2).$$

What happens when (i) $V = c$, (ii) $V < c$? (M.T. 1930.)

21. If water of depth h be flowing with velocity proportional to the distance from the bottom, V being the velocity of the stream at the surface prove that the velocity U of propagation of waves in the direction of the stream is given by

$$(U - V)^2 + V(U - V)\, W^2/gh - W^2 = 0,$$

where W is the velocity of propagation in still water. (M.T. 1881.)

22. A stream of water is running steadily with uniform velocity U in a horizontal canal of depth h of which the bottom is slightly undulating: shew that there will be a depression η_2 in the steady free surface, above each elevation η_1 in the bottom, and *vice versa*, given by

$$\eta_2 = \eta_1 \left(\frac{gh}{U^2} - 1\right)^{-1}.$$

What happens as U^2 approaches and passes the value gh? Explain the general principle of which this is an example. (St John's Coll. 1899.)

23. Prove that, if a canal of rectangular section is terminated by two rigid vertical walls whose distance apart is $2a$, and if the water is initially at rest and has its surface plane and inclined at a small angle β to the length of the canal, the altitude η of the wave at any time t is given by

$$\eta = \frac{8a\beta}{\pi^2} \Sigma \frac{1}{s^2} (-1)^{\frac{s-1}{2}} \sin \frac{s\pi x}{2a} \cos \frac{s\pi ct}{2a},$$

where c is the velocity of a wave of length $4a/s$ on an infinitely long canal, and Σ implies summation for all odd integral values of s. (M.T. 1893.)

24. Find the possible periods of standing oscillations in a trough of depth h and length l, and shew that, if initially the water be at rest with its free surface plane and inclined at a small angle α to the horizontal, the velocity potential and the stream function at any time are given by

$$\phi + i\psi = -\alpha \sum_{s=0}^{\infty} \frac{4l^2}{\pi^3} \frac{p_s \sin p_s t}{(2s+1)^3} \frac{\cos\{(2s+1)\pi(x+iy)/l\}}{\sinh\{(2s+1)\pi h/l\}},$$

where $p_s/2\pi$ is the frequency for the vibration of type s.

(Trinity Coll. 1908.)

25. The free undisturbed surface of a liquid of great depth is the plane $y=0$, and it extends to infinity in both directions of the axis of x. In the surface there is a shallow depression, bounded by the planes $\pm x/a=1+y/\epsilon$, due to the presence of a floating body. Everything being at rest, the floating body is suddenly removed. Shew that after the lapse of a time t the equation to the free surface is

$$y=-\frac{4\epsilon}{a\pi}\int_0^\infty \frac{\cos kx . \sin^2 \frac{1}{2}ka . \cos(t\sqrt{gk})}{k^2}\,dk. \qquad \text{(M.T. 1902.)}$$

26. A rectangular trough containing water of given depth is slightly tilted at one end, and then let fall again into the horizontal position: find the period of the to-and-fro oscillations of the water that are thus set up.

Shew that, if the tilt is removed suddenly in comparison with this period, but without jarring, the surface of the water will assume, at the end of each swing, the form of an inclined plane, until friction and other causes modify the motion; and also that, if the water is shallow, its surface will at any intermediate time be in part horizontal, and in part a plane of constant slope. (St John's Coll. 1896.)

27. Shew that, if water is flowing with velocity V along a horizontal canal of rectangular section and depth h, and the bottom of the canal is agitated so that its form is given by $a\cos m(x-vt)$, where a is small, the form of the free surface is given by

$$y=a'\cos m(x-vt),$$

where
$$a=a'\left\{\cosh mh-\frac{g+m^2T/\rho}{m(V-v)^2}\sinh mh\right\},$$

T is the surface tension of the water and ρ its density. (M.T. 1898.)

28. The bottom of a straight uniform canal of rectangular section has the form $y=\alpha\sin(2\pi x/\lambda)$ referred to horizontal and vertical axes Ox and Oy through a point O in itself, and is moving with uniform velocity V in the direction Ox, α being small. If the mean depth of the liquid in the canal be h, find the velocity potential of the wave motion generated, and shew that the form of the free surface is given by

$$y=h+\alpha\sinh\frac{2\pi H}{\lambda}\operatorname{cosech}\frac{2\pi(H-h)}{\lambda}\sin\frac{2\pi(x-Vt)}{\lambda},$$

referred to fixed axes originally coinciding with Ox and Oy, H being the depth of the liquid corresponding to the free propagation under gravity, with velocity V, of waves of length λ. (M.T. 1900.)

29. A stream is running with mean velocity U in the plane xy between a horizontal bottom $y=0$ and a fixed upper boundary $y=h+a\cos mx$, where a is small. Find the character of the motion by determining its velocity potential or stream function.

Prove that, if U^2 exceeds a critical value $\frac{g}{m}\tanh mh$, the pressure on the upper boundary is in excess of the mean in its higher parts and in defect in its lower parts: and *vice versa*. (M.T. 1919.)

30. Shew how to take account of a variable pressure acting on the surface of a uniform canal; and in particular examine the effect of a travelling distribution of surface-pressure of the type

$$A + B \cos k(ct - x),$$

where x is the longitudinal coordinate, the canal being supposed infinitely long. (M.T. 1911.)

31. Find, at any time, the form of the free surface of an infinite canal, of uniform breadth, and uniform equilibrium depth h, if the initial conditions are $\eta = a \sin kx$ and $\dot{\eta} = 0$.

If the variations of pressure on the surface of such a canal are given by $b \sin kx \sin kv_0 t$, where b is small, then the form of the surface at any time will be

$$\eta = \frac{b}{g\rho\left(\dfrac{v_0^2}{v^2} - 1\right)} \sin kx \sin kv_0 t,$$

where v is the velocity of propagation of waves of length $2\pi/k$.

(Coll. Exam. 1906.)

32. An estuary extending from $x = 0$ to $x = a$ has at x a rectangular cross section of uniform depth Hx and a breadth Bx, where H and B are constants. The estuary meets the open sea at $x = a$, in which a tidal oscillation given by $\eta = \eta_0 \cos(\sigma t + \epsilon)$ is maintained. Prove that in the estuary

$$\eta = \eta_0 \cos(\sigma t + \epsilon) \frac{\sqrt{a} J_1(k\sqrt{x})}{\sqrt{x} J_1(k\sqrt{a})},$$

where $k^2 = 4\sigma^2/gH$. (M.T. 1926.)

33. A canal of uniform rectangular section and length l is closed at one end by a vertical wall, while the other end communicates with the sea. The velocity u may be supposed the same at all depths, but friction produces a resisting force $\kappa\rho h u$ per unit area of the bottom, where ρ is the density, h the depth, and κ a constant. A harmonic oscillation of period $2\pi/\sigma$ takes place in the level of the sea. Shew that the motion of the water in the canal may be represented by two waves, one travelling away from the sea and one towards it; and that the amplitudes of these waves are equal at the landward end, but in the ratio $\exp\left(-\dfrac{2\sigma l}{\sqrt{(gh)}} \sec^{\frac{1}{2}} \alpha \sin \tfrac{1}{2}\alpha\right)$ at the seaward end; where $\tan \alpha = \dfrac{\kappa}{\sigma}$. (M.T. 1925.)

CHAPTER XI

VIBRATIONS OF STRINGS

11·1. IN the last chapter we considered some cases of small oscillations of fluids regarded as *incompressible*. The theory of the oscillations of *elastic* fluids is also a branch of Hydrodynamics and it includes the theory of sound or waves in the atmosphere. The theory of sound is too extensive a subject to receive adequate treatment in an elementary text-book on hydrodynamics; but we propose in this chapter and the following to give a short account of some of the elements of the theory of sound waves together with the kindred subject of the vibrations of stretched strings.

11·12. Transverse Vibrations of a Stretched String. By transverse vibration we mean a motion in which each point is displaced at right angles to the equilibrium position of the string, and the slight extension of any element of the string is of the second order compared to the displacement. In fact the string is regarded as inextensible 'or rather the elastic modulus of extension is indefinitely great. The very beginnings of a local disturbance of tension will then be equalized along the string with speed practically infinite*'; and we may take it that the tension P remains constant along the string and throughout the motion. Let the string be of uniform line density ρ. Take the x axis in the equilibrium position of the string, and let y be the displacement at the point x at time t. If ψ be the inclination to the x axis of the tangent to the string we shall suppose that ψ is small.

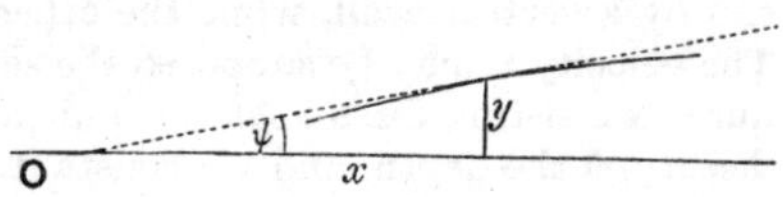

The equation of transverse motion of the element δx is

$$\rho\,\delta x\,\ddot{y} = -P\sin\psi + P\sin\psi + \delta(P\sin\psi),$$

for the forces acting on the element in the direction of motion are the components of the tension at its ends; viz. $P\sin\psi$ at one

* See a paper 'On the Dynamics of Radiation' by Sir Joseph Larmor, *International Congress*, 1912, *Proceedings*, vol. I, where the motion of a string is used as an illustration.

end and $P\sin\psi+\delta(P\sin\psi)$ at the other, and $\sin\psi=\dfrac{\partial y}{\partial s}=\dfrac{\partial y}{\partial x}$ approximately, neglecting $(\partial y/\partial x)^3$; therefore

$$\ddot{y}=\frac{P}{\rho}\frac{\partial^2 y}{\partial x^2}.$$

If we put $P=\rho c^2$ we may write the result

$$\frac{\partial^2 y}{\partial t^2}=c^2\frac{\partial^2 y}{\partial x^2}\quad\text{..........................(1).}$$

This is the same equation as we obtained in the theory of long waves in shallow water and as in **10·2** the solution is

$$y=f(ct-x)+F(ct+x)\quad\text{..................(2),}$$

where f and F are arbitrary functions.

If, for the moment, we take F to be zero, we have

$$y=f(ct-x)\quad\text{......................(3).}$$

This represents a wave form travelling with velocity c in the positive direction of the x axis. For, if we increase x and ct by the same amount, we leave y unaltered, which means that the displacement which exists at the instant t at the place x will at time $t+\tau$ be found at the place $x+c\tau$.

In the same way the equation

$$y=F(ct+x)\quad\text{......................(4)}$$

represents a wave form travelling with velocity c in the negative direction of the x axis.

Referring again to equation (3) we find by differentiation

$$\frac{\partial y}{\partial t}=-c\frac{\partial y}{\partial x}\quad\text{......................(5),}$$

which is a relation connecting the velocity at any point with the slope of the string. It is obvious that motion might be begun with arbitrary velocity and arbitrary slope but unless the two are connected by equation (5) the resulting motion cannot be given by a relation of the form (3). In the same way a motion represented by (4) implies a relation

$$\frac{\partial y}{\partial t}=c\frac{\partial y}{\partial x}\quad\text{.........................(6)}$$

between velocity and slope.

The general motion of the string may be regarded as the result of the superposition of two such wave systems travelling in opposite directions; and in this case the initial values of $\partial y/\partial t$ and $\partial y/\partial x$ may be regarded as composed of two parts which separately satisfy equations (5) and (6).

11·13. Unlimited String with given initial conditions. Suppose that, when $t=0$, we have

$$y=\phi(x) \quad \text{.........................(1)},$$

and

$$\dot{y}=\psi(x) \quad \text{.........................(2)}.$$

Taking for the general solution

$$y=f(ct-x)+F(ct+x) \quad \text{.................(3)},$$

we have, when $t=0$,

$$\phi(x)=f(-x)+F(x) \quad \text{....................(4)},$$

and

$$\psi(x)=cf'(-x)+cF'(x) \quad \text{..............(5)}.$$

By integrating the last equation we get

$$\int^x \psi(z)\,dz=-cf(-x)+cF(x) \quad \text{..............(6)};$$

and then from (4) and (6)

$$f(-x)=\tfrac{1}{2}\phi(x)-\frac{1}{2c}\int^x \psi(z)\,dz,$$

and

$$F(x)=\tfrac{1}{2}\phi(x)+\frac{1}{2c}\int^x \psi(z)\,dz;$$

so that

$$y=\tfrac{1}{2}\{\phi(x-ct)+\phi(x+ct)\}+\frac{1}{2c}\int_{x-ct}^{x+ct}\psi(z)\,dz \quad \text{......(7)}.$$

11·14. A given initial Displacement. In the special case in which there is no initial velocity but merely an initial displacement, the last result reduces to

$$y=\tfrac{1}{2}\{\phi(x-ct)+\phi(x+ct)\},$$

in which the two component waves resemble the initial form of the string but are of half the height at corresponding points.

The form of the string at any subsequent time may be constructed by drawing a curve in which the ordinate of each point is half the initial displacement of the point, imagining that two such curves initially occupy the same position and then moving them in opposite directions along the x axis with velocity c. The sum of the ordinates of the two curves at any point at any instant will give the displacement of the point at that instant.

11·15. Energy. The kinetic energy of any portion of the string is given by

$$T = \tfrac{1}{2}\rho\int \dot{y}^2 dx \quad \ldots\ldots\ldots\ldots\ldots\ldots\ldots\ldots(1).$$

For the potential energy V it is necessary to calculate the work done in the slight extension of the string against the tension P.

The increase in length in the element δx

$$= \delta s - \delta x = \delta x\left\{1 + \left(\frac{\partial y}{\partial x}\right)^2\right\}^{\frac{1}{2}} - \delta x$$

$$= \tfrac{1}{2}\left(\frac{\partial y}{\partial x}\right)^2 \delta x.$$

Therefore
$$V = \tfrac{1}{2}P\int\left(\frac{\partial y}{\partial x}\right)^2 dx \quad \ldots\ldots\ldots\ldots\ldots\ldots\ldots(2).$$

Now $P = \rho c^2$, and in either component wave from **11·12** (5) and (6)
$$\partial y/\partial t = \mp c\,\partial y/\partial x,$$
hence in any *single* progressive wave the kinetic and potential energies are equal.

11·16. String of Limited Length. Suppose that the origin is a fixed point on the string. In this case we must have $y=0$ when $x=0$, for all values of t. Hence, in the equation

$$y = f(ct - x) + F(ct + x),$$

we have
$$0 = f(ct) + F(ct),$$
or
$$F(z) = -f(z).$$

The general solution in this case is therefore

$$y = f(ct - x) - f(ct + x).$$

As applied to the string on the left of the origin this means the superposition of an *incident wave*, represented by the first term, travelling towards the origin, and a *reflected wave*, represented by the second term, and travelling away from the origin. The waves are similar in shape, their amplitudes being equal in magnitude and opposite in sign.

Let us consider the case of a disturbance represented by

$$y = f(ct - x) \quad \ldots\ldots\ldots\ldots\ldots\ldots\ldots\ldots(1)$$

advancing towards the origin, the disturbance being confined to a length l of the string, and suppose the string to be fixed at the origin O. Until the head of the disturbance reaches O the motion is represented completely by (1), but when this instant arrives we must take as the equation that represents the motion

$$y = f(ct - x) - f(ct + x) \quad \ldots\ldots\ldots\ldots\ldots\ldots(2).$$

The terms of this equation do not both apply to the same range of the string continuously. Thus if $t=0$ when the head of the disturbance reaches O, then when $0<t<l/c$ the first term applies between $x=0$ and $x=-l+ct$, and the second term between $x=0$

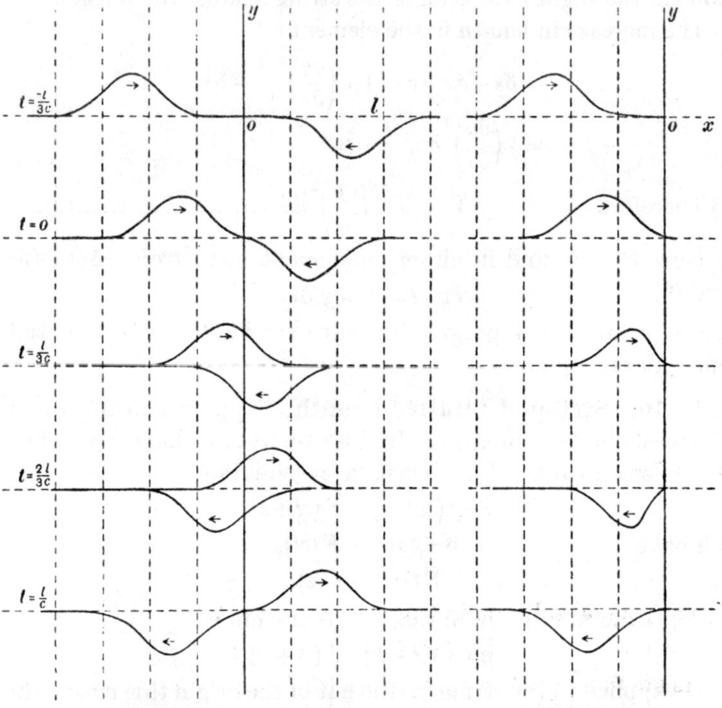

Components Resultant

and $x=-ct$. When $t=l/c$ the first term ceases to apply and the subsequent motion is represented by

$$y=-f(ct+x) \quad \text{......................(3)}$$

alone, or the reflection of the wave is complete.

When the initial form of the disturbance is given the form of the string at any time can be constructed graphically. Thus in the accompanying diagram the figures on the left represent the components of the displacement at intervals $l/3c$ before and after the head of the disturbance reaches O. They are obtained by drawing the curve that represents the disturbance with its head at O and drawing a similar curve so that the two are antisymmetrical with regard to O, and then displacing these curves to the right and left respectively with velocity c. The resultant form of the string, as shewn on the right, is obtained by taking the sum of the ordinates of the component waves on the left.

Later this will be seen to represent also the reflection of a sound wave at the *closed* end of a straight pipe.

11·17. If however the end of the string at the origin is capable of free transverse motion—it might, for example, be attached to a ring of negligible mass free to slide on a smooth wire along the y axis—the condition is $\partial y/\partial x = 0$, when $x = 0$, for all values of t. This follows from the equation of motion of the massless ring along the wire, which shews that there can be no component of tension along the y axis.

Taking $$y = f(ct - x)$$

for the incident wave, and

$$y = f(ct - x) + F(ct + x)$$

for the complete disturbance, we have

$$0 = -f'(ct) + F'(ct)$$

for all values of t.

Therefore $$F'(z) = f'(z),$$

or $$F(z) = f(z),$$

so that $$y = f(ct - x) + f(ct + x).$$

The reflected wave is therefore exactly the same in form as the incident wave, the amplitude being unchanged in sign.

This case corresponds to the reflection of a sound wave at the *open* end of a straight pipe.

11·18. String Fixed at both Ends. Let the fixed points be at $x = 0$ and $x = l$. Then we have

$$y = f(ct - x) + F(ct + x),$$

and the condition that $y = 0$ when $x = 0$, for all values of t, makes $F = -f$, as before.

Hence $$y = f(ct - x) - f(ct + x).$$

Also $y = 0$ when $x = l$, for all values of t, so that

$$0 = f(ct - l) - f(ct + l);$$

or, putting z for $ct - l$, $\quad f(z + 2l) = f(z)$.

Therefore $f(z)$ is a periodic function with a period $2l$ in z. Hence the motion of the string is periodic with respect to t, the period being $2l/c$, or twice the time taken by a wave to travel the length l.

It is otherwise evident that if a disturbance starts from any point A of the string, and moves with velocity c in either direction, it will after successive reflections at the two ends pass the point A again in the same direction with its original amplitude and sign in time $2l/c$.

11·19. Plucked String. When the string starts from rest with a given displacement, as for example when the string is drawn aside at one or more points and then set free, we have initially

$$y=\phi(x), \text{ say, and } \dot{y}=0.$$

And by substituting in the general solution

$$y=f(ct-x)+F(ct+x),$$

we get

$$\phi(x)=f(-x)+F(x),$$

and

$$0=cf'(-x)+cF'(x).$$

Therefore, by integrating the last equation,

$$0=-f(-x)+F(x);$$

whence

$$f(-x)=F(x)=\tfrac{1}{2}\phi(x).$$

Hence

$$y=\tfrac{1}{2}\phi(x-ct)+\tfrac{1}{2}\phi(x+ct),$$

as might have been written down from **11·14**.

Again y vanishes when $x=0$ and when $x=l$ for all values of t, so that

$$0=\phi(-ct)+\phi(ct),$$

and

$$0=\phi(l-ct)+\phi(l+ct).$$

Therefore

$$\phi(-z)=-\phi(z);$$

and, putting $ct=z+l$, we have also

$$\phi(z+2l)=-\phi(-z)=\phi(z).$$

Hence we get the following method for constructing the successive forms of the string: draw the curve $y=\phi(x)$ between $x=0$ and $x=l$ and continue it in both directions subject to the foregoing conditions, i.e. draw a similar curve in the third quadrant between $x=0$ and $x=-l$ and then repeat the whole figure in every successive space of length $2l$. Imagine curves of this type to travel in both directions with velocity c and take the arithmetic mean of their ordinates at any instant. The resulting curve represents the form at that instant of an unlimited string moving in such a manner that the points $x=0$, $\pm l$, $\pm 2l$, etc. are at rest, and therefore the portion between $x=0$ and $x=l$ satisfies all the required conditions. See the figure below.

In the case of a string plucked at one point and then set free the string at any instant consists of either two or three straight portions, generally three; and the two outer portions are always

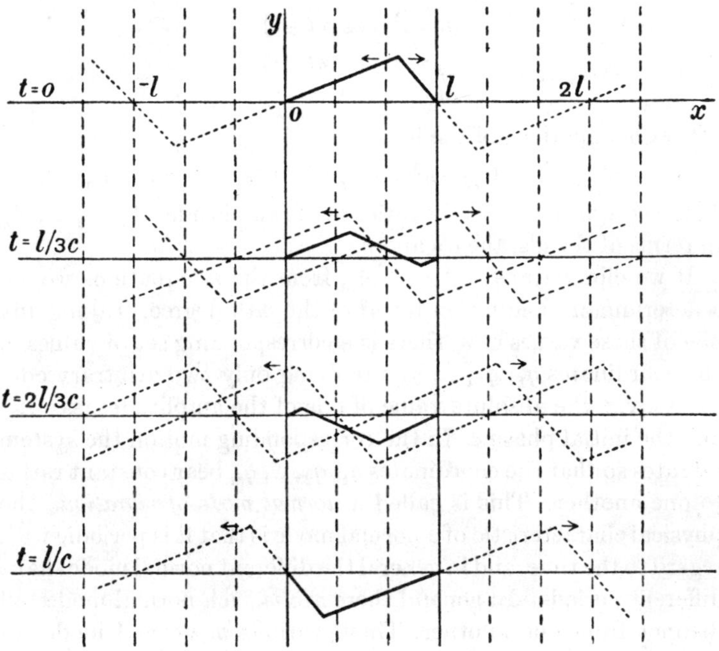

in the directions of the two portions in the initial position, while the gradient of the intermediate portion is a mean between the gradients of the other two having due regard to sign. Thus the figure shews the form of a string of length l, plucked at one point, after three intervals of time $l/3c$.

11·2. Normal Modes of Vibration. The position of a system which possesses m degrees of freedom and vibrates about a position of stable equilibrium can be defined by the values of m parameters or coordinates $q_1, q_2, \ldots q_m$. The kinetic energy T is given by

$$2T = a_{11}\dot{q}_1^2 + a_{22}\dot{q}_2^2 + \ldots + 2a_{12}\dot{q}_1\dot{q}_2 + \ldots;$$

where the a's are generally functions of the q's, but in small vibrations they may be regarded as constants.

And the potential energy is given by

$$2V = c_{11}q_1^{\,2} + c_{22}q_2^{\,2} + \ldots + 2c_{12}q_1q_2 + \ldots.$$

In the case of free vibrations, Lagrange's equations give

$$a_{11}\ddot{q}_1 + a_{12}\ddot{q}_2 + \ldots + a_{1m}\ddot{q}_m + c_{11}q_1 + c_{12}q_2 + \ldots + c_{1m}q_m = 0$$

and $m-1$ similar equations.

If, to solve these equations, we substitute

$$q_1 = A_1 \cos(nt+\epsilon),$$
$$q_2 = A_2 \cos(nt+\epsilon),$$
etc.,

we get m equations of the form

$$(c_{11} - n^2a_{11})A_1 + (c_{12} - n^2a_{12})A_2 + \ldots + (c_{1m} - n^2a_{1m})A_m = 0.$$

These m equations give the ratios of the amplitudes $A_1, A_2, \ldots A_m$ in terms of the a's, the c's and n.

If we eliminate $A_1, A_2, \ldots A_m$ from the m equations we get a determinantal equation for n^2 of the mth degree. Taking any one of these values of n, there is a corresponding set of values of the coordinates $q_1, q_2, \ldots q_m$ involving only two arbitrary constants, viz. the absolute value of one of the amplitudes, say A_1, and the initial phase ϵ. In the corresponding motion the system vibrates so that the coordinates $q_1, q_2, \ldots q_m$ bear constant ratios to one another. This is called a *normal mode of vibration*. The physical characteristic of a normal mode is that it is periodic with regard to the time, and in general the different normal modes have different periods. In general there are m such normal modes all distinct from one another. These various m normal modes of motion each with its arbitrary absolute amplitude and phase may be superposed; and the complete solution is given by m equations of the form

$$q_1 = B_1\cos(n_1t+\epsilon_1) + B_2\cos(n_2t+\epsilon_2) + \ldots + B_m\cos(n_mt+\epsilon_m),$$

and contains $2m$ arbitrary constants, namely $B_1, B_2, \ldots B_m$ and $\epsilon_1, \epsilon_2, \ldots \epsilon_m$. These are all the arbitrary constants because the quantities corresponding to the B's in the expressions for the other $m-1$ coordinates are all constant multiples of the B's.

It is shewn in books on Dynamics* that it is possible to choose the coordinates of a system so that the expressions for kinetic and

* Whittaker, *Analytical Dynamics*, 1904, § 77.

potential energy only contain squares and not products of the $\dot{q}$'s and the q's. When the coordinates are so chosen they are called the *normal coordinates* or *principal coordinates of the system* and each normal mode of vibration affects one and only one coordinate. For we have

$$2T = a_{11}\dot{q}_1^2 + a_{22}\dot{q}_2^2 + \dots,$$

and

$$2V = c_{11}q_1^2 + c_{22}q_2^2 + \dots,$$

so that by Lagrange's equation we get m equations

$$a_{11}\ddot{q}_1 + c_{11}q_1 = 0, \quad a_{22}\ddot{q}_2 + c_{22}q_2 = 0, \quad \text{etc.},$$

and the *complete* solution is

$$q_1 = A_1 \cos(n_1 t + \epsilon_1), \quad q_2 = A_2 \cos(n_2 t + \epsilon_2), \quad \text{etc.},$$

where

$$n_1^2 = c_{11}/a_{11}, \text{ etc.},$$

containing as before $2m$ arbitrary constants $A_1, A_2, \dots A_m$ and $\epsilon_1, \epsilon_2, \dots \epsilon_m$.

11·21. Normal Modes of Vibration of a Finite String. Since a string has an infinite number of degrees of freedom it has an infinite number of normal modes of vibration. To find modes let us assume that the displacement of every point of the string is proportional to $\cos(nt + \epsilon)$.

The differential equation to be satisfied is

$$\frac{\partial^2 y}{\partial t^2} = c^2 \frac{\partial^2 y}{\partial x^2} \quad \text{.........................(1)};$$

and if $y \propto \cos(nt + \epsilon)$, we have $\ddot{y} = -n^2 y$, therefore

$$\frac{\partial^2 y}{\partial x^2} + \frac{n^2}{c^2} y = 0.$$

The complete solution of this equation, including the time factor, is

$$y = \left(A \cos\frac{nx}{c} + B \sin\frac{nx}{c}\right) \cos(nt + \epsilon) \quad \text{...........(2)}.$$

If the ends of the string are fixed at the points $x = 0$ and $x = l$, we must have $A = 0$ and $\sin nl/c = 0$.

Hence

$$n = \frac{\pi c}{l}, \frac{2\pi c}{l}, \frac{3\pi c}{l}, \text{ etc.} \quad \text{..................(3)}.$$

This gives the infinitely many values of n that correspond to the different normal modes, and the solution corresponding to the sth normal mode may be written

$$y = B_s \sin\frac{s\pi x}{l} \cos\left(\frac{s\pi ct}{l} + \epsilon_s\right) \quad \text{..............(4)}.$$

The gravest or fundamental note of the string is that for which $s=1$. Its frequency is

$$\frac{n}{2\pi}=\frac{c}{2l}=\frac{1}{2l}\sqrt{\frac{P}{\rho}}.$$

The facts embodied in this formula, namely that the frequency varies inversely as the length and the square root of the density and directly as the square root of the tension, are known as *Mersenne's Laws.* They are capable of experimental verification by fixing one end of a string and then passing the string over two edges or 'bridges', whose distance apart can be varied and measured, and suspending a weight from the other end of the string.

In the next normal mode to the fundamental $s=2$ and the middle point of the string $x=\frac{1}{2}l$ remains at rest throughout the motion. In the sth normal mode of which the frequency is $sc/2l$, the $(s-1)$ points

$$x=\frac{l}{s},\ \frac{2l}{s},\ \dots\ \frac{(s-1)\,l}{s}$$

are at rest throughout the motion. These points are called *nodes*; the points midway between them are the points of maximum amplitude and are called *loops.* Each segment into which the $s-1$ nodes divide the string vibrates like the fundamental mode of a string of length l/s.

A general vibration of the string is obtained by the superposition of the several normal modes with amplitudes and phase constants chosen to suit whatever may be the given initial conditions. The equation that represents this motion is therefore

$$y=\Sigma B_s \sin\frac{s\pi x}{l}\cos\left(\frac{s\pi ct}{l}+\epsilon_s\right),$$

where B_s and ϵ_s are chosen to suit the initial conditions and the summation extends to all integral values of s.

11·22. Two special cases. (1) If the string starts from rest at time $t=0$, then $\dot{y}=0$ when $t=0$ for all values of x, so that all the ϵ's are zero, and

$$y=\Sigma B_s \sin\frac{s\pi x}{l}\cos\frac{s\pi ct}{l}.$$

(2) If the string starts from the equilibrium position at time $t=0$, then $y=0$ when $t=0$ for all values of x, so that all the ϵ's are odd multiples of $\frac{1}{2}\pi$, and

$$y=\Sigma B_s \sin\frac{s\pi x}{l}\sin\frac{s\pi ct}{l}.$$

11·23. Plucked String. Let the string be drawn aside through a small distance β at a distance b from the end $x=0$ and then released.

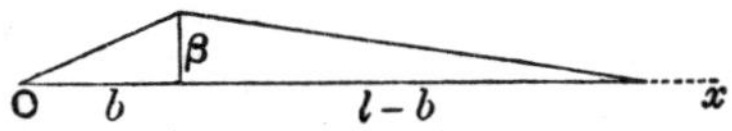

We have to determine the coefficients B_s in the solution

$$y=\Sigma B_s \sin\frac{s\pi x}{l}\cos\frac{s\pi ct}{l} \quad\text{.....................(1).}$$

The initial values of y are

$$y=\beta x/b,\ (0<x<b);\ \text{and}\ y=\beta(l-x)/(l-b),\ (b<x<l).$$

Multiply both sides of equation (1) by $\sin\dfrac{s\pi x}{l}$ and integrate between the values 0 and l of x, giving y its proper values in terms of x for each part of the range, and taking $t=0$. Then, since, when $r\neq s$,

$$\int_0^l \sin\frac{s\pi x}{l}\sin\frac{r\pi x}{l}\,dx=0,$$

therefore

$$\int_0^b \frac{\beta x}{b}\sin\frac{s\pi x}{l}\,dx+\int_b^l \frac{\beta(l-x)}{l-b}\sin\frac{s\pi x}{l}\,dx=\int_0^l B_s\sin^2\frac{s\pi x}{l}\,dx;$$

which gives
$$B_s=\frac{2\beta l^2}{s^2\pi^2 b(l-b)}\sin\frac{s\pi b}{l},$$

so that
$$y=\frac{2\beta l^2}{\pi^2 b(l-b)}\Sigma\frac{1}{s^2}\sin\frac{s\pi b}{l}\sin\frac{s\pi x}{l}\cos\frac{s\pi ct}{l}.$$

11·24. Energy of a String with Fixed Ends. If the string be vibrating in its sth normal mode we have from **11·21**

$$y=B_s\sin\frac{s\pi x}{l}\cos\left(\frac{s\pi ct}{l}+\epsilon_s\right).$$

The kinetic energy T is given by

$$\begin{aligned}T&=\tfrac{1}{2}\rho\int_0^l \dot{y}^2dx\\&=\frac{s^2\pi^2c^2\rho}{2l^2}\int_0^l B_s{}^2\sin^2\frac{s\pi x}{l}\sin^2\left(\frac{s\pi ct}{l}+\epsilon_s\right)dx\\&=\frac{s^2\pi^2c^2\rho}{4l}B_s{}^2\sin^2\left(\frac{s\pi ct}{l}+\epsilon_s\right) \quad\text{.......................(1).}\end{aligned}$$

And the potential energy, as in **11·15**, by

$$\begin{aligned}V&=\tfrac{1}{2}P\int_0^l\left(\frac{dy}{dx}\right)^2dx\\&=\frac{s^2\pi^2P}{4l}B_s{}^2\cos^2\left(\frac{s\pi ct}{l}+\epsilon_s\right) \quad\text{...............(2),}\end{aligned}$$

in the same way.

Also since $P=c^2\rho$, (**11·12**), therefore

$$T+V=\frac{s^2\pi^2c^2\rho}{4l}B_s^{\,2} \quad\text{.....................(3)}$$

gives the whole energy of the vibration in the sth mode.

Again if the motion be of the general type

$$y=\Sigma B_s\sin\frac{s\pi x}{l}\cos\left(\frac{s\pi ct}{l}+\epsilon_s\right),$$

we have

$$T=\tfrac{1}{2}\rho\int_0^l \dot{y}^2\,dx=\frac{\pi^2c^2\rho}{2l^2}\int_0^l\left[\Sigma\left\{sB_s\sin\frac{s\pi x}{l}\sin\left(\frac{s\pi ct}{l}+\epsilon_s\right)\right\}\right]^2dx.$$

Now
$$\int_0^l \sin\frac{s\pi x}{l}\sin\frac{r\pi x}{l}\,dx=0,$$

and
$$\int_0^l \sin^2\frac{s\pi x}{l}=\frac{l}{2}.$$

Therefore
$$T=\frac{\pi^2c^2\rho}{4l}\Sigma s^2B_s^{\,2}\sin^2\left(\frac{s\pi ct}{l}+\epsilon_s\right) \quad\text{............(4).}$$

Similarly we get

$$V=\frac{\pi^2P}{4l}\Sigma s^2B_s^{\,2}\cos^2\left(\frac{s\pi ct}{l}+\epsilon_s\right) \quad\text{............(5),}$$

and
$$T+V=\frac{\pi^2c\rho}{4l}\Sigma s^2B_s^{\,2} \quad\text{.....................(6).}$$

In these results it appears that the whole kinetic energy, containing square terms but no product terms, is the sum of the kinetic energy due to each separate normal mode of vibration, and similarly in regard to the potential energy, which is of course in accordance with the general theory of normal modes as explained in **11·2**.

11·3. Normal functions and coordinates*. When a vibrating system has a finite number (m) of degrees of freedom, we saw (**11·2**) that its position could be specified in terms of m normal coordinates each corresponding to a normal mode of vibration, and that the kinetic and potential energies contained only squares and not products of these normal coordinates. A vibrating string has, however, an infinite number of degrees of

* This use of normal coordinates is due to Lord Rayleigh, see *Theory of Sound*, I, § 128.

freedom and therefore infinitely many normal coordinates, and when we express the form by the equation

$$y = \Sigma B_s \sin \frac{s\pi x}{l} \cos \left(\frac{s\pi ct}{l} + \epsilon_s \right) \quad \text{..............(1)},$$

the coefficients of $\sin \frac{s\pi x}{l}$ for all integral values of s are the normal coordinates and the typical one may be denoted by ϕ_s, so that

$$y = \Sigma \phi_s \sin \frac{s\pi x}{l} \quad \text{.......................(2)}.$$

Taking the ϕ's as the coordinates that determine the position and motion of the string we may use Lagrange's equations. As in **11·24** we have

$$T = \tfrac{1}{4}\rho l \sum_1^\infty \dot{\phi}_s^2 \text{ and } V = \tfrac{1}{4}\frac{\rho c^2 \pi^2}{l} \sum_1^\infty s^2 \phi_s^2 \quad \text{.........(3)}.$$

And if Φ_s is the force tending to cause a displacement $\delta\phi_s$ (using the word force in a generalized sense) we have

$$\frac{d}{dt}\left(\frac{\partial T}{\partial \dot{\phi}_s}\right) - \frac{\partial T}{\partial \phi_s} + \frac{\partial V}{\partial \phi_s} = \Phi_s.$$

That is

$$\ddot{\phi}_s + \frac{s^2\pi^2c^2}{l^2}\phi_s = \frac{2}{\rho l}\Phi_s \quad \text{....................(4)}.$$

If we write this equation

$$\ddot{\phi}_s + n^2\phi_s = \frac{2}{\rho l}\Phi_s \quad \text{....................(5)},$$

for a particular integral, using D for d/dt, we have

$$\begin{aligned}
\phi_s &= \frac{2}{\rho l}\frac{1}{D^2+n^2}\Phi_s \\
&= \frac{1}{in\rho l}\left\{\frac{1}{D-in} - \frac{1}{D+in}\right\}\Phi_s \\
&= \frac{1}{in\rho l}\left\{e^{int}\int_0^t e^{-int}\Phi_s dt - e^{-int}\int_0^t e^{int}\Phi_s dt\right\} \\
&= \frac{1}{in\rho l}\int_0^t (e^{in(t-t')} - e^{-in(t-t')})\Phi_s dt' \\
&= \frac{2}{n\rho l}\int_0^t \sin n(t-t')\Phi_s dt',
\end{aligned}$$

and adding the complementary function, the complete solution is

$$\phi_s = (\phi_s)_0 \cos nt + (\dot{\phi}_s)_0 \frac{\sin nt}{n} + \frac{2}{n\rho l}\int_0^t \sin n(t-t')\Phi_s dt' \text{...(6)},$$

where the zero suffixes denote values when $t = 0$.

If the impressed force is a single force Y at the point $x=b$, then

$$\Sigma\Phi_s\delta\phi_s = Y\delta y,$$

so that
$$\Phi_s = Y\left(\frac{\partial y}{\partial\phi_s}\right)_{x=b} = Y\sin\frac{s\pi b}{l} \quad\text{..............(7).}$$

11·31. Examples of use of normal coordinates. Plucked String. Taking the case considered in **11·23**, Φ_s is zero except when $t=0$, and then its value is $Y\sin\frac{s\pi b}{l}$, where Y is the force by which the string is held. Since the string starts from rest $(\dot{\phi}_s)_0=0$ and **11·3** (5) gives

$$n^2(\phi_s)_0 = \frac{2}{\rho l}\Phi_s = \frac{2}{\rho l}Y\sin\frac{s\pi b}{l}.$$

And at time t we have from (6)

$$\phi_s = (\phi_s)_0\cos nt = \frac{2Y}{\rho l n^2}\sin\frac{s\pi b}{l}\cos nt.$$

Therefore
$$\begin{aligned} y &= \Sigma\phi_s\sin\frac{s\pi x}{l} \\ &= \frac{2Y}{\rho l}\Sigma\sin\frac{s\pi b}{l}\sin\frac{s\pi x}{l}\frac{\cos nt}{n^2} \\ &= \frac{2lY}{\rho\pi^2c^2}\Sigma\frac{1}{s^2}\sin\frac{s\pi b}{l}\sin\frac{s\pi x}{l}\cos\frac{s\pi ct}{l}, \end{aligned}$$

which agrees with the result of **11·23**, if we note that Y is equal to the resolved part of the tensions perpendicular to the x axis; that is

$$\begin{aligned} Y &= P\left(\frac{\beta}{b}+\frac{\beta}{l-b}\right), \text{ to the first order of } \beta \\ &= \frac{c^2\rho l\beta}{b(l-b)}. \end{aligned}$$

11·32. String set in Motion by an Impulse. Let an impulse I be applied at the point $x=b$. We may regard this as the limit of $\int_0^\tau I'dt'$, where I' is a force that begins to act when the string is at rest and ceases to act after a short time τ. Then using (6) of **11·3**, $(\phi_s)_0=0$ and $(\dot{\phi}_s)_0=0$ so that

$$\begin{aligned} \phi_s &= \frac{2}{n\rho l}\int_0^\tau \sin n(t-t')\Phi_s dt' \\ &= \frac{2}{n\rho l}\sin nt\int_0^\tau \Phi_s dt', \end{aligned}$$

neglecting the term $\sin nt'$ for the range $t'=0$ to $t'=\tau$ since τ is small. But from **11·3** (7)

$$\Phi_s = I' \sin \frac{s\pi b}{l},$$

therefore
$$\int_0^\tau \Phi_s dt' = \sin \frac{s\pi b}{l} \int_0^\tau I' dt' = I \sin \frac{s\pi b}{l}.$$

Hence
$$\phi_s = \frac{2I}{s\pi c\rho} \sin \frac{s\pi b}{l} \sin \frac{s\pi ct}{l},$$

and
$$y = \frac{2I}{\pi c\rho} \sum_1^\infty \frac{1}{s} \sin \frac{s\pi b}{l} \sin \frac{s\pi x}{l} \sin \frac{s\pi ct}{l}.$$

11·4. Forced Vibrations of a String. There are two cases to be considered; the first, when a given point $x=b$ is given an arbitrary transverse periodic motion; the second when a given periodic force acts at $x=b$.

In the first case let the given motion at $x=b$ be represented by

$$y = \gamma \cos(pt+\alpha).$$

We have to satisfy the equation

$$\frac{\partial^2 y}{\partial t^2} = c^2 \frac{\partial^2 y}{\partial x^2},$$

and if we assume that y varies as $\cos(pt+\alpha)$, this equation becomes

$$\frac{\partial^2 y}{\partial x^2} + \frac{p^2}{c^2} y = 0 \quad \text{.......................(1).}$$

Now we cannot assume that the same solution will represent the form of both portions into which the string is divided at the point $x=b$ which is given a forced motion; so we assume that there are two distinct solutions of (1) corresponding to the two parts of the string, viz.

when $\quad 0 < x < b, \; y_1 = \left(A \cos \frac{px}{c} + B \sin \frac{px}{c}\right) \cos(pt+\alpha)$ ...(2),

and $\quad b < x < l, \; y_2 = \left(C \cos \frac{px}{c} + D \sin \frac{px}{c}\right) \cos(pt+\alpha)$ ...(3).

Then we have $y_1 = 0$ when $x=0$, so that $A=0$,

and
$$y_1 = \gamma \cos(pt+\alpha) \text{ when } x=b,$$

so that
$$B \sin \frac{pb}{c} = \gamma.$$

Hence when $0 < x < b$,

$$y_1 = \gamma \frac{\sin px/c}{\sin pb/c} \cos(pt+\alpha) \quad \text{...............(4).}$$

Similarly $y_2 = 0$ when $x = l$, so that

$$C\cos\frac{pl}{c} + D\sin\frac{pl}{c} = 0 \qquad\text{.....................(5)},$$

and $y_2 = \gamma\cos(pt+\alpha)$ when $x = b$, so that

$$C\cos\frac{pb}{c} + D\sin\frac{pb}{c} = \gamma \qquad\text{..................(6)}.$$

Then by solving (5) and (6) for C and D and substituting in (3) we find that when $b < x < l$,

$$y_2 = \gamma\frac{\sin p(l-x)/c}{\sin p(l-b)/c}\cos(pt+\alpha) \qquad\text{............(7)}.$$

In the second case, let there be a force $F\cos(pt+\alpha)$ at the point $x = b$. We may deduce the solution for this case from the last by the consideration that the resultant of the tensions at the point must balance the impressed force.

That is, if P denotes the tension

$$F\cos(pt+\alpha) = P\frac{\partial y_1}{\partial x} - P\frac{\partial y_2}{\partial x}, \text{ at } x = b.$$

Therefore, by differentiating (4) and (7)

$$F = \frac{P\gamma p}{c}\,\frac{\sin pl/c}{\sin pb/c\sin p(l-b)/c};$$

whence we get

$$y_1 = \frac{F}{P}\,\frac{\sin p(l-b)/c\sin px/c}{p/c\sin pl/c}\cos(pt+\alpha),\quad 0 < x < b \text{ ...(8)},$$

and $$y_2 = \frac{F}{P}\,\frac{\sin pb/c\sin p(l-x)/c}{p/c\sin pl/c}\cos(pt+\alpha),\quad b < x < l \text{ ...(9)}.$$

This is an example of a reciprocal theorem that the motion at x when the force acts at b is the same as would be the motion at b if the force acted at x.

We notice that in the first case the motion of either portion of the string is independent of the length of the other portion and depends only on the forced motion at the point $x = b$; also that if pb/c is an integral multiple of π, i.e. if $p/2\pi$ is a natural frequency for a string of length b, the amplitude in (4) appears to be infinite. This is a case of 'resonance' in which we have a forced oscillation of the same period as free oscillations. In actuality the amplitude is not infinite as our equations cease to represent the motion when the displacement is other than small, also there are small frictional forces which oppose the motion and damp out the free oscillations. In the second case the same phenomenon of 'resonance' occurs when pl/c is an integral multiple of π, i.e. when the frequency of the applied force is a natural frequency of the whole string.

11·41. Vibrations of a String carrying a Load. Let a particle of mass M be attached at the point $x=b$.

If we assume that the motion of the particle M is given by

$$y=\gamma\cos(pt+\alpha) \quad \text{.........................(1)},$$

then the motions of the two parts into which it divides the string are given by (4) and (7) of **11·4**. And the frequency $p/2\pi$ is to be found from the equation of motion of M; namely

$$M\ddot{y}=-P\frac{\partial y_1}{\partial x}+P\frac{\partial y_2}{\partial x}$$

at $x=b$, P denoting the tension of the string.

Substituting from (4) and (7) of **11·4** and from (1) above, we get

$$-p^2M=-P\frac{p}{c}\cot\frac{pb}{c}-P\frac{p}{c}\cot\frac{p(l-b)}{c}.$$

Therefore
$$pM=\frac{P}{c}\frac{\sin pl/c}{\sin pb/c\sin p(l-b)/c} \quad \text{...............(2)}.$$

This equation must be satisfied by p, and the form of the string at time t is then given by (4) and (7) of **11·4**, γ and α being arbitrary constants depending on initial conditions. Since those normal modes of motion which have a node at $x=b$ could exist without causing the motion of this point, it is clear that the presence of M will not affect these normal modes. Thus if M be at the middle point of the string, the normal modes of even order are unchanged, and we can shew that the frequencies of the odd components are diminished. For, in this case

$$b=l-b=\tfrac{1}{2}l,$$

so that (2) becomes
$$pM=\frac{2P}{c}\cot\frac{pl}{2c},$$

or
$$\frac{pl}{2c}\tan\frac{pl}{2c}=\frac{Pl}{c^2M}=\frac{\rho l}{M}.$$

The frequencies of the normal modes concerned are therefore given by

$$pl/2c=x_1,\ x_2,\ x_3,\ \ldots,$$

where $x_1, x_2, x_3, \ldots$ are the successive roots of the equation

$$x\tan x=\frac{\rho l}{M}.$$

By drawing the curves $y=\tan x$ and $y=\rho l/Mx$ it is easily seen that the roots lie between zero and $\frac{1}{2}\pi$, π and $\frac{3}{2}\pi$, 2π and $\frac{5}{2}\pi$ and so on.

But the natural frequencies of the unloaded string are given by

$$pl/2c=\tfrac{1}{2}\pi,\ \pi,\ \tfrac{3}{2}\pi,\ 2\pi,\ \tfrac{5}{2}\pi,\ \ldots \qquad \textbf{(11·21)}.$$

Hence it follows that frequencies of the normal modes of odd order are decreased.

11·42. Finite String with Ends not rigidly Fastened. We will consider two cases, namely when one end of the string is attached to a mass M capable of moving transversely, either (i) as a bead on a smooth wire, or (ii) under the control of a spring of strength μ, the other end of the string being fixed.

As solution of $$\frac{\partial^2 y}{\partial t^2}=c^2\frac{\partial^2 y}{\partial x^2},$$

we take $$y=\left(A\cos\frac{nx}{c}+B\sin\frac{nx}{c}\right)\cos(nt+\epsilon) \quad\text{.............(1)};$$

the terminal conditions being

(i) $$M\ddot{y}=P\,\partial y/\partial x \text{ when } x=0,$$

and $$y=0 \text{ when } x=l.$$

Therefore $$-n^2AM=PB\,n/c,$$

and $$A\cos nl/c+B\sin nl/c=0;$$

whence $$\frac{nl}{c}\tan\frac{nl}{c}=\frac{Pl}{c^2M}=\frac{\rho l}{M} \quad\text{........................(2)},$$

which is the same equation for the frequencies as if the particle were at the middle point of a string of length $2l$; as is otherwise obvious.

(ii) The terminal conditions in this case are

$$M\ddot{y}+\mu y=P\,\partial y/\partial x \text{ when } x=0,$$

and $$y=0 \text{ when } x=l.$$

Therefore $$(\mu-n^2M)A=PBn/c,$$

and $$A\cos nl/c+B\sin nl/c=0;$$

whence $$\tan\frac{nl}{c}=\frac{Pn}{c(n^2M-\mu)}=\frac{\rho cn}{n^2M-\mu} \quad\text{.................(3)}.$$

In either case equation (1) takes the form

$$y=C\sin\frac{n(l-x)}{c}\cos(nt+\epsilon) \quad\text{....................(4)};$$

and equations (2) and (3) both have an infinite number of solutions so that the motion in general will be given by equating y to the sum of an infinite number of terms like (4).

11·5. Damped Oscillations. If the motion of the string be retarded by a force acting on each element of mass and proportional to its velocity, the equation of motion of **11·12** becomes

$$\frac{\partial^2 y}{\partial t^2}=c^2\frac{\partial^2 y}{\partial x^2}-\kappa\frac{\partial y}{\partial t} \quad\text{....................(1)}.$$

If we put $y=ze^{-\frac{1}{2}\kappa t}$, this reduces to

$$\frac{\partial^2 z}{\partial t^2}-\tfrac{1}{4}\kappa^2 z=c^2\frac{\partial^2 z}{\partial x^2} \quad\text{....................(2)},$$

and we may obtain solutions of this equation to suit particular cases. Thus to find the frequency $p/2\pi$ of waves of length $2\pi/m$, if we assume that

$$z\propto e^{imx},$$

we get $$\frac{\partial^2 z}{\partial t^2}+p^2z=0,$$

where $p^2=c^2m^2-\frac{1}{4}\kappa^2$.

And the solution is

$$y = Ae^{-\frac{1}{2}\kappa t + imx}\cos\{(c^2m^2 - \tfrac{1}{4}\kappa^2)^{\frac{1}{2}}t + \alpha\},$$

or, rejecting the imaginary part,

$$y = Ae^{-\frac{1}{2}\kappa t}\cos mx\cos\{(c^2m^2 - \tfrac{1}{4}\kappa^2)^{\frac{1}{2}}t + \alpha\} \quad ......(3).$$

This represents a vibration whose amplitude diminishes continuously because of the factor $e^{-\frac{1}{2}\kappa t}$. The time $2/\kappa$ in which the amplitude is reduced to e^{-1} of its former value is called the *modulus of decay*.

11·51. If the resistance is so small that κ^2 may be neglected, (2) becomes

$$\frac{\partial^2 z}{\partial t^2} = c^2\frac{\partial^2 z}{\partial x^2},$$

the solution of which, as in **11·12**, is

$$z = f(ct - x) + F(ct + x),$$

and therefore $y = e^{-\frac{1}{2}\kappa t}f(ct - x) + e^{-\frac{1}{2}\kappa t}F(ct + x)$ ....(4).

Since the functions are arbitrary we may write

$$e^{\frac{1}{2}\kappa\left(t - \frac{x}{c}\right)}f(ct - x) \text{ instead of } f(ct - x),$$

and $e^{\frac{1}{2}\kappa\left(t + \frac{x}{c}\right)}F(ct + x)$ instead of $F(ct + x)$;

so that $y = e^{-\kappa x/2c}f(ct - x) + e^{\kappa x/2c}F(ct + x)$ ....(5)

is also a solution.

For example, suppose that the string is of infinite length and is subject to a forced motion $E\cos pt$ at a particular point, which we may take to be the origin, the motion will be represented by

$$y = Ee^{-\kappa x/2c}\cos p(t - x/c) \quad ....(6)$$

on the positive side of the origin; and by

$$y = Ee^{\kappa x/2c}\cos p(t + x/c) \quad ....(7)$$

on the negative side; these equations representing a progressive wave whose amplitude decreases in the ratio $1 : e$ as the distance from the origin increases by $2c/\kappa$, i.e. at intervals of time $2/\kappa$, since c is the wave velocity.

11·6. Reflection and Transmission of Waves. Consider a uniform stretched string of great length. Let a train of simple harmonic transverse waves travelling along the string encounter a massive particle attached to the string at a particular point. There will be a reflected and a transmitted train of waves.

Let M be the mass of the particle, T the tension of the string, ρ its line density and let $c^2 = T/\rho$ give the velocity of propagation of the waves.

Take the origin at the particle and let the incident wave train, coming from the part of the string for which x is negative, be represented by

$$y = A\cos(mx - nt) \qquad \text{.........(1)}.$$

For facility in working it is simpler to take this to be the real part of $Ae^{i(mx-nt)}$. If we assume similar expressions $A'e^{i(m'x-n't)}$ and $A_1e^{i(m_1x-n_1t)}$ for the displacement in the reflected and transmitted wave, then continuity requires that the period shall be the same for all, i.e. $n' = n_1 = n$; and as the velocity of propagation c has the same numerical value for all waves on the string, therefore $m' = -m$ in the reflected wave and $m_1 = m$ in the transmitted wave. Hence for the total displacements we have when $x < 0$,

$$y_1 = Ae^{i(mx-nt)} + A'e^{-i(mx+nt)} \qquad \text{.........(2)}$$

and when $x > 0$,

$$y_2 = A_1e^{i(mx-nt)} \qquad \text{.........(3)},$$

where the ratios of A' and A_1 to A may be complex numbers.

For the motion of the particle at the origin we have

$$M\ddot{y} = -T\frac{\partial y_1}{\partial x} + T\frac{\partial y_2}{\partial x} \quad \text{where } x = 0,$$

i.e.

$$-Mn^2A_1 = -iTm(A - A' - A_1),$$

or, since $T = c^2\rho$ and $c^2 = n^2/m^2$,

$$MmA_1 = i\rho(A - A' - A_1) \qquad \text{.........(4)}.$$

Also, when $x = 0$, we have $y_1 = y_2$, so that

$$A + A' = A_1 \qquad \text{.........(5)}.$$

From (4) and (5) we find that

$$\frac{A'}{iMm} = \frac{A_1}{2\rho} = \frac{A}{2\rho - iMm} \qquad \text{.........(6)},$$

or

$$\frac{A'}{i\sin\epsilon} = \frac{A_1}{\cos\epsilon} = \frac{A}{\cos\epsilon - i\sin\epsilon} \qquad \text{.........(7)}$$

if $\tan\epsilon = Mm/2\rho$.

Hence the reflected and transmitted waves differ both in amplitude and phase from the incident wave, these differences being exhibited in the formulae

$$A' = A\sin\epsilon \, . \, e^{i(\frac{1}{2}\pi+\epsilon)} \text{ and } A_1 = A\cos\epsilon \, . \, e^{i\epsilon} \qquad \text{.........(8)}.$$

By using the method of **11·15** it is easy to shew that the energy per wave length of a simple harmonic wave is proportional to the square of he amplitude, and from (7) it is clear that

$$|A'|^2 + |A_1|^2 = |A|^2$$

so that the energies of the reflected and transmitted waves are together equal to the energy of the incident wave.

11·7. Longitudinal Vibrations. Suppose the string to be elastic and stretched and to obey Hooke's Law. If P, Q are two points whose coordinates are x, $x + \delta x$ in the equilibrium position and these are displaced to P', Q' where the coordinate of P' is $x + \xi$ then that of Q' is

$$x + \delta x + \xi + \frac{\partial \xi}{\partial x}\delta x.$$

If T be the tension at P' and E the modulus of elasticity

$$T = E\,\frac{P'Q' - P_0Q_0}{P_0Q_0},$$

where $P_0Q_0\,(=\delta x_0)$ is the unstretched length of PQ.

Therefore $$T = E\left(\delta x + \frac{\partial \xi}{\partial x}\,\delta x - \delta x_0\right)\Big/\,\delta x_0$$

$$= E\,\frac{\delta x}{\delta x_0}\,\frac{\partial \xi}{\partial x} + E\,\frac{\delta x - \delta x_0}{\delta x_0}.$$

Now $E\,\dfrac{\delta x - \delta x_0}{\delta x_0} = T_0$ is the tension in equilibrium.

Also $\delta x/\delta x_0$ is the ratio of the equilibrium stretched length to the natural stretched length for the whole string $= l/l_0$ say, and if we put $El/l_0 = E'$, a definite constant for the string in its equilibrium position, we have

$$T = E'\,\frac{\partial \xi}{\partial x} + T_0.$$

Let ρ be the line density in the equilibrium state and X the external impressed force per unit mass at P' acting on the string; then the equation of motion of the element PQ is

$$\rho\,\delta x\,\frac{\partial^2 \xi}{\partial t^2} = -T + \left(T + \frac{\partial T}{\partial x}\,\delta x\right) + \rho X\,\delta x,$$

or

$$\frac{\partial^2 \xi}{\partial t^2} = \frac{E'}{\rho}\,\frac{\partial^2 \xi}{\partial x^2} + X \quad\ldots\ldots\ldots\ldots\ldots\ldots\ldots\ldots(1).$$

If there be no impressed force, and we write $E' = \rho c^2$ the equation takes the form

$$\frac{\partial^2 \xi}{\partial t^2} = c^2\,\frac{\partial^2 \xi}{\partial x^2} \quad\ldots\ldots\ldots\ldots\ldots\ldots\ldots\ldots(2).$$

This is the same differential equation as for transverse vibrations and its solutions may be interpreted in a similar manner when applied to the propagation of longitudinal vibrations, but it is important to observe a difference in the form of terminal conditions. Thus, at a fixed end we have $\xi = 0$, and $\partial\xi/\partial t = 0$, for all values of t.

If a free end were possible we should have at that end $T = 0$ and therefore $\partial\xi/\partial x = 0$.

It is to be observed that c is the velocity with which waves travel along the string stretched to length l and that

$$c^2 = \frac{E'}{\rho} = \frac{E}{\rho}\,\frac{l}{l_0}.$$

But if ρ_0 be the line density of the unstretched string then $\rho l = \rho_0 l_0$, so that

$$c^2 = \frac{E}{\rho_0}\left(\frac{l}{l_0}\right)^2,$$

and the period $\dfrac{2l}{c} = 2l_0\sqrt{\dfrac{\rho_0}{E}}$ is independent of the amount of stretching.

Instead of the equilibrium stretched length x in (2) we might use the corresponding unstretched length x_0 as independent variable, since $x/x_0 = l/l_0$, then (2) becomes

$$\frac{\partial^2 \xi}{\partial t^2} = \frac{E}{\rho_0}\frac{\partial^2 \xi}{\partial x_0^2} = c_0^2 \frac{\partial^2 \xi}{\partial x_0^2} \quad \text{.....................(3)}$$

where $c_0^2 = E/\rho_0$.

The foregoing arguments also apply to the longitudinal vibrations of bars.

Longitudinal waves are reflected and transmitted in the same way as transverse waves. Thus if the method of **11·6** is applied to solve the same problem for longitudinal waves, similar results are obtained.

11·8. Transverse Oscillations of an inextensible Chain hanging from one End. Let l be the length of the chain. Take the origin at the equilibrium position of the free end and the axis Ox vertically upwards.

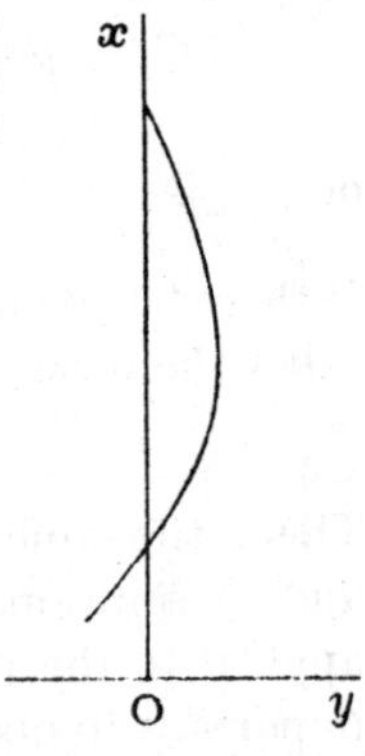

Neglecting the vertical motion the tension at the point (x, y) is $T = g\rho x$, where ρ is the line density supposed uniform. The equation of motion of an element δx at (x, y) is

$$\rho\,\delta x\,\ddot{y} = -T\frac{\partial y}{\partial x} + \left\{T\frac{\partial y}{\partial x} + \frac{\partial}{\partial x}\left(T\frac{\partial y}{\partial x}\right)\delta x\right\},$$

or

$$\rho\ddot{y} = \frac{\partial}{\partial x}\left(T\frac{\partial y}{\partial x}\right)$$

or

$$\ddot{y} = g\frac{\partial}{\partial x}\left(x\frac{\partial y}{\partial x}\right) \quad \text{..................(1).}$$

To find the normal modes assume that $y \propto e^{int}$, so that

$$\frac{\partial}{\partial x}\left(x\frac{\partial y}{\partial x}\right) + \kappa y = 0 \quad \text{...............(2),}$$

where $\kappa = n^2/g$.

If we now substitute $y = \sum_0^\infty c_r x^r$ in (2), it is easy to obtain the relation between successive coefficients in the series and shew that (2) has a solution

$$y = c_0\left\{1 - \kappa x + \frac{\kappa^2 x^2}{(2\,!)^2} - \frac{\kappa^3 x^3}{(3\,!)^2} + \dots\right\} \quad \text{..................(3),}$$

subject to the condition that $y = 0$ when $x = l$.

We may also transform (2) by the substitution $x = \frac{1}{4}gz^2$*, giving

$$\frac{\partial^2 y}{\partial z^2} + \frac{1}{z}\frac{\partial y}{\partial z} + n^2 y = 0 \quad \text{.......................(4),}$$

* Lamb's *Dynamical Theory of Sound*, 1910, § 31.

which is Bessel's Equation of order zero and it has a solution

$$y=c_0\left(1-\frac{n^2z^2}{2^2}+\frac{n^4z^4}{2^2.4^2}\cdots\right) \quad\text{.....................(5)}$$

identical with (3). The series in brackets is denoted by $J_0(nz)$, i.e. Bessel's Function of order zero. Hence when we introduce the time factor we have

$$y=c_0J_0\left(2n\sqrt{\frac{x}{g}}\right)\cos(nt+\epsilon) \quad\text{..................(6)}$$

subject to the condition that $y=0$ when $x=l$, so that possible values of n are given by the equation

$$J_0\left(2n\sqrt{\frac{l}{g}}\right)=0 \quad\text{..........................(7).}$$

11·9. Transverse Vibrations of a Stretched Membrane.

We shall suppose the membrane to be perfectly flexible and of uniform material and thickness and so stretched that the tension at every point is the same in every direction and constant throughout the motion. If T_1 denote this tension, then, as in *Hydrostatics*, Art. 101, there is a normal force on an element of area dS surrounding a point P equal to

$$T_1dS\left(\frac{1}{\rho}+\frac{1}{\rho'}\right),$$

where ρ, ρ' are the principal radii of curvature of the surface at P. If x, y, z are the coordinates of this point in the displaced position, the xy plane coinciding with the equilibrium position, and the displacement is such that squares of $\partial z/\partial x$ and $\partial z/\partial y$ can be neglected, we have

$$\frac{1}{\rho}+\frac{1}{\rho'}=\frac{\partial^2z}{\partial x^2}+\frac{\partial^2z}{\partial y^2}.$$

Hence if m is the mass of unit area

$$m\frac{\partial^2z}{\partial t^2}dS=T_1dS\left(\frac{\partial^2z}{\partial x^2}+\frac{\partial^2z}{\partial y^2}\right)$$

or

$$\frac{\partial^2z}{\partial t^2}=c^2\left(\frac{\partial^2z}{\partial x^2}+\frac{\partial^2z}{\partial y^2}\right) \quad\text{.....................(1),}$$

where $c^2=T_1/m$.

When the membrane is circular it is convenient to change x, y into polar coordinates and the equation becomes

$$\frac{\partial^2z}{\partial t^2}=c^2\left(\frac{\partial^2z}{\partial r^2}+\frac{1}{r}\frac{\partial z}{\partial r}\right) \quad\text{.....................(2),}$$

which is the form suitable for a drum head.

The hypothesis $z \propto e^{ipt}$ reduces the equations to the form

$$\frac{\partial^2 z}{\partial x^2}+\frac{\partial^2 z}{\partial y^2}+\frac{p^2}{c^2}z=0 \quad\text{.....................(3)},$$

and

$$\frac{\partial^2 z}{\partial r^2}+\frac{1}{r}\frac{\partial z}{\partial r}+\frac{p^2}{c^2}z=0 \quad\text{.....................(4)}.$$

If the membrane be rectangular and bounded by the axes and $x=a$, $y=b$, a particular integral is clearly

$$z=\sin\frac{m\pi x}{a}\sin\frac{n\pi y}{b}\cos pt,$$

where

$$p^2=c^2\pi^2\left(\frac{m^2}{a^2}+\frac{n^2}{b^2}\right),$$

and m and n are integers; and the general solution is

$$z=\sum_{m=1}^{\infty}\sum_{n=1}^{\infty}\sin\frac{m\pi x}{a}\sin\frac{n\pi y}{b}(A_{m,n}\cos pt+B_{m,n}\sin pt).$$

The solution of (4) involves the use of Bessel's Functions.

EXAMPLES

1. Shew that, if a string is of infinite length and the disturbance at time $t=0$ is given by

$$\eta=\chi(x) \text{ and } \dot{\eta}=\theta(x),$$

then

$$\eta=\tfrac{1}{2}\{\chi(x+ct)+\chi(x-ct)\}+\frac{1}{2c}\int_{x-ct}^{x+ct}\theta(z)\,dz.$$

Prove further, that if the initial disturbance is confined to a finite portion between the points $x=\pm\alpha$ and be such that $\eta=0$ and $\dot{\eta}=\theta(x)$, then, for any time t greater than α/c, there will be a portion of length $2ct-2\alpha$ which will be straight and parallel to the axis of x and at a distance $\frac{1}{2c}\int_{-\alpha}^{\alpha}\theta(z)\,dz$ from it. (Coll. Exam. 1908.)

2. A stretched string is drawn aside at $n-1$ points and let go from rest. Shew that generally the string consists of $2n-1$ straight portions; and in the case where the two points of trisection are drawn aside equal distances in the same direction, draw the shape of the string after three intervals each one-twelfth of a complete oscillation. (M.T. 1896.)

3. A uniform string is stretched between two points. Shew that if the middle point is plucked aside it will move to and fro with a constant velocity, and describe the motion of any other point of the string. (M.T. 1915.)

4. A uniform stretched string of length l, density ρ and tension $a^2\rho$ is initially at rest and the displacement of any point at a distance x from one end is $\frac{1}{2}\epsilon x(l-x)$ where ϵ is small, so that the curvature is constant and equal

to ϵ. Prove that at any subsequent time t less than $l/2a$ it consists of an arc of constant curvature ϵ and length $l-2at$ and two straight pieces, which are tangents at the ends of the arc. (Coll. Exam.)

5. A uniform string whose length is $2l$ and mass $2lm$ is stretched at tension T between two fixed points, the middle point of the string being displaced a small distance b perpendicular to the string and then released, shew that the subsequent motion of the string, referred to axes through its middle point, along and perpendicular to the string, is given by the equation

$$y=\frac{8b}{\pi^2}\sum_{r=0}^{r=\infty}\frac{1}{(2r+1)^2}\cos\frac{(2r+1)\pi x}{2l}\cos\frac{(2r+1)\pi ct}{2l},$$

where c is given by the equation $mc^2=T$. (M.T. 1900.)

6. A string of length $l+l'$ is stretched with tension P between two fixed points. The length l has mass m per unit of length, the length l' has mass m' per unit of length. Prove that the possible periods t of transverse vibration are given by the equation

$$\frac{\tan\left(\dfrac{2\pi l}{t}\sqrt{\dfrac{m}{P}}\right)}{\tan\left(\dfrac{2\pi l'}{t}\sqrt{\dfrac{m'}{P}}\right)}+\sqrt{\frac{m}{m'}}=0. \quad \text{(Coll. Exam. 1898.)}$$

7. If a slightly elastic string is stretched between two fixed points and motion is started by drawing aside through a distance b a point on the string distant one-fifth of the length l of the string from one end, the displacement at any instant will be given by the equation

$$y=\frac{25b}{2\pi^2}\sum_{n=1}^{\infty}\left(\frac{1}{n^2}\sin\frac{n\pi}{5}\sin\frac{n\pi x}{l}\cos\frac{n\pi ct}{l}\right).$$

Find the energy of the vibrating string. (Coll. Exam. 1895.)

8. A stretched string of length l has one end fixed and the other attached to a massless ring free to slide on a smooth rod. If the ring is displaced a small distance b from the position of equilibrium and the system start from rest, shew that the displacement at time t of any point of the string at distance x from the fixed end is

$$\frac{8b}{\pi^2}\sum_{0}^{\infty}\frac{(-1)^s}{(2s+1)^2}\sin\frac{(2s+1)\pi x}{2l}\cos\frac{(2s+1)\pi ct}{2l},$$

where c is the velocity of transverse vibrations.

Shew that, if $ct<l$, the shape of the string is given by

$$y=bx/l \text{ from } x=0 \text{ to } l-ct,$$
$$y=b(l-ct)/l \text{ beyond.} \quad \text{(Trinity Coll. 1905.)}$$

9. One end of a string of length l is fixed at A and the other end is fastened to the end B of a rod BC of length b which can turn freely about C. Shew that the period of a principal transverse oscillation is $2\pi/c\xi$, where ξ is a root of the equation

$$M\xi^2/3\rho-1/b=\xi\cot l\xi,$$

ρ being the line density of the string, M the mass of the rod, and c the wave velocity for the given tension. (M.T. 1899.)

10. If a stretched string be acted on at two points equidistant from the two ends by equal transverse forces Y, prove that the modes of vibration of even order are not excited and the modes of odd order are excited in the same way as if a single force $2Y$ had acted at one of the points. (M.T. 1895.)

11. A string is stretched between two given points and a given point of the string is (1) drawn aside and then let go, (2) struck by a sharp point; shew that the relative intensity of any upper partial tone to the fundamental tone is greater in the second case than in the first. (M.T. 1897.)

12. A stretched cord is held displaced from the natural straight position at a number of points, so that it assumes the form of a series of straight lines: shew that when it is let go, the form assumed at each instant in the ensuing transverse vibration will be a series of straight lines.

In the particular case when the two points of trisection of the cord are held displaced transversely by equal amounts, compute the ratios in which the harmonics of the fundamental enter into the tone of the note emitted by the cord when released. (St John's Coll. 1896.)

13. A string of length $2a$ is fixed at the two ends. The left-hand half of the string is of uniform density ρ per unit length, and the right-hand half of density ρ'. Find an equation whose roots are the frequencies of the normal modes of vibration of the string, and shew that if $m^2\rho' = n^2\rho$, where m and n are integers having no common factor, then the frequencies may be put in the form

$$\frac{Nm}{2a}\sqrt{\left(\frac{T}{\rho}\right)} + \nu_s, \quad N = 0, 1, 2, 3, \ldots,$$

where ν_s may take $n+m$ values. (M.T. 1933.)

14. A uniformly stretched string, of which the extremities are fixed, starts from rest in the form $y = A \sin \frac{m\pi x}{l}$, where m is an integer and l the distance between the fixed extremities. Prove that, if the resistance of the air be taken into account and be assumed to be $2k$ times the momentum per unit length, the displacement after any time t is

$$y = Ae^{-kt}\left(\cos m't + \frac{k}{m'} \sin m't\right) \sin \frac{m\pi x}{l},$$

where $m'^2 = \frac{m^2\pi^2c^2}{l^2} - k^2$ and c is the velocity of waves of transverse vibration. (Coll. Exam.)

15. A uniform string of length $2(l + l')$ and line density ρ is stretched between two fixed points; a length $2l'$ in the middle is uniformly wrapped with wire so that its line density becomes ρ'. Prove that, if the tension $T = c^2\rho = c'^2\rho'$, the periods of the notes which can be sounded are $2\pi/p$, where p satisfies either of the equations

$$c' \tan(pl'/c') + c \tan(pl/c) = 0 \quad \text{and} \quad \tan(pl'/c') \tan(pl/c) = c'/c.$$

(Coll. Exam. 1901.)

16. If a stretched string be held at its middle point, drawn aside at a point of quadrisection, and released from rest, prove that in the ensuing vibration the energy in the harmonic of order r is proportional to

$$r^{-2} \sin^2(r\pi/4) \sin^4(r\pi/8).$$ (St John's Coll. 1908.)

17. Find the periods of the normal modes of vibration of a tense string fixed at the ends. Prove that the period of the gravest mode is almost exactly nine-tenths of that of a simple pendulum whose length is equal to the sag in the middle (due to gravity) if the string be horizontal.

If the string consist of two portions of lengths a_1, a_2, and different densities ρ_1, ρ_2, prove that the periods $(2\pi/p)$ are determined by the equation

$$k_1 \cot k_1 a_1 + k_2 \cot k_2 a_2 = 0,$$

provided $$k_1^2 = p^2\rho_1/T, \quad k_2^2 = p^2\rho_2/T,$$

T being the tension.

Examine the case of $\rho_2 = 0$, and explain how the resulting period-equation may be solved graphically. (M.T. 1911.)

18. A uniform extensible string is stretched, at tension T, between two points A and B, distance l apart; and the wave velocity for small transverse vibrations is a. At the middle point a particle of mass m is attached. The ends A and B are given small inexorable transverse vibrations, the displacement of each at any time being $\kappa \sin mat$. Find the corresponding forced motion of the particle. (Trinity Coll. 1898.)

19. The ends of a stretched uniform string, of length l, are attached to small rings without mass which can slide on two parallel rods at right angles to the string. The middle point of the string is acted on by the transverse force $F \sin pt$. Prove that the forced vibration at a distance ξ from either end is given by

$$y = -\frac{cF}{2pT} \operatorname{cosec} \frac{pl}{2c} \cos \frac{p\xi}{c} \sin pt,$$

where c is the wave velocity and T is the tension. (Trinity Coll. 1902.)

20. Two uniform strings are attached together and stretched in a straight line between two fixed points with tension T and carry a particle of mass M attached at the point of junction. Their line-densities are ρ and ρ' and their lengths l and l'. Shew that, if $T = c^2\rho = c'^2\rho'$, the periods $2\pi/n$ of transverse vibration are given by

$$Mn = c\rho \cot \frac{nl}{c} + c'\rho' \cot \frac{nl'}{c'}.$$ (Coll. Exam. 1905.)

21. An infinitely long tense string has a mass M attached to it at one point. The string being initially straight and at rest, a transverse impulse P is given to M. Find the form of the string at any subsequent instant, and prove that the ultimate displacement of M is $P/2\rho c$, where ρ is the line density and c the velocity of transverse waves. (M.T. 1922.)

22. A transverse force $\gamma \sin pt$ acts at the point of junction of two strings of different mass per unit length which are joined at this point and stretched between two points at distance l apart, the lengths of the strings being b and $l-b$. Prove that, if c_1 and c_2 be the velocities of transverse waves in the two strings, the displacement of the point of junction of the strings at the time t is

$$\gamma \sin pt \Big/ \left\{ \frac{pT}{c_1} \cot \frac{pb}{c_1} + \frac{pT}{c_2} \cot \frac{p(l-b)}{c_2} \right\},$$

where T is the tension. (Trinity Coll. 1896.)

23. A long stretched string has a portion (of length l) in the centre whose density is $\mu^2\rho$, the density of the rest of the string being ρ. A train of simple harmonic waves approaches this portion from one end; prove that the energy in the reflected wave is to the energy in the transmitted wave in the ratio

$$(\mu^2-1)\sin^2\theta : 4\mu^2,$$

where $\theta = 2\pi l/\lambda$, the wave-length in the central part being equal to λ.

Shew also that the sum of these energies is equal to the energy in the incident wave. (St John's Coll. 1914.)

24. If the density of a stretched string be m/x^2, where x is measured from a point in the line of prolongation of the string, the ends of the string being $x=l_1$, $x=l_2$, shew that the frequency equation is

$$4p^2/c^2 = 1 + \{2n\pi/(\log l_2/l_1)\}^2,$$

where $c^2 = T/m$ and T is the tension in equilibrium, the vibrations being transversal. (M.T. 1905.)

25. If a string of length l and tension T_0 stretched between two fixed points be not uniform but of line density $\rho_0/(1+\kappa x)^2$, where x is the distance from one end, shew that the transverse vibrations are of period $2\pi/n$ when

$$\sqrt{4n^2-\kappa^2c^2}\log(1+\kappa l) = 2ic\kappa\pi,$$

where $c^2 = T_0/\rho_0$ and i is a positive integer. Examine the case of $i=0$.

(Coll. Exam. 1898.)

26. A tight string of length l hangs in the catenary $y = c\cosh x/c$, under the action of gravity, from two points, distant l apart, in the same horizontal line. If gravity be supposed suddenly to cease to act, prove that after a time t the form of the string will be given by the equation

$$y = -\frac{4cl^2}{\pi}\cosh\left(\frac{l}{2c}\right)\sum_1^\infty \frac{1}{r}\frac{\sin^2\frac{1}{2}r\pi}{l^2+r^2c^2\pi^2}\sin r\pi\left(\frac{x}{l}+\frac{1}{2}\right)\cos\frac{r\pi t\sqrt{cg}}{l} + c\cosh\frac{l}{2c},$$

c being very large compared with l. (Coll. Exam. 1898.)

27. A particle of mass M is suspended by a string whose mass is m. Shew that if the particle be slightly displaced in a vertical direction the periods of the vibration are the values of $\frac{2\pi}{z}\sqrt{\frac{ml}{\lambda}}$, where z is given by the equation $z\tan z = \frac{m}{M}$; l being the natural length and λ the modulus of elasticity of the string. (M.T. 1899.)

28. Investigate the free transverse vibrations of a tense string, taking account of the lateral yielding of the supports. Assume that each support has inertia M, and is urged towards its equilibrium position by a force equal to Mn^2 times the displacement. Taking the case of the symmetrical displacements prove that the periods $(2\pi/p)$ are given by the equation

$$x\tan x = \frac{M}{2m}\left(\frac{n^2l^2}{c^2} - 4x^2\right),$$

where m is the total mass of the string, l is the length, c is the wave velocity, and $x = pl/2c$.

Shew how to solve this equation graphically, and find approximately the change of frequency of the gravest mode due to the yielding, on the assumption that nl/c is relatively negligible and M/m is large. (M.T. 1923.)

29. A very long uniform flexible string is stretched in a straight line, the tension being T, and the line density m. A portion of the string of length l, far from the ends, receives a small transverse displacement, and is released from rest. Describe the ensuing motion, and find an expression for the displacement at any point of the string at any subsequent time, the given displacement being denoted by $f(x)$, where $0<x<l$. Shew that the ratio of the kinetic energy to the potential energy of the string changes in time $\frac{1}{2}l(m/T)^{\frac{1}{2}}$ from 0 to 1, and afterwards remains equal to 1.

A bead of mass M is fastened to the string at a point $x=0$, and a train of waves in which the displacement is $A\sin\frac{2\pi}{\lambda}(x-ct)$ advances towards the bead. Shew that after passing the bead the energy per unit length of the waves is diminished in the ratio

$$1:1+(M\pi/\lambda m)^2;$$

and find the change of phase on passing the bead. (M.T. 1910.)

30. A uniform string of great length and of line density Tc^{-2} has one end fixed, carries a mass M at a distance a from the fixed end, and is stretched with tension T. A train of transverse waves of period $2\pi/p$ is coming along the string and is being reflected; prove that the change of phase that accompanies the reflection at M is

$$2\tan^{-1}\left\{\frac{Mpc}{T}-\cot\frac{pa}{c}\right\}.$$ (St John's Coll. 1905.)

31. A uniform string is of indefinite length, stretching from $x=-\infty$ to $x=0$, and is at tension T; at its end $(x=0)$ it is tied to two strings of similar make to the first, each at tension $\frac{1}{2}T$, which stretch from $x=0$ to $x=+\infty$ nearly parallel to each other. A harmonic train of waves of transverse vibrations perpendicular to the plane of the string, is continually advancing on the first string along the axis of x towards the junction; its amplitude is k. Prove that the amplitude of the transmitted trains and that of the reflected train are $2(\sqrt{2}-1)k$ and $(\sqrt{2}-1)^2k$ respectively, where the mass of the knot is neglected. (Trinity Coll. 1908.)

32. If a stretched elastic string is of great length and its end A is fastened to one end of an elastic string of different material, whose other end B is fixed, shew that if a train of longitudinal waves of period $2\pi/p$ advances upon A, the reflected train is of equal amplitude. Shew also that each portion of the string forms stationary waves, the amplitudes of the waves in AB and in the rest of the string being in the ratio $\sin\alpha:\sin\frac{pl}{c'}$, where m', c' are the line mass and wave velocity for the portion AB, m, c are the corresponding quantities for the rest of the string, l is the length AB and

$$\tan\alpha=\frac{mc}{m'c'}\tan\frac{pl}{c'}.$$ (M.T. 1908.)

33. Longitudinal waves come from infinity along the string (0), are transmitted through a string of length l and proceed to infinity along the string (1), shew that the amplitude is lessened in the ratio

$$\left\{\left(1+\frac{\rho_1 c_1}{\rho_0 c_0}\right)^2 \cos^2(nl/c)+\left(\frac{\rho_1 c_1}{\rho c}+\frac{\rho c}{\rho_0 c_0}\right)^2 \sin^2(nl/c)\right\}^{\frac{1}{2}} : 2,$$

where $n/2\pi$ is the frequency. (St John's Coll. 1895.)

34. A stretched string, infinite in both directions, is of density ρ, when undisturbed, and has attached to it a single particle of mass m. The velocity of waves of longitudinal displacement in the string is c. An infinite harmonic train of such waves, such that the period of the displacement of each point of the string is $2\pi/p$, impinges on the particle. Prove that the train is partly transmitted and partly reflected: that the energies per wave length of the incident, the reflected and transmitted trains are as $m^2p^2+4\rho^2c^2$ to m^2p^2 to $4\rho^2c^2$; and that the change of phase of the transmitted train is $\tan^{-1}\dfrac{mp}{2\rho c}$. (Trinity Coll. 1897.)

35. A stretched string is in equilibrium with its ends fixed; shew that, on being slightly disturbed from its position of equilibrium, the potential energy of deformation per unit length of stretched string is

$$m\left[b^2\frac{\partial \xi}{\partial x}+\tfrac{1}{2}\left\{a^2\left(\frac{\partial \xi}{\partial x}\right)^2+b^2\left[\left(\frac{\partial \eta}{\partial x}\right)^2+\left(\frac{\partial \zeta}{\partial x}\right)^2\right]\right\}\right],$$

where m is the equilibrium line mass, and a, b the longitudinal and transverse wave velocities. Deduce the equations of vibration. (M.T. 1905.)

36. A uniform extensible string is stretched with its ends fixed and simultaneously executes in a plane free longitudinal motions, which are not necessarily small, and transverse vibrations which are small. The coordinates of any point in the string when undisturbed are $(\xi, 0)$ and at the time t $(\xi+z, y)$, prove that

$$\frac{\partial^2 z}{\partial t^2}=\frac{T_1+\lambda}{\rho_1}\frac{\partial^2 z}{\partial \xi^2},$$

$$\frac{\partial^2 y}{\partial t^2}=\frac{T_1}{\rho_1}\frac{\partial^2 y}{\partial \xi^2}+\frac{\lambda}{\rho_1}\frac{\partial}{\partial \xi}\left\{\frac{\partial y}{\partial \xi}\frac{\partial z}{\partial \xi}\Big/\left(1+\frac{\partial z}{\partial \xi}\right)\right\},$$

where T_1, ρ_1 are the undisturbed tension and line density, λ is the coefficient of elasticity and y, $\dfrac{\partial y}{\partial \xi}$ are assumed to be always small.

(Trinity Coll. 1903.)

37. A uniform rod of mass M is freely pivoted at its mid-point, and its ends are fastened to the mid-points of two stretched strings, one elastic, the other inextensible. There is equilibrium when the rod is vertical, and the strings are straight, horizontal and perpendicular to one another. Shew that the period $2\pi/p$ of a small oscillation of the system satisfies the equation

$$\tfrac{1}{6}Mp=\frac{T}{\alpha}\cot\frac{p}{\alpha}l+\frac{E}{\beta}\cot\frac{p}{\beta}l',$$

where T, $2l$, $2lT/\alpha^2$, are the tension, length and mass of the inextensible string, and E, $2l'$, $2l'E/\beta^2$, the modulus, equilibrium length and mass of the other. (St John's Coll. 1903.)

38. A uniform extensible string has its two ends fixed, and is stretched when in equilibrium to a length $l_1 + l_2$. At a distance l_1 from one end a ring of mass m is attached, which can slide on a smooth fixed rod making an angle α with the undisturbed string which is straight. Prove that the periods $2\pi/p$ of small oscillations of the system are given by

$$mp = \rho b \cos^2 \alpha (\cot pl_1/b + \cot pl_2/b) + \rho a \sin^2 \alpha (\cot pl_1/a + \cot pl_2/a);$$

where ρ is the density per unit length and a and b are respectively the wave velocities of transverse and longitudinal disturbances of the string as thus stretched. (Trinity Coll. 1899.)

39. If a membrane be a rectangle of edges a and b shew that

$$z = A \sin pt \sin \frac{m\pi x}{a} \sin \frac{n\pi y}{b}$$

is a possible form of stationary vibrations, where

$$\left(\frac{p}{\pi}\right)^2 = c^2 \left(\frac{m^2}{a^2} + \frac{n^2}{b^2}\right),$$

the origin being at a corner, and c being the velocity of propagation of a rectilinear disturbance across the membrane. If $b = a/\sqrt{2}$, shew that there are two such modes of vibration of period $\tau/\sqrt{11}$, τ being the period of vibration. (Univ. of London, 1911.)

40. If a stretched membrane be of the shape of a sector of a circle of angle 72°, shew how to calculate its natural tones.

(Univ. of London, 1907.)

CHAPTER XII

SOUND WAVES

12·1. A FEW simple appeals to experience shew that sound is transmitted by waves in the atmosphere. If a bell is rung under the receiver of an air pump from which the air is gradually exhausted the sound becomes fainter and soon ceases to affect the organs of the ear; shewing that atmospheric communication is necessary between the ear and the disturbance that causes the sound. We infer that sound is accompanied by the motion of the intervening medium from the fact that a musical note sounded on any instrument may produce a vibration, in unison with it, in another body not in contact with it. That the motions of the medium are small is evident from the fact that sound will travel through a dust-laden atmosphere without perceptible motion of the dust.

In this chapter we shall consider the propagation of waves in an elastic fluid, confining our attention for the most part to plane waves.

12·11. General equations. In considering the propagation of sound waves we shall regard the velocities of the elements of fluid as so small that their squares may be neglected. In the kinetic theory of gases, a mass of gas is regarded as composed of a large number of separate molecules moving in different directions with velocities which undergo frequent changes owing to the collisions of the molecules; but the hypothesis that we now make about the magnitude of the velocity of a fluid element in wave propagation does not contravene this conception of a gas, because what we take to be the velocity of a fluid element in a given direction is the average velocity in that direction of the molecules composing the element; and there is nothing in the molecular hypothesis to prevent this average velocity from being small, since molecules may move in opposite directions.

Neglecting friction, the motion being due to natural causes must be irrotational, so that the pressure equation is

$$\int \frac{dp}{\rho} = \frac{\partial \phi}{\partial t} - \tfrac{1}{2}q^2 - V + F(t) \quad \text{..............(1).}$$

If ρ_0 denotes the equilibrium density of a mass of fluid which is compressed until its density becomes

$$\rho=\rho_0(1+s),$$

s is called the *condensation*.

When the condensation s and the velocities u, v, w are small, the equation of continuity

$$\frac{\partial\rho}{\partial t}+\frac{\partial\rho u}{\partial x}+\frac{\partial\rho v}{\partial y}+\frac{\partial\rho w}{\partial z}=0$$

becomes

$$\frac{\partial s}{\partial t}+\frac{\partial u}{\partial x}+\frac{\partial v}{\partial y}+\frac{\partial w}{\partial z}=0,$$

or

$$\frac{\partial s}{\partial t}=\nabla^2\phi \quad \text{..........................(2).}$$

Again, if $p=p_0+\delta p$ denotes the pressure when the density is ρ, p_0 being the equilibrium pressure, and if we neglect q^2 and all impressed forces, (1) may be written

$$\frac{\delta p}{\rho_0}=\frac{\partial\phi}{\partial t} \quad \text{..........................(3).}$$

But if we assume that p is a function of ρ we have

$$\delta p=p-p_0=\left(\frac{dp}{d\rho}\right)_0(\rho-\rho_0)+\ldots$$

$$=\left(\frac{dp}{d\rho}\right)_0\rho_0 s \text{ to the first power of } s,$$

or

$$\delta p=c^2\rho_0 s, \text{ where } c^2=(dp/d\rho)_0 \quad \text{.................(4).}$$

Hence (3) becomes

$$c^2 s=\frac{\partial\phi}{\partial t} \quad \text{..........................(5);}$$

and by eliminating s between (2) and (4) we get

$$\frac{\partial^2\phi}{\partial t^2}=c^2\nabla^2\phi \quad \text{.......................(6).}$$

12·12. The simplest case is that in which the wave fronts are planes. If we take the x axis perpendicular to the wave fronts the last equation reduces to

$$\frac{\partial^2\phi}{\partial t^2}=c^2\frac{\partial^2\phi}{\partial x^2} \quad \text{.......................(7),}$$

the solution of which is

$$\phi=f(x-ct)+F(x+ct) \quad \text{.................(8),}$$

representing the propagation of independent waves in the positive and negative directions with the same velocity c.

12·13. The Velocity of Sound. The quantity c of **12·12** which represents the velocity of propagation of sound waves, within the limits of the approximations which led to (6), is clearly independent of the form of the waves. It was defined in **12·11** by the relation $c^2=(dp/d\rho)_0$; and it is possible to calculate a numerical value when the relation connecting p and ρ is known. Newton adopted Boyle's Law $p=\kappa\rho$ as the basis of his investigation. This makes $c=\sqrt{\kappa}=\sqrt{(p_0/\rho_0)}=279{\cdot}945$ metres per second at 0° C., falling short of the result of observation by about one-sixth part*. The discrepancy is due to the fact that Boyle's Law requires the compressions and rarefactions to take place isothermally, whereas it is a matter of observation that the compression of a gas is always accompanied by a rise in temperature. The hypothesis that the vibrations are so rapid that there is no time for a gain or loss of quantity of heat, i.e. that the relation between p and ρ is the adiabatic one $p=\kappa\rho^{\gamma}$†, leads to a result more in accordance with observation. This makes

$$c^2=(dp/d\rho)_0=\gamma p_0/\rho_0 \quad \text{..................(1)},$$

and if we take $\gamma=1{\cdot}41$, we get $c=332$ metres per second at 0° C., which agrees with the result of experiment.

12·14. Plane Waves. Instead of using the velocity potential we may obtain the equation for plane waves directly in terms of the displacement ξ of a layer of particles whose abscissa is x when undisturbed. Thus the stratum which in equilibrium is of density ρ_0 between the planes x and $x+\delta x$ becomes at time t a stratum of density ρ between the planes $x+\xi$ and $x+\delta x+\xi+\frac{\partial\xi}{\partial x}\delta x$, so that, from constancy of mass,

$$\rho_0\delta x=\rho\left(\delta x+\frac{\partial\xi}{\partial x}\delta x\right)$$

or

$$\rho_0=\rho\left(1+\frac{\partial\xi}{\partial x}\right) \quad \text{......................(1)}.$$

The equation of motion of unit area of this stratum is

$$\rho_0\delta x\,\ddot{\xi}=-\frac{\partial p}{\partial x}\delta x,$$

or

$$\rho_0\ddot{\xi}=-\frac{\partial p}{\partial x} \quad \text{........................(2)}.$$

* Rayleigh, *Theory of Sound*, II, p. 19. † *Hydrostatics*, Art. 94.

But $$\frac{p}{p_0}=\left(\frac{\rho}{\rho_0}\right)^\gamma=\left(1+\frac{\partial\xi}{\partial x}\right)^{-\gamma} \text{ from (1);}$$

hence (2) becomes $$\ddot{\xi}=\gamma\frac{p_0}{\rho_0}\frac{\partial^2\xi}{\partial x^2}\left(1+\frac{\partial\xi}{\partial x}\right)^{-\gamma-1} \quad\text{.................(3).}$$

This is an exact equation giving ξ in terms of x and t; but from (1) we note that $\partial\xi/\partial x=-s$, so that if we take the last factor in (3) to be unity and use **12·13** (1) we get an equation

$$\ddot{\xi}=c^2\frac{\partial^2\xi}{\partial x^2} \quad\text{..............................(4)}$$

which is correct to the same order of smallness as the equation **12·12** (7), and its solution (**10·2**) is

$$\xi=f(ct-x)+F(ct+x) \quad\text{.................(5).}$$

12·2. Energy. *In a plane progressive wave the energy is half kinetic and half potential.*

The *kinetic energy* is $$\tfrac{1}{2}\rho_0\int\dot{\xi}^2 dv \quad\text{..........................(1)}$$

integrated over the space occupied by the disturbed air when in its equilibrium state.

The *potential energy* of an element is the work stored up in compression, or the work that it would do in expanding from its compressed to its equilibrium state. Consider an element which in the equilibrium state has volume dv_0 and density ρ_0 and in the compressed state has density $\rho=\rho_0(1+s)$ and therefore volume $dv=dv_0/(1+s)$. Let this element expand from the compressed state. At any stage of the expansion the volume $dv'=dv_0/(1+s')$, where s' is the condensation, and an infinitesimal increment in this volume is

$$\delta(dv')=-\frac{dv_0}{(1+s')^2}\delta s'.$$

The effective part of the pressure at this stage is $\delta p=c^2\rho_0 s'$ (**12·11**); therefore the work done in this small expansion is

$$-c^2\rho_0 dv_0\frac{s'\delta s'}{(1+s')^2}.$$

And as the condensation decreases from s to 0 the work done by the element dv_0

$$=-c^2\rho_0 dv_0\int_s^0\frac{s'ds'}{(1+s')^2}$$

$$=\tfrac{1}{2}\rho_0 c^2 s^2 dv_0 \text{ to the second order of } s.$$

Hence the potential energy of the whole mass of gas is

$$\tfrac{1}{2}\rho_0\int c^2 s^2\, dv_0 \quad \text{......................(2)},$$

the integration extending over the space occupied by the disturbed air when in its equilibrium state.

For a plane progressive wave

$$\xi = f(ct - x)$$

so that
$$\dot{\xi} = cf'(ct-x) = -c\frac{\partial \xi}{\partial x} = cs, \qquad \textbf{12·14}\ (1),$$

and by comparing (1) and (2) it follows that the kinetic and potential energies are equal.

12·21. Intensity of Sound. The rate at which energy is transmitted across unit area of a plane parallel to the front of a progressive wave may be taken as a measure of the intensity of the radiation.

If W is the energy transmitted in time t then

$$\frac{dW}{dt} = p\dot{\xi} = (p_0 + \delta p)\,\dot{\xi}.$$

For a simple harmonic wave

$$\xi = A\cos\frac{2\pi}{\lambda}(x - ct) \quad \text{......................(1)},$$

where λ is the wave length and c, the wave velocity, is the same for all wave lengths. And from **12·11** (4) and **12·14** (1)

$$\delta p = \rho_0 c^2 s = -\rho_0 c^2 \frac{\partial \xi}{\partial x}.$$

Therefore

$$\frac{dW}{dt} = \left\{p_0 + \rho_0 c^2 A\frac{2\pi}{\lambda}\sin\frac{2\pi}{\lambda}(x - ct)\right\} A\frac{2\pi c}{\lambda}\sin\frac{2\pi}{\lambda}(x - ct)$$

$$= \tfrac{1}{2}\rho_0 A^2 c^3\left(\frac{2\pi}{\lambda}\right)^2 + \text{periodic terms} \quad \text{......................(2)}.$$

This is the required measure of the intensity, and by integration the energy transmitted in any given time is found; and for any number of periods or for any interval of time so long that a fraction of a period is negligible we have

$$\frac{W}{t} = \tfrac{1}{2}\rho_0 A^2 c^3\left(\frac{2\pi}{\lambda}\right)^2 \quad \text{......................(3)}.$$

This represents the average intensity and it may also be expressed in the form

$$\frac{W}{t}=\tfrac{1}{2}\rho_0\xi_1{}^2c\left(\frac{2\pi}{T}\right)^2 \quad\ldots\ldots\ldots\ldots\ldots\ldots\ldots(4),$$

where T is the period λ/c and ξ_1 is the maximum displacement.

Again, since $$s=-\frac{\partial\xi}{\partial x}=\frac{2\pi A}{\lambda}\sin\frac{2\pi}{\lambda}(x-ct),$$

the maximum condensation $s_1=2\pi A/\lambda$, and the average intensity may also be written

$$\frac{W}{t}=\tfrac{1}{2}\rho_0 s_1{}^2c^3 \quad\ldots\ldots\ldots\ldots\ldots\ldots\ldots(5).$$

Though these formulae for the rate of transmission of energy across a unit of area of the wave front have been obtained for plane waves of harmonic type, they will also hold good for all harmonic waves at a sufficient distance from the disturbing source.

We note that if the wave be given by a velocity potential

$$\phi=A\cos\frac{2\pi}{\lambda}(x-ct)$$

the foregoing formula (3) needs slight modification, for since

$$\xi=-\frac{\partial\phi}{\partial x},$$

therefore $$\xi=\frac{A}{c}\cos\frac{2\pi}{\lambda}(x-ct) \quad\ldots\ldots\ldots\ldots\ldots\ldots\ldots(6),$$

but formulae (4) and (5) are unaltered.

12·3. Exact Equation and its solution. Change of type. We may obtain the differential equation which gives the actual position of a layer at time t in terms of t and the equilibrium position x, by writing $y=x+\xi$ in **12·14** (3) which takes the form

$$\frac{\partial^2 y}{\partial t^2}=c^2\frac{\partial^2 y}{\partial x^2}\left(\frac{\partial y}{\partial x}\right)^{-\gamma-1} \quad\ldots\ldots\ldots\ldots\ldots\ldots\ldots(1).$$

To solve this equation let $$\frac{\partial y}{\partial t}=f\left(\frac{\partial y}{\partial x}\right);$$

therefore $$\frac{\partial^2 y}{\partial t^2}=f'\left(\frac{\partial y}{\partial x}\right)\frac{\partial^2 y}{\partial x\,\partial t}=\left\{f'\left(\frac{\partial y}{\partial x}\right)\right\}^2\frac{\partial^2 y}{\partial x^2},$$

and by comparing this equation with (1) we get

$$f'\left(\frac{\partial y}{\partial x}\right)=\pm c\left(\frac{\partial y}{\partial x}\right)^{-\frac{1}{2}(\gamma+1)},$$

so that $$\frac{\partial y}{\partial t}=f\left(\frac{\partial y}{\partial x}\right)=A\pm\frac{2c}{\gamma-1}\left(\frac{\partial y}{\partial x}\right)^{-\frac{1}{2}(\gamma-1)} \quad\ldots\ldots\ldots\ldots\ldots\ldots(2).$$

Again for a progressive wave with no translation of the medium as a whole $\partial y/\partial t = 0$ when $\rho = \rho_0$, that is when $\partial y/\partial x = 1$; therefore

$$A = \mp 2c/(\gamma - 1),$$

and
$$\frac{\partial y}{\partial t} = \mp \frac{2c}{\gamma - 1}\left\{1 - \left(\frac{\partial y}{\partial x}\right)^{\frac{1}{2}(1-\gamma)}\right\} \quad \text{......(3).}$$

The complete integral of this differential equation is of the form

$$y = \alpha x + \beta t + C,$$

provided the constants α, β are chosen so as to satisfy (3), that is provided

$$\beta = \mp \frac{2c}{\gamma - 1}\{1 - \alpha^{\frac{1}{2}(1-\gamma)}\}.$$

Hence the complete integral is

$$y = \alpha x \mp \frac{2c}{\gamma - 1}\{1 - \alpha^{\frac{1}{2}(1-\gamma)}\}\, t + C \quad \text{......(4),}$$

and the general integral is the result of eliminating α between

$$\left.\begin{aligned} y &= \alpha x \mp \frac{2c}{\gamma - 1}\{1 - \alpha^{\frac{1}{2}(1-\gamma)}\}\, t + \phi(\alpha) \\ \text{and} \quad 0 &= x \mp c\alpha^{-\frac{1}{2}(1+\gamma)}\, t + \phi'(\alpha) \end{aligned}\right\} \quad \text{......(5),}$$

where ϕ is an arbitrary function.

Taking the upper sign, if u denote the velocity $\dot{y}$, we have

$$u = -\frac{2c}{\gamma - 1}\{1 - \alpha^{\frac{1}{2}(1-\gamma)}\},$$

and, eliminating x from (5),

$$y = -\frac{ct}{\gamma - 1}\{2 - (\gamma + 1)\,\alpha^{\frac{1}{2}(1-\gamma)}\} + \phi(\alpha) - \alpha\phi'(\alpha),$$

so that
$$y - \{c + \tfrac{1}{2}(\gamma + 1)\, u\}\, t = \phi(\alpha) - \alpha\phi'(\alpha).$$

Hence $y - \{c + \frac{1}{2}(\gamma + 1)\, u\}\, t$ is an arbitrary function of α and therefore of u, and conversely u is an arbitrary function of $y - \{c + \frac{1}{2}(\gamma + 1)\, u\}\, t$, and we may write

$$u = f\,[y - \{c + \tfrac{1}{2}(\gamma + 1)\, u\}\, t] \quad \text{......(6),}$$

where f is an arbitrary function.

This equation was given by Poisson for the special case $\gamma = 1$*. The equation shews that a progressive wave in air cannot be propagated without change of type. A relation $u = f(y - ct)$ would represent the propagation of u with uniform velocity c, and relation (6) shews that if we draw a curve whose ordinate represents the value of u corresponding to the abscissa y at any instant, then the form of the curve at time t later is got by moving each point of the original curve a distance $\{c + \frac{1}{2}(\gamma + 1)\, u\}\, t$ in the direction of propagation, and as this is a different quantity for the different points of the curve it follows that the curve is continually changing shape and a discontinuity will occur as soon as the velocity curve has a vertical tangent, after which we cannot infer that the integral has a real application.

* *Journal de l'École Polytechnique*, VII, p. 319.

12·31. Condition for permanence of type. To find the condition that a train of plane waves may be propagated unchanged in type, we impose on the whole mass of air a velocity equal and opposite to that of propagation so that if the wave form is permanent it becomes stationary in space and the motion becomes steady.

If u_0, p_0, ρ_0 denote the velocity, pressure and density in the undisturbed state of the fluid and u, p, ρ are the corresponding quantities at a point in the wave, the equation of continuity is

$$\rho u = \rho_0 u_0 \quad \text{.......................(1)},$$

and the pressure equation is

$$\int_{p_0}^{p} \frac{dp}{\rho} = \tfrac{1}{2}u_0^2 - \tfrac{1}{2}u^2 \quad \text{.......................(2)}.$$

If we eliminate u we get $\displaystyle\int_{p_0}^{p} \frac{dp}{\rho} = \tfrac{1}{2}u_0^2 (1 - \rho_0^2/\rho^2)$ ....(3);

so that

$$\frac{dp}{d\rho} = u_0^2 \rho_0^2/\rho^3 \quad \text{.......................(4)},$$

or

$$p = \text{const.} - u_0^2 \rho_0^2/\rho \quad \text{.......................(5)}.$$

This relation must exist between pressure and density in order that the wave may maintain itself. As this relation between the pressure and density of the atmosphere is an impossibility a train of waves cannot maintain itself unchanged in form. If however the variations in density are small, the condition is approximately satisfied by taking $u_0 = \sqrt{(dp/d\rho)}$, and this hypothesis is the basis of our theory to the order of approximation to which it is carried.

12·4. Vibrations in Tubes. Using ξ to denote displacement the general solution for a plane wave is, as in **12·14**,

$$\xi = f(ct - x) + F(ct + x) \quad \text{.................(1)}.$$

If there be a *fixed barrier* at the origin parallel to the wave fronts then $\xi = 0$ when $x = 0$ for all values of t; therefore

$$0 = f(ct) + F(ct),$$

or $F = -f$; so that $\xi = f(ct - x) - f(ct + x)$ ....(2).

The first term may be regarded as a wave system approaching the origin from the left and the second term as the reflected system. The two have equal amplitudes, the velocity $\dot{\xi}$ is reversed in the reflected system, but the condensation s $(= -\partial\xi/\partial x)$ has its sign unchanged.

Another type of boundary condition is the hypothesis of a *surface of constant pressure*, i.e. $\delta p = 0$, but $\delta p = c^2 \rho_0 s$ (**12·11**), therefore $s = 0$, or $\partial\xi/\partial x = 0$. If this condition holds at the origin for all values of t we have

$$-f'(ct) + F'(ct) = 0.$$

Hence f and F differ only by a constant which we may omit as it would merely imply a displacement of the whole mass; therefore in this case

$$\xi = f(ct - x) + f(ct + x) \quad \text{.................(3)},$$

and as before the first term may be taken to represent an incident train on the left of the origin and the second term the reflected train. The velocity $\dot{\xi}$ is now reflected unchanged but the condensation $s\ (= -\partial\xi/\partial x)$ has its sign reversed.

The condition $s = 0$ is realized approximately at the open end of a tube whose diameter is negligible compared to the wave length.

12·41. Normal Modes for a uniform straight Tube. The equation to be solved is

$$\frac{\partial^2 \xi}{\partial t^2} = c^2 \frac{\partial^2 \xi}{\partial x^2} \quad \text{.........................(1)}$$

and as in **11·21**, to find the normal modes we assume that $\xi \propto \cos(nt + \epsilon)$, so that $\ddot{\xi} = -n^2 \xi$, and the equation becomes

$$\frac{\partial^2 \xi}{\partial x^2} + \frac{n^2}{c^2}\xi = 0 \quad \text{.......................(2)},$$

and the complete solution including the time factor is

$$\xi = (A \cos nx/c + B \sin nx/c) \cos(nt + \epsilon) \quad \text{........(3)},$$

representing stationary waves, the corresponding progressive waves in free air being of length $\lambda = 2\pi c/n$.

(1) *Tube closed at both ends $x = 0$ and $x = l$.* We have $\xi = 0$ when $x = 0$ and $x = l$. Therefore

$$A = 0 \text{ and } \sin nl/c = 0.$$

Hence $nl/c = m\pi$, where m is an integer, gives the frequencies of the normal modes, and

$$\xi = \sum_{m=1}^{\infty} B_m \sin \frac{m\pi x}{l} \cos\left(\frac{m\pi ct}{l} + \epsilon_m\right).$$

The frequency of the gravest tone is $n/2\pi$ or $c/2l$; that is, the period $2l/c$ is twice the time taken for a pulse to travel the length of the tube. In any particular normal mode, say the mth, there is a series of *nodes*, or points for which $\xi = 0$, at intervals l/m along the tube, and a series of *loops* or points of zero condensation $(\partial\xi/\partial x = 0)$ half-way between the nodes.

(2) *Tube open at both ends.* We have $-\partial\xi/\partial x = s = 0$ when $x = 0$ and $x = l$. Therefore

$$B = 0 \text{ and } \sin nl/c = 0,$$

so that the frequencies of the normal modes are the same as in the last case, and

$$\xi = \sum_{m=1}^{\infty} A_m \cos \frac{m\pi x}{l} \cos\left(\frac{m\pi ct}{l} + \epsilon_m\right).$$

The nodes and loops are distributed at the same distances as in the closed tube, but in the open tube the ends of the tube are loops.

(3) *Tube open at $x = l$ and closed at $x = 0$.* Now we have $\xi = 0$ when $x = 0$, and $-\partial\xi/\partial x = 0$ when $x = l$. Therefore

$$A = 0 \text{ and } \cos nl/c = 0.$$

Hence $nl/c = m\pi/2$, where m is an *odd* integer, gives the frequencies of the normal modes, and

$$\xi = \sum_{p=0}^{\infty} B_{2p+1} \sin \frac{(2p+1)\pi x}{2l} \cos\left(\frac{(2p+1)\pi ct}{2l} + \epsilon_{2p+1}\right).$$

The frequency of the gravest mode is now $n/2\pi$ or $c/4l$, so that the period $4l/c$ is in this case four times the time taken by a pulse to travel the length of the tube. In the pth normal mode the nodes will be at distances $2l/(2p-1)$ apart and there is of course a node at one end of the tube and a loop at the other.

The period of the gravest mode in each of the foregoing cases may also be obtained from the considerations of **12·4** by considering a pulse of condensation to start from a point P in the tube and travel towards the end A, if A is a closed end in the reflected wave the sign of s is unaltered and that of ξ is reversed, and the same happens when the reflected wave reaches B, and after time $2l/c$ the wave is passing P again under the same conditions as at first. A similar argument holds for a tube open at both ends.

For a 'stopped tube', i.e. a tube open at one end A and closed at the other B, under similar circumstances, at the reflections at A the sign of s is changed and that of ξ is unchanged and at the reflections at B the sign of s is unchanged and that of ξ reversed, so that it is not until after four reflections or an interval $4l/c$ that the pulse passes through P again under exactly the same conditions as initially.

Hence in every case the frequency of the gravest mode varies inversely as the length of the tube and for a stopped tube the gravest mode has half the frequency or is an octave lower than for an open or closed tube of the same length.

12·42. Since the velocity potential ϕ satisfies (1) of **12·41** its value is also given by

$$\phi=(A\cos nx/c+B\sin nx/c)\cos(nt+\epsilon)$$

with the conditions $\partial\phi/\partial x=0$ at a closed end of the tube and $\partial\phi/\partial t=0$ at an open end, since $c^2s=\partial\phi/\partial t$. This method of course leads to the same results as are obtained in **12·41**.

12·43. Forced Vibrations in a Tube. Let a vibration of given frequency $n/2\pi$ be maintained at one end of a straight tube. The motion may be due for example to the inexorable motion of a piston at the origin, so that $\xi=C\cos(nt+\epsilon)$ when $x=0$. Taking the solution

$$\xi=(A\cos nx/c+B\sin nx/c)\cos(nt+\epsilon)$$

we must have $A=C$, and if the tube be closed at $x=l$,

$$0=C\cos nl/c+B\sin nl/c,$$

so that
$$\xi=C\frac{\sin n(l-x)/c}{\sin nl/c}\cos(nt+\epsilon)\quad\text{.................(1).}$$

But if the tube be open at $x=l$ so that $\partial\xi/\partial x=0$ at this end, then

$$0=-C\sin nl/c+B\cos nl/c,$$

and
$$\xi=C\frac{\cos n(l-x)/c}{\cos nl/c}\cos(nt+\epsilon)\quad\text{.................(2).}$$

In the first case the amplitude of the displacements is a minimum if $\sin nl/c=\pm 1$, i.e. if l is an odd multiple of $\pi c/2n$ or $\frac{1}{4}\lambda$, and as, in this case, $x=l$ is a closed end this makes $x=0$ a loop. If l is an even multiple of $\frac{1}{4}\lambda$, the amplitude appears to be infinite, but the origin would have to be a node which is precluded by the conditions of the forced motion at the origin.

In the second case, in like manner, if l is an odd multiple of $\pi c/2n$ or $\frac{1}{4}\lambda$, the amplitude according to (2) is infinite, but if $x=l$ were really a loop the origin in this case would have to be a node and so the solution again fails.

In cases in which $\sin nl/c$ or $\cos nl/c$ is small the amplitude will be large, and if the tube contains a little fine sand, or lycopodium powder the positions of the nodes will be rendered visible. This method was used by Kundt* in experiments for comparing the velocity of sound in different gases.

12·44. Piston controlled by a Spring. As another example let us find the frequency equation when the end $x=l$ of the tube is closed and, at the end $x=0$, there is a piston of mass M controlled by a spring of strength μ.

Assuming that $\xi\propto e^{int}$, equation (1) of **12·41** takes the form

$$\frac{\partial^2\xi}{\partial x^2}+\frac{n^2}{c^2}\xi=0,\quad\text{..............................(1)}$$

and has a solution
$$\xi=\left(A\cos\frac{nx}{c}+B\sin\frac{nx}{c}\right)e^{int}\quad\text{.................(2).}$$

At $x=l$ we have $\xi=0$, so that

$$A\cos\frac{nl}{c}+B\sin\frac{nl}{c}=0,$$

* *Pogg. Ann.* CXXXV, 1868, p. 337. See also Rayleigh's *Theory of Sound*, II, Art. 260.

and (2) may be written $\xi = C \sin \dfrac{n(l-x)}{c} e^{int}$.

For the motion of the piston, supposed to be of unit area

$$M\ddot{\xi} + \mu\xi = -\delta p = -c^2\rho s = c^2\rho \frac{\partial \xi}{\partial x} \quad \text{at } x=0.$$

Therefore $$(\mu - Mn^2)\, C \sin\frac{nl}{c} = -c^2\rho C \frac{n}{c} \cos\frac{nl}{c},$$

and the frequency is given by

$$\tan nl/c = \frac{\rho nc}{Mn^2 - \mu} \qquad \text{.......................(3)}.$$

12·45. Sound Waves in a Branching Pipe. A solution may be obtained by assuming expressions of the form

$$(E \cos nx/c + F \sin nx/c)\, e^{int}$$

for the velocity potential in each branch A, B, C and determining the constants so as to satisfy the conditions at the junction O, viz.

(i) the pressure at O must be the same in each branch, i.e. $\partial\phi/\partial t$ has the same value at O for each branch;

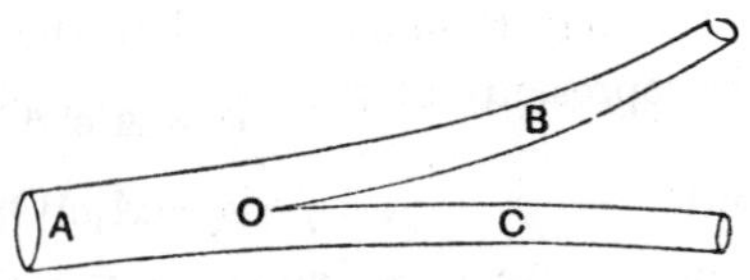

(ii) velocity × cross section in A = sum of velocity × cross section in B and C.

These conditions together with the conditions obtained from data as to whether the ends of the pipe are open or closed will suffice to give the ratios of the constants and an equation for the frequency.

12·5. Reflection and Refraction of Plane Waves.

When a train of plane waves reaches the surface of separation of two distinct media, there is a reflected and a transmitted train of waves. Let the plane yz separate the two media and let the wave fronts be oblique to this plane, the z axis being taken parallel to the line of intersection of the wave fronts with the yz plane.

Let the x axis be drawn into the first medium and suppose c, c_1 to be the velocities of sound in the two media.

The equations for the velocity potentials in the two media are

$$\frac{\partial^2\phi}{\partial t^2} = c^2\left(\frac{\partial^2\phi}{\partial x^2} + \frac{\partial^2\phi}{\partial y^2}\right) \qquad \text{....................(1)},$$

and $$\frac{\partial\phi}{\partial t} = c^2 s \qquad \text{..............................(2)},$$

for the first; and $$\frac{\partial^2\phi_1}{\partial t^2} = c_1^2\left(\frac{\partial^2\phi_1}{\partial x^2} + \frac{\partial^2\phi_1}{\partial y^2}\right) \qquad \text{..................(3)},$$

and $$\frac{\partial\phi_1}{\partial t} = c_1^2 s_1 \qquad \text{...........................(4)},$$

for the second.

The special conditions to be satisfied at the boundary $x=0$ are

(i) continuity of velocity normal to the boundary, i.e.

$$\partial\phi/\partial x = \partial\phi_1/\partial x \quad \text{.....................(5)};$$

(ii) continuity of pressure, i.e. $\delta p = \delta p_1$, if these denote the small increments of pressure due to the wave motion.

But $\delta p = c^2\rho s$ and $\delta p_1 = c_1{}^2\rho_1 s_1$ (**12·11**), where ρ, ρ_1 are the equilibrium densities of the two media; hence from (2) and (4) we must have

$$\rho\,\partial\phi/\partial t = \rho_1\,\partial\phi_1/\partial t \text{ when } x=0 \quad \text{..............(6)}.$$

To represent waves of harmonic type we take for the incident train

$$\phi = Ae^{i(ax+by+\omega t)} \quad \text{........................(7)},$$

so that $ax+by=$ const. gives the direction of the wave fronts.

We may then assume that the reflected and refracted trains are represented by

$$\phi' = A'e^{i(a'x+by+\omega t)} \quad \text{.....................(8)},$$

and

$$\phi_1 = A_1 e^{i(a_1 x+by+\omega t)} \quad \text{.....................(9)}.$$

The coefficient of t must be the same in all because all the waves must have the same period, and the coefficient of y must be the same because an incident, reflected and refracted wave front will all have the same trace on the yz plane.

The velocity potential of the whole motion in the first medium is $\phi+\phi'$ and by substituting the values from (7) and (8) in (1) and observing that the result must be true for all values of x, y and t we get

$$\omega^2 = c^2(a^2+b^2) = c^2(a'^2+b^2) \quad \text{...........(10)};$$

and in like manner from (3) and (9)

$$\omega^2 = c_1{}^2(a_1{}^2+b^2) \quad \text{.....................(11)}.$$

It follows that $a'^2=a^2$, and we take $a'=-a$ for a reflected wave so that the reflected and incident waves are equally inclined to the surface of separation.

Again if θ, θ_1 are the angles that the normals to the incident and refracted waves make with the x axis,

$$\sin\theta = b/\sqrt{a^2+b^2} \text{ and } \sin\theta_1 = b/\sqrt{a_1{}^2+b^2};$$

and therefore

$$c/\sin\theta = c_1/\sin\theta_1 \quad \text{.....................(12)}.$$

This is the law of refraction.

If $c>c_1$, there will be a real value of θ_1 for all angles of incidence so that sound can pass at all angles from a rarer to a denser medium. But if $c_1 > c^*$, then θ_1 is imaginary when $\theta > \sin^{-1}(c/c_1)$, and for such angles of incidence the waves are totally reflected.

It remains to find the relations between the amplitudes A, A', A_1 of the waves, by means of the boundary conditions (5), (6). From these we get

$$\left.\begin{aligned} a(A-A') &= a_1 A_1 \\ \rho(A+A') &= \rho_1 A_1 \end{aligned}\right\} \quad \text{.....................(13).}$$

and

Therefore
$$\frac{A}{a\rho_1 + a_1\rho} = \frac{A'}{a\rho_1 - a_1\rho} = \frac{A_1}{2a\rho} \quad \text{..............(14).}$$

But
$$a_1 \tan\theta_1 = b = a\tan\theta \quad \text{..................(15)}$$

so that (14) may be written

$$\frac{A}{\dfrac{\rho_1}{\rho} + \dfrac{\cot\theta_1}{\cot\theta}} = \frac{A'}{\dfrac{\rho_1}{\rho} - \dfrac{\cot\theta_1}{\cot\theta}} = \frac{A_1}{2} \quad \text{..............(16).}$$

It follows that there is no reflected wave when

$$\rho_1/\rho = \cot\theta_1/\cot\theta;$$

but from (12) $\quad (1+\cot^2\theta_1)/(1+\cot^2\theta) = c^2/c_1^{\,2}$,

so that, by eliminating $\cot\theta_1$, we get

$$\cot^2\theta\left(\frac{\rho_1^{\,2}}{\rho^2} - \frac{c^2}{c_1^{\,2}}\right) = \frac{c^2}{c_1^{\,2}} - 1 \quad \text{..............(17).}$$

Hence there is a real θ for which there is no reflected wave if, and only if, c/c_1 lies between ρ_1/ρ and unity.

12·51. Energy. The energy transmitted in any time across any area of the incident wave must be equal to the energy transmitted in the same time across the corresponding areas of the reflected and refracted waves. These three corresponding areas are in the ratio

$$\cos\theta : \cos\theta : \cos\theta_1,$$

and taking the expression for energy transmitted from **12·21** (3) and (6), the frequency being the same for all the waves, we have

$$\cos\theta \,.\, \rho A^2/c = \cos\theta \,.\, \rho A'^2/c + \cos\theta_1 \,.\, \rho_1 A_1^{\,2}/c_1;$$

or, using
$$c/\sin\theta = c_1/\sin\theta_1,$$

$$\rho(A^2 - A'^2)\cot\theta = \rho_1 A_1^{\,2}\cot\theta_1.$$

This is the energy condition and it can be verified at once by using **12·5** (16).

* If in this case we write $-ia_1'$ for a_1 we find a wave travelling along the surface of separation with amplitude decreasing exponentially. *V.* Rayleigh, *Theory of Sound*, II, p. 84.

12·52. Impact of Plane Waves on a flexible Membrane. Let the membrane of surface density σ and uniform tension T separate media of densities ρ, ρ_1. Take the yz plane to coincide with the undisturbed membrane and the z axis parallel to the intersection of the wave fronts with the membrane, and draw the x axis into the first medium.

If, following the line of argument of **12·5**, we assume as the velocity potentials of the incident, reflected and refracted waves the expressions

$$\phi = Ae^{i(ax+by+\omega t)} \quad \text{.........................(1)},$$

$$\phi' = A'e^{i(-ax+by+\omega t)} \quad \text{.....................(2)},$$

and

$$\phi_1 = A_1 e^{i(a_1x+by+\omega t)} \quad \text{.........................(3)},$$

we may take for the displacement of the membrane at time t

$$\xi = Be^{i(by+\omega t)} \quad \text{.................................(4)},$$

where a, a_1, b, ω are connected with the velocities of sound in the two media as in **12·5**.

From the continuity of normal velocity, we get

$$-\dot{\xi} = \frac{\partial}{\partial x}(\phi + \phi') = \frac{\partial \phi_1}{\partial x}, \text{ when } x = 0,$$

or

$$-\omega B = a(A - A') = a_1 A_1 \quad \text{.....................(5)}.$$

The equation of motion of the membrane is

$$\sigma \ddot{\xi} = T\partial^2\xi/\partial y^2 + \delta p_1 - \delta p, \text{ when } x = 0 \quad \text{..............(6)},$$

where

$$\delta p_1 = \rho_1 \partial\phi_1/\partial t \text{ and } \delta p = \rho\partial(\phi + \phi')/\partial t.$$

Substituting from (1), (2), (3), (4) we get

$$B(Tb^2 - \sigma\omega^2) = i\omega\{\rho_1 A_1 - \rho(A + A')\},$$

and eliminating B by means of (5), and writing n for b/ω

$$i\rho(A + A') - A_1\{a_1(Tn^2 - \sigma) + i\rho_1\} = 0 \quad \text{..............(7)}.$$

From (5) and (7) we find

$$\frac{A}{a\rho_1 + a_1\rho - iaa_1(Tn^2 - \sigma)} = \frac{A'}{a\rho_1 - a_1\rho - iaa_1(Tn^2 - \sigma)} = \frac{A_1}{2a\rho} \quad \text{...(8)};$$

which may also be written

$$\frac{A}{\{(a\rho_1 + a_1\rho)^2 + a^2a_1^2(Tn^2 - \sigma)^2\}^{\frac{1}{2}} e^{i\epsilon}}$$

$$= \frac{A'}{\{(a\rho_1 - a_1\rho)^2 + a^2a_1^2(Tn^2 - \sigma)^2\}^{\frac{1}{2}} e^{i\epsilon'}} = \frac{A_1}{2a\rho} \quad \text{...(9)},$$

where

$$\tan\epsilon = -aa_1(Tn^2 - \sigma)/(a\rho_1 + a_1\rho),$$

and

$$\tan\epsilon' = -aa_1(Tn^2 - \sigma)/(a\rho_1 - a_1\rho).$$

The amplitudes of the incident, reflected and transmitted waves are therefore in the ratio

$$\{(a\rho_1 + a_1\rho)^2 + a^2a_1^2(Tn^2 - \sigma)^2\}^{\frac{1}{2}} : \{(a\rho_1 - a_1\rho)^2 + a^2a_1^2(Tn^2 - \sigma)^2\}^{\frac{1}{2}} : 2a\rho;$$

while the phases of the reflected and incident waves differ by $\epsilon' - \epsilon$ and those of the transmitted and incident waves differ by ϵ. From (5) it follows that the vibrations of the membrane are in the same phase as the transmitted wave, as is otherwise obvious.

12·6. Spherical Waves. When there is symmetry about a point, the general equation

$$\frac{\partial^2\phi}{\partial t^2}=c^2\nabla^2\phi$$

takes the form
$$\frac{\partial^2\phi}{\partial t^2}=c^2\left(\frac{\partial^2\phi}{\partial r^2}+\frac{2}{r}\frac{\partial\phi}{\partial r}\right),$$

or
$$\frac{\partial^2(r\phi)}{\partial t^2}=c^2\frac{\partial^2(r\phi)}{\partial r^2} \qquad \text{.....................(1)}$$

which has a general solution

$$r\phi=f(ct-r)+F(ct+r) \qquad \text{..................(2);}$$

the two terms representing two wave systems one diverging from the origin and the other converging on the origin with velocity c.

The velocity and condensation are given by

$$u=-\partial\phi/\partial r, \text{ and } c^2s=\partial\phi/\partial t \qquad \text{..............(3).}$$

In the case of a diverging wave we have

$$r\phi=f(ct-r) \qquad \text{........................(4),}$$

so that from (3)
$$crs=f'(ct-r) \qquad \text{........................(5).}$$

This shews that any value of rs is propagated unchanged so that the condensation s decreases like $1/r$ as the wave advances.

In this case the velocity, from (3) and (4), is given by

$$u=\frac{1}{r}f'(ct-r)+\frac{1}{r^2}f(ct-r) \qquad \text{..............(6).}$$

As the wave spreads outwards the second term in u becomes negligible in comparison with the first, and ultimately for large values of r, from (5) and (6), $u=cs$ as in a plane wave.

From (5) and (6) we get

$$f(ct-r)=r^2(u-cs) \qquad \text{....................(7).}$$

It follows that if the disturbance is confined to a spherical shell within and without which there is neither velocity nor condensation, then $f(ct-r)$ is zero both inside and outside the shell to which the disturbance is limited. Hence, if α, β are radii less and greater than the bounding radii of this shell, we have from (5)

$$c\int_\alpha^\beta srdr=\int_\alpha^\beta f'(ct-r)\,dr=-\Big[f(ct-r)\Big]_\alpha^\beta=0.$$

This shews that s cannot be of the same sign throughout the region occupied by the wave, so that a wave of condensation or of rarefaction cannot exist alone.

12·61. Given Initial Conditions. Suppose that at time $t=0$ the values of u and s are given for all positive values of r and represented by u_0 and s_0.

The total flux at any time across a sphere of radius r is $4\pi r^2 u$, and from **12·6** (3) and (2) this is

$$4\pi\{f(ct-r)+F(ct+r)\}+4\pi r\{f'(ct-r)-F'(ct+r)\};$$

and, if the origin is not a source at which fluid is produced or absorbed, this expression must vanish with r.

Hence we must have $f(ct)+F(ct)=0$ ....(1)

for all positive values of t.

Again from **12·6** (3) $u_0=-(\partial\phi/\partial r)_0$, so that from **12·6** (2)

$$f(-r)+F(r)=r\phi_0=-r\int u_0 dr \quad \text{...........(2)}.$$

Also from **12·6** (3) and (2)

$$c^2 rs=\frac{\partial}{\partial t}(r\phi)=cf'(ct-r)+cF'(ct+r),$$

so that
$$crs_0=f'(-r)+F'(r),$$

and
$$f(-r)-F(r)=-c\int s_0 r\,dr \quad \text{..............(3)}.$$

Equations (2) and (3) then determine f for all negative arguments and F for all positive arguments, and the form of f for positive arguments then follows from (1). The form of F for negative arguments is not required.

Assuming the initial disturbance to be confined to a sphere or a spherical shell in open space, it breaks into two parts which travel in opposite directions outwards and inwards and the inward wave is continually reflected at the centre. In both waves $r\phi$ is propagated with constant velocity c. From (1) we see that $r\phi$ vanishes at the origin, so that the case is somewhat similar to that of a straight tube with an open end.

12·62. Harmonic Waves diverging from a Source. If we assume that $\phi\propto e^{int}$, then **12·6** (1) becomes

$$\frac{\partial^2(r\phi)}{\partial r^2}+\frac{n^2}{c^2}(r\phi)=0 \quad \text{....................(1)},$$

so that
$$r\phi=(Ae^{-\frac{in}{c}r}+Be^{\frac{in}{c}r})e^{int} \quad \text{..............(2)}.$$

The first term represents a wave diverging from the origin, and in real form we may write

$$\phi=\frac{A}{4\pi r}\cos n\left(t-\frac{r}{c}\right)\ \ldots\ldots\ldots\ldots\ldots\ldots\ldots(3).$$

The flux across a sphere of radius r is

$$-4\pi r^2\frac{\partial\phi}{\partial r}=A\left\{\cos n\left(t-\frac{r}{c}\right)-\frac{nr}{c}\sin n\left(t-\frac{r}{c}\right)\right\}$$

$$=A\cos nt, \text{ if } r \text{ is small enough,}$$

so that A denotes the maximum rate of introduction of fluid.

The existence of such a source in unlimited space implies an expenditure of energy which can be measured by the average rate at which work is done at the surface of a sphere of radius r by pressure exerted on the fluid outside the sphere, i.e. by the mean value of

$$-4\pi r^2 p\frac{\partial\phi}{\partial r}\ \ldots\ldots\ldots\ldots\ldots\ldots\ldots\ldots(4),$$

where the pressure $p=p_0+\delta p=p_0+\rho\frac{\partial\phi}{\partial t}$, p_0, ρ denoting equilibrium pressure and density.

Substituting from (3) in (4), we find for the mean value

$$W=\frac{\rho n^2A^2}{8\pi c}\ \ldots\ldots\ldots\ldots\ldots\ldots\ldots\ldots(5).$$

This result is only valid for an isolated source in free space. Thus it has been remarked by Lamb* that the emission of energy may be greatly modified by the neighbourhood of an obstacle. Thus a simple source near a plane rigid boundary will have an equal source as image on the other side of the boundary; the result of the reflection as from this image is to double the amplitude at any point, so that the intensity is quadrupled, and the emission on one side of the plane is therefore twice that of an equal source in free space.

12·63. Doublets. Such simple sources cannot be realized in practice. But a vibrating body such as the prong of a tuning fork since it produces alternate condensations and rarefactions may be considered to be a pair of simple sources in opposite phase at a small distance apart, i.e. *a doublet.*

* *Dynamical Theory of Sound*, § 76.

If we consider two such simple sources $\pm A$ at a small distance δx apart and put $A\,\delta x = C$, the velocity potential is

$$-\phi + \left(\phi + \frac{\partial \phi}{\partial x}\delta x\right) = \frac{\partial \phi}{\partial x}\delta x,$$

where $$\phi = \frac{A}{4\pi r}\cos n\left(t - \frac{r}{c}\right),$$

and $$\partial r/\partial x = -\cos\theta.$$

Hence the velocity potential due to the doublet

$$= \left\{-\frac{A}{4\pi r^2}\cos n\left(t-\frac{r}{c}\right) + \frac{An}{4\pi rc}\sin n\left(t-\frac{r}{c}\right)\right\}(-\cos\theta\,\delta x).$$

At a great distance from the doublet this approximates to a velocity potential

$$\phi = -\frac{nC}{4\pi rc}\sin n\left(t-\frac{r}{c}\right)\cos\theta \quad \ldots\ldots\ldots\ldots\ldots(1).$$

Then, as in **12·62**, the flux of energy across a unit of area

$$= -\left(p_0 + \rho\frac{\partial \phi}{\partial t}\right)\frac{\partial \phi}{\partial r},$$

and substituting from (1) gives a mean value

$$\frac{\rho n^4 C^2 \cos^2\theta}{32\pi^2 r^2 c^3}$$

The total average rate of emission of energy from a doublet of strength $C\cos nt$ is therefore

$$W = \frac{\rho n^4 C^2}{32\pi^2 c^3}\int_0^\pi 2\pi\cos^2\theta\sin\theta\,d\theta = \frac{\rho n^4 C^2}{24\pi c^3} \quad \ldots\ldots\ldots(2).$$

The effect produced by a vibrating sphere may be represented by that of an equivalent doublet*.

12·7. Musical Sounds. Musical sounds as distinct from noises possess three main characteristics: (1) pitch, (2) intensity, (3) timbre.

The *pitch* of a note depends on the rapidity with which the successive waves impinge upon the ear, that is on the frequency of the vibration or on the wave length. For the velocity of propagation is the same for waves of all lengths so that the frequency varies inversely as the wave length. A siren is the instrument used for experiments on the pitch of sounds. It is an apparatus by which air under pressure escapes through a hole which has as a shutter a revolving disc pierced with holes at

* Lamb, *Dynamical Theory of Sound*, § 77.

regular intervals. When the disc revolves with sufficient rapidity the vibrations caused by the escaping air produce a note of definite pitch, and it is found that increasing the frequency of the vibrations raises the pitch of the note. If the frequency of one note is double that of another the former is an octave higher than the latter. Notes whose frequencies are multiples of that of a given note are called its harmonics.

The *intensity* of a note depends on the amplitude of the vibrations. The loudness of notes can only be compared when they are of approximately the same pitch and then the square of the amplitude gives a physical measure of the intensity.

The *timbre* of a note is a quality dependent on the method by which the note is produced; for example, there is a marked difference in quality between notes of the same pitch produced from the pianoforte and the violin, this quality is called timbre and experiment shews that it is dependent on the *form* of the wave produced*.

12·71. Beats. When two notes of nearly the same frequency are sounded together a phenomenon known as 'beats' occurs, that is a succession of intervals in which the resultant vibration gradually increases to a maximum and then dies away. Let the vibrations have equal amplitudes and be in the same phase so that the resultant vibration may be represented by

$$y = a\cos(nt) + a\cos(mt),$$

where m and n are nearly equal.

Hence $\quad y = 2a\cos\tfrac{1}{2}(n-m)t\cos\tfrac{1}{2}(n+m)t,$

which may be regarded as a simple harmonic vibration of frequency $(n+m)/4\pi$ with amplitude $2a\cos\frac{1}{2}(n-m)t$, and as the amplitude varies between 0 and $2a$ with a period $4\pi/(n-m)$ the phenomenon will be as described. For example, if two tuning forks of frequencies 500 and 501 be equally excited there is a rise and fall of sound once a second corresponding to the coincidence or opposition of the vibrations.

12·8. For further information on the subject of the last two chapters, reference should be made to Donkin's *Acoustics*, Lord Rayleigh's *Theory of Sound* and Sir H. Lamb's *Dynamical Theory of Sound.*

* A paper on 'The Graphical Recording of Sound Waves' was read by D. C. Miller at the International Congress, 1912, *Proceedings*, II, p. 245.

EXAMPLES

1. Prove that the velocity potential of the one-dimensional motion of a gas, for which $p=\kappa\rho$, satisfies the equation

$$\frac{\partial}{\partial t}\left\{\frac{\partial\phi}{\partial t}-\left(\frac{\partial\phi}{\partial x}\right)^2\right\}=\frac{\partial}{\partial x}\left\{\kappa\frac{\partial\phi}{\partial x}-\frac{1}{3}\left(\frac{\partial\phi}{\partial x}\right)^3\right\},$$

where $\partial/\partial t$ denotes differentiation at a fixed point. (Trinity Coll. 1897.)

2. Prove that in a fluid medium in which the pressure (p) and the density (ρ) are connected by an equation $p=\phi(\rho)$, where $\phi'(\rho)$ is positive and increases when ρ increases, a plane wave of finite amplitude cannot be propagated indefinitely without the occurrence of discontinuity. (M.T. 1897.)

3. In an organ pipe of length l, closed at one end, the pressure at the other end is made to vary according to the law $\delta p=p_0\sin nt$. Find the velocity potential of the motion of the air inside. (Trinity Coll. 1897.)

4. Taking γ as 1·41 and the height of the homogeneous atmosphere as 8000 metres, calculate the velocity of sound in air in metres per second. Find also the length of an organ pipe which with one end open and the other stopped will sound the middle C (frequency 256). (Univ. of London, 1911.)

5. What is the difference between the overtones present in an 8-ft. stopped organ pipe and a 16-ft. open pipe? (M.T. 1913.)

6. Find the length of a stopped pipe with a fundamental frequency of 64. Assume the air to be under a pressure of $1{\cdot}013\times10^6$ dynes per sq. cm. and to have a density of 1·293 grams per litre, the ratio of the specific heats being 1·41. (M.T. 1915.)

7. Assuming the atmosphere to be in convective equilibrium (i.e. in equilibrium according to the law of pressure $p=\kappa\rho^\gamma$) under the action of gravity, prove that the equation of propagation of sound vertically upwards is

$$\frac{\partial^2\xi}{\partial t^2}=g\left\{(\gamma-1)(\lambda-x)\frac{\partial^2\xi}{\partial x^2}-\gamma\frac{\partial\xi}{\partial x}\right\},$$

where $g\lambda(\gamma-1)/\gamma$ is the ratio of the pressure to the density at the surface of the earth and ξ is the displacement at a height x. (Coll. Exam. 1899.)

8. A tube containing air has one end rigidly closed, and the other end stopped by a plug of mass M, which can move without friction in the tube. If the length of tube filled with air be l, prove that the periodicity of the free vibrations is given by

$$\frac{pl}{c}\tan\frac{pl}{c}=\frac{M'}{M},$$

where c is the velocity of sound in the enclosed air, and M' the mass of the air. (Coll. Exam. 1906.)

9. A tube of unit cross section open at both ends is divided into two parts of lengths l, l' by a thin piston of mass M attached to a spring such

that $2\pi/m$ is its natural period of vibration. When the air waves are taken into account prove that the period of vibration is $2\pi/n$, where

$$M(m^2 - n^2) = \rho cn [\tan nl'/c + \tan nl/c].$$

(Coll. Exam. 1907.)

10. A piston of mass M is supported by a spring of strength Mn^2, and separates two gases of densities ρ, ρ' in a long tube; the area of the section is S and c, c' are the velocities of sound in the two gases. Shew that the free oscillations of the piston are given by

$$\frac{d^2\xi}{dt^2} + \frac{S}{M}(c\rho + c'\rho')\frac{d\xi}{dt} + n^2\xi = 0.$$

(St John's Coll. 1910.)

11. Air is confined in a straight tube of unit section between two pistons, one of which is made to vibrate with velocity $a \cos nct$, and the other is of mass M and is constrained by a spring of strength μ. Shew that the velocity potential for the air vibrations is

$$-\frac{a}{n}\frac{\cos n(l-x+\epsilon)}{\sin n(l+\epsilon)}\cos nct,$$

where $\tan n\epsilon = \dfrac{n\rho c^2}{\mu - Mn^2c^2}$, l being the distance between the pistons and ρ the density in equilibrium. (M.T. 1903.)

12. A tube of length l is closed at one end and open at the other, and is filled with a gas of mean density ρ_0. A pressure disturbance $A\rho_0 \sin \sigma t$ is maintained at the open end by waves passing outside the tube. Prove that the velocity at any point within the tube is

$$u = A\frac{\cos \sigma t \sin \{\sigma(l-x)/c\}}{c \cos(\sigma l/c)},$$

where $c^2 = dp/d\rho$ evaluated for $\rho = \rho_0$ and the origin is at the open end. Find how the pressure varies at the closed end.

Explain the physical significance of the vanishing of the denominator for certain values of σ. (M.T. 1929.)

13. A straight pipe of length l is closed at one end and open at the other. Prove that, if the air extend only from the open end to the middle point, the other half being occupied by a gas of density ρ_1, then the frequencies of the natural modes of the pipe are the values of p satisfying the equation

$$\tan\frac{\pi pl}{c}\tan\frac{\pi pl}{c_1} = \rho_1 c_1/\rho c,$$

where ρ is the density of the air, and c, c_1 are respectively the velocities of sound in air and in the gas. (M.T. 1895.)

14. In a cylindrical pipe, open at one end, closed at the other, it is found experimentally that, when the fundamental note is being sounded, the pressure at the closed end varies on either side of its mean value by one nth of that value. Prove that at the open end the amplitude of vibration of the particles of air is $2l/n\pi\gamma$, where l is the length of the pipe and γ the ratio of the specific heats of air at constant pressure and constant volume.

(Coll. Exam. 1911.)

15. A horizontal uniform tube closed at both ends and containing air is divided into two parts of equal length l by a tightly fitting piston of mass m which can move without friction in the tube. The piston is displaced a small distance d from its equilibrium position in the middle of the tube and then released, the air being initially in equilibrium. Shew that the motion of the piston is made up of harmonic components whose periods $(2\pi/n)$ are given by $(nl/c)\tan(nl/c) = m'/m$, where c is the velocity of propagation of small disturbances in the enclosed air whose mass is m'.

Assuming the possibility of the expansion

$$f(x) = A_1 \cos(n_1 x/c) + A_2 \cos(n_2 x/c) + \dots, \qquad 0 \leqslant x \leqslant l,$$

where $n_1, n_2, \dots$ are the roots of the period equation, prove that the initial condensation in either part of the tube can be expressed in the form

$$\pm \frac{2d}{l} \sum_{r=1}^{\infty} \left\{ \frac{\sin \frac{n_r l}{c}}{\frac{n_r l}{c} + \sin \frac{n_r l}{c} \cos \frac{n_r l}{c}} \cos \frac{n_r x}{c} \right\}. \qquad \text{(M.T. 1920.)}$$

16. A long straight tube of cross section α has at one point a close-fitting piston controlled by a spring but otherwise free to move in the tube. The mass of the piston is m and its period of oscillation *in vacuo* would be $2\pi/n$. The tube is open to the atmosphere at both ends and initially the piston and the air are at rest. Prove that, if a velocity u is suddenly given to the piston, the displacement of a layer of air at a distance x from the piston after a time t is

$$\frac{u}{\sqrt{(n^2 - k^2)}} e^{-k(t - x/c)} \sin\{\sqrt{(n^2 - k^2)}\,(t - x/c)\},$$

where $k = c\rho_0 \alpha/m$, ρ_0 being the equilibrium density of the air and c the velocity of sound. (M.T. 1932.)

17. In a uniform straight tube of length $2l$ and sectional area ω, closed at one end, a quantity of gas is imprisoned by a thin movable piston of mass M. Under the pressure of the external atmosphere of density ρ the equilibrium position of the piston is at the middle of the tube, and the density of the enclosed gas is then σ. Prove that the periodic times $2\pi/p$ of the oscillations of the piston about its position of equilibrium are given by the equation

$$Mp/\omega = c\sigma \cot(pl/c) - c'\rho \tan(pl/c'),$$

c and c' being the velocities of propagation of sound in the enclosed gas and in the atmosphere respectively. (St John's Coll. 1900.)

18. A long straight speaking-tube is obstructed in the middle by a uniform rigid plug with plane ends, of length z and density equal to N times that of the air. The plug fits the tube accurately, but is free to move in it without friction. Prove that, if sound of wave length λ is advancing along the tube, the intensity of the sound transmitted beyond the plug will be less in the ratio $1 : 1 + \pi^2 N^2 z^2/\lambda^2$, and its phase retarded by

$$\{\tan^{-1}(\pi N z/\lambda) - 2\pi z/\lambda\}. \qquad \text{(M.T. 1901.)}$$

19. A closed pipe of length $2l$ contains air whose density is slightly greater than that of the outside air, in the ratio $1 + \epsilon : 1$. Everything being

at rest, the discs closing the ends of the pipe are suddenly drawn aside. Shew that after a time t the velocity potential is

$$\phi=\frac{8\epsilon cl}{\pi^2}\sum_{s=0}^{s=\infty}\frac{(-1)^s}{(2s+1)^2}\cos\frac{(2s+1)\pi x}{2l}\sin\frac{(2s+1)\pi ct}{2l},$$

the origin being taken at the middle of the pipe, and c denoting the velocity of sound. (St John's Coll. 1903.)

20. A straight pipe of length l is open at one end and the disc closing the other end executes small inexorable oscillations, its displacement at any time t being $A\sin pt$. Prove that at any time the kinetic energy of the air in the pipe is

$$\tfrac{1}{4}MA^2\left(p^2\sec^2\frac{pl}{c}+\frac{cp}{l}\tan\frac{pl}{c}\right)\cos^2 pt,$$

where c is the velocity of sound in air and M is the mass of air contained in the pipe. Investigate also the potential energy of the air in the pipe. (Trinity Coll. 1900.)

21. Plane waves of sound represented by $\phi=A\cos m(x+ct)$ impinge perpendicularly on a rigid screen and are continuously reflected by it. Prove that the increment of the pressure per unit area on the screen lies between $\pm 2Amc\rho_0$, where ρ_0 is the density of the air. (Coll. Exam.)

22. Determine the velocity potential ϕ of a plane wave of sound, of small amplitude, for all x, t, given that when $t=0$

$$\phi=\phi_0(x),\quad \frac{\partial\phi}{\partial t}=\psi_0(x).$$

Fluid is contained in a long straight tube closed at one end $x=0$. When $t=0$ the fluid is everywhere at rest while the condensation s is s_0 (constant) for values of x between 0 and a and zero elsewhere. Determine s for all x, t; draw (s, t) graphs for the values $\tfrac{1}{2}a$ and $\tfrac{3}{2}a$ of x, and explain the difference between them. (M.T. 1933.)

23. Shew that the form of the equation

$$\frac{\partial^2\xi}{\partial t^2}=c^2\frac{\partial^2\xi}{\partial x^2}$$

for plane waves of sound in a pipe remains unaltered when the velocity of sound c is a function of x.

Sound waves are set up in a closed pipe of length k in which the absolute temperature of the gas varies as the square of the distance x from a point at a distance l outside one end of the pipe. The velocity of sound at this end is c_0. Shew that the form of the sound waves in the pipe must be

$$\xi=A\sqrt{x}\sin\left\{q\log\frac{x}{l}\right\}e^{ipt},$$

where $q^2=\dfrac{p^2l^2}{c_0{}^2}-\tfrac{1}{4}$ and that the possible frequencies p are the roots of

$$\frac{p^2l^2}{c_0{}^2}=\tfrac{1}{4}+\frac{n^2\pi^2}{\{\log(1+k/l)\}^2},$$

where n is any positive integer. Compare these frequencies with those when the velocity of sound has the constant value c_0. (M.T. 1924.)

24. A long straight pipe of unit sectional area extends to infinity in one direction (taken as that of the z axis) and is closed at $z=-l$; at $z=0$ there is a freely movable piston of mass m, on both sides of which there is air of density ρ. Waves in which the displacement is given by $a \sin \kappa (z+ct)$ impinge on the positive side of the piston. Shew that if the reflected waves are expressed by $a_1 \sin\{\kappa(z-ct)+\epsilon\}$, then $a_1{}^2 = a^2$; also shew that if a_1 is taken equal to a

$$\cot \tfrac{1}{2}\epsilon = \cot \kappa l - \kappa m/\rho.$$

(M.T. 1928.)

25. If a straight tube of indefinite length be occupied by two different gases with the section $x=0$ for surface of contact; shew that the displacements in an incident wave together with those of the corresponding reflected and refracted waves may be represented by

$$f(t-x/c_1), \quad Af(t+x/c_1), \quad Bf(t-x/c_2),$$

where

$$A : B : 1 :: \rho_1 c_1 - \rho_2 c_2 : 2\rho_1 c_1 : \rho_1 c_1 + \rho_2 c_2,$$

and determine the distribution of the primitive energy between the reflected and refracted systems. (St John's Coll. 1906.)

26. Two plugs, each of mass M, fit closely into a long straight tube, and can slide without friction in the tube. They are kept apart by a light spring, the tube, except the part between the plugs, being filled with air. A train of simple harmonic sound waves of amplitude a impinges on one of the plugs. Shew that the amplitude of the wave transmitted beyond the other plug is $a\alpha\beta/\sqrt{(a^2+1)\{\alpha^2+(1-\beta)^2\}}$. Here $2\pi\alpha = M'/M$, where M' is the mass of air in a length of the tube equal to a wave length of the incident wave, and $\beta = 2k/n^2M$, where $2\pi/n$ is the period of the incident wave, and kx is the increase in the thrust of the spring when its length is decreased by x. (M.T. 1931.)

27. An infinite long straight tube of unit cross section contains gases of densities ρ, ρ' separated by a smoothly fitting piston of mass M. An harmonic train of sound waves of length $2\pi/m$ is incident on the piston in the gas of density ρ. Shew that in transmission beyond the piston the amplitude of the waves is changed in the ratio

$$2c\rho : \sqrt{\{(c\rho + c'\rho')^2 + M^2m^2c^2\}},$$

and the phase by

$$\tan^{-1} Mmc/(c\rho + c'\rho'),$$

where c, c' are the velocities of sound in the two gases. (M.T. 1930.)

28. An endless tube of uniform cross section contains two pistons; the intervening portions of the tube, of lengths l_1, l_2 respectively, containing air at atmospheric pressure. If one of the pistons be found to vibrate so that its displacement at time t is $X \cos pt$, shew that the displacement of the other is

$$\frac{c\{\operatorname{cosec}(pl_1/c) + \operatorname{cosec}(pl_2/c)\}}{c\{\cot(pl_1/c) + \cot(pl_2/c)\} - mp} X \cos pt,$$

where m is the ratio of the mass of the piston to that of the air contained in unit length of the tube and c is the velocity of sound in air.

(Trinity Coll. 1898.)

29. The period of the fundamental note of a flue pipe, open at one end and closed at the other, would be T, if the closed end were rigid. But the

barrier at the closed end is replaced by a piston of mass M, controlled by a strong spring of strength μ. Prove that the period of the fundamental note is approximately

$$T\left\{1+\frac{16m}{\mu T^2-4\pi^2 M}\right\},$$

where m is the mass of the air in the pipe and $16m/(\mu T^2-4\pi^2 M)$ is assumed to be small. (Trinity Coll. 1902.)

30. Plane sound waves of length λ in a medium of density ρ impinge normally on a plane membrane which separates the medium from another of density ρ'. The membrane is such that a pressure difference δp on opposite sides of it causes a displacement $\mu\,\delta p$. Shew that the phase of the transmitted wave differs from those of the incident and reflected waves by $\cot^{-1} 2\pi\mu\left(\frac{c'^2\rho'}{\lambda'}\pm\frac{c^2\rho}{\lambda}\right)$, where c, c' are the wave velocities in the two media and λ' is the length of the transmitted wave.

Compare the amplitudes of the three waves. (M.T. 1925.)

31. A train of waves of air, velocity potential $=A\cos\frac{2\pi(ct-x)}{\lambda}$, is advancing in a straight pipe infinite in both directions, and at $x=0$ impinges on a movable piston of mass M which separates the air of the pipe into two portions. Prove that the velocity potential of the train of waves transmitted to the air beyond the piston is

$$A\cos\frac{2\pi\epsilon}{\lambda}\cos\frac{2\pi(ct-x-\epsilon)}{\lambda},$$

where m is the mass of the air in a wave length of the pipe, and

$$\cos\frac{2\pi\epsilon}{\lambda}=m\{\pi^2M^2+m^2\}^{-\frac{1}{2}}. \quad \text{(Trinity Coll. 1903.)}$$

32. A long straight tube, of cross section ω, is obstructed in the middle by a piston of mass M, whose ends are plane, fitting the tube accurately but free to move in it. To the right of the piston is gas of density ρ, to the left gas of density ρ', and the velocities of propagation of sound in the gases are c and c'. Sound of wave length λ is advancing through the tube from the right, and undergoes partial reflection at the piston. Shew that the intensities of the reflected and incident waves are in the ratio

$$\left\{\left(\frac{\rho'}{\rho}-\frac{c}{c'}\right)^2+\left(\frac{2\pi Mc}{\omega\rho\lambda c'}\right)^2\right\}:\left\{\left(\frac{\rho'}{\rho}+\frac{c}{c'}\right)^2+\left(\frac{2\pi Mc}{\omega\rho\lambda c'}\right)^2\right\}.$$

(St John's Coll. 1901.)

33. An infinite long straight tube of unit section contains gases of densities ρ and ρ', at the same pressure p, separated by a piston of mass M which can vibrate under the action of a spring of strength μ. Sound waves of harmonic type and amplitude A travelling in medium ρ are incident on the piston. Shew that if A_1 and A' are the amplitudes of the reflected and transmitted waves

$$A^2:A_1^2:A'^2=(\mu-n^2M)^2+\gamma^2p^2(m+m')^2:$$
$$(\mu-n^2M)^2+\gamma^2p^2(m-m')^2:4\gamma^2p^2m^2,$$

where $2\pi/m$, $2\pi/m'$ are the wave lengths in the media ρ and ρ' and n/m is the velocity of the waves in medium ρ. (M.T. 1898.)

34. Plane waves of sound are travelling normally from a gas of density ρ_1 into one of density ρ_3. Shew that the mean transmission of energy into the latter gas is increased by interposing between the gases a layer of a different gas of density ρ_2, provided that ρ_2 is intermediate in value between ρ_1 and ρ_3, the ratio of the specific heats having the same value in each gas
(M.T. 1911.)

35. Two media of different densities have a plane surface of separation, one medium extends to infinity and the other is bounded by a rigid plane at a distance l from their common plane of separation. Plane waves of sound travelling in the first medium are refracted into the second medium and, after reflection at the rigid boundary and another refraction, emerge into the first medium again; prove that the amplitudes of the incident and emergent waves are equal, and that there is a loss of phase of amount

$$2\tan^{-1}\left\{\frac{\sin 2\alpha'}{\sin 2\alpha}\tan\left(\frac{2\pi l\cos\alpha'}{\lambda'}\right)\right\},$$

where α and α' are the angles of incidence and refraction at the surface separating the two media and λ' is the wave length in the second medium.
(M.T. 1906.)

36. A train of plane waves of sound of a type given by a velocity potential

$$\phi = A\sin\frac{2\pi}{\lambda}(x-ct)$$

is incident at an angle α on an infinite plane rigid surface. Find the velocity potential of the reflected system of waves, and shew that the pressure on a square area in this plane, whose side is $2a$, differs from its equilibrium value by the quantity

$$\frac{8Aa\rho_0 c}{\sin\alpha}\sin\frac{2\pi a\sin\alpha}{\lambda}\cos\theta,$$

where θ is the phase at the centre of the square, ρ_0 being the mean density of the fluid, and the sides of the square being parallel and perpendicular to the intersections of the wave fronts and the rigid surface. (M.T. 1900.)

37. A plane wave of sound of wave length λ travelling with velocity V in an infinite medium of density ρ is transmitted through a plane plate of thickness l and density ρ_1 in which the velocity of sound is V_1 into another infinite medium of density ρ_2 in which the velocity of sound is V_2. Shew that the phase of the transmitted disturbance is the same as that of the original disturbance if

$$\frac{\tan E_2\rho_2}{\tan E_1\rho_1} = \frac{EE_2+E_1^2}{E_1(E+E_2)},$$

where

$$E = \frac{2\pi l\cos\theta}{\lambda\rho},$$

$$E_1 = 2\pi l\,(V^2/V_1^2-\sin^2\theta)^{\frac{1}{2}}/\lambda\rho_1,\quad E_2 = 2\pi l\,(V^2/V_2^2-\sin^2\theta)^{\frac{1}{2}}/\lambda\rho_2,$$

θ is the angle of incidence at the first surface of the plate and it is supposed that $V/V_2 > V/V_1 > \sin\theta$. What would be the physical nature of the disturbance within and beyond the plate if $V/V_2 > \sin\theta > V/V_1$? (M.T. 1896.)

38. In the case of refraction of plane waves of sound at a plane surface of separation of two media of densities ρ, ρ_1, the ratio of the energy transmitted per unit time into the second medium through a given area of the boundary to the energy of the train incident per unit time on that area is

$$(4\rho\rho_1 \cot\alpha \cot\alpha_1)/(\rho_1 \cot\alpha + \rho \cot\alpha_1)^2,$$

α, α_1 being the angles of incidence and refraction. (M.T. 1894.)

39. An infinite plane membrane of uniform surface density σ and uniform tension T, coinciding with the plane xOz, separates two gases of densities ρ and ρ' in which the velocities of propagation of sound are V and V'. The infinitesimal motion of the membrane being given by

$$y = A \cos mx \sin pt,$$

shew that the velocity potentials in the gases are

$$\phi = -Apn^{-1}e^{-ny} \cos mx \cos pt \quad \text{and} \quad \phi' = Apn'^{-1}e^{n'y} \cos mx \cos pt$$

where $m^2 - n^2 = p^2/V^2$, $m^2 - n'^2 = p^2/V'^2$ and $Tm^2/p^2 = \sigma + \rho/n + \rho'/n'$, all the quantities concerned being supposed real. (Coll. Exam. 1898.)

40. A gas extends everywhere to a distance a from a plane rigid wall and is separated from a second gas by a light perfectly flexible membrane from which the second gas extends to a great distance. Shew that, if c_1, c_2 be the velocities of sound in the two media, the displacements perpendicular to the wall for plane waves of period $2\pi/p$ are respectively of the form

$$\xi_1 = A \cos(pt + \alpha) \sin\frac{px}{c_1} \operatorname{cosec}\frac{pa}{c_1}$$

$$\xi_2 = A \cos(pt + \alpha) \cos\left(\frac{px}{c_2} - \epsilon\right) \sec\left(\frac{pa}{c_2} - \epsilon\right)$$

and determine the necessary value of ϵ. (Coll. Exam. 1903.)

41. A tube of small uniform section S and length l has one end closed while the other end branches into two tubes of small uniform sections S', S'' and lengths l', l'' respectively with their ends closed. Shew that the periods of the notes which the air in the tubes can sound are the values of T satisfying the equation

$$S \tan\frac{2\pi l}{cT} + S' \tan\frac{2\pi l'}{cT} + S'' \tan\frac{2\pi l''}{cT} = 0,$$

where c is the velocity of sound in air. (M.T. 1899.)

42. Determine the periods of the fundamental tone and overtones (i) of a conical pipe open at both ends, (ii) of an open wedge-shaped pipe whose walls are formed of two planes inclined to each other and two other planes perpendicular to both of them. (St John's Coll. 1899.)

43. A point source of sound of strength $C \cos nt$ is at a point O at a perpendicular distance h from an infinite rigid plane which is the only boundary of the medium. Shew that at time t the velocity potential at a point at distances r_1 and r_2 from O and the image of O in the boundary is

$$\frac{C}{4\pi}\left[\frac{1}{r_1} \cos n\left(t - \frac{r_1}{c}\right) + \frac{1}{r_2} \cos n\left(t - \frac{r_2}{c}\right)\right],$$

and by considering the rate at which energy is transmitted across the surface of a large sphere centre O, or otherwise, shew that in maintaining the source work must be done at twice the rate which would be necessary if the medium were unbounded. (M.T. 1934.)

44. An infinite train of divergent waves is set up by the pulsation of the spherical internal boundary, of radius $R(1+\alpha\sin\kappa ct)$, where α is small, in an otherwise unlimited mass of uniform fluid. Find the velocity potential of the motion, and prove that the mean energy-density at radius r is

$$\rho\kappa^2c^2\alpha^2R^6(1+2\kappa^2r^2)/4r^4(1+\kappa^2R^2). \qquad \text{(M.T. 1928.)}$$

45. Explain the characters of the sources of sound which give at a distance velocity potentials of the forms

$$\frac{d}{dx}\frac{\sin\kappa(t-r/c)}{r} \quad \text{and} \quad \frac{d^2}{dx^2}\frac{\sin\kappa(t-r/c)}{r}$$

respectively. Which of them would most suitably represent the action of an ordinary tuning fork?

Explain the alternations of sound and silence that occur when a vibrating fork is rotated on its axis near the ear. (St John's Coll. 1897.)

CHAPTER XIII

VISCOSITY

13·1. THE **Viscosity** of a fluid is that property in virtue of which it is able to offer resistance to shearing stress. It is a kind of molecular frictional resistance. All known fluids whether liquid or gaseous possess the property of viscosity but in greatly varying degree.

There is a distinction between plastic solids and viscous fluids which was defined by Maxwell thus: when the smallest stress if continued long enough will cause a constantly increasing change of form the body must be regarded as a viscous fluid however hard it may be; but if the continuous alteration of form is only produced by stresses exceeding a certain value, the substance is called a solid however soft (or plastic) it may be*.

13·11. Measurement of Viscosity. The method of measuring the effects of viscosity may be illustrated by considering a

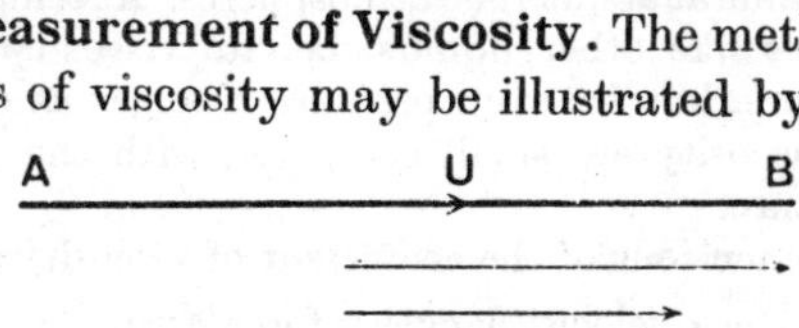

simple example: Suppose that fluid is bounded above and below by horizontal planes AB, CD of which the upper has a uniform horizontal velocity U while the lower is at rest.

We assume for the moment that a fluid in contact with a solid does not slip on the surface. The fluid between the planes will then move in horizontal strata with velocities which decrease as we go downwards from U in contact with AB to zero in contact with CD. If d be the distance between the planes, the velocity gradient (assumed to be uniform) is U/d, and to maintain the motion of the plane AB will require a horizontal force proportional to U/d per unit area of AB. If we denote this force per unit area by $\mu U/d$, then μ is called the **coefficient of viscosity** of the fluid under consideration.

* *Theory of Heat*, p. 303.

If we consider any horizontal plane in the fluid, the portions of fluid above and below it will exert on one another a horizontal traction $\mu U/d$ per unit area; i.e. μ times the velocity gradient. And in a more general case, where we do not assume the velocity gradient to be uniform, if z be an axis at right angles to the planes and u the velocity at any point, the tractive force on either portion into which a horizontal plane through the point divides the fluid is measured by $\mu du/dz$ per unit area.

The foregoing hypothesis of a frictional resistance proportional to the relative velocity of the fluid elements was introduced by Newton* in a discussion of the circular motion of a fluid produced by a revolving solid cylinder, and it has been found to constitute a satisfactory basis for a theory which has been frequently tested by experiment.

13·12. The coefficient of viscosity of a fluid is not a constant, but depends in general on pressure and temperature. For gases μ, as deduced from the kinetic theory, is independent of the pressure, but increases rapidly with increase of temperature. For liquids in general, water being an exception, μ increases with the pressure. At temperatures below 30° the viscosity of water at first decreases with increasing pressure and has a minimum value at about 1000 atmospheres. At temperatures above 30° water behaves like other liquids, i.e. its viscosity increases with increase of pressure, but for such pressure changes as ordinarily occur the changes in viscosity are small compared with the changes due to varying temperature.

The physical dimensions of the coefficient of viscosity are given by

$$\mu \times \text{velocity/length} = \text{force/area},$$

or

$$\mu = \mathrm{ML^{-1}T^{-1}}.$$

13·2. Stresses in a Fluid in Motion.

The essential distinction between a real fluid and the ideal perfect fluid of the previous chapters is that while the stress across any plane surface in the latter is always normal to the surface this is not true of real fluids, and when these are in motion tangential components of stress always exist unless the rate of deformation is zero. The immediate consequence of the existence of tangential stresses is that the theorem of equality of pressure in every direction at a point, true for a perfect fluid, no longer exists. We need therefore, in the first place, a mode of specifying the components of stress at a point in a fluid, and secondly to determine what relations exist between these components; and our ultimate

* *Principia Mathematica*, 2nd edition, 1713, Bk. II, Sec. IX, 'Hypothesis: *Resistentiam, quae oritur ex defectu lubricitatis partium Fluidi, caeteris paribus, proportionalem esse velocitati, qua partes Fluidi separantur ab invicem*'.

object is to determine what expressions are to enter into the equations of motion of a viscous fluid in place of the pressure terms $\partial p/\partial x$, $\partial p/\partial y$, $\partial p/\partial z$ which occur in the equations of motion of a perfect fluid.

13·21. Definitions. Imagine a small plane surface, whose area we take as the unit, placed in an arbitrary direction at a point P in a fluid. The direction of the area may be indicated by a vector $\boldsymbol{h}$ at right angles to it. Take any set of rectangular axes $Pxyz$; then the stress across the surface may be resolved into three rectangular components in the directions of the axes and these will be denoted by

$$p_{hx},\quad p_{hy},\quad p_{hz}.$$

In this notation the first suffix indicates the direction of the plane surface (*not* of the force upon it), and the second suffix indicates the direction of the component stress.

The resultant stress across the surface has of course direction cosines proportional to p_{hx}, p_{hy} and p_{hz}.

Now let a small plane area centred at (x, y, z) be placed at right angles to each of the coordinate axes in turn, then in accordance with the above symbolism the components of stress per unit area parallel to the axes in the three cases are

$$p_{xx},\quad p_{xy},\quad p_{xz},$$
$$p_{yx},\quad p_{yy},\quad p_{yz},$$
and
$$p_{zx},\quad p_{zy},\quad p_{zz},$$

where the components p_{xx}, p_{yy}, p_{zz} are clearly normal to the surfaces on which they act, while the other six symbols denote tangential components; e.g. p_{xy} is a force in the direction y on an area perpendicular to x.

We shall consider the symbols p_{xx}, p_{yy}, p_{zz} to be positive numbers when they represent tensions, so that a pressure is to be regarded as a negative stress. In a non-viscous fluid we have

$$p_{xx} = p_{yy} = p_{zz} = -p,$$
$$p_{xy} = p_{yx} = p_{xz}, \text{ etc.} = 0.$$

13·22. Relations between rectangular components of Stress. Consider a small rectangular parallelepiped with its centre at $(x, y. z)$ and edges of lengths δx, δy, δz parallel to the axes.

In relation to a surface at right angles to the axis of x, the stresses per unit area at (x, y, z) are p_{xx}, p_{xy}, p_{xz}. The corresponding stresses at the centre of the face $\delta y\,\delta z$ remote from the origin are

$$p_{xx}+\frac{1}{2}\frac{\partial p_{xx}}{\partial x}\delta x,\quad p_{xy}+\frac{1}{2}\frac{\partial p_{xy}}{\partial x}\delta x,\quad p_{xz}+\frac{1}{2}\frac{\partial p_{xz}}{\partial x}\delta x,$$

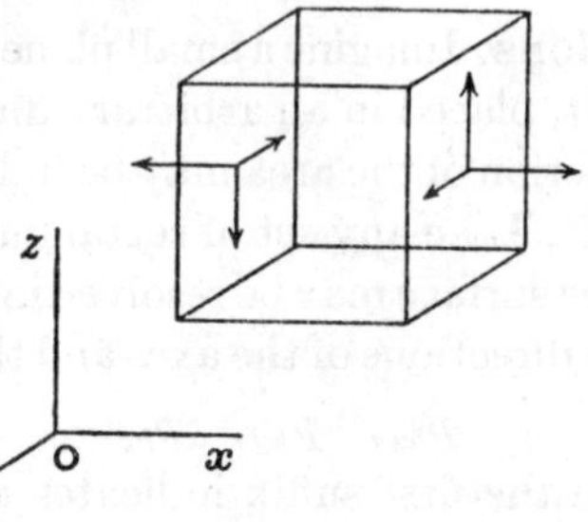

and the senses in which they act on the fluid in the parallelepiped are indicated in the diagram. At the centre of the opposite face the corresponding stresses are

$$p_{xx}-\frac{1}{2}\frac{\partial p_{xx}}{\partial x}\delta x,\quad p_{xy}-\frac{1}{2}\frac{\partial p_{xy}}{\partial x}\delta x,\quad p_{xz}-\frac{1}{2}\frac{\partial p_{xz}}{\partial x}\delta x,$$

acting on the fluid in the parallelepiped in the opposite senses to the former. Proceeding as in **1·3** we may shew that the stresses on this pair of opposite faces may be compounded into forces

$$\frac{\partial p_{xx}}{\partial x}\delta x\,\delta y\,\delta z,\quad \frac{\partial p_{xy}}{\partial x}\delta x\,\delta y\,\delta z,\quad \frac{\partial p_{xz}}{\partial x}\delta x\,\delta y\,\delta z$$

acting at (x, y, z) parallel to Ox, Oy, Oz respectively, and couples $-p_{xz}\,\delta x\,\delta y\,\delta z$, $p_{xy}\,\delta x\,\delta y\,\delta z$ about Oy, Oz respectively.

The stresses on the other two pairs of opposite forces may be compounded into similar forces at (x, y, z) parallel to the axes, and couples

$$-p_{yx}\,\delta x\,\delta y\,\delta z,\quad p_{yz}\,\delta x\,\delta y\,\delta z$$

about Oz, Ox and $-p_{zy}\,\delta x\,\delta y\,\delta z,\quad p_{zx}\,\delta x\,\delta y\,\delta z$

about Ox, Oy respectively.

It follows that if, as in **2·1**, we write down equations of motion for the fluid in the parallelepiped by resolving parallel to the axes, we get

$$\left.\begin{aligned}\rho\frac{Du}{Dt}&=\rho X+\frac{\partial p_{xx}}{\partial x}+\frac{\partial p_{yx}}{\partial y}+\frac{\partial p_{zx}}{\partial z}\\ \rho\frac{Dv}{Dt}&=\rho Y+\frac{\partial p_{xy}}{\partial x}+\frac{\partial p_{yy}}{\partial y}+\frac{\partial p_{zy}}{\partial z}\\ \rho\frac{Dw}{Dt}&=\rho Z+\frac{\partial p_{xz}}{\partial x}+\frac{\partial p_{yz}}{\partial y}+\frac{\partial p_{zz}}{\partial z}\end{aligned}\right\}\quad\text{...........(1).}$$

And further if we take moments about lines through (x, y, z) parallel to the axes for the kinetic reactions and the forces acting on the fluid in the parallelepiped, we get

$$(p_{yz}-p_{zy})\,\delta x\,\delta y\,\delta z + \text{terms of the fourth degree in } \delta x, \delta y, \delta z = 0$$

and two similar equations.

So that on dividing by $\delta x\,\delta y\,\delta z$ and making the edges shrink to zero, we get $p_{yz}=p_{zy}$, and similarly $p_{zx}=p_{xz}$ and $p_{xy}=p_{yx}$; so that the nine rectangular components of stress at a point are reduced to six.

We have now to find the connection between these six components of stress and the gradients of the velocity of the fluid.

13·23. Connection between Stresses and Gradients of Velocity. It is the relative motion of fluid particles which give rise to the tangential stresses just described. The stresses in an element of fluid are not affected by its translation or rotation but only by its distortion, i.e. by the relative motion of its parts. We have seen in **4·1** that the relative motion can be analysed into pure strain and rotation, and the state of stress depends only on the state of strain. At every point there is a rate of strain quadric whose axes are in the directions in which the lines joining particles are undergoing elongation at uniform rates. It follows from considerations of symmetry that the stresses across the axial planes of the strain quadric at any point are normal to these planes. We may call these the **principal stresses** at the point and denote them by p_1, p_2, p_3. We shall use the principal stresses as a connecting link between the six general components of stress and the gradients of the velocity.

Let $Px'y'z'$ be the axes of the rate of strain quadric at P, and let l_1, m_1, n_1; l_2, m_2, n_2; l_3, m_3, n_3 be their direction cosines referred to any other rectangular axes $Pxyz$. The velocity components at P are u, v, w referred to $Pxyz$, and u', v', w' referred to $Px'y'z'$, with the distinction that while u, v, w are in general all functions of x, y and z, yet u' is a function of x' alone, v' a function of y' alone and w' a function of z' alone (**4·1**).

Then with the notation of **4·1**

$$a=\frac{\partial u}{\partial x}=\left(l_1\frac{\partial}{\partial x'}+l_2\frac{\partial}{\partial y'}+l_3\frac{\partial}{\partial z'}\right)(l_1u'+l_2v'+l_3w')$$

$$=l_1^2\frac{\partial u'}{\partial x'}+l_2^2\frac{\partial v'}{\partial y'}+l_3^2\frac{\partial w'}{\partial z'},$$

$$\left.\begin{aligned} \text{or} \quad & a = l_1^2 a' + l_2^2 b' + l_3^2 c'; \\ \text{similarly} \quad & b = m_1^2 a' + m_2^2 b' + m_3^2 c' \\ \text{and} \quad & c = n_1^2 a' + n_2^2 b' + n_3^2 c', \end{aligned}\right\} \quad \ldots\ldots\ldots\ldots\ldots\ldots(1)$$

and, incidentally, $a + b + c = a' + b' + c'$,

each side representing the dilatation (**1·3**).

Again, from **4·1**,

$$2f = \frac{\partial w}{\partial y} + \frac{\partial v}{\partial z} = \left(m_1 \frac{\partial}{\partial x'} + m_2 \frac{\partial}{\partial y'} + m_3 \frac{\partial}{\partial z'}\right)(n_1 u' + n_2 v' + n_3 w')$$

$$+ \left(n_1 \frac{\partial}{\partial x'} + n_2 \frac{\partial}{\partial y'} + n_3 \frac{\partial}{\partial z'}\right)(m_1 u' + m_2 v' + m_3 w')$$

$$= 2m_1 n_1 \frac{\partial u'}{\partial x'} + 2m_2 n_2 \frac{\partial v'}{\partial y'} + 2m_3 n_3 \frac{\partial w'}{\partial z'},$$

$$\left.\begin{aligned} \text{or} \quad & f = m_1 n_1 a' + m_2 n_2 b' + m_3 n_3 c'; \\ \text{similarly} \quad & g = n_1 l_1 a' + n_2 l_2 b' + n_3 l_3 c' \\ \text{and} \quad & h = l_1 m_1 a' + l_2 m_2 b' + l_3 m_3 c'. \end{aligned}\right\} \quad \ldots\ldots\ldots\ldots(2)$$

Now let a plane at right angles to Px cut Px', Py', Pz' in A, B, C, forming with the coordinate planes $Px'y'z'$ a tetrahedron of small dimensions. Let Δ denote the area ABC, then the areas PBC, PCA, PAB are $l_1\Delta$, $l_2\Delta$, $l_3\Delta$ and the only stresses on them are the normal stresses

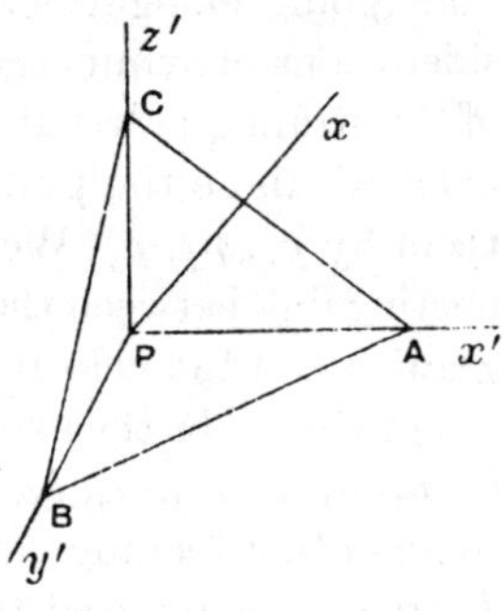

$$p_1 l_1 \Delta, \quad p_2 l_2 \Delta, \quad p_3 l_3 \Delta.$$

But the stresses on ABC are

$$p_{xx}\Delta, \quad p_{xy}\Delta, \quad p_{xz}\Delta$$

parallel to Px, Py, Pz. So if we resolve in the direction Px for the fluid in the tetrahedron and note that the resolved parts of the kinetic reactions and external forces will be of higher order of small quantities, we have

$$p_{xx}\Delta = p_1 l_1 \Delta . l_1 + p_2 l_2 \Delta . l_2 + p_3 l_3 \Delta . l_3,$$

$$\left.\begin{aligned} \text{or} \quad & p_{xx} = p_1 l_1^2 + p_2 l_2^2 + p_3 l_3^2; \\ \text{similarly} \quad & p_{yy} = p_1 m_1^2 + p_2 m_2^2 + p_3 m_3^2 \\ \text{and} \quad & p_{zz} = p_1 n_1^2 + p_2 n_2^2 + p_3 n_3^2. \end{aligned}\right\} \quad \ldots\ldots\ldots\ldots\ldots(3)$$

Whence $\quad p_{xx} + p_{yy} + p_{zz} = p_1 + p_2 + p_3 \quad \ldots\ldots\ldots\ldots(4).$

Thus the sum of the normal stresses across any three perpendicular planes at a point is the same. We denote this sum by $-3p$, so that p denotes the *mean pressure* at a point.

Again, if for the same tetrahedron we resolve parallel to Py, we get

$$p_{xy}\Delta = p_1 l_1 \Delta . m_1 + p_2 l_2 \Delta . m_2 + p_3 l_3 \Delta . m_3$$

or

$$\left.\begin{aligned} p_{xy} &= p_1 l_1 m_1 + p_2 l_2 m_2 + p_3 l_3 m_3; \\ \text{similarly}\quad p_{yz} &= p_1 m_1 n_1 + p_2 m_2 n_2 + p_3 m_3 n_3 \\ \text{and}\quad p_{zx} &= p_1 n_1 l_1 + p_2 n_2 l_2 + p_3 n_3 l_3. \end{aligned}\right\} \quad \text{.........(5)}$$

We have thus expressed the six stresses of **13·22** in terms of the principal stresses, but we cannot proceed further without an assumption.

We assume that these principal stresses p_1, p_2, p_3 differ from their mean value $-p$ by linear functions of the rates of distortion a', b', c' of the fluid element, and write

$$\left.\begin{aligned} p_1 &= -p + \lambda(a' + b' + c') + 2\mu a' \\ p_2 &= -p + \lambda(a' + b' + c') + 2\mu b' \\ p_3 &= -p + \lambda(a' + b' + c') + 2\mu c' \end{aligned}\right\} \quad \text{.........(6).}$$

Since $p_1 + p_2 + p_3 = -3p$, it follows by addition that

$$3\lambda + 2\mu = 0 \quad \text{.........(7).}$$

Hence, from (3) and (1),

$$p_{xx} = -p - \tfrac{2}{3}\mu(a + b + c) + 2\mu a,$$

or

$$\left.\begin{aligned} p_{xx} &= -p - \frac{2}{3}\mu\left(\frac{\partial u}{\partial x} + \frac{\partial v}{\partial y} + \frac{\partial w}{\partial z}\right) + 2\mu\frac{\partial u}{\partial x}; \\ \text{similarly}\quad p_{yy} &= -p - \frac{2}{3}\mu\left(\frac{\partial u}{\partial x} + \frac{\partial v}{\partial y} + \frac{\partial w}{\partial z}\right) + 2\mu\frac{\partial v}{\partial y} \\ \text{and}\quad p_{zz} &= -p - \frac{2}{3}\mu\left(\frac{\partial u}{\partial x} + \frac{\partial v}{\partial y} + \frac{\partial w}{\partial z}\right) + 2\mu\frac{\partial w}{\partial z}. \end{aligned}\right\} \quad \text{......(8)}$$

Again, since $m_1 n_1 + m_2 n_2 + m_3 n_3 = 0$, therefore, from (5), (6) and (2) we have

$$\left.\begin{aligned} p_{yz} &= 2\mu(m_1 n_1 a' + m_2 n_2 b' + m_3 n_3 c') \\ &= 2\mu f \\ &= \mu\left(\frac{\partial w}{\partial y} + \frac{\partial v}{\partial z}\right) = p_{zy} \quad [\mathbf{4{\cdot}1}\ (2)]. \\ \text{Similarly}\quad p_{zx} &= p_{xz} = 2\mu g = \mu\left(\frac{\partial u}{\partial z} + \frac{\partial w}{\partial x}\right) \\ \text{and}\quad p_{xy} &= p_{yx} = 2\mu h = \mu\left(\frac{\partial v}{\partial x} + \frac{\partial u}{\partial y}\right). \end{aligned}\right\} \quad \text{.........(9)}$$

Thus (8) and (9) express the six components of stress at a point in terms of the mean pressure at the point and the gradients of velocity.

13·24. We notice that in the special case to which reference was made in **13·1**, viz. that of a steady flow with velocity u at right angles to the axis of z, we have $v=w=0$, and the tangential stress in the direction of flow, from **13·23** (9), is

$$p_{zx}=\mu\,\partial u/\partial z,$$

so that the symbol μ introduced in **13·23** (6) is what we have already defined as the coefficient of viscosity.

13·3. Equations of Motion. We now obtain the equations of motion in their most general form by substituting in **13·22** (1) from **13·23** (8), (9), viz.

$$\left.\begin{aligned}\rho\frac{Du}{Dt}&=\rho X-\frac{\partial p}{\partial x}+\frac{1}{3}\mu\frac{\partial}{\partial x}\left(\frac{\partial u}{\partial x}+\frac{\partial v}{\partial y}+\frac{\partial w}{\partial z}\right)+\mu\nabla^2 u\\ \rho\frac{Dv}{Dt}&=\rho Y-\frac{\partial p}{\partial y}+\frac{1}{3}\mu\frac{\partial}{\partial y}\left(\frac{\partial u}{\partial x}+\frac{\partial v}{\partial y}+\frac{\partial w}{\partial z}\right)+\mu\nabla^2 v\\ \rho\frac{Dw}{Dt}&=\rho Z-\frac{\partial p}{\partial z}+\frac{1}{3}\mu\frac{\partial}{\partial z}\left(\frac{\partial u}{\partial x}+\frac{\partial v}{\partial y}+\frac{\partial w}{\partial z}\right)+\mu\nabla^2 w\end{aligned}\right\}\ \ldots(1).$$

In the case of incompressible fluid

$$\frac{\partial u}{\partial x}+\frac{\partial v}{\partial y}+\frac{\partial w}{\partial z}=0,$$

and we may write the equations

$$\left.\begin{aligned}\frac{Du}{Dt}&=X-\frac{1}{\rho}\frac{\partial p}{\partial x}+\nu\nabla^2 u\\ \frac{Dv}{Dt}&=Y-\frac{1}{\rho}\frac{\partial p}{\partial y}+\nu\nabla^2 v\\ \frac{Dw}{Dt}&=Z-\frac{1}{\rho}\frac{\partial p}{\partial z}+\nu\nabla^2 w\end{aligned}\right\}\ \ldots\ldots\ldots\ldots\ldots\ldots(2)^*,$$

where $\nu=\mu/\rho$ is called the *kinematic coefficient of viscosity.*

* These equations were first obtained by Navier (1822), but the mode of investigation given here follows that of Sir H. Lamb (see *Hydrodynamics*, §§ 323–328 (1932)) and is based upon a paper of Stokes, 'On the Theories of the Internal Friction of Fluids in Motion and of the Equilibrium and Motion of Elastic Solids', *Trans. Camb. Phil. Soc.* VIII, 1845, or *Math. and Phys. Papers*, I, p. 75.

13·31. Boundary Conditions. At the surface of separation of two fluids, consider an element of a thin stratum whose opposite faces are close together but one in each fluid. Since the kinetic reactions and the external forces on the element are of higher order of small quantities than the stresses on its surface, therefore the resultant stresses on opposite sides of the surface of separation must be equal and opposite.

Let h be a vector normal to the surface of separation and l, m, n its direction cosines, and as in **13·23** construct a small tetrahedron $PABC$ with the face ABC perpendicular to h and the corners A, B, C on Px, Py, Pz; then we find that the stress components in one fluid are given by

$$\left.\begin{aligned} p_{hx} &= lp_{xx} + mp_{yx} + np_{zx} \\ p_{hy} &= lp_{xy} + mp_{yy} + np_{zy} \\ p_{hz} &= lp_{xz} + mp_{yz} + np_{zz} \end{aligned}\right\} \quad \text{.................}(1),$$

with similar expressions for the components p_{hx}', p_{hy}', p_{hz}' in the other fluid; and, omitting the effects of capillarity, we must have

$$p_{hx} = p_{hx}', \quad p_{hy} = p_{hy}', \quad p_{hz} = p_{hz}' \quad \text{...........}(2).$$

We shall assume in what follows that at the surface of separation of a solid and a fluid no slipping takes place. Different theories were put forward by earlier students of the subject allowing for the possibility of slip, but strong evidence that in most cases no slipping takes place exists in the fact that mathematical results based upon this hypothesis are in general accord with calculations based upon experiment where such can be made*.

13·32. Equations of Motion in Cylindrical and Polar Coordinates. As we shall have occasion to use equations of motion and stress components in other than rectangular coordinates we proceed to obtain the required forms. The transformations may readily be effected by the tensor calculus†, but without assuming the necessary knowledge of this subject we proceed as follows: The acceleration components have already been found in **1·52**; hence for an incompressible fluid we have only to calculate the terms in cylindrical and polar coordinates which are to take the place of the terms $\nu\nabla^2 u$, $\nu\nabla^2 v$, $\nu\nabla^2 w$ in the cartesian equations.

With *cylindrical coordinates* r, θ, z let v_r, v_θ, v_z denote the components of velocity and f_r, f_θ, f_z those of acceleration. Then

$$u = v_r \cos\theta - v_\theta \sin\theta, \quad v = v_r \sin\theta + v_\theta \cos\theta,$$

or

$$u + iv = e^{i\theta}(v_r + iv_\theta) \quad \text{and} \quad w = v_z \quad \text{.................}(1).$$

* There are exceptional cases and there is a physical explanation based on the kinetic theory of gases. On this subject see *The Physics of Solids and Fluids*, Ewald, Pöschl and Prandtl, 1930, p. 271.

† See e.g. *Handbuch der Physik*, VII, p. 94, J. Springer, Berlin, 1927.

Also $$f_r = \cos\theta \frac{Du}{Dt} + \sin\theta \frac{Dv}{Dt},\quad f_\theta = -\sin\theta \frac{Du}{Dt} + \cos\theta \frac{Dv}{Dt} \quad\text{......(2).}$$

Therefore $$f_r = -\frac{\partial V}{\partial r} - \frac{1}{\rho}\frac{\partial p}{\partial r} + \nu(\cos\theta\,\nabla^2 u + \sin\theta\,\nabla^2 v)$$

and $$f_\theta = -\frac{\partial V}{r\,\partial\theta} - \frac{1}{\rho}\frac{\partial p}{r\,\partial\theta} + \nu(-\sin\theta\,\nabla^2 u + \cos\theta\,\nabla^2 v) \quad\text{.........(3),}$$

where V is the potential of the external forces.

Now from (1)

$$e^{-i\theta}\nabla^2(u+iv) = e^{-i\theta}\nabla^2 e^{i\theta}(v_r + iv_\theta) = \left(\nabla^2 + \frac{2i}{r^2}\frac{\partial}{\partial\theta} - \frac{1}{r^2}\right)(v_r + iv_\theta) \quad\text{......(4),}$$

where, on the right, $$\nabla^2 \equiv \frac{\partial^2}{\partial r^2} + \frac{1}{r}\frac{\partial}{\partial r} + \frac{1}{r^2}\frac{\partial^2}{\partial\theta^2} + \frac{\partial^2}{\partial z^2} \quad\text{.....................(5).}$$

Whence, by equating real and imaginary parts

$$\cos\theta\,\nabla^2 u + \sin\theta\,\nabla^2 v = \nabla^2 v_r - \frac{2}{r^2}\frac{\partial v_\theta}{\partial\theta} - \frac{v_r}{r^2}$$

and $$-\sin\theta\,\nabla^2 u + \cos\theta\,\nabla^2 v = \nabla^2 v_\theta + \frac{2}{r^2}\frac{\partial v_r}{\partial\theta} - \frac{v_\theta}{r^2} \quad\text{...........(6).}$$

Hence, by taking the acceleration components from **1·52**, and substituting (6) in (3) we have for **cylindrical coordinates**

$$\frac{Dv_r}{Dt} - \frac{v_\theta^2}{r} = -\frac{\partial V}{\partial r} - \frac{1}{\rho}\frac{\partial p}{\partial r} + \nu\left(\nabla^2 v_r - \frac{2}{r^2}\frac{\partial v_\theta}{\partial\theta} - \frac{v_r}{r^2}\right)$$

$$\frac{Dv_\theta}{Dt} + \frac{v_r v_\theta}{r} = -\frac{\partial V}{r\,\partial\theta} - \frac{1}{\rho}\frac{\partial p}{r\,\partial\theta} + \nu\left(\nabla^2 v_\theta + \frac{2}{r^2}\frac{\partial v_r}{\partial\theta} - \frac{v_\theta}{r^2}\right) \quad\text{........(7),}$$

$$\frac{Dv_z}{Dt} = -\frac{\partial V}{\partial z} - \frac{1}{\rho}\frac{\partial p}{\partial z} + \nu\nabla^2 v_z$$

where $$\frac{D}{Dt} \equiv \frac{\partial}{\partial t} + v_r\frac{\partial}{\partial r} + v_\theta\frac{\partial}{r\,\partial\theta} + v_z\frac{\partial}{\partial z} \quad\text{.....................(8),}$$

and ∇^2 has the value (5).

Again with *polar coordinates* r, θ, ϕ, let q_r, q_θ, q_ϕ denote the components of velocity and F_r, F_θ, F_ϕ those of acceleration. Then by comparing the velocities with those in cylindrical coordinates and taking account of the different meanings of r, θ in the two systems, viz. that the former r becomes $r\sin\theta$, and the former θ becomes ϕ, we have

$$v_r = q_r\sin\theta + q_\theta\cos\theta,\quad v_\theta = q_\phi,\quad v_z = q_r\cos\theta - q_\theta\sin\theta,$$

so that $$v_z + iv_r = e^{i\theta}(q_r + iq_\theta) \quad\text{......................(9).}$$

Also

$F_r = f_z\cos\theta + f_r\sin\theta$, which from (3) and (6) or (7)

$$= -\frac{\partial V}{\partial r} - \frac{1}{\rho}\frac{\partial p}{\partial r} + \nu\cos\theta\,\nabla^2 v_z + \nu\sin\theta\left(\nabla^2 v_r - \frac{2}{r^2\sin^2\theta}\frac{\partial v_\theta}{\partial\phi} - \frac{v_r}{r^2\sin^2\theta}\right)$$

$$F_\theta = -f_z\sin\theta + f_r\cos\theta$$

$$= -\frac{\partial V}{r\,\partial\theta} - \frac{1}{\rho}\frac{\partial p}{r\,\partial\theta} - \nu\sin\theta\,\nabla^2 v_z + \nu\cos\theta\left(\nabla^2 v_r - \frac{2}{r^2\sin^2\theta}\frac{\partial v_\theta}{\partial\phi} - \frac{v_r}{r^2\sin^2\theta}\right)$$

and

$$F_\phi = f_\theta = -\frac{\partial V}{r\sin\theta\,\partial\phi} - \frac{1}{\rho}\frac{\partial p}{r\sin\theta\,\partial\phi} + \nu\left(\nabla^2 v_\theta + \frac{2}{r^2\sin^2\theta}\frac{\partial v_r}{\partial\phi} - \frac{v_\theta}{r^2\sin^2\theta}\right)$$

$$\text{.....(10).}$$

Now from (9) $e^{-i\theta}\nabla^2(v_z+iv_r)=e^{-i\theta}\nabla^2 e^{i\theta}(q_r+iq_\theta)$,
where, in polars,

$$\nabla^2\equiv\frac{\partial^2}{\partial r^2}+\frac{2}{r}\frac{\partial}{\partial r}+\frac{\cot\theta}{r^2}\frac{\partial}{\partial\theta}+\frac{1}{r^2}\frac{\partial^2}{\partial\theta^2}+\frac{1}{r^2\sin^2\theta}\frac{\partial^2}{\partial\phi^2}\;\ldots\ldots\ldots(11);$$

therefore $$e^{-i\theta}\nabla^2(v_z+iv_r)=\left(\nabla^2+\frac{i\cot\theta}{r^2}+\frac{2i}{r^2}\frac{\partial}{\partial\theta}-\frac{1}{r^2}\right)(q_r+iq_\theta)\;\ldots\ldots(12),$$

and by equating real and imaginary parts we get

$$\left.\begin{aligned}\cos\theta\,\nabla^2 v_z+\sin\theta\,\nabla^2 v_r&=\nabla^2 q_r-\frac{q_r}{r^2}-\frac{\cot\theta}{r^2}q_\theta-\frac{2}{r^2}\frac{\partial q_\theta}{\partial\theta}\\ \text{and}\quad -\sin\theta\,\nabla^2 v_z+\cos\theta\,\nabla^2 v_r&=\nabla^2 q_\theta-\frac{q_\theta}{r^2}+\frac{\cot\theta}{r^2}q_r+\frac{2}{r^2}\frac{\partial q_r}{\partial\theta}\end{aligned}\right\}\;\ldots(13).$$

Then substituting in (10) and taking the acceleration components from **1·52** we have for **polar coordinates**

$$\left.\begin{aligned}&\frac{Dq_r}{Dt}-\frac{q_\theta^2+q_\phi^2}{r}=-\frac{\partial V}{\partial r}-\frac{1}{\rho}\frac{\partial p}{\partial r}\\ &\qquad+\nu\left(\nabla^2 q_r-\frac{2q_r}{r^2}-\frac{2\cot\theta}{r^2}q_\theta-\frac{2}{r^2}\frac{\partial q_\theta}{\partial\theta}-\frac{2}{r^2\sin\theta}\frac{\partial q_\phi}{\partial\phi}\right)\\ &\frac{Dq_\theta}{Dt}-\frac{q_\phi^2\cot\theta}{r}+\frac{q_r q_\theta}{r}=-\frac{\partial V}{r\partial\theta}-\frac{1}{\rho}\frac{\partial p}{r\partial\theta}\\ &\qquad+\nu\left(\nabla^2 q_\theta-\frac{q_\theta}{r^2\sin^2\theta}+\frac{2}{r^2}\frac{\partial q_r}{\partial\theta}-\frac{2\cos\theta}{r^2\sin^2\theta}\frac{\partial q_\phi}{\partial\phi}\right)\\ &\frac{Dq_\phi}{Dt}+\frac{q_r q_\phi}{r}+\frac{q_\theta q_\phi\cot\theta}{r}=-\frac{\partial V}{r\sin\theta\,\partial\phi}-\frac{1}{\rho}\frac{\partial p}{r\sin\theta\,\partial\phi}\\ &\qquad+\nu\left(\nabla^2 q_\phi-\frac{q^\phi}{r^2\sin^2\theta}+\frac{2}{r^2\sin\theta}\frac{\partial q_r}{\partial\phi}+\frac{2\cos\theta}{r^2\sin^2\theta}\frac{\partial q_\theta}{\partial\phi}\right)\end{aligned}\right\}\;\ldots\ldots(14),$$

where in polars $$\frac{D}{Dt}\equiv\frac{\partial}{\partial t}+q_r\frac{\partial}{\partial r}+q_\theta\frac{\partial}{r\partial\theta}+q_\phi\frac{\partial}{r\sin\theta\,\partial\phi}$$

and ∇^2 has the form (11).

13·33. Components of stress in cylindrical and polar coordinates. We shall merely state the results in cylindrical coordinates as the reader will be able to adapt to cylindrical coordinates the method used below for obtaining the formulae in polars.

With the notation of **13·32** we have for **cylindrical coordinates** r, θ, z:

$$\left.\begin{aligned}&p_{rr}=-p+2\mu\frac{\partial v_r}{\partial r},\quad p_{\theta\theta}=-p+2\mu\left(\frac{\partial v_\theta}{r\partial\theta}+\frac{v_r}{r}\right),\quad p_{zz}=-p+2\mu\frac{\partial v_z}{\partial z}\\ &p_{\theta z}=\mu\left(\frac{\partial v_z}{r\partial\theta}+\frac{\partial v_\theta}{\partial z}\right),\quad p_{zr}=\mu\left(\frac{\partial v_r}{\partial z}+\frac{\partial v_z}{\partial r}\right),\quad p_{r\theta}=\mu\left(\frac{\partial v_r}{r\partial\theta}-\frac{v_\theta}{r}+\frac{\partial v_\theta}{\partial r}\right)\end{aligned}\right\}\;\ldots(1).$$

For *polar coordinates* we proceed thus: take a set of rectangular axes $Oxyz$ and let r, θ, ϕ be polar coordinates measured in the usual way. Take a second set of rectangular axes $Ox'y'z'$ such that the plane $x'z'$ contains Oz and makes an angle ϕ' with xz, and let the angle $zOz'=\theta'$. Then the direction cosines of the axes $Ox'y'z'$ referred to $Oxyz$ are

$$\begin{array}{lll}\cos\theta'\cos\phi', & \cos\theta'\sin\phi', & -\sin\theta'\\ -\sin\phi', & \cos\phi', & 0\\ \sin\theta'\cos\phi', & \sin\theta'\sin\phi', & \cos\theta'.\end{array}$$

And with the notation of **13·32**

$$\left.\begin{aligned} u&=q_r\sin\theta\cos\phi+q_\theta\cos\theta\cos\phi-q_\phi\sin\phi\\ v&=q_r\sin\theta\sin\phi+q_\theta\cos\theta\sin\phi+q_\phi\cos\phi\\ w&=q_r\cos\theta-q_\theta\sin\theta \end{aligned}\right\}\quad\ldots\ldots\ldots\ldots(2).$$

And if u', v', w' are components of velocity referred to $Ox'y'z'$ we have from (2)

$$\left.\begin{aligned} u'&=(q_r\sin\theta\cos\phi+q_\theta\cos\theta\cos\phi-q_\phi\sin\phi)\cos\theta'\cos\phi'\\ &\quad+(q_r\sin\theta\sin\phi+q_\theta\cos\theta\sin\phi+q_\phi\cos\phi)\cos\theta'\sin\phi'\\ &\quad-(q_r\cos\theta-q_\theta\sin\theta)\sin\theta'\\ v'&=-(q_r\sin\theta\cos\phi+q_\theta\cos\theta\cos\phi-q_\phi\sin\phi)\sin\phi'\\ &\quad+(q_r\sin\theta\sin\phi+q_\theta\cos\theta\sin\phi+q_\phi\cos\phi)\cos\phi'\\ w'&=(q_r\sin\theta\cos\phi+q_\theta\cos\theta\cos\phi-q_\phi\sin\phi)\sin\theta'\cos\phi'\\ &\quad+(q_r\sin\theta\sin\phi+q_\theta\cos\theta\sin\phi+q_\phi\cos\phi)\sin\theta'\sin\phi'\\ &\quad+(q_r\cos\theta-q_\theta\sin\theta)\cos\theta' \end{aligned}\right\}\quad\ldots(3).$$

Now we find the required expressions for the stresses from the values of $\partial u'/\partial x'$, etc. when the axes $Ox'y'z'$ are so moved that $\theta'=\theta$ and $\phi'=\phi$, and then

$$\frac{\partial}{\partial x'}=\frac{\partial}{r\,\partial\theta},\quad \frac{\partial}{\partial y'}=\frac{\partial}{r\sin\theta\,\partial\phi}\quad\text{and}\quad \frac{\partial}{\partial z'}=\frac{\partial}{\partial r}.$$

Hence by differentiating (3) and putting $\theta'=\theta$ and $\phi'=\phi$ after differentiation we get

$$\left.\begin{aligned} &\frac{\partial u'}{\partial x'}=\frac{\partial q_\theta}{r\,\partial\theta}+\frac{q_r}{r},\quad \frac{\partial u'}{\partial y'}=\frac{\partial q_\theta}{r\sin\theta\,\partial\phi}-\frac{q_\phi\cot\theta}{r},\quad \frac{\partial u'}{\partial z'}=\frac{\partial q_\theta}{\partial r}\\ &\frac{\partial v'}{\partial x'}=\frac{\partial q_\phi}{r\,\partial\theta},\quad \frac{\partial v'}{\partial y'}=\frac{\partial q_\phi}{r\sin\theta\,\partial\phi}+\frac{q_r}{r}+\frac{q_\theta\cot\theta}{r},\quad \frac{\partial v'}{\partial z'}=\frac{\partial q_\phi}{\partial r}\\ &\frac{\partial w'}{\partial x'}=\frac{\partial q_r}{r\,\partial\theta}-\frac{q_\theta}{r},\quad \frac{\partial w'}{\partial y'}=\frac{\partial q_r}{r\sin\theta\,\partial\phi}-\frac{q_\phi}{r},\quad \frac{\partial w'}{\partial z'}=\frac{\partial q_r}{\partial r} \end{aligned}\right\}\quad\ldots(4).$$

Thus we have for the components of stress in **polar coordinates**

$$\left.\begin{aligned} p_{\theta\theta}&=-p+2\mu\left(\frac{\partial q_\theta}{r\,\partial\theta}+\frac{q_r}{r}\right)\\ p_{\phi\phi}&=-p+2\mu\left(\frac{\partial q_\phi}{r\sin\theta\,\partial\phi}+\frac{q_r}{r}+\frac{q_\theta\cot\theta}{r}\right)\\ p_{rr}&=-p+2\mu\frac{\partial q_r}{\partial r}\\ p_{\theta\phi}&=\mu\left(\frac{\partial q_\phi}{r\,\partial\theta}+\frac{\partial q_\theta}{r\sin\theta\,\partial\phi}-\frac{q_\phi\cot\theta}{r}\right)\\ p_{r\theta}&=\mu\left(\frac{\partial q_r}{r\,\partial\theta}-\frac{q_\theta}{r}+\frac{\partial q_\theta}{\partial r}\right)\\ p_{r\phi}&=\mu\left(\frac{\partial q_r}{r\sin\theta\,\partial\phi}-\frac{q_\phi}{r}+\frac{\partial q_\phi}{\partial r}\right) \end{aligned}\right\}\quad\ldots\ldots\ldots\ldots(5).$$

13·4. Dissipation of Energy. If T denotes the kinetic energy at time t of a limited portion of fluid bounded by a surface S, then

$$2T=\iiint \rho\,(u^2+v^2+w^2)\,dx\,dy\,dz \quad\ldots\ldots\ldots\ldots\ldots(1)$$

and following the motion of the same portion of fluid

$$\frac{DT}{Dt}=\iiint \rho\left(u\frac{Du}{Dt}+v\frac{Dv}{Dt}+w\frac{Dw}{Dt}\right)dx\,dy\,dz \quad\ldots\ldots(2).$$

Substituting from the equations of motion of **13·22** (1), we get

$$\frac{DT}{Dt}=\iiint \rho\,(uX+vY+wZ)\,dx\,dy\,dz$$
$$+\iiint\left\{u\left(\frac{\partial p_{xx}}{\partial x}+\frac{\partial p_{yx}}{\partial y}+\frac{\partial p_{zx}}{\partial z}\right)+v\left(\frac{\partial p_{xy}}{\partial x}+\frac{\partial p_{yy}}{\partial y}+\frac{\partial p_{zy}}{\partial z}\right)\right.$$
$$\left.+w\left(\frac{\partial p_{xz}}{\partial x}+\frac{\partial p_{yz}}{\partial y}+\frac{\partial p_{zz}}{\partial z}\right)\right\}dx\,dy\,dz \quad\ldots\ldots\ldots\ldots\ldots\ldots\ldots(3).$$

The first integral represents the rate at which the external forces are doing work throughout the mass of the fluid.

The second integral may be integrated by parts and gives

$$-\iint\{u\,(lp_{xx}+mp_{yx}+np_{zx})+v\,(lp_{xy}+mp_{yy}+np_{zy})$$
$$+w\,(lp_{xz}+mp_{yz}+np_{zz})\}\,dS$$
$$-\iiint\left\{\frac{\partial u}{\partial x}p_{xx}+\frac{\partial v}{\partial y}p_{yy}+\frac{\partial w}{\partial z}p_{zz}+\left(\frac{\partial v}{\partial z}+\frac{\partial w}{\partial y}\right)p_{yz}+\ldots+\right\}dx\,dy\,dz$$
$$\ldots\ldots(4),$$

where l, m, n are the direction cosines of the inward drawn normal to dS. By **13·31** (1) the surface integral may be written

$$-\iint(up_{hx}+vp_{hy}+wp_{hz})\,dS \quad\ldots\ldots\ldots\ldots\ldots(5),$$

where the suffix h indicates a normal to dS, and this integral represents the rate at which the kinetic energy is being increased by the action of the stresses on the boundary of the fluid.

If in the remaining volume integral we substitute from **13·23** (8) and (9) we get

$$\iiint p\left(\frac{\partial u}{\partial x}+\frac{\partial v}{\partial y}+\frac{\partial w}{\partial z}\right)dx\,dy\,dz$$
$$+\frac{2}{3}\mu\iiint\left(\frac{\partial u}{\partial x}+\frac{\partial v}{\partial y}+\frac{\partial w}{\partial z}\right)^2 dx\,dy\,dz$$
$$-\mu\iiint\left\{2\left(\frac{\partial u}{\partial x}\right)^2+2\left(\frac{\partial v}{\partial y}\right)^2+2\left(\frac{\partial w}{\partial z}\right)^2+\left(\frac{\partial w}{\partial y}+\frac{\partial v}{\partial z}\right)^2\right.$$
$$\left.+\left(\frac{\partial u}{\partial z}+\frac{\partial w}{\partial x}\right)^2+\left(\frac{\partial v}{\partial x}+\frac{\partial u}{\partial y}\right)^2\right\}dx\,dy\,dz \quad\ldots\ldots\ldots\ldots\ldots(6).$$

In this expression p denotes, in the case of an elastic fluid, the pressure statically corresponding to the density of the fluid at (x, y, z), and the first integral represents the rate at which the various elements of fluid are losing intrinsic energy in consequence of internal expansion*.

The remaining integrals on the whole are negative or at least never positive and represent the rate of dissipation of energy in consequence of internal friction. This cannot vanish unless

$$\frac{\partial u}{\partial x}=\frac{\partial v}{\partial y}=\frac{\partial w}{\partial z},$$

and
$$\frac{\partial w}{\partial y}+\frac{\partial v}{\partial z}=\frac{\partial u}{\partial z}+\frac{\partial w}{\partial x}=\frac{\partial v}{\partial x}+\frac{\partial u}{\partial y}=0;$$

i.e. in the notation of **4·1** unless $a=b=c$ and $f=g=h=0$.

Since for an incompressible fluid $a+b+c=0$ therefore there must be dissipation of energy in a liquid unless

$$a=b=c=f=g=h=0$$

at every point; i.e. no extension or contraction of linear elements. It follows that only when the motion consists of a translation or rotation of the mass as a whole can there be no dissipation of energy†.

It follows that in the special case of a liquid the rate of dissipation of energy is represented by

$$\mu\iiint\left\{2\left(\frac{\partial u}{\partial x}\right)^2+2\left(\frac{\partial v}{\partial y}\right)^2+2\left(\frac{\partial w}{\partial z}\right)^2\right.$$
$$\left.+\left(\frac{\partial w}{\partial y}+\frac{\partial v}{\partial z}\right)^2+\left(\frac{\partial u}{\partial z}+\frac{\partial w}{\partial x}\right)^2+\left(\frac{\partial v}{\partial x}+\frac{\partial u}{\partial y}\right)^2\right\}dx\,dy\,dz \;\ldots\ldots(7).$$

* In the case of an elastic fluid, from the equation of continuity

$$\iiint p\left(\frac{\partial u}{\partial x}+\frac{\partial v}{\partial y}+\frac{\partial w}{\partial z}\right)dx\,dy\,dz=\iiint -p\,.\,\frac{1}{\rho}\frac{D\rho}{Dt}\,dx\,dy\,dz$$
$$=\iiint p\,\frac{D}{Dt}\left(\frac{1}{\rho}\right).\,\rho\,dx\,dy\,dz$$
$$=-\iiint \frac{DE}{Dt}\,\rho\,dx\,dy\,dz,$$

where $E=-\int p\,d\left(\frac{1}{\rho}\right)$ denotes the intrinsic energy, or the work done by a unit mass of the fluid against external pressure as it passes from its actual volume to some standard volume.

† See Stokes 'On the Effect of the Internal Friction of Fluid on the Motion of Pendulums', *Trans. Camb. Phil. Soc.* IX, p. 8, or *Math. and Phys. Papers*, III, p. 69, where a detailed discussion of the conclusion is given.

13·5. The Reynolds Number. Experimental work concerning the motion of solid bodies through fluids is usually performed with models and it is of importance to know to what extent the full-sized body will behave like the model. A like consideration arises in connection with the flow of fluid through tubes of different diameters.

If we regard an external field of force as producing a hydrostatic pressure and take p to be the dynamical part of the pressure, i.e. the amount by which the total pressure exceeds the hydrostatic pressure, then in the equations of motion of a fluid, omitting the external field of force and the hydrostatic pressure there are three types of forces, viz. (i) the reversed effective forces or inertia terms of the type $\rho\, \partial u/\partial t$ or $\rho u\, \partial u/\partial x$, (ii) pressure terms $\partial p/\partial x$, etc., and (iii) terms arising from frictional forces of the type $\mu\, \partial^2 u/\partial x^2$. In order that the flow may be geometrically similar in two corresponding motions it is necessary that the ratios of the forces represented by these sets of terms shall be the same in both motions. Since however the forces (including reversed effective forces, as in D'Alembert's Principle) which enter into any equation balance one another, it will suffice to consider the ratio of two of the types, e.g. (i) and (iii).

In the case of a body moving through a fluid the velocities are all proportional to the velocity of the body, say U; and we can choose a length l associated with the body to represent the linear scale of measurement. Terms of type (i) are then of dimensions $\rho U^2/l$, and terms of type (iii) are of dimensions $\mu U/l^2$, and for similarity we require that the ratio $\rho U^2/l \div \mu U/l^2$ or $R = \rho U l/\mu$ shall be the same in both motions.

The expression $\rho U l/\mu$ or $U l/\nu$, where ν is the kinematic coefficient of viscosity is called the **Reynolds number**, after Osborne Reynolds who first investigated the question of similarity*. It represents the ratio of the inertia terms to the frictional terms in the equation of motion and it is clearly non-dimensional. A necessary condition for the dynamical similarity of two fluid motions in which the systems are geometrically similar, is that they must have the same Reynolds number and the boundary conditions must be the same.

Further, if F represents the component in an assigned direction of the force on the body due to the stresses on its surface, then $F/\rho U^2 l^2$ is non-

* *Phil. Trans. R.S.* CLXXIV, 1883, or *Scientific Papers*, II, p. 51.

dimensional and must therefore depend on a non-dimensional combination of the data ρ, U, l, μ, and R is the only such combination so that

$$F = \rho U^2 l^2 f(R).$$

For a particular case a series of values of $f(R)$ is determined by experiment by varying U, l or ν as may be convenient.

The importance of the Reynolds number lies in the fact that it gives some indication of the nature of the corresponding fluid motion. Thus a small Reynolds number implies that viscosity is predominant, and a large number implies that viscosity is small or that the effects of inertia outweigh the effects of friction.

Also it was found by Reynolds* that in the case of flow through a tube the steady laminar flow breaks down and the flow becomes turbulent when Ul/ν exceeds a definite limiting value, where U denotes the mean velocity of the fluid and l the diameter of the tube. It follows that for small values of l the dynamical equations represent the actual motion for a wide range of velocities, but if l is large then either U must be small or the viscosity large for otherwise the motion will be turbulent.

The simpler problems of fluid motion which can be considered are divided into two classes according as the corresponding Reynolds number is small or large. In the former case viscosity is predominant and the inertia terms in the equations may be regarded as negligible. In the latter case the frictional terms are small, and we shall see later that in the case of relative motion of a fluid and solid boundaries this means that at a distance from the boundaries the frictional terms in the equations are negligible so that the conditions there approximate to the motion of a 'perfect' fluid, but near the boundaries there is a thin layer of fluid in which viscosity is effective and in which the velocity of the fluid varies from that of the solid in contact with it to that of the frictionless motion outside the layer, and the smaller the viscosity the thinner is this layer†.

13·6. We shall next consider some problems of steady motion either of such a special kind that the inertia terms vanish identically or such that the Reynolds number is so small that the inertia terms are negligible in comparison with the frictional terms.

13·61. Steady Motion between Parallel Planes. Let viscous fluid of uniform density ρ fill the space between parallel

* *Loc. cit.* p. 377.

† L. Prandtl, *Verh. d. 3. intern. Math. Ver.* Heidelberg, 1904.

plane boundaries $z=0$, $z=h$. Let the former boundary have a velocity U in the x direction and the latter be at rest. We assume that there is a steady motion with a velocity which at any point is denoted by u parallel to x, the components v and w being zero. From the equation of continuity we must therefore have $\partial u/\partial x=0$, and as u is clearly independent of y it must be a function of z alone. It is now apparent that the inertia terms in the equations of motion are all zero. At a distance z from the plane $z=0$, there is a

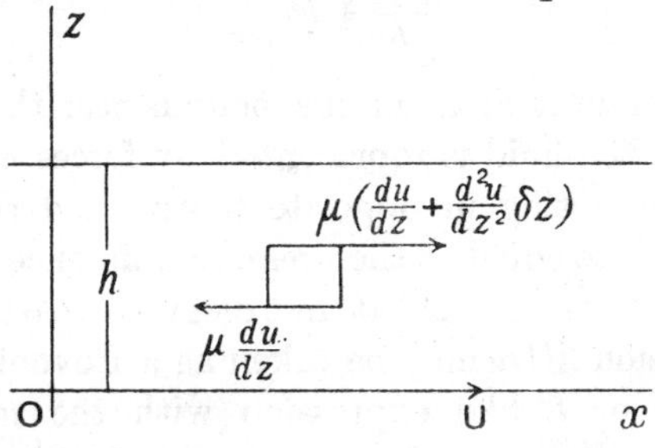

tractive force $\mu du/dz$ per unit area in the plane parallel to the boundaries opposing relative motion (**13·23** (9)). Considering an element of the fluid with faces parallel to the coordinate planes and of linear dimensions δx, 1, δz, the tractive forces on its faces parallel to the boundary planes give a resultant $\mu\frac{d^2u}{dz^2}\delta z\,\delta x$ in the x direction, and the resultant of the mean pressures on the faces parallel to yz is $-\frac{\partial p}{\partial x}\delta x\,\delta z$ in the same direction. There being no acceleration these forces have a zero sum, so that

$$\mu\frac{d^2u}{dz^2}=\frac{\partial p}{\partial x} \quad\text{.........................(1)},$$

which on the hypotheses stated might have been written down directly from **13·3** (1).

Since there is no motion save in the x direction therefore $\partial p/\partial y$ and $\partial p/\partial z$ are zero, and since u is independent of x, (1) shews that the pressure gradient dp/dx is a constant.

Integrating (1) we get

$$\mu u=\tfrac{1}{2}z^2\frac{dp}{dx}+Az+B \quad\text{....................(2)}$$

and since $u=U$ when $z=0$, and $u=0$ when $z=h$, therefore

$$u=\left(1-\frac{z}{h}\right)U-\frac{1}{2\mu}z(h-z)\frac{dp}{dx} \quad\text{..............(3).}$$

The total flux per unit breadth across a plane perpendicular to x is

$$\int_0^h u\,dz = \tfrac{1}{2}hU - \frac{h^3}{12\mu}\frac{dp}{dx} \quad\text{..................(4).}$$

The tangential stress at any point is

$$\mu\frac{du}{dz} = -\frac{\mu U}{h} - \tfrac{1}{2}(h-2z)\frac{dp}{dx} \quad\text{..................(5)}$$

which gives

$$\frac{\mu U}{h} \pm \tfrac{1}{2}h\frac{dp}{dx}$$

as the drag per unit area on the boundaries; the drag of the boundaries on the fluid is represented by forces equal and opposite to these acting in opposite senses, and their resultant $h\,dp/dx$ is of course equal to the pressure difference on two planes at right angles to the stream at unit distance apart.

The expression hU/ν may be taken as a Reynolds number in this problem, or a like expression with the mean velocity instead of U.

13·611. Theory of Lubrication. It is a familiar fact that parallel or nearly parallel surfaces can slide over one another with great ease if a film of viscous fluid is maintained between them. The mathematical theory is due to O. Reynolds*. A necessary condition is that the opposing surfaces should be slightly inclined to one another and that the relative motion should tend to drag the fluid from the wider to the narrower part of the intervening space. The following discussion is based upon a paper by Lord Rayleigh†.

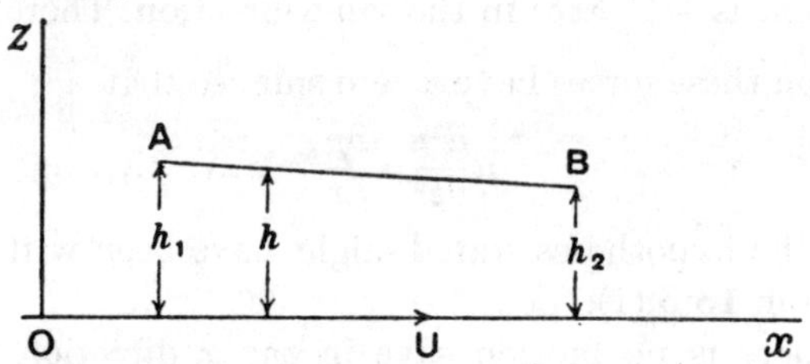

Consider a fixed block with a plane face AB nearly parallel to another plane $z=0$, which has a uniform velocity U in the x direction. Let the block be so wide in the y direction that the problem may be treated as two-dimensional.

Let a, h_1; b, h_2 and x, h be the coordinates of A, B and any other point on AB. Since the inclination of the plane faces is small, the velocity u at any

* 'On the Theory of Lubrication, etc.', *Phil. Trans.* CLXXVII, p. 157, 1886, or *Scientific Papers*, II, p. 228.

† 'Notes on the Theory of Lubrication', *Phil. Mag.* (6), XXXV, p. 1, 1918, or *Scientific Papers*, VI, p. 523.

point may be determined as in **13·61**; and as the addition of a constant pressure throughout the fluid will make no difference to the solution, we may for convenience assume that $p=0$ beyond the ends of the block.

The condition of continuity is that the total flux

$$\int_0^h u\,dz \equiv \tfrac{1}{2}hU - \frac{h^3}{12\mu}\frac{dp}{dx} \quad \text{.......................(1)}$$

must be independent of x.

Therefore
$$\frac{dp}{dx} = 6\mu U \frac{(h-h_0)}{h^3} \quad \text{.......................(2),}$$

where h_0 is the value of h at points of maximum pressure.

But
$$h(b-a) = h_1(b-x) + h_2(x-a),$$

so that
$$l\frac{dh}{dx} = h_2 - h_1 \quad \text{.......................(3),}$$

where l is the length of the block.

Hence, from (2),
$$\frac{dp}{dh} = \frac{6\mu Ul}{h_2-h_1}\cdot\frac{h-h_0}{h^3} \quad \text{.......................(4),}$$

and, by integration,
$$p = \frac{3\mu Ul}{h_2-h_1}\left\{\frac{h_0-2h}{h^2}+C\right\} \quad \text{.......................(5);}$$

and we must now determine h_0 and C so that $p=0$ when $h=h_1$ and when $h=h_2$.

This gives
$$h_0 = 2h_1h_2/(h_1+h_2) \quad \text{.......................(6)}$$

and
$$p = \frac{6\mu Ul}{h_1^2-h_2^2}\,\frac{(h_1-h)(h-h_2)}{h^2} \quad \text{.......................(7).}$$

It follows that p cannot be positive unless $h_1 > h_2$, i.e. unless the stream contracts in the direction of U.

The total pressure is given by

$$P = \int_a^b p\,dx = \int_{h_1}^{h_2} \frac{pl}{h_2-h_1}\,dh$$
$$= \frac{6\mu Ul^2}{(k-1)^2h_2^2}\left\{\log k - \frac{2(k-1)}{k+1}\right\} \quad \text{.......................(8),}$$

where $k = h_1/h_2$.

Again from **13·61** (5) the tangential stress on either surface is

$$(p_{zx})_{z=0} = -\frac{\mu U}{h} - \tfrac{1}{2}h\frac{dp}{dx},$$

so that, using (2) and (3), we get for the total frictional force

$$F = \int_a^b -(p_{zx})_{z=0}\,dx$$
$$= \frac{2\mu Ul}{(k-1)h_2}\left\{2\log k - \frac{3(k-1)}{k+1}\right\} \quad \text{.......................(9).}$$

Comparing (8) and (9) we see that the ratio F/P of the total friction to the total load is independent of both μ and U, but proportional to h if the scale of h is altered.

The position of the centre of pressure may be calculated from

$$\bar{x}P = \int_a^b px\,dx,$$

leading to
$$\bar{x} - a = \tfrac{1}{2}l\,\frac{2k(k+2)\log k - 5k^2 + 4k + 1}{(k^2-1)\log k - 2(k-1)^2} \quad \text{............(10).}$$

It has been shewn by Reynolds and Rayleigh that the value of k which makes P a maximum is 2·2; that this makes $P = 0{\cdot}1602\mu Ul^2/h_2^2$ and $F = 0{\cdot}75\mu Ul/h_2$.

Since the film of fluid is thin the above arguments would hold good if the surfaces were cylindrical instead of plane, provided h is everywhere small compared to the radii of curvature.

Where there is a flow in the direction of y as well as x, we shall obtain as in **13·61** for the total flow in the y direction

$$\int_0^h v\,dz \equiv \tfrac{1}{2}hV - \frac{h^3}{12\mu}\frac{\partial p}{\partial y} \quad \text{.....................(11),}$$

and the equation of continuity is now

$$\frac{\partial}{\partial x}\int_0^h u\,dz + \frac{\partial}{\partial y}\int_0^h v\,dz = 0,$$

or
$$\frac{\partial}{\partial x}\left(h^3\frac{\partial p}{\partial x}\right) + \frac{\partial}{\partial y}\left(h^3\frac{\partial p}{\partial y}\right) = 6\mu\left\{\frac{\partial}{\partial x}(hU) + \frac{\partial}{\partial y}(hV)\right\} \quad \text{......(12)*.}$$

13·62. Steady Motion in a Tube of Circular Section. Let a be the internal radius of the tube and w the velocity along the tube at a distance r from the axis Oz. The other components of velocity are assumed to be zero so that by reference to the equation of continuity $\partial w/\partial z$ must be zero, therefore and by symmetry w is a function of r alone. It now appears, from the equations of motion in cylindrical coordinates (**13·32** (7)), that in steady motion the inertia terms are all zero, and the equations, in the absence of external forces reduce to

$$\left.\begin{aligned} &\frac{\partial p}{\partial r} = 0, \quad \frac{\partial p}{r\,\partial \theta} = 0 \\ \text{and} \quad &\frac{\partial p}{\partial z} = \mu\left(\frac{\partial^2 w}{\partial r^2} + \frac{1}{r}\frac{\partial w}{\partial r}\right)\end{aligned}\right\} \quad \text{.....................(1).}$$

It follows that the pressure is a function of z alone and that the pressure gradient is constant, and equal to $-(p_1 - p_2)l$ where p_1, p_2 denote the values of p at the ends of a length l of the tube.

We might also argue directly that since there is no motion at right angles to the axis p does not vary over a cross section, and

* Michell, *Zeits. f. Math.* LIII, 1905, p. 123.

taking the tangential stress along a plane perpendicular to r as $p_{rz}=\mu\frac{dw}{dr}$ (**13·33** (1)), the frictional forces on the fluid between two cylinders of length δz and radii r and $r+\delta r$ are

$$-\mu\frac{dw}{dr}2\pi r\,\delta z \quad\text{and}\quad \mu\frac{dw}{dr}2\pi r\,\delta z+\mu\frac{d}{dr}\left(\frac{dw}{dr}2\pi r\,\delta z\right)\delta r$$

and in steady motion the sum of these must be balanced by the pressure difference on the plane ends of the fluid mass, viz.

$$\frac{dp}{dz}\delta z\,.\,2\pi r\,\delta r,$$

so that
$$\mu\frac{d}{dr}\left(r\frac{dw}{dr}\right)=r\frac{dp}{dz}, \text{ as above}$$

or
$$\frac{d}{dr}\left(r\frac{dw}{dr}\right)=-\frac{p_1-p_2}{\mu l}r \quad\text{.................(2).}$$

Integrating, we get

$$w=-\frac{p_1-p_2}{4\mu l}r^2+A\log r+B \quad\text{..............(3).}$$

The velocity along the axis must be finite, so that $A=0$, and since there is no slipping on the tube, $w=0$ when $r=a$, therefore

$$w=\frac{(p_1-p_2)}{4\mu l}(a^2-r^2)\text{....................(4).}$$

The total flux across any section is

$$\int_0^a w\,.\,2\pi r\,dr=\frac{p_1-p_2}{l}\cdot\frac{\pi a^4}{8\mu} \quad\text{...............(5).}$$

The drag on the cylinder is $2\pi a l\left(\mu\frac{dw}{dr}\right)_{r=a}$, which is easily shewn to be $\pi a^2(p_1-p_2)$, as is otherwise obvious. The result that the flux is proportional to the pressure gradient and to the fourth power of the radius of the tube was discovered experimentally by G. Hagen* and was rediscovered independently by Poiseuille†. It is a result of fundamental importance in connection with the law of frictional resistance as it can be confirmed by experiment with great accuracy. It also provides a method of measuring μ. But as stated in **13·5** it is only in narrow tubes that the result is true for all such velocities as are likely to

* *Pogg. Ann.* XLVI, 1839, p. 423.

† *Comptes Rendus*, XI, XII, 1840–1; *Mém. des Savants Étrangers*, IX, 1846.

occur in experiment. There is a definite Reynolds number which determines in every case a critical velocity in relation to the viscosity and the diameter of the tube. For smaller velocities the flow is 'stream line' or 'laminar' but when the critical velocity is exceeded the regular stream line character of the motion is destroyed, small eddies appear in the fluid and the motion becomes 'turbulent'. Then the relation (5) ceases to be true, and instead of the pressure gradient being proportional to the flux it is found to be approximately proportional to the square of the flux*.

13·621. Steady Flow between Coaxial Circular Cylinders. With the same notation, let the flow take place between two coaxial cylinders of radii a, b. Let the inner boundary have a velocity W while the outer is at rest. Then in **13·62** (3), we have $w=W$ when $r=a$, and $w=0$ when $r=b$, so that

$$w=W\frac{\log(r/b)}{\log(a/b)}-\frac{p_1-p_2}{4\mu l}\left\{r^2-\frac{b^2\log(r/a)-a^2\log(r/b)}{\log(b/a)}\right\} \quad \text{......(1)}$$

giving a flux relative to the fixed boundary of $\int_a^b w\,.\,2\pi r\,dr$

$$=\pi W\left\{\frac{\frac{1}{2}(b^2-a^2)}{\log(b/a)}-a^2\right\}+\frac{\pi}{8\mu}\cdot\frac{p_1-p_2}{l}\left\{b^4-a^4-\frac{(b^2-a^2)^2}{\log(b/a)}\right\} \quad \text{...(2).}$$

The drag on the cylinders can be calculated as in **13·62.**

13·63. Steady Flow in Tubes of cross section other than Circular. If we assume that w is a function of x, y but not of z, and that $u=v=0$, then in steady motion the inertia terms disappear and in the absence of external forces the equations **13·3** (1) reduce to

$$\left.\begin{aligned}&\frac{\partial p}{\partial x}=0,\quad \frac{\partial p}{\partial y}=0\\ \text{and}\qquad &\mu\left(\frac{\partial^2 w}{\partial x^2}+\frac{\partial^2 w}{\partial y^2}\right)=\frac{\partial p}{\partial z}\end{aligned}\right\} \quad \text{........................(1).}$$

It follows that in steady flow along a tube the pressure gradient $\partial p/\partial z$ is constant, and denoting this by $-P$, the velocity w has to satisfy the equation

$$\frac{\partial^2 w}{\partial x^2}+\frac{\partial^2 w}{\partial y^2}=-\frac{P}{\mu} \quad \text{........................(2)}$$

with a boundary condition $w=0$ on the surface of the tube. If we write $w=\psi-\frac{P}{4\mu}(x^2+y^2)$, then ψ has to satisfy the equation

$$\frac{\partial^2\psi}{\partial x^2}+\frac{\partial^2\psi}{\partial y^2}=0,$$

with a boundary condition $\psi=\frac{P}{4\mu}(x^2+y^2)$ on the surface of the tube. But from **5·1** (3) these are the conditions which have to be satisfied by the

* *The Physics of Solids and Fluids*, Ewald, Pöschl and Prandtl, 1930, p. 277.

stream function when a like tube containing frictionless fluid rotates about its length with angular velocity $P/2\mu$.

This analogy was pointed out by Greenhill *.

For example in the case of a tube whose cross section is an **equilateral triangle** by analogy from **5·5** (3) we write

$$w = A(x^3 - 3xy^2) + B - \frac{P}{4\mu}(x^2 + y^2) \quad \text{..............(3)},$$

where this expression is to vanish at all points of the boundary. If we determine A and B so as to make $x = a$ part of the boundary, we find that

$$w = -\frac{P}{12a\mu}\{x^3 - 3xy^2 + 3a(x^2 + y^2) - 4a^3\}$$

or

$$w = -\frac{P}{12a\mu}(x - a)(x - \sqrt{3}y + 2a)(x + \sqrt{3}y + 2a) \quad \text{......(4)};$$

from which we see that w vanishes on the three sides of an equilateral triangle, and since it also satisfies (2), it represents the required velocity.

By evaluating $\iint w\,dx\,dy$ over the cross section we find that the flux of liquid is $\dfrac{27}{20\sqrt{3}}\dfrac{Pa^4}{\mu}$.

For an **elliptic section** by analogy from **5·5** (2) we write

$$w = A(x^2 - y^2) + B - \frac{P}{4\mu}(x^2 + y^2) \quad \text{..................(5)},$$

and determine A and B so that w vanishes on $x^2/a^2 + y^2/b^2 = 1$. This requires that

$$a^2\left(A - \frac{P}{4\mu}\right) = -b^2\left(A + \frac{P}{4\mu}\right) = -B,$$

from which we find that

$$w = \frac{P}{2\mu}\frac{a^2b^2}{a^2 + b^2}\left(1 - \frac{x^2}{a^2} - \frac{y^2}{b^2}\right) \quad \text{.....................(6)}.$$

By evaluating $\iint w\,dx\,dy$ over the area of the ellipse we find that the flux of fluid is $\dfrac{\pi a^3b^3}{a^2 + b^2}\cdot\dfrac{P}{4\mu}$.

13·64. Steady Rotatory Motion.

Let a fluid have a steady rotation about the axis of z, the angular velocity ω being a function of the distance r from the axis. Then in the notation of **13·32**, $v_r = 0$, $v_\theta = \omega r$ and $v_z = 0$, and, in the absence of external forces, by symmetry the mean pressure p is independent of θ, so that the equations of motion **13·32** (7) reduce to

$$\omega^2 r = \frac{1}{\rho}\frac{\partial p}{\partial r}, \quad r\frac{\partial^2 \omega}{\partial r^2} + 3\frac{\partial \omega}{\partial r} = 0 \quad \text{and} \quad \frac{\partial p}{\partial z} = 0 \quad \text{........(1)}.$$

* 'On the Flow of a Viscous Liquid in a Pipe or Channel', *Proc. L.M.S.* (1), XIII, 1881, p. 43.

We might also obtain the equation for ω from the consideration that the only stress in the fluid is the tangential force

$$p_{r\theta}=\mu r\frac{d\omega}{dr} \quad \textbf{(13·33 (1))}$$

and the moment about the axis of the tangential force on a cylinder of radius r is therefore $2\pi\mu r^3 d\omega/dr$ per unit length of cylinder. Then since, in the steady motion, the moments on the inner and outer surfaces of an annulus must be equal and opposite, it follows that the last expression must be constant; i.e.

$$r^3\frac{d\omega}{dr}=A \quad \text{..........................(2),}$$

this being the integral of the second of equations (1).

Hence
$$\omega=-\frac{A}{2r^2}+B \quad \text{........................(3).}$$

If the fluid has no internal boundary we must have $A=0$, since the angular velocity cannot become infinite as $r\to 0$, and therefore ω is constant and there is no relative motion.

But if the fluid is bounded by coaxial cylinders of radii a and b, the second of which, the outer, rotates with angular velocity Ω while the inner is at rest, we have $\omega=0$ when $r=a$, and $\omega=\Omega$ when $r=b$, so that

$$\omega=\Omega\frac{b^2}{r^2}\cdot\frac{r^2-a^2}{b^2-a^2} \quad \text{.......................(4).}$$

The couple per unit length on either cylinder is $2\pi\mu A$ or

$$\frac{4\pi\mu a^2b^2}{b^2-a^2}\Omega \quad \text{..........................(5).}$$

Experimental work has been done on this basis, the couple on the inner cylinder being measured by the torsion of a suspending wire.

We have not imposed any limitations on the angular velocity, and it has been shewn by Taylor* that the above steady motion of a liquid is stable for all speeds of the outer cylinder; and also that when the outer cylinder is fixed there is stability for sufficiently small angular velocities of the inner.

* 'Stability of a Viscous Liquid contained between two Rotating Cylinders', *Phil. Trans.* A, CCXXIII, 1922, p. 289.

13·65. Steady Motion of a Viscous Fluid due to a slowly Rotating Sphere. Writing

$$u = -\omega y, \quad v = \omega x \quad \text{and} \quad w = 0,$$

where ω is a function of r alone ($r^2 = x^2 + y^2 + z^2$), and neglecting squares of velocities, for steady motion equations **13·3** (1) become

$$\left.\begin{aligned} 0 &= -\frac{\partial p}{\partial x} - \mu y\left(\frac{d^2\omega}{dr^2} + \frac{4}{r}\frac{d\omega}{dr}\right) \\ 0 &= -\frac{\partial p}{\partial y} + \mu x\left(\frac{d^2\omega}{dr^2} + \frac{4}{r}\frac{d\omega}{dr}\right) \\ 0 &= -\frac{\partial p}{\partial z} \end{aligned}\right\} \quad \text{..................(1).}$$

These are satisfied by $p = \text{const.}$ and

$$\frac{d^2\omega}{dr^2} + \frac{4}{r}\frac{d\omega}{dr} = 0,$$

or
$$r^4 \frac{d\omega}{dr} = \text{const.}$$

i.e.
$$\omega = \frac{A}{r^3} + B \quad \text{..............................(2).}$$

If the motion is produced by a solid sphere of radius a rotating with angular velocity Ω and the fluid extends to infinity and is at rest there, we have

$$\omega = \frac{a^3}{r^3}\Omega \quad \text{.................................(3).}$$

And if there is an outer fixed concentric spherical boundary of radius b, then

$$\omega = \frac{a^3}{r^3} \cdot \frac{b^3 - r^3}{b^3 - a^3}\Omega \quad \text{..........................(4).}$$

The couple on the sphere may be found from the formulae for the stresses in **13·33** (5). Here $q_r = 0$, $q_\theta = 0$ and $q_\phi = \omega r \sin\theta$ and the only stress which has a moment about the axis is

$$p_{r\phi} = \mu r \sin\theta \frac{d\omega}{dr},$$

and its moment is $\mu r^2 \sin^2\theta \dfrac{d\omega}{dr}$.

Hence the couple on the sphere of radius a is

$$\int_0^\pi \mu a^2 \sin^2\theta \left(\frac{d\omega}{dr}\right)_{r=a} . 2\pi a^2 \sin\theta\, d\theta = -\frac{8\pi\mu\Omega a^3 b^3}{b^3 - a^3} \quad \text{........(5).}$$

The same result might be found from the consideration that if N is the couple which must be applied to the moving sphere to maintain the rotation, then

$$N\Omega = \text{rate of dissipation of energy.}$$

It should be noted that the results of this article are of little value for experimental purposes because of the necessary limitation about the smallness of the velocity. Unless the motion is so slow that the squares of velocities are negligible steady annular motion is impossible*.

* Stokes, *loc. cit.* p. 370.

13·7. Vorticity in Viscous Fluids. The equations of motion **13·3** (1) may be written

$$\left.\begin{aligned}\frac{Du}{Dt}&=-\frac{\partial Q}{\partial x}+\tfrac{1}{3}\nu\frac{\partial\theta}{\partial x}+\nu\nabla^2 u\\ \frac{Dv}{Dt}&=-\frac{\partial Q}{\partial y}+\tfrac{1}{3}\nu\frac{\partial\theta}{\partial y}+\nu\nabla^2 v\\ \frac{Dw}{Dt}&=-\frac{\partial Q}{\partial z}+\tfrac{1}{3}\nu\frac{\partial\theta}{\partial z}+\nu\nabla^2 w\end{aligned}\right\}\quad\text{..............(1)},$$

where $\quad Q=V+\int\frac{dp}{\rho}\quad$ and $\quad\theta=\frac{\partial u}{\partial x}+\frac{\partial v}{\partial y}+\frac{\partial w}{\partial z}.$

Proceeding exactly as in **2·6** these equations take the form

$$\left.\begin{aligned}\frac{D}{Dt}\left(\frac{\xi}{\rho}\right)&=\frac{\xi}{\rho}\frac{\partial u}{\partial x}+\frac{\eta}{\rho}\frac{\partial u}{\partial y}+\frac{\zeta}{\rho}\frac{\partial u}{\partial z}+\frac{\nu}{\rho}\nabla^2\xi\\ \frac{D}{Dt}\left(\frac{\eta}{\rho}\right)&=\frac{\xi}{\rho}\frac{\partial v}{\partial x}+\frac{\eta}{\rho}\frac{\partial v}{\partial y}+\frac{\zeta}{\rho}\frac{\partial v}{\partial z}+\frac{\nu}{\rho}\nabla^2\eta\\ \frac{D}{Dt}\left(\frac{\zeta}{\rho}\right)&=\frac{\xi}{\rho}\frac{\partial w}{\partial x}+\frac{\eta}{\rho}\frac{\partial w}{\partial y}+\frac{\zeta}{\rho}\frac{\partial w}{\partial z}+\frac{\nu}{\rho}\nabla^2\zeta\end{aligned}\right\}\quad\text{.........(2)}.$$

In the case of an incompressible fluid the equations take the simpler form

$$\left.\begin{aligned}\frac{D\xi}{Dt}&=\xi\frac{\partial u}{\partial x}+\eta\frac{\partial u}{\partial y}+\zeta\frac{\partial u}{\partial z}+\nu\nabla^2\xi\\ \frac{D\eta}{Dt}&=\xi\frac{\partial v}{\partial x}+\eta\frac{\partial v}{\partial y}+\zeta\frac{\partial v}{dz}+\nu\nabla^2\eta\\ \frac{D\zeta}{Dt}&=\xi\frac{\partial w}{\partial x}+\eta\frac{\partial w}{\partial y}+\zeta\frac{\partial w}{\partial z}+\nu\nabla^2\zeta\end{aligned}\right\}\quad\text{...........(3)}.$$

As in **9·21** the first three terms on the right of these equations represent the rates at which ξ, η, ζ vary for a given particle, when the vortex lines move with the fluid and the strengths of the vortices remain constant.

In any case of slow motion the first three terms on the right of each of the equations are negligible and the remaining terms exhibit the variations of the vorticity, and the equations are the same in form as the equations of the conduction of heat, so by analogy vorticity cannot originate in the interior of a viscous fluid but must be transmitted from the surface.

13·71. Circulation in Viscous Fluids. The conclusion of the preceding article may also be reached from considerations of the circulation in a circuit moving with the fluid. Thus if I denote the circulation, then as in **4·23**

$$\frac{DI}{Dt}=\frac{D}{Dt}\int_C (u\,dx+v\,dy+w\,dz)$$
$$=\int_C\left(\frac{Du}{Dt}dx+\frac{Dv}{Dt}dy+\frac{Dw}{Dt}dz\right)+\int_C(u\,du+v\,dv+w\,dw).$$

The last integral vanishes for a closed circuit, and therefore from **13·22** (1)

$$\frac{DI}{Dt}=\int_C (X\,dx+Y\,dy+Z\,dz)$$
$$+\int_C\left\{\frac{1}{\rho}\left(\frac{\partial p_{xx}}{\partial x}+\frac{\partial p_{yx}}{\partial y}+\frac{\partial p_{zx}}{\partial z}\right)dx+\quad+\right\}.$$

For a conservative field of force the first integral vanishes and from **13·23** (8) and (9)

$$\frac{DI}{Dt}=\int_C\left(\frac{1}{\rho}\left[\frac{\partial}{\partial x}\left(-p-\tfrac{2}{3}\mu\theta+2\mu\frac{\partial u}{\partial x}\right)+\frac{\partial}{\partial y}\left\{\mu\left(\frac{\partial v}{\partial x}+\frac{\partial u}{\partial y}\right)\right\}+\frac{\partial}{\partial z}\left\{\mu\left(\frac{\partial u}{\partial z}+\frac{\partial w}{\partial x}\right)\right\}\right]dx\right.$$
$$\left.+\frac{1}{\rho}\left[\qquad\right]dy+\frac{1}{\rho}\left[\qquad\right]dz\right),$$
$$=\int_C\frac{1}{\rho}d(-p-\tfrac{2}{3}\mu\theta)$$
$$+2\int_C\frac{1}{\rho}\left[\frac{\partial}{\partial x}\left(\mu\frac{\partial u}{\partial x}\right)-\frac{\partial}{\partial y}\left\{\mu\left(\zeta-\frac{\partial v}{\partial x}\right)\right\}+\frac{\partial}{\partial z}\left\{\mu\left(\eta+\frac{\partial w}{\partial x}\right)\right\}\right]dx$$
$$+2\int_C\frac{1}{\rho}\left[\frac{\partial}{\partial x}\left\{\mu\left(\zeta+\frac{\partial u}{\partial y}\right)\right\}+\frac{\partial}{\partial y}\left(\mu\frac{\partial v}{\partial y}\right)-\frac{\partial}{\partial z}\left\{\mu\left(\xi-\frac{\partial w}{\partial y}\right)\right\}\right]dy$$
$$+2\int_C\frac{1}{\rho}\left[-\frac{\partial}{\partial x}\left\{\mu\left(\eta-\frac{\partial u}{\partial z}\right)\right\}+\frac{\partial}{\partial y}\left\{\mu\left(\xi+\frac{\partial v}{\partial z}\right)\right\}+\frac{\partial}{\partial z}\left(\mu\frac{\partial w}{\partial z}\right)\right]dz$$
$$=-\int_C\frac{dp}{\rho}-\tfrac{2}{3}\int\frac{1}{\rho}d(\mu\theta)$$
$$+2\int_C\frac{1}{\rho}\left\{\left(\frac{\partial\mu\eta}{\partial z}-\frac{\partial\mu\zeta}{\partial y}\right)dx+\left(\frac{\partial\mu\zeta}{\partial x}-\frac{\partial\mu\xi}{\partial z}\right)dy+\left(\frac{\partial\mu\xi}{\partial y}-\frac{\partial\mu\eta}{\partial x}\right)dz\right\}$$
$$+2\int_C\frac{1}{\rho}\left(\frac{\partial\mu}{\partial x}du+\frac{\partial\mu}{\partial y}dv+\frac{\partial\mu}{\partial z}dw\right)+2\int\frac{\mu}{\rho}d\theta \quad\text{.......................(1).}$$

This result was obtained by Jeffreys by the shorter process of the tensor calculus. He remarks on the different terms that the first vanishes if ρ is a function of p alone, as in many cases; but if there are variations of density not due to pressure, but to temperatures or composition this term will not vanish. The second and fifth terms involve products of the viscosity and the divergence, and are probably unimportant in most cases. The fourth shews that circulation may arise in a moving fluid in which the viscosity varies from place to place. The third term is the only one which survives in a uniform incompressible fluid, and it shews that in such a fluid changes in circulation depend only on the vorticity in the neighbourhood of the contour*.

In fact when μ and ρ are constant, the equation reduces to

$$\frac{DI}{Dt} = \nu \int_C (\nabla^2 u\,dx + \nabla^2 v\,dy + \nabla^2 w\,dz)$$

$$= \nu \nabla^2 I \qquad \text{.......................................(2)},$$

shewing the same law of conduction of heat for variations in the circulation in a moving circuit.

13·72. Further Special Cases. Diffusion of Vorticity.

(i) *Laminar Motion or flow in parallel planes.* When the motion is *not* steady but uniform in a set of parallel planes, so that we may take $v = w = 0$ and u a function of z only, then the equations of motion **13·3** (1) are satisfied by taking $p = \text{const.}$ and

$$\frac{\partial u}{\partial t} = \nu \frac{\partial^2 u}{\partial z^2} \qquad \text{..................................(1)}.$$

This is also the equation for the linear conduction of heat and the appropriate solution depends on the initial and boundary conditions.

Thus, when the fluid extends to infinity on both sides of the plane $z = 0$ and the initial velocity throughout the fluid is given by $u = \pm U$ according as z is positive or negative, the solution of (1) is

$$u = \frac{2U}{\sqrt{\pi}} \int_0^{z/2\sqrt{(\nu t)}} e^{-\theta^2} d\theta \qquad \text{.....................(2)†}.$$

For it can be shewn that the solution of (1) which fits given boundary conditions is unique‡, and it is easy to verify that (2) satisfies (1), and that as $z \to 0$, $u \to \pm U$ according as z is positive or negative.

The vorticity is given by

$$2\eta = \frac{\partial u}{\partial z} = \frac{U}{\sqrt{(\pi \nu t)}} e^{-z^2/4\nu t} \qquad \text{........................(3)}.$$

The initial conditions imply the existence of a vortex sheet in the plane $z = 0$. Equation (3) represents the manner in which the vorticity is diffused into the fluid on both sides of the sheet, and (2) enables us to measure the decay of velocity, throughout the fluid.

* H. Jeffreys, 'The Equations of Viscous Motion and the Circulation Theorem', *Proc. Camb. Phil. Soc.* XXIV, 1928, p. 477.

† Carslaw, *Conduction of Heat*, 1921, p. 34, where notes will be found on the tabular value of the integral (2).

‡ *Ibid.* p. 14.

(ii) *Periodic Laminar Motion.* With the same notation let us suppose that the fluid lies on the positive side only of the plane $z=0$ and that its motion is due to the motion of a plane rigid boundary oscillating in the plane $z=0$ with velocity

$$u=a\cos(nt+\epsilon) \quad \text{.........................(4).}$$

Taking this as the real part of $ae^{i(nt+\epsilon)}$ the general velocity will now have the time factor $e^{i(nt+\epsilon)}$ so that (1) becomes

$$\frac{\partial^2 u}{\partial z^2}=\frac{in}{\nu}u \quad \text{...............................(5).}$$

The solution of this equation is

$$u=Ae^{(1+i)\sqrt{\left(\frac{n}{2\nu}\right)}z}+Be^{-(1+i)\sqrt{\left(\frac{n}{2\nu}\right)}z} \quad \text{..............(6).}$$

If the fluid extends to infinity in the positive direction of z, we must have $A=0$, and, introducing the time factor in (6) and using (4) to determine the value of B, we get

$$u=ae^{i(nt+\epsilon)-(1+i)\sqrt{\left(\frac{n}{2\nu}\right)}z} \quad \text{.......................(7)}$$

and the real part corresponding to (4) is

$$u=ae^{-\sqrt{\left(\frac{n}{2\nu}\right)}z}\cos\left\{nt-\sqrt{\left(\frac{n}{2\nu}\right)}z+\epsilon\right\} \quad \text{.............(8).}$$

This represents a transverse wave of length $2\pi\sqrt{\left(\frac{2\nu}{n}\right)}$ propagated in the z direction with velocity $\sqrt{(2\nu n)}$, but with a rapidly decreasing amplitude, in consequence of the exponential factor, the decrease in amplitude in a wave length being in the ratio $e^{-2\pi}:1$ or as $1:535$, so that the motion is propagated only a short distance into the fluid.

The drag of the fluid on the boundary per unit area is measured by

$$-\mu\left(\frac{\partial u}{\partial z}\right)_{z=0}=\rho a\sqrt{(\tfrac{1}{2}n\nu)}\{\cos(nt+\epsilon)-\sin(nt+\epsilon)\}$$
$$=\rho a\sqrt{(n\nu)}\cos(nt+\epsilon+\tfrac{1}{4}\pi) \quad \text{.................(9)*.}$$

The foregoing represents the forced oscillations on which any normal modes of free oscillations may be superposed.

If the fluid were bounded by a rigid plane $z=h$, both terms in (6) would be required in the solution, with the further condition that $u=0$ when $z=h$.

(iii) *Diffusion of vorticity from a line vortex.* Let there be initially a vortex filament of strength κ along the axis of z in an infinite mass of fluid. The motion will be in circles about the z axis, the velocity at distance r from the axis being a function of r.

We have therefore $w=0$, $\xi=\eta=0$, and it is easy to verify by putting $u=-(y/r)f(r)$ and $v=(x/r)f(r)$, $(r^2=x^2+y^2)$ that in this case

$$D\zeta/Dt=\partial\zeta/\partial t,$$

so that equations **13·7** (3) reduce to

$$\frac{\partial\zeta}{\partial t}=\nu\left(\frac{\partial^2\zeta}{\partial x^2}+\frac{\partial^2\zeta}{\partial y^2}\right) \quad \text{.......................(10),}$$

* Stokes, *loc. cit.* p. 376, see also Lamb, *Hydrodynamics*, 1932, p. 660.

or, in terms of r, $$\frac{\partial \zeta}{\partial t} = \nu\left(\frac{\partial^2 \zeta}{\partial r^2} + \frac{1}{r}\frac{\partial \zeta}{\partial r}\right) \quad \text{..........................(11).}$$

This corresponds to the radial flow of heat from a line source and the solution is

$$\zeta = \frac{\kappa}{8\pi\nu t}\, e^{-r^2/4\nu t} \quad \text{..........................(12)*}$$

for it is easy to verify that this satisfies (11), and it makes the circulation in a circle of radius r

$$= \int_0^r 2\zeta \,.\, 2\pi r\, dr = \kappa\,(1 - e^{-r^2/4\nu t})$$

and this $\rightarrow \kappa$ as $t \rightarrow 0$.

Regarded as a function of t the vorticity ζ at a distance r from the axis increases to a maximum $\kappa/2\pi r^2 e$ and then decreases asymptotically to zero.

The velocity is found by dividing the circulation by $2\pi r$ and

$$= \kappa\,(1 - e^{-r^2/4\nu t})/2\pi r;$$

and as t increases from 0 to ∞ the velocity decreases from $\kappa/2\pi r$ to zero.

13·8. Motion the same in all Planes through an Axis.

Here it is convenient to use Stokes's Stream Function ψ as defined in **7·3**. Since the resultant velocity and vorticity at a point are independent of the azimuthal angle ϕ, the former lies in the meridian plane and the latter is at right angles to the same plane (**4·25**).

Hence, taking the x axis as the given axis, we may denote the components of velocity by

$$u, \quad v = V\cos\phi, \quad w = V\sin\phi \quad \text{...........(1)}$$

and the components of vorticity by

$$0, \quad \eta = -\omega\sin\phi, \quad \zeta = \omega\cos\phi \quad \text{...........(2),}$$

where, as in **9·82**, $$2\omega = \frac{1}{\varpi}\left\{\frac{\partial^2\psi}{\partial x^2} + \frac{\partial^2\psi}{\partial \varpi^2} - \frac{1}{\varpi}\frac{\partial\psi}{\partial\varpi}\right\} \quad \text{.................(3).}$$

For a motion which is so slow that the squares of velocities can be neglected, equations **13·7** (3) reduce to

$$\frac{\partial\xi}{\partial t} = \nu\nabla^2\xi, \quad \frac{\partial\eta}{\partial t} = \nu\nabla^2\eta, \quad \frac{\partial\zeta}{\partial t} = \nu\nabla^2\zeta \quad \text{...........(4).}$$

Now $$\nabla^2\eta = -\left(\frac{\partial^2}{\partial x^2} + \frac{\partial^2}{\partial\varpi^2} + \frac{1}{\varpi}\frac{\partial}{\partial\varpi} + \frac{1}{\varpi^2}\frac{\partial^2}{\partial\phi^2}\right)\omega\sin\phi,$$

or, since ω is independent of ϕ,

$$\nabla^2\eta = -\sin\phi\left(\frac{\partial^2}{\partial x^2} + \frac{\partial^2}{\partial\varpi^2} + \frac{1}{\varpi}\frac{\partial}{\partial\varpi} - \frac{1}{\varpi^2}\right)\omega$$

$$= -\frac{\sin\phi}{\varpi}\left(\frac{\partial^2}{\partial x^2} + \frac{\partial^2}{\partial\varpi^2} - \frac{1}{\varpi}\frac{\partial}{\partial\varpi}\right)\varpi\omega \quad \text{...........(5).}$$

* Carslaw, *Conduction of Heat*, 1921, p. 152.

Similarly $$\nabla^2\zeta = \frac{\cos\phi}{\varpi}\left(\frac{\partial^2}{\partial x^2}+\frac{\partial^2}{\partial\varpi^2}-\frac{1}{\varpi}\frac{\partial}{\partial\varpi}\right)\varpi\omega \quad\text{.............(6).}$$

Now if we put $D \equiv \frac{\partial^2}{\partial x^2}+\frac{\partial^2}{\partial\varpi^2}-\frac{1}{\varpi}\frac{\partial}{\partial\varpi}$ from (2), (4), (5) and (6) we get

$$\left(D-\frac{1}{\nu}\frac{\partial}{\partial t}\right)\varpi\omega = 0 \quad\text{.....................(7).}$$

But, from (3), $$2\varpi\omega = D\psi,$$

therefore the equation for the stream function is

$$\left(D-\frac{1}{\nu}\frac{\partial}{\partial t}\right)D\psi = 0 \quad\text{..................(8).*}$$

The operator D may be expressed in polar coordinates r, θ where $x = r\cos\theta$ and $\varpi = r\sin\theta$ by writing

$$\frac{\partial^2}{\partial x^2}+\frac{\partial^2}{\partial\varpi^2} \equiv \frac{\partial^2}{\partial r^2}+\frac{1}{r}\frac{\partial}{\partial r}+\frac{1}{r^2}\frac{\partial^2}{\partial\theta^2},$$

and $$\frac{1}{\varpi}\frac{\partial}{\partial\varpi} \equiv \frac{1}{r}\frac{\partial}{\partial r}+\frac{\cot\theta}{r^2}\frac{\partial}{\partial\theta},$$

so that $$D \equiv \frac{\partial^2}{\partial r^2}+\frac{1}{r^2}\frac{\partial^2}{\partial\theta^2}-\frac{\cot\theta}{r^2}\frac{\partial}{\partial\theta} \quad\text{.................(9),}$$

or $$D \equiv \frac{\partial^2}{\partial r^2}+\frac{1-\mu^2}{r^2}\frac{\partial^2}{\partial\mu^2} \quad\text{.................(10),}$$

where $\mu \equiv \cos\theta$.

In a case of steady motion (8) becomes

$$D^2\psi = 0 \quad\text{........................(11).}$$

This has a solution $$\psi = (1-\mu^2)\,f(r) \quad\text{....................(12),}$$

provided that $$\left(\frac{d^2}{dr^2}-\frac{2}{r^2}\right)^2 f(r) = 0 \quad\text{....................(13).}$$

The solution of this equation is

$$f(r) = \frac{A}{r}+Br+Cr^2+Er^4 \quad\text{.............(14),}$$

so that $$\psi = \left(\frac{A}{r}+Br+Cr^2+Er^4\right)\sin^2\theta \quad\text{...........(15).}$$

In the notation of **13·32** the velocity components (**7·31**) are

$$q_r = -\frac{1}{r^2\sin\theta}\frac{\partial\psi}{\partial\theta} \quad\text{and}\quad q_\theta = \frac{1}{r\sin\theta}\frac{\partial\psi}{\partial r} \quad\text{........(16)}$$

* Stokes, *loc. cit.* p. 376.

along and perpendicular to the radius vector; so that a finite velocity at infinity requires that the constant $E=0$, and a steady flow U at infinity requires that $\psi=-\frac{1}{2}Ur^2\sin^2\theta$, so that $C=-\frac{1}{2}U$.

If the liquid is streaming past a fixed sphere of radius a, the boundary conditions require that $q_r=0$ and $q_\theta=0$ when $r=a$, i.e.

$$\frac{A}{a}+Ba+Ca^2=0 \quad\text{and}\quad -\frac{A}{a^2}+B+2Ca=0,$$

giving

$$A=-\tfrac{1}{4}Ua^3, \quad B=\tfrac{3}{4}Ua \qquad\text{.....................(17)},$$

and

$$\psi=-\tfrac{1}{2}U\left(1-\tfrac{3}{2}\frac{a}{r}+\tfrac{1}{2}\frac{a^3}{r^3}\right)r^2\sin^2\theta \qquad\text{..................(18)}.$$

For a sphere moving with velocity U through a liquid at rest, we reverse the velocity U by adding $\frac{1}{2}Ur^2\sin^2\theta$ to ψ and have

$$\psi=\tfrac{3}{4}Uar\left(1-\tfrac{1}{3}\frac{a^2}{r^2}\right)\sin^2\theta \qquad\text{.....................(19)}.$$

Hence from (16)

$$q_r=-\tfrac{3}{2}U\cos\theta\left(\frac{a}{r}-\tfrac{1}{3}\frac{a^3}{r^3}\right) \quad\text{and}\quad q_\theta=\tfrac{3}{4}U\sin\theta\left(\frac{a}{r}+\tfrac{1}{3}\frac{a^3}{r^3}\right) \qquad\text{...(20)}.$$

To determine the force exerted on the sphere we might find the components of stress from **13·33** (5) and integrate over the surface of the sphere, but this would involve finding an expression for the variable pressure p^*, and it is simpler to use the expressions for the velocity gradients in **13·33** (4) and evaluate the rate of dissipation of energy (**13·4** (7)).

Thus substituting from (20) in **13·33** (4) and putting $q_\phi=0$, the relevant terms are

$$\left.\begin{aligned}&\frac{\partial u'}{\partial x'}=\frac{\partial v'}{\partial y'}=-\tfrac{3}{4}U\cos\theta\left(\frac{a}{r^2}-\frac{a^3}{r^4}\right),\quad \frac{\partial w'}{\partial z'}=\tfrac{3}{2}U\cos\theta\left(\frac{a}{r^2}-\frac{a^3}{r^4}\right)\\ &\frac{\partial w'}{\partial x'}+\frac{\partial u'}{\partial z'}=-\tfrac{3}{2}U\sin\theta\frac{a^3}{r^4}\end{aligned}\right\} \qquad\text{(21)},$$

so that the dissipation function **13·4** (7)

$$=\mu U^2\int_0^\pi\int_a^\infty\left\{\tfrac{27}{4}\left(\frac{a}{r^2}-\frac{a^3}{r^4}\right)^2\cos^2\theta+\tfrac{9}{4}\frac{a^6}{r^8}\sin^2\theta\right\}2\pi r^2\sin\theta\, d\theta\, dr=6\pi\mu aU^2$$

......(22).

This represents the rate at which the retarding forces are doing work, but the body is moving with velocity U, so that the resultant retarding force

$$=6\pi\mu aU \qquad\text{.............................(23)}.$$

This is **Stokes's formula**† for the resistance to the motion of a sphere in viscous liquid. It must not be overlooked however that the squares of velocities have been neglected. Attempts have been made to obtain a more complete solution. In particular it has been pointed out by Oseen ‡ that if we write

$$u=-U+u', \quad v=v', \quad w=w',$$

then the term $\rho u\frac{\partial u}{\partial x}$ which we neglected contains a term $\rho U\frac{\partial u}{\partial x}$ which may

* Evaluated by Stokes, *loc. cit.* p. 376. † Stokes, *loc. cit.*

‡ 'Ueber die Stokes'sche Formel, und...', *Arkiv för matematik*, VI, 1910, p. 29.

be comparable with terms retained. Proceeding in this way he obtained the result

$$6\pi\mu aU\left(1+\tfrac{3}{8}\frac{\rho a\,|\,U\,|}{\mu}\right) \quad\text{.....................(24)*.}$$

The formula (23) was employed by Stokes to determine the terminal velocity of a sphere falling vertically in a fluid, by equating the resistance to the excess of the weight of the sphere over the force of buoyancy; thus

$$6\pi\mu aU=\tfrac{4}{3}\pi\,(\sigma-\rho)\,a^3g,$$

where σ is the density of the sphere, so that

$$U=\tfrac{2}{9}\frac{g}{\nu}\left(\frac{\sigma}{\rho}-1\right)a^2 \quad\text{.......................(25)†,}$$

but the result is subject to the limitation stated above.

13·9. Prandtl's Boundary Layer.

We now return to the case mentioned in **13·5** of fluids of small viscosity such as water or air, in which motion is approximately that of a perfect fluid save in the immediate neighbourhood of solid boundaries, where it is affected by the boundary condition that the fluid clings to the boundary. In the case of flow past a solid obstacle there is a thin layer of fluid in the immediate neighbourhood of the solid in which friction is effective and in which, as explained below, vortices are produced; there is also the region containing vortices which are thrown off from such a body and constitute its 'wake'; and there is a region which contains all the remaining fluid and in this the motion may be regarded as irrotational.

It must not be assumed however that the solution for a 'perfect fluid' is the limiting form of the solution for a viscous fluid when the viscosity tends to zero or $R\to\infty$. The differential equations for viscous fluid are of a higher order than for perfect fluid and so require more definition in the way of boundary conditions for a complete solution of a particular problem; and there is the essential difference in the boundary conditions in the two cases, that a perfect fluid slips freely over a solid boundary while a viscous fluid clings to the boundary and has no velocity relative to it. It has indeed been shewn by Jeffreys‡ that if we assume the existence of a velocity potential and that a fluid has no motion relative to a solid in contact with it, then the only possible motion is one such that the fluid, the solid and the containing vessel have a single velocity of translation, no solid can rotate and a 'classical fluid' is more rigid than any solid.

* See also Lamb's *Hydrodynamics*, 1932, § 340. † Stokes, *loc. cit.* p. 376.

‡ *Proc. Roy. Soc.* A, CXXVIII, 1930, p. 376.

Prandtl's theory* simplifies problems by considering the effects of friction only in a thin layer round solid boundaries and treating the rest of the fluid as frictionless, but it assumes that the distribution of pressure outside the layer can be ascertained. If the layer does not separate from the surface the distribution of pressure can be found with fair approximation from the assumption of irrotational motion; but otherwise the distribution of pressure is affected by the 'wake', and it cannot be found save by experiment.

13·91. Differential Equations for the Boundary Layer. Consider a two-dimensional case, take the x axis along the surface, assuming at first that the boundary is rectilinear, and let the layer have a small thickness δ. Then if we suppose the velocity u and its derivative $\partial u/\partial x$ to be of order 1 within the layer, $\partial u/\partial y$ is large being of order $1/\delta$, u decreasing from a finite value at the outer boundary of the layer to zero at the solid surface.

From the equation of continuity

$$\frac{\partial u}{\partial x}+\frac{\partial v}{\partial y}=0 \qquad (1),$$

since $\partial u/\partial x$ is of order 1 so is $\partial v/\partial y$; but $v=0$ at the surface of the solid, therefore v will only be of order δ at the outer boundary of the layer.

Then considering the equations of motion

$$\frac{\partial u}{\partial t}+u\frac{\partial u}{\partial x}+v\frac{\partial u}{\partial y}=-\frac{1}{\rho}\frac{\partial p}{\partial x}+\nu\left(\frac{\partial^2 u}{\partial x^2}+\frac{\partial^2 u}{\partial y^2}\right) \qquad (2)$$

and

$$\frac{\partial v}{\partial t}+u\frac{\partial v}{\partial x}+v\frac{\partial v}{\partial y}=-\frac{1}{\rho}\frac{\partial p}{\partial y}+\nu\left(\frac{\partial^2 v}{\partial x^2}+\frac{\partial^2 v}{\partial y^2}\right) \qquad (3),$$

in the bracket on the right of (2) $\partial^2 u/\partial x^2$ is of order 1 and $\partial^2 u/\partial y^2$ is of order $1/\delta^2$, so the former is negligible compared with the latter and the frictional terms in (2) are of order ν/δ^2. If ν/δ^2 is large the frictional terms preponderate and the terms involving squares of velocities are negligible—the case of slow motion. If ν/δ^2 is small, the frictional terms are negligible and we have the equation for a perfect fluid. It is only when ν is of the same order as δ^2 that the conditions apply to the case under consideration. We therefore conclude that the thickness of the layer is proportional to $\sqrt{\nu}$.

* *Loc. cit.* p. 378.

Equation (3) then reduces to $\partial p/\partial y = 0$, so that the pressure in the boundary layer is a function of x alone.

Hence we have two equations for u and v in the layer, viz.

$$\frac{\partial u}{\partial t} + u\frac{\partial u}{\partial x} + v\frac{\partial u}{\partial y} = -\frac{1}{\rho}\frac{dp}{dx} + \nu\frac{\partial^2 u}{\partial y^2} \quad \text{............(4)}$$

$$\frac{\partial u}{\partial x} + \frac{\partial v}{\partial y} = 0 \quad \text{........................(5);}$$

and the pressure has the same value through the layer as on its external surface and is continuous with its value in the frictionless motion outside the layer, so that it may be regarded as a known externally applied force in the layer, determined in a case of steady motion by Bernoulli's Equation

$$p + \tfrac{1}{2}\rho U^2 = \text{const.} \quad \text{....................(6),}$$

where U is the velocity in the irrotational motion, so that

$$\frac{dp}{dx} = -\rho U\frac{dU}{dx} \quad \text{........................(7).}$$

Similar conditions apply in steady motion when the boundary is not rectilinear, provided that its radius of curvature is large compared with δ.

Equations (4) and (5) may then be written

$$\left.\begin{aligned} q\frac{\partial q}{\partial s} + w\frac{\partial q}{\partial n} &= -\frac{1}{\rho}\frac{\partial p}{\partial s} + \nu\frac{\partial^2 q}{\partial n^2} \\ \frac{\partial q}{\partial s} + \frac{\partial w}{\partial n} &= 0 \end{aligned}\right\} \quad \text{...............(8),}$$

where s denotes arc measured along the surface and n distance measured in the normal direction, and q, w are the components of velocity in the directions s and n*. Instead of equation (3) we now have an equation connecting the centrifugal force with the normal gradient of p, i.e. $\partial p/\partial n$, which in this case is of order 1, the total change in p across a section of the layer being of order δ†. The solution must be such that $q = w = 0$ for $n = 0$, and q, w must take their values in the given stream outside the layer.

* See Bairstow on 'Skin Friction', *Journal of the Royal Aero. Soc.* XXIX, 1925, p. 3.

† This remark is due to Dr Goldstein.

13·92. Kármán's Integral Condition. On the hypothesis that the effects of friction are confined to a thin layer of fluid whose thickness δ is some function of x, and that in this layer we abandon the attempt to determine the actual value of u but assume that u is a definite function of y which vanishes at the rigid boundary and takes the value which belongs to the external irrotational motion at the surface $y=\delta$, we can integrate the equation **13·91** (4) with regard to y between the limits 0 and δ, and by substituting from the equation of continuity we obtain

$$\int_0^\delta \rho \frac{\partial u}{\partial t} dy + \int_0^\delta \rho u \frac{\partial u}{\partial x} dy + [\rho v u]_0^\delta + \int_0^\delta \rho u \frac{\partial u}{\partial x} dy = -\delta . \frac{dp}{dx} + \left[\mu \frac{\partial u}{\partial y}\right]_0^\delta \quad \text{......(1)}.$$

Then

$$\int_0^\delta \rho \frac{\partial u^2}{\partial x} dy = \frac{\partial}{\partial x}\int_0^\delta \rho u^2 dy - \frac{\partial \delta}{\partial x}[\rho u^2]^\delta = \frac{\partial}{\partial x}\int_0^\delta \rho u^2 dy - \rho U^2 \frac{\partial \delta}{\partial x},$$

where U is the value of u in the irrotational motion. The term $[\rho v u]_0^\delta$ vanishes at the lower limit and is therefore equal to

$$\begin{aligned} U[\rho v]_0^\delta = U\int_0^\delta \rho \frac{\partial v}{\partial y} dy &= -U\int_0^\delta \rho \frac{\partial u}{\partial x} dy \\ &= -U\frac{\partial}{\partial x}\int_0^\delta \rho u\, dy + U\frac{\partial \delta}{\partial x}[\rho u]^\delta \\ &= -U\frac{\partial}{\partial x}\int_0^\delta \rho u\, dy + \rho U^2 \frac{\partial \delta}{\partial x}. \end{aligned}$$

Substituting in (i) and observing also that the tangential stress $\mu\, \partial u/\partial y$ vanishes at the outer boundary of the layer, the equation takes the form

$$\frac{\partial}{\partial t}\int_0^\delta \rho u\, dy + \frac{\partial}{\partial x}\int_0^\delta \rho u^2 dy - U\frac{\partial}{\partial x}\int_0^\delta \rho u\, dy = -\frac{dp}{dx}\delta - \mu\left(\frac{\partial u}{\partial y}\right)_{y=0} \quad \text{......(2)}.$$

If when considering a stationary state we make any plausible hypothesis as to a functional form for u satisfying the boundary conditions, then δ is the only unknown quantity in the equation and so a corresponding form for the thickness of the layer as a function of x is determined.

Equation (2) has also been obtained by Kármán* from considerations of momentum, thus: let NP, $N'P'$ be two ordinates to the curve $y=\delta(x)$ at a small distance dx apart, and consider the change of momentum which takes place in unit time in the area $NPP'N'$.

* *Zeits. f. angew. Math. u. Mech.* I, 1921, p. 233.

The time increase of x momentum is $\frac{\partial}{\partial t}\int_0^\delta \rho u\,dy\,.\,dx$ and this must be equal to the x component of applied force plus the gain of momentum by flux across the boundaries. Now the amount by which the momentum crossing NP exceeds that crossing $N'P'$ is $-\frac{\partial}{\partial x}\int_0^\delta \rho u^2 dy\,.\,dx$; and the mass of fluid crossing $N'P'$ exceeds that which crosses NP by $\frac{\partial}{\partial x}\int_0^\delta \rho u\,dy\,.\,dx$, and this must represent the fluid crossing PP' where the velocity parallel to x is sensibly U, so that the momentum entering across PP' is

$$U\frac{\partial}{\partial x}\int_0^\delta \rho u\,dy\,.\,dx.$$

The applied forces are the pressure excess $-\frac{dp}{dx}dx\,.\,\delta$ and the tangential stress along $N'N$, viz. $-\mu\left(\frac{\partial u}{\partial y}\right)_{y=0}.\,dx$. Equating the terms as stated and dividing by dx we get equation (2).

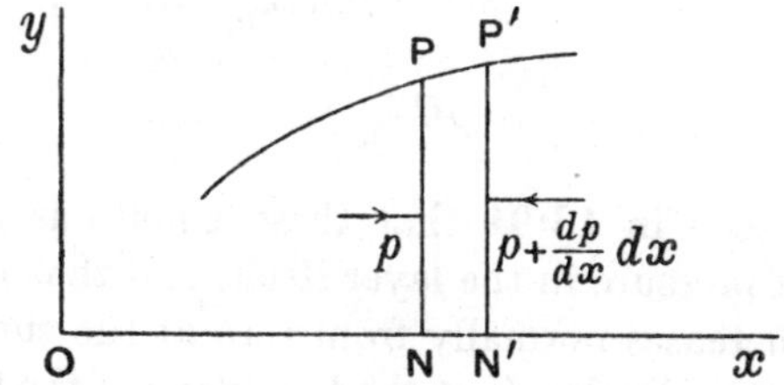

13·93. One of the simplest applications of the boundary layer theory is to the case of a plane plate of finite length immersed in a steady uniform stream of velocity U in the direction of the length of the plate. The problem was solved by Blasius* who found that the thickness of the sheet is proportional to $\sqrt{x}$, where x is measured from the leading edge, and that the tangential drag on either side of the lamina per unit area is

$$\cdot 332\rho U^2\sqrt{\left(\frac{\nu}{Ux}\right)}.$$

It was shewn by Lamb† that if in **13·92** (2) in steady motion we assume that

$$u = U\sin\frac{\pi y}{2\delta}$$

we get for the drag per unit area

$$\cdot 328\rho U^2\sqrt{\left(\frac{\nu}{Ux}\right)},$$

a result differing but little from that of Blasius.

13·94. Generation of Vortices in Fluids of small Viscosity. The existence of the Prandtl layer round a body immersed in a stream of fluid of small viscosity affords an explanation of the generation of vortices in the fluid. According

* *Zeits. f. Math. u. Phys.* LVI, 1908, p. 13.
† *Hydrodynamics*, 1932, § 687.

to the Helmholtz-Kelvin laws of vortex motion, if an element of a perfect fluid is once at rest it can never acquire rotation nor can a rotating element ever cease to rotate, but, as we remarked in **13·9**, we cannot regard the perfect fluid solution of problems as the limiting case of a real fluid solution in which the viscosity tends to zero because of the essential difference in the boundary conditions. And we saw, without considerations of viscosity, in **9·72**, that a trail of vortices may arise in the wake of a solid with a compensating circulation round the solid. But the theory of the boundary layer in fluid of small viscosity such as water or air greatly simplifies the considerations. The layer is thin and its thickness decreases with the viscosity; outside it the motion is irrotational and, in a steady state, we have Bernoulli's Equation

$$p+\tfrac{1}{2}\rho U^2=\text{const.} \qquad \text{.....................(1)},$$

so that

$$\frac{\partial p}{\partial s}+\rho U\frac{\partial U}{\partial s}=0 \qquad \text{.....................(2)};$$

and we have seen in **13·91** that these equations govern the distribution of pressure in the layer itself, and that in the layer the velocity increases normally from zero at the surface of the solid to the external value U at the boundary of the layer.

Again, in the boundary layer, in the notation of **13·91** the vorticity is given by

$$\zeta=\frac{1}{2}\left(\frac{\partial v}{\partial x}-\frac{\partial u}{\partial y}\right),$$

where $v=0$ at the boundary of the solid and is only of order δ at the boundary of the layer, while $\partial u/\partial y$ is large being of order $1/\delta$ in the layer. Consequently in such a layer friction is active in producing vorticity. Such fluid elements as do not enter the layer remain without rotation and those which enter the layer acquire vorticity. Moreover the manner in which the tangential velocity u increases along the normal from 0 at the solid face to U at the outside of the layer depends on the fall of pressure only. Assuming U to depend in a definite way upon s, then in a case in which U increases with s, we see from (2) that $\partial p/\partial s$ is negative and the fluid in the boundary layer is therefore accelerated owing to the falling pressure in the direction of motion. The velocity graphs at successive sections of the layer will be as in fig. (i), and the particles in the layer will all continue to move along the surface of the body. On the other hand if U decreases as s increases, then, from (2), $\partial p/\partial s$ is positive and the particles in the layer are retarded by the pressure gradient as well as by friction. When their kinetic energy is destroyed their direction of motion is reversed and the velocity graphs at successive sections of the layer in this

Fig. (i)

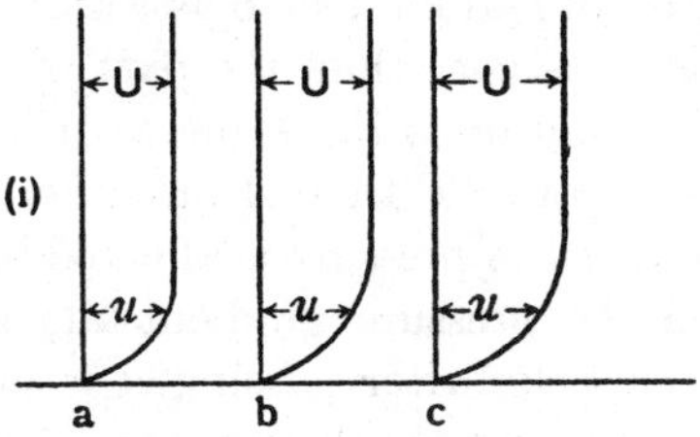

Fig. (ii)

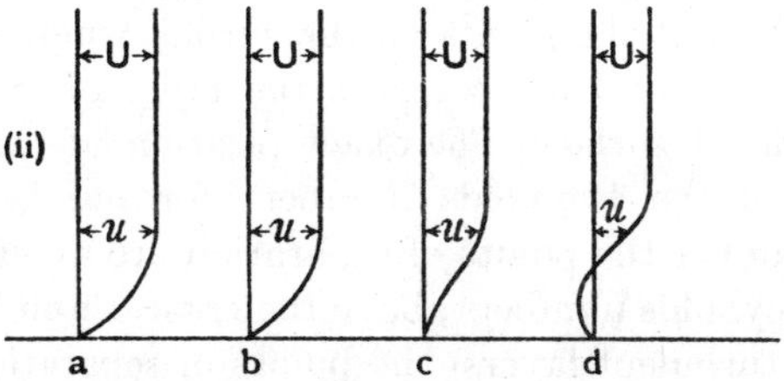

case are as in fig. (ii), the flow being forwards near the points a and b and backwards near d, with an intermediate point c at which $\partial u/\partial y = 0$, and at such a point the advancing and receding streams meet and break away from the body throwing off vortices.

To take a concrete case: consider a steady streaming about a long circular cylinder.

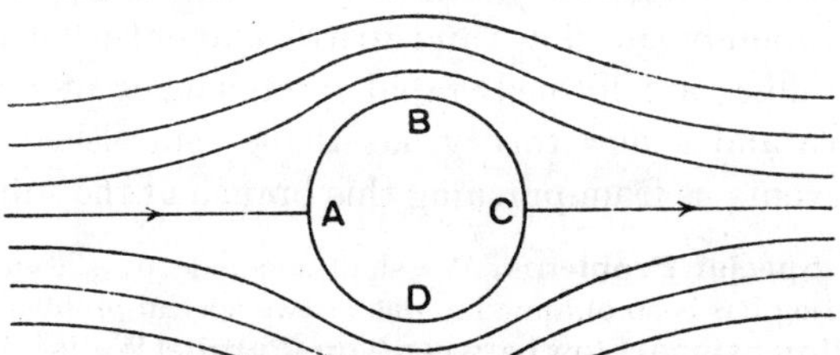

In the perfect fluid solution the stream divides at A and unites again at C and the stream lines are symmetrical about the diametral plane at right angles to AC. The velocity is zero at A and C and the pressure there is greatest. The velocity is greatest at B and D and the pressure there is least. There being no friction the kinetic energy acquired by particles moving along AB is sufficient to carry them from B to C against the increasing pressure. When the upper and lower streams unite again at C they have the same velocity and there is no vortex sheet.

But in the case of real fluid such as water, the layer in which friction is effective is very thin, the part of the layer in which the pressure distribution is the same as in irrotational motion extends at most up to 30° from A on either side and round the rest of the cylinder the pressure is affected by the wake; owing to friction and the pressure gradient the kinetic energy acquired by the particles after passing A is insufficient to carry them round the cylinder, back currents set in from C towards B and D, and at the points on the surface where the oncoming current meets the back currents the layer of vorticity breaks away from the surface. The exact position of the break away depends on the Reynolds number; for numbers below the critical number the points of separation are about 80° from A, and for Reynolds numbers above the critical number (turbulent or partly turbulent layers) the points of separation lie between 120° and 130° from A*.

13·95. Turbulence. We have several times made reference to the fact that the solutions of certain problems only hold good for definite ranges of values of the Reynolds number. Thus in the case of the flow of viscous fluid through a tube the motion is only regular laminar motion provided a certain value of the Reynolds number is not exceeded. When the critical value is passed the regular motion breaks down and an irregular or turbulent motion ensues. Eddies are formed, rapid interchanges of momentum take place and a new theory has to be established. Lack of space prevents us from pursuing this branch of the subject.

13·96. Special Problems. We shall conclude by shewing how an exact solution has been obtained in one or two special problems.

Two-Dimensional Flow between Non-Parallel Walls†. Using polar

* For these data I am indebted to Dr Goldstein. See also Prandtl, *Journal of the Royal Aero. Soc.* XXXI, 1927, p. 720, containing many photographic reproductions of experiments, some of which may also be found in *The Physics of Solids and Fluids*, *loc. cit.* p. 384.

† G. Hamel, *Jahresb. der deutsch. Math. Verein.* XXV, 1916, p. 34. See also K. Pohlhausen, *Zeits. f. angew. Math. u. Mech.* I, 1921, p. 266, and v. Kármán, *Vorträge aus dem Gebiete der Hydro. u. Aerodyn.* p. 146, Innsbruck, 1920.

coordinates in two dimensions, the only component of velocity is the radial component v_r, and, in steady motion, we have from **13·32** (7)

$$v_r \frac{\partial v_r}{\partial r} = -\frac{1}{\rho}\frac{\partial p}{\partial r} + \nu\left(\frac{\partial^2 v_r}{\partial r^2} + \frac{1}{r}\frac{\partial v_r}{\partial r} + \frac{1}{r^2}\frac{\partial^2 v_r}{\partial \theta^2} - \frac{v_r}{r^2}\right) \quad \text{.........(1)},$$

$$0 = -\frac{1}{\rho}\frac{\partial p}{r\,\partial \theta} + \nu\left(\frac{2}{r^2}\frac{\partial v_r}{\partial \theta}\right) \quad \text{.........(2)},$$

and the equation of continuity $\frac{\partial v_r}{\partial r} + \frac{v_r}{r} = 0$ ....(3).

The last equation is clearly satisfied by

$$r v_r = f(\theta) \quad \text{.........(4)},$$

and substituting in (1) and (2) and eliminating p by differentiating the equations we get

$$2ff' + \nu(f''' + 4f') = 0 \quad \text{.........(5)},$$

or, on integration,

$$f^2 + \nu(f'' + 4f) = A \quad \text{.........(6)}.$$

The solution of this equation must be subject to the condition that the velocity vanishes along the boundary walls, so that $f=0$ when $\theta = \pm\alpha$ say.

An exact solution can be obtained by the help of elliptic integrals. For Reynolds number we may take the mean flux over a circular cross section multiplied by the length of that section divided by ν—this will be the same for all sections. Then for small R the velocity distribution is analogous to that between parallel walls. But for large R there are two cases to be considered, according as the stream is converging to a sink or diverging from a source. In the former case there is a uniform distribution of velocity the only defect being close to the walls, i.e. within an angular distance proportional to $1/\sqrt{R}$. In the latter case the differences of velocity are more marked, most of the flow is concentrated in the centre, and there is a critical value of R for which $dv_r/d\theta$ vanishes at the walls, beyond which value of R there is a backflow along the walls and as R increases so does the number of angular regions in which backflow is possible.

13·97. Motion of a Viscous Fluid produced by the Uniform Rotation of a Disc. Let the disc lie in the plane $z=0$ and rotate with uniform angular velocity ω about the z axis. Neglecting edge effects and taking the equations of motion in cylindrical coordinates (**13·32** (7)), we put

$$v_r = rf(z), \quad v_\theta = rg(z) \quad \text{and} \quad v_z = h(z) \quad \text{.........(1)}.$$

Substituting in the equations of motion and taking account of the fact that p is clearly a function of z alone, we get

$$f^2 - g^2 + h\frac{df}{dz} = \nu\frac{d^2 f}{dz^2} \quad \text{.........(2)},$$

$$2fg + h\frac{dg}{dz} = \nu\frac{d^2 g}{dz^2} \quad \text{.........(3)},$$

$$h\frac{dh}{dz} = -\frac{1}{\rho}\frac{dp}{dz} + \nu\frac{d^2 h}{dz^2} \quad \text{.........(4)},$$

and the equation of continuity is

$$\frac{dh}{dz} + 2f = 0 \quad \text{.........(5)}.$$

Equations (2), (3) and (5) determine f, g, h and (4) determines the pressure. The boundary conditions are

$$\left.\begin{array}{lll} f(0)=0, & g(0)=\omega, & h(0)=0 \\ f(\infty)=0, & g(\infty)=0 & \end{array}\right\} \qquad (6).$$

As $z \to \infty$, $h(z)$ tends to a finite limit. In the immediate neighbourhood of the disc there will be a radial flow of fluid and this will be compensated by an axial flow towards the disc.

If we change the independent variable to $z_1 = z\sqrt{(\omega/\nu)}$ and write $f(z) = \omega f_1(z_1)$, $g(z) = \omega g_1(z_1)$ and $h(z) = \sqrt{(\nu\omega)}\, h_1(z_1)$ equations (2), (3) and (5) become

$$f_1^2 - g_1^2 + h_1 \frac{df_1}{dz_1} = \frac{d^2 f_1}{dz_1^2} \qquad (2'),$$

$$2f_1 g_1 + h_1 \frac{dg_1}{dz_1} = \frac{d^2 g_1}{dz_1^2} \qquad (3'),$$

and

$$\frac{dh_1}{dz_1} + 2f_1 = 0 \qquad (5')$$

with boundary conditions

$$\left.\begin{array}{lll} f_1(0)=0, & g_1(0)=1, & h_1(0)=0 \\ f_1(\infty)=0, & g_1(\infty)=0 & \end{array}\right\} \qquad (6'),$$

so that the equations are now independent of the data of any special problem. Since $g_1(z_1)$ is the ratio of the angular velocity at a distance $z_1\sqrt{(\nu/\omega)}$ from the disc to the angular velocity of the disc, it is clear that with increasing angular velocity of the disc and decreasing viscosity angular velocity will only exist within a short distance of the disc. And the relation $h = \sqrt{(\nu\omega)}\, h_1$ shews that the axial velocity at an infinite distance increases with $\sqrt{(\nu\omega)}$.

If $h_1 \to -c$ as $z_1 \to \infty$, expansions have been obtained for f_1, g_1, h_1 in powers of e^{-cz_1} satisfying the differential equations and the conditions at infinity, and there are also expansions for f_1, g_1, h_1 in powers of z_1 which satisfy the differential equations and the conditions at $z_1 = 0$. And the constants in the two sets of expansions can be determined so as to make $f_1, g_1, h_1, f_1', g_1'$ continuous*.

The couple on a rotating disc of radius a can be deduced if the effect of the finite boundary can be neglected, which will probably be the case if a is large compared with the thickness of the boundary layer.

The tangential stress is by **13·33** (1)

$$p_{z\theta} = \mu \left(\frac{\partial v_\theta}{\partial z}\right)_{z=0} = \mu r g'(0)$$

$$= \rho (\nu\omega^3)^{\frac{1}{2}} r g_1'(0);$$

and the couple on one side of the disc

$$= \int_0^a 2\pi r^2 p_{z\theta}\, dr = \tfrac{1}{2}\pi\rho a^4 (\nu\omega^3)^{\frac{1}{2}} g_1'(0).$$

* See W. G. Cochran, *Proc. Camb. Phil. Soc.* XXX (1934), p. 365, for the complete solution.

Or, if we put $R = aU/\nu = a^2\omega/\nu$ as the Reynolds number, where U is the velocity at the edge of the disc, the couple is

$$\tfrac{1}{2}\pi\rho a^5\omega^2 g_1'(0)/R^{\frac{1}{2}}.$$

The value obtained by Cochran for g_1' is $-0{\cdot}616$. In the paper to which reference is given on the opposite page he indicates an error in an earlier solution by v. Kármán.

EXAMPLES

1. Prove that, for liquid filling a closed vessel which is at rest, the rate of dissipation of energy due to viscosity is

$$4\mu\iiint(\xi^2+\eta^2+\zeta^2)\,dx\,dy\,dz,$$

where $\xi = \frac{1}{2}\left(\frac{\partial w}{\partial y} - \frac{\partial v}{\partial z}\right)$, $\eta =\quad$, $\zeta =\quad$ is the vorticity.

If the vessel has the form of a surface of revolution and is rotating about its axis (the axis of z) with angular velocity ω, prove that the rate of dissipation of energy has an additional term

$$2\mu\omega\iint(lDu + mDv)\,dS,$$

where $D = y\frac{\partial}{\partial x} - x\frac{\partial}{\partial y}$, and (l, m, n) are the direction cosines of the inward drawn normal at the element dS of the surface of the vessel.

(M.T. 1926.)

2. A circular disc of radius a is at a small distance h_0 from a fixed parallel plane and can rotate about an axis perpendicular to its plane through its centre O. The space between the disc and the plane is filled with viscous liquid. Shew that the traction on the disc gives rise to a couple

$$\tfrac{1}{2}\pi\mu\omega a^4/h_0,$$

where ω is the angular velocity of the disc.

If the plane of the disc is inclined at a small angle α to the fixed plane, and it is assumed that the pressure in the film of liquid round the boundary of the disc is the atmospheric pressure Π, shew that, to the order α^2, the pressure in the film is

$$\Pi + \frac{3}{4h_0^3}\mu\alpha\omega(a^2-x^2-y^2)\,y - \frac{15}{8h_0^4}\mu\alpha^2\omega(a^2-x^2-y^2)\,xy,$$

where Ox, Oy are axes in the plane of the disc such that the thickness of the film at the point (x, y) is $h_0 + \alpha x$.

Obtain a formula for the value of the couple to this order of approximation.

(M.T. 1925.)

3. A thin film of viscous liquid is contained between two long plane strips of breadth $2d$. One strip is fixed and the other can turn about its medial line parallel to its long edges. The distance between the strips in the position in which they are parallel is h_0. Examine the nature of a slow

steady motion of the liquid produced by displacing the movable strip and shew that, if θ be the small angle between the strips at any time, the pressure in the film at a point where the distance between the strips is h, is

$$\frac{6\mu}{\theta^3}\frac{d\theta}{dt}\left(\log h+\frac{2h_0}{h}-\frac{h_0^2}{2h^2}\right)+\frac{A}{h^2}+B,$$

where A, B are constants determined by the condition that at the edges of the film the pressure is atmospheric. (M.T. 1924.)

4. One surface (nearly plane) is fixed and another near surface (plane) rotates with angular velocity ω about an axis perpendicular to its plane and there is a film of viscous fluid between them. Prove that the pressure p in the film satisfies the equation

$$h^3\left(\frac{\partial^2 p}{\partial r^2}+\frac{1}{r}\frac{\partial p}{\partial r}+\frac{1}{r^2}\frac{\partial^2 p}{\partial \theta^2}\right)+\frac{\partial h^3}{\partial r}\frac{\partial p}{\partial r}+\frac{1}{r^2}\frac{\partial h^3}{\partial \theta}\frac{\partial p}{\partial \theta}=6\mu\omega\frac{\partial h}{\partial \theta},$$

where (r, θ) are polar coordinates in the plane of the film, the origin being in the axis of rotation, and h is the thickness of the film. (M.T. 1927.)

5. A cylindrical shaft with a plane end is made to rotate about its axis with angular velocity ω, the end of the shaft standing on a fixed plane plate from which it is separated by a film of viscous liquid of thickness h, and a uniform flow U of fluid is supplied through a perforation in the plate in line with the axis of the shaft. Prove that the rate of working necessary to maintain the rotation and the supply of lubricant is

$$\frac{\pi\mu\omega^2}{2h}(r_1^4-r_0^4)+\frac{2h^3P^2}{3\pi\mu r_1^4}\log\frac{r_1}{r_0},$$

where r_0, r_1 are the radii of the perforation and the shaft and P is the load on the shaft. (Rayleigh.)

6. A circular disc of radius c is suspended in a horizontal position midway between two fixed horizontal planes by a wire passing through a small hole in the upper plane and having its upper end fixed. The space between the planes contains viscous fluid. The disc is given a slow oscillatory motion and it is assumed that the motion of the fluid is laminar and that variations in pressure are negligible. Neglecting the effects at the edge of the disc, shew that its equation of motion is of the form

$$I\left(\frac{\partial^2\theta}{\partial t^2}+2\kappa\frac{\partial\theta}{\partial t}+\omega^2\theta\right)+\pi c^4\mu\frac{\partial^2\theta}{\partial z\,\partial t}=0.$$

Give the physical meaning of the constants in the equation.

Find also the equation of motion of the fluid and the solution which fits the boundary conditions; and assuming that the motion of the disc is represented by $\theta=ae^{-lt}\cos(\sigma t+\epsilon)$, shew that the constants are connected by the relations

$$I(\omega^2+l^2-\sigma^2-2\kappa l)=\pi c^4\mu\frac{(l\alpha+\sigma\beta)\sinh 2\alpha h+(\sigma\alpha-l\beta)\sin 2\beta h}{\cosh 2\alpha h-\cos 2\beta h},$$

$$2I\sigma(\kappa-l)=\pi c^4\mu\frac{(l\alpha+\sigma\beta)\sin 2\beta h-(\sigma\alpha-l\beta)\sinh 2\alpha h}{\cosh 2\alpha h-\cos 2\beta h},$$

where $\beta^2-\alpha^2=\rho l/\mu$, $2\alpha\beta=\rho\sigma/\mu$, $2h$ is the distance between the fixed planes and ρ is the density of the fluid.

What practical use has been made of these results? (Maxwell.)

7. Incompressible liquid is flowing steadily through a circular pipe. Prove that the mean pressure is constant over the cross section and that the rate of flow is

$$\pi a^4 (p_1 - p_2)/8\mu l,$$

where p_1, p_2 are the pressures over sections at distance l apart.

In the case of steady flow of compressible fluid, shew that the mass which crosses any section per unit time is

$$\pi a^4 (p_1 - p_2)(\rho_1 + \rho_2)/16\mu l,$$

where ρ_1 and ρ_2 are the densities at the two sections at distance l apart. It is assumed that the temperature is constant, and that the gradient of the velocity in the direction of the axis may be neglected in comparison with its gradient in the direction of a radius. (M.T. 1924.)

8. Viscous liquid is flowing steadily under pressure through an infinitely long rectangular tube whose axis is parallel to the axis of z. The sides $x=0$, $x=a$ are smooth, and the sides $y=0$, $y=a$ do not permit of slipping of the liquid in contact with them. The pressure gradient maintaining the motion is suddenly annulled, shew that the total flux across any section is $Qa^2/10\nu$, where Q is the flux per unit time across a section in the initial steady motion.

[In obtaining the above result it may be assumed that $\sum\limits_0^\infty \frac{1}{(2n+1)^6} = \frac{\pi^6}{960}$.]

(M.T. 1925.)

9. Incompressible viscous liquid is moving steadily under pressure between planes $y=0$, $y=h$. The plane $y=0$ has a constant velocity U in the direction of the axis of x, and the plane $y=h$ is fixed. The planes are porous, and liquid is sucked in uniformly over one and ejected uniformly over the others. Shew that a possible solution is given by

$$u = \frac{(Ue^{h/a} + Ah) - (U + Ah)\, e^{y/a}}{e^{h/a} - 1} + Ay, \quad v = \nu/a,$$

where ν is the coefficient of viscosity.

Determine the meanings of the constants A and a. (M.T. 1923.)

10. A liquid occupying the space between two coaxial circular cylinders is acted upon by a force C/r per unit mass, where r is the distance from the axis, the lines of force being circles round the axis. Prove that in the steady motion the velocity at any point is given by the formula

$$\frac{1}{2}\frac{C}{\mu}\left\{\frac{b^2}{r}\frac{r^2 - a^2}{b^2 - a^2}\log\frac{b}{a} - r\log\frac{r}{a}\right\},$$

where μ is the coefficient of viscosity, and a, b are the two radii.

How could this state of motion be realised experimentally?

(M.T. 1895.)

11. The space between two coaxial cylinders of radii a and b is filled with viscous fluid, and the cylinders are made to rotate with angular velocities ω_1, ω_2. Prove that in steady motion the angular velocity of the fluid is given by

$$\omega = \frac{a^2(b^2 - r^2)\,\omega_1 - b^2(r^2 - a^2)\,\omega_2}{r^2(b^2 - a^2)}.$$

12. Two rigid circular discs of radius a are separated by a thin layer of liquid of viscosity μ and initial thickness ζ_0. At time zero a weight W is placed on the upper disc so that the liquid is squeezed out at the edges, the discs remaining horizontal. Prove that at any instant the pressure in the liquid is proportional to a^2-r^2, and that the thickness of the layer is given by

$$\frac{1}{\zeta^2}=\frac{1}{\zeta_0^{\ 2}}+\frac{4}{3}\frac{Wt}{\pi\mu a^4}. \qquad \text{(M.T. 1931.)}$$

13. Prove that $\left(\nu\nabla^2-\frac{\partial}{\partial t}\right)\nabla^2\psi=\frac{\partial(\psi, \nabla^2\psi)}{\partial(x, y)}$, where ψ is the stream function for a two-dimensional motion of a viscous liquid.

A circular cylinder of internal radius a can rotate freely without friction about its axis. It is filled with viscous liquid and the whole system is rotating as if solid about the axis of the cylinder with angular velocity ω_0. The cylinder is instantaneously brought to rest at time $t=0$, and then immediately released. Shew that the angular velocity of the cylinder at time t is

$$\omega_1+\Sigma A_k J_1(ka)\, e^{-k^2\nu t},$$

where ω_1 is the final angular velocity of the system when it is again rotating as if solid, and the values of k are the roots of

$$\{k^2a^2(\omega_0-\omega_1)/4\omega_1+2\}\, J_1(ka)+kaJ_0(ka)=0.$$

State other necessary conditions.

It may be assumed that the cylinder is so long that the disturbing effect of the plane ends may be neglected. (M.T. 1926.)

14. Prove that, in the slow steady motion of a viscous liquid in two dimensions,

$$\nu\nabla^4\psi=\frac{\partial Y}{\partial x}-\frac{\partial X}{\partial y},$$

where (X, Y) is the impressed force per unit area.

If the fluid is bounded by the circle $r=a$, and a concentrated force P acts at the origin in the positive direction of x, prove that

$$4\pi\mu\psi=Pr\sin\theta\left\{\log\frac{r}{a}+\frac{1}{2}\left(1-\frac{r^2}{a^2}\right)\right\}. \qquad \text{(Lamb.)}$$

15. Find the stream function of the motion due to a sphere of radius a performing rectilinear oscillations in a viscous liquid, the period of an oscillation being $2\pi/n$, and prove that the resultant force at any time on the sphere is

$$M\left(\frac{1}{2}+\frac{9}{4\beta a}\right)\dot{w}+\tfrac{9}{4}Mn\left(\frac{1}{\beta a}+\frac{1}{\beta^2a^2}\right)w,$$

where w is the velocity at that time, M the mass of liquid displaced and $\beta=(n/2\nu)^{\frac{1}{2}}$. (M.T. 1900.)

16. A sphere of radius a, surrounded by viscous fluid, is oscillating by the torsion of a suspending wire, the angular velocity being

$$\omega=\omega_0\cos(pt+\epsilon);$$

investigate the motion of the fluid.

Prove that energy is being dissipated by viscosity at the mean rate

$$\tfrac{4}{3}\pi\mu a^3\,\frac{3+6\beta a+6\beta^2a^2+2\beta^3a^3}{1+2\beta a+2\beta^2a^2}\,\omega_0^{\ 2},$$

where a is the radius of the sphere, and $\beta=(p\rho/2\mu)^{\frac{1}{2}}$. (M.T. 1902.)

17. Shew that, for a plane plate of length l placed lengthwise in a uniform stream, the assumption

$$u = Uy\,(2\delta - y)/\delta^2$$

in Kármán's Integral Condition **13·92** leads to

$$\delta = \sqrt{\{30\mu x/\rho U\}},$$

with a frictional resistance $\quad 8\sqrt{\{\mu\rho l\,U^3/30\}}$. (Kármán.)

18. A plane solid surface is wetted with viscous liquid and set up vertically to drain. If h is the thickness of the liquid layer adhering to it at any point, prove that h satisfies the equation

$$\frac{\partial h}{\partial t} + \frac{gh^2}{\nu}\frac{\partial h}{\partial z} = 0,$$

where the coordinate z is measured vertically from the upper edge. The motion is slow and inertia is neglected. Find a solution of the form

$$h = Az^{\alpha}t^{\beta},$$

where A, α, β are constants.

If the original thickness is H, prove that this solution is approximately correct at depth z when a time exceeding $3\nu z/gH^2$ has elapsed from the start. (M.T. 1929.)

19. Shew that, at a distance x from the leading edge of a flat plate parallel to a stream of unbounded fluid moving outside the boundary layer with velocity U, the tangential stress on the plate is $\frac{1}{4}a\,(\mu\rho U^3/x)^{\frac{1}{2}}$, where

$$2a^{-\frac{2}{3}} = \lim_{\xi\to\infty} F'(\xi),$$

and $F(\xi)$ is the solution of the equation

$$\frac{d^3F}{d\xi^3} + F\frac{d^2F}{d\xi^2} = 0,$$

for which $F(0) = F'(0) = 0$, $F''(0) = 1$. (M.T. 1934.)

INDEX OF AUTHORS

The numbers refer to the pages

GENERAL INDEX

The numbers refer to the pages